ENCYCLOPEDIA
of
WORLD AIRCRAFT

ROYAL NAVY

ENCYCLOPEDIA
of
WORLD AIRCRAFT

Christopher Chant

Aircraft Specifications Research
Hugh Cowin

Brian Trodd Publishing House Limited

Published in 1990 by
Brian Trodd Publishing House Limited
27 Swinton Street, London
WC1X 9NW

ISBN 1 85361 106 9

Printed in Italy

Right: A BAe 146-200 of Air Zimbabwe.

Acknowledgments
All artwork by John Batchelor with the exception of pages 46-7 (upper), 56 (lower), 79, 119, 139, 151 (lower), 165 (upper), 167 (lower), 179, 193, 226, 227, 238, 241, 243, 244 (lower), 276, 279 (lower), 280 (lower), 306-7, 312, 326-7, 328, 340-1, 342-3, 343 (Brian Trodd Publishing House Limited).

Photographs from the Hugh W. Cowin Collection are used on the following pages: 49, 50, 51 (lower), 52, 64 (upper), 69, 105, 112 (upper), 122, 125, 127 (upper), 130, 134 (lower), 138 (lower), 142 (lower), 144, 156 (lower), 157 (upper), 161 (lower), 170, 172, 173, 182 (upper), 182 (lower), 183 (upper), 186 (upper), 186 (lower), 191 (upper), 201 (upper), 211, 213, 214 (upper), 216 (upper), 217, 286, 292 (lower), 296, 297, 298, 308, 310, 315 (upper), 315 (lower), 318 (upper), 338 (upper), 339 (upper), 348.

The publisher also wishes to thank the following organizations for their generous help in providing photographs: Aeritalia, Aermacchi, Aerospatiale (Aircraft and Helicopter Divisions), Agusta, Airbus Industrie, American Airlines, Beech, Bell Helicopter, Boeing (Commercial Airplane, De Havilland Canada and Helicopter Divisions), British Aerospace, British Airways, Canadair, CASA, Cessna, Dassault-Breguet, Dornier, Embraer, European Helicopter Industries, Fabrica Militar de Aviones, Fokker, General Dynamics, Grumman Aerospace, Gulfstream Aerospace, Israel Aircraft Industries, Kaman Aerospace, Kawasaki, KLM, Learjet, Lockheed (California and Georgia Divisions), McDonnell Douglas Corporation, MBB, Mitsubishi, Northrop, Panavia, Piaggio, Pilatus, Piper, Rockwell North American, Rolls-Royce, Royal Air Force Strike Command, SAAB, Shorts, Sikorsky Division of United Technologies, TASS, United Airlines, United States Air Force, United States Navy, Vought Aircraft Products Group of LTV Corporation, Westland Helicopters.

air zimbabwe

INTRODUCTION

One hundred years ago the powered aeroplane had yet to fly, but even so the world of aviation was a thriving scene of practical flight with balloons and all manner of theoretical and experimental work with dirigibles and gliders. And behind this practical work lay a vast reservoir of imagination, some of it possessing practical applications but much of it fanciful and even dangerous.

Out of this comparatively tranquil background there rapidly developed motorized dirigibles, rudimentary gliders of considerable performance and, in 1903, the world's first successful powered aeroplane. This is the scope of the *Encyclopedia of World Aircraft*, which is a concise overview of the aircraft, small and large, which have come to dominate a large part of 20th-century history. The book does not essay a thorough coverage of aviation in this century, but seeks instead to create an impression of aviation through presentation of carefully selected aircraft types arranged for ease of reference alphabetically rather than chronologically. Text entries concentrate briefly on the history of a particular aeroplane type, with artworks and photographs plus the specification of a typical variant helping to amplify the reader's insight.

From about 1910, progress in aviation was extraordinarily rapid. World War I spurred great technological advances in airframe and powerplant development. Then, after a slow start in the 1920s, the period between the world wars witnessed the first emergence of commercial aviation and the evolution of all-metal aircraft, and increasing use

Right: A typical 'flight line' scene of 727-200s prior to delivery from Boeing's Renton plant.

of the 'modern' low-wing monoplane configuration with a stressed-skin structure and advanced features such as flaps, retractable landing gear and the enclosed cockpit. Next, the almost incredible demands of World War II pushed aerodynamic and propulsive technology forward some 25 years in a mere six years. The period after World War II witnessed the flowering of aviation as the capabilities of the gas turbine engine over the piston engine came to be appreciated and turned into increasingly sophisticated hardware, making use of modern aerodynamic and structural technologies in features such as swept wings. Finally, the present age has almost without a second glance progressed to the stage in which aircraft are now accepted as just another means of transport for recreational as well as military purposes, and increasingly available to all as the market in mass travel booms. In these circumstances aviation has developed into a multi-faceted giant with military, commercial, social, business and recreational aspects.

Aviation has now branched upwards into space, with current vehicles such as the Space Shuttle combining attributes of the aeroplane and the rocket, but paving the way for hypersonic machines equally at home in the air and in the outer fringes of the atmosphere.

AEG G IV
(Germany)

This was one of Germany's more important bombers of World War I, and appeared in late 1916 as a development of the earlier G I, II and III with greater power, balanced control surfaces and an increased maximum bombload. However, this could be carried only over short ranges, so the type was used mainly over the Western Front for tactical work, in which the interconnection of the cockpits for the four crew was most useful. Production totalled about 500, including the G IVb with three-bay wings of increased span, and the G IVk with a 20mm Becker cannon in the nose position.

AEG G IV

Role: Medium bomber
Crew/Accommodation: Three
Power Plant: Two 260 hp Mercedes D.IVa water-cooled inline
Dimensions: Span 18.4m (60.4 ft); length 9.7m (31.8 ft); wing area 67.0m^2 (721 sq ft)
Weights: Empty 2,400kg (5,291 lb); MTOW 3,630kg (8,003 lb)
Performance: Maximum speed 165km/h (103 mph) at sea level; operational ceiling 4,500m (14,764 ft); range 700km (435 miles) with full bombload
Load: Two 7.9mm machine guns, plus up to 800kg (1,764 lb) of bombs

AERO L-39 ALBATROS
(Czechoslovakia)

This attractive aeroplane was produced in succession to the L-29 Delfin from the same company, and since 1973 has become the standard basic and advanced trainer of most communist air arms. Unswept flying surfaces curtail outright performance, but positive features are the stepped seats, tractable handling and a fuel-economical turbofan. Variants are the basic L-39C, the L-39ZO weapons trainer, the L-39ZA attack/reconnaissance type with an underfuselage gun pod and four underwing hardpoints, the L-39V target-tug and the L-39MS updated trainer with greater power and a modernized cockpit.

AERO L-39Z ALBATROS

Role: Light strike/advanced trainer
Crew: Two
Power Plant: One 1,500kgp (3,307 lb s.t.) Walter Titan turbofan
Dimensions: Span 9.46m (31 ft); length 12.13m (39.76 ft); wing area 18.80m^2 (202.4 sq ft)
Weights: Empty 3,565kg (7,859 lb); MTOW 5,650kg (12,450 lb)
Performance: Maximum speed 700km/h (378 knots) at sea level; operational ceiling 11,000m (36,090 ft); range 1,100km (594 naut. miles) on internal fuel only
Load: One 23mm cannon, plus up to 2,000kg (3,300 lb) of externally underslung ordnance

AERITALIA G222
(Italy)

Though it stemmed from a NATO requirement for a V/STOL tactical transport, the G222 appeared in 1970 as a conventional type with blistered main landing gear units, a rear ramp/door providing straight-in access to the 44-man hold, and two General Electric turboprops. In its basic form the type serves with the Italian and other air forces, and a few special-mission versions have also been produced. To get round an embargo on the export of U.S. items to Libya, the company produced the G222L version with 3,536kW (3,400 shp) Rolls-Royce Tyne turboprops.

AERITALIA G222

Role: Medium-lift, rough-field transport
Crew/Accommodation: Three, plus up to 44 troops
Power Plant: Two 3,400 shp General Electric T64-GE-P4D turboprops, or two 4,860 shp Rolls-Royce Tyne RTy 20 Mk 801 turboprops on G222T
Dimensions: Span 28.7m (94.16 ft); length 22.7m (74.48 ft); wing area 82m^2 (882.64 sq ft)
Weights: Empty15,400kg (33,950 lb); MTOW 28,000kg (61,730 lb)
Performance: Maximum speed 540km/h (291 knots) at 4,575m (15,000 ft); operational ceiling 7,620m (25,000 ft); range 2,200km (1,198 naut. miles) with 44 troops
Load: Up to 9,000kg (19,841 lb)
Note: above characteristics apply to standard T64 powered aircraft

AERMACCHI MB.326
(Italy)

One of Macchi's most successful designs, the MB.326 was planned as a basic and advanced trainer. The type reveals its 1950s' design age in its unswept flying surfaces and vertically unstaggered seats, but in its time it was an excellent aeroplane that scored considerable export success in a number of different armed and unarmed variants, including the Impala Mk 1 built under licence in South Africa and the Xavante in Brazil. The basic design was also adapted as the commercially successful MB.326K single-seat light attack aeroplane, also built in South Africa as the Impala Mk 2.

AERMACCHI MB.326H

Role: Light/strike trainer
Crew: Two
Power Plant: One 1,134kgp (2,500 lb s.t.) Rolls-Royce Viper II turbojet
Dimensions: Span 10.85m (35.71 ft); length 10.67m (35 ft); wing area 19.4m^2 (208.3 sq ft)
Weights: Empty 2,685kg (6,920 lb); MTOW 5,216kg (11,500 lb)
Performance: Maximum speed 867km/h (468 knots) at sea level; operational ceiling 14,325m (47,000 ft); range 1,850km (998 naut. miles)
Load: Up to 1,814kg (4,000 lb) of externally carried armament

AERMACCHI MB.339
(Italy)

The MB.339A first flew in 1976 as an updated version of the MB.326 with essentially the same unswept flying surfaces and turbojet power plant married to an airframe with larger vertical tail and a deeper nose to allow the incorporation of the vertically staggered seating now thought essential. The MB.339B has an uprated engine, larger wingtip tanks and greater weapon-carrying capability. The latest version is the MB.339C with an advanced cockpit and weapons for use in the light attack/maritime roles. These sophisticated nav/attack and weapon systems are features of the single-seat dedicated attack version, the MB.339K.

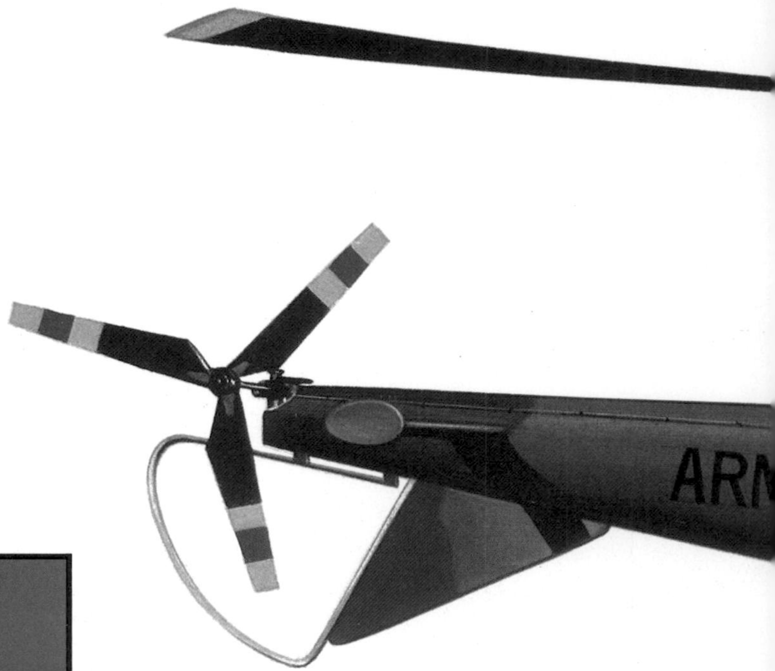

AERMACCHI MB.339

Role: Light strike/trainer
Crew/Accommodation: Two
Power Plant: One 1,814kgp (4,000 lb s.t.) Rolls-Royce Viper 632-43 turbojet
Dimensions: Span 10.86m (35.6 ft); length 10.97m (36 ft); wing area 19.3m^2 (207.7 sq ft)
Weights: Empty 3,125kg (6,883 lb); MTOW 5,897kg (13,000 lb)
Performance: Maximum speed 898km/h (485 knots) at sea level; operational ceiling 14,630m (47,500 ft); range 1,760km (950 naut. miles)
Warload: Up to 1,815kg (4,000 lb) of externally carried weapons

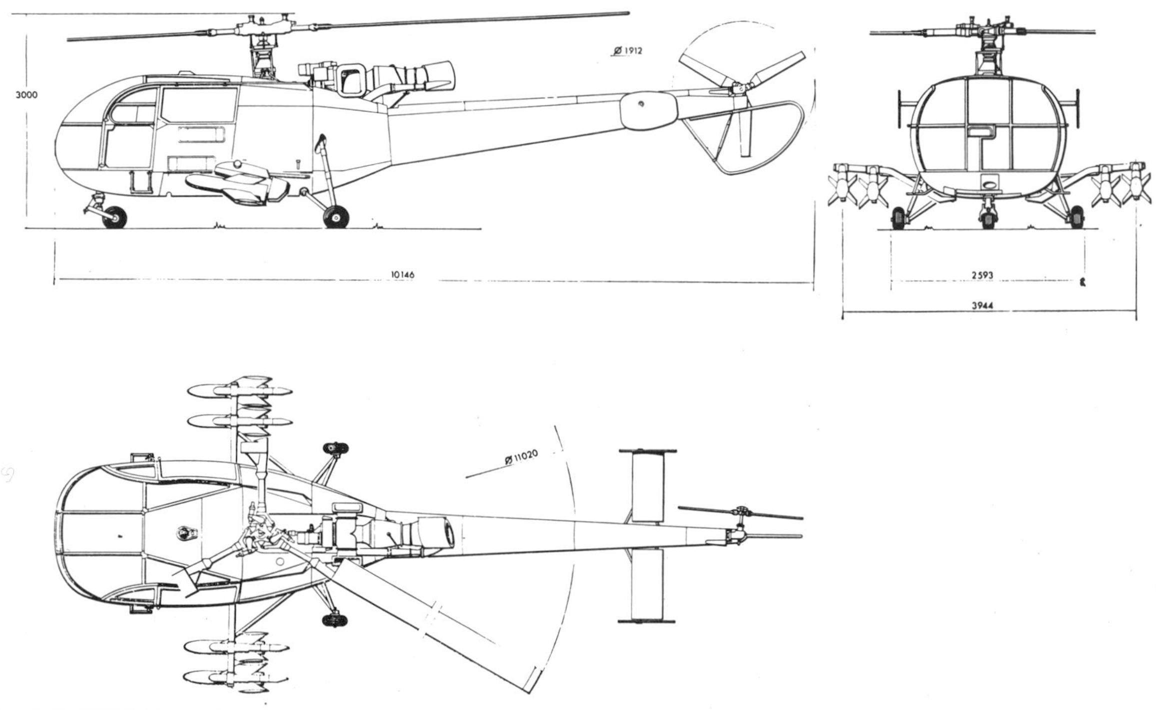

AEROMARINE 39
(U.S.A.)

In 1917 this was the first trainer ordered into large-scale production for the U.S. Navy, and was an orthodox two-bay biplane that could be fitted with wheel, float or even ski landing gear. The 50 Model 39-A aircraft had the 75kW (100 hp) Hall-Scott A-7A engine and, in the maritime role, twin floats; the 150 Model 39-Bs had the similarly rated Curtiss OXX-6 engine and, in the maritime role, the arrangement of one long central float and two underwing stabilizing floats that became standard for U.S. Navy floatplanes. After World War I the type was the pioneer of U.S. Navy deck landing experiments.

AEROMARINE 39-B

Role: Trainer
Crew: Two
Power Plant: One 100 hp Curtiss OXX-6 water-cooled inline
Dimensions: Span 14.33m (47 ft); length 9.27m (30.4 ft)
Weights: Empty 800kg (1,939 lb); MTOW 1,143kg (2,520 lb)
Performance: Maximum speed 109km/h (68 mph) at sea-level; operational ceiling 2,134m (7,000 ft); radius 440km (273 miles)
Load: None

AEROSPATIALE SA 316/319 ALOUETTE III
(France)

Developed from the Alouette II in the late 1950s with a larger cabin, improved equipment, greater power and an all-round improvement in performance, the Alouette III remains a classic helicopter in tasks as diverse in the civil role as air-taxi work and mountain rescue, and in the military role as anti-submarine and observation work. Just under 1,400 were produced in France, India, Romania and Switzerland in two basic versions: the SA 316B with the 425kW (570 shp) Artouste IIIB turboshaft, and the SA 319B with the 448kW (600 shp) Astazou XIV turboshaft.

AEROSPATIALE SA 319 ALOUETTE III

Role: Light utility/communications helicopter
Crew/Accommodation: One, plus up to six passengers
Power Plant: One 789 shp Turbomeca Astazou XIV turboshaft
Dimensions: Rotor diameter 11.02m (36.1 ft); length 12.84m (42.1 ft) rotors turning
Weights: Empty 1,090kg (2,403 lb); MTOW 2,205kg (4,960 lb)
Performance: Maximum speed 220km/h (136 mph) at sea level; range 605km (375 miles) with six passengers
Load: Variants can carry up to 500kg (1,102 lb) of ordnance, including two lightweight anti-submarine torpedoes

▲ AEROSPATIALKE SA 330 PUMA and AS 332 SUPER PUMA
(France)

The Puma was designed to meet a French army requirement for an all-weather tactical and logistic transport helicopter, but since its first flight in 1965 the type has matured as an extremely versatile civil and military craft, with Turmo III and IV turboshafts, and accommodation for 20 passengers. In 1978 the company first flew the upgraded Super Puma with more powerful Makila 1 engines, a lightweight rotor head and improved landing gear. The SA 330L was stretched for 24-seat capacity, and in the latest Super Puma Mk II a revised rotor, uprated gearbox and improved avionics give a marked improvement in overall capabilities.

AEROSPATIALE AS 332M SUPER PUMA

Role: Medium-lift transport helicopter
Crew/Accommodation: Two crew, plus up to 25 troops
Power Plant: Two 1,877shp Turbomeca Makila 1A1 turboshafts
Dimensions: Rotor diameter 15.6m (51.8 ft); length 18.70m (61.35 ft) rotors turning
Weights: Empty 4,420kg (9,745 lb); MTOW 9,000kg (19,840 lb)
Performance: Maximum speed 278km/h (150 knot) at sea level; operational ceiling 4,100m (13,448 ft); range 842km (455 naut. miles) on standard fuel tanks
Load: Can lift external underslung loads of up to 4,500kg (9,920 lb)

AEROSPATIALE SA 341/342 GAZELLE
(France)

Resulting from a French army requirement for a light observation helicopter, the Gazelle uses the same power plant and transmission as the SA 318C Alouette II. The type first flew in 1967 and has been bought mainly by military operators for the observation, liaison and, in special armed versions, anti-tank and anti-helicopter operations. The Gazelle's most distinctive feature is the *fenestron* tail rotor housed in the fin, and while the SA 341 has the Astazou III turboshaft, the SA 342 has the more powerful Astazou XIV for greater performance and payload, especially in 'hot and high' conditions.

◄ AEROSPATIALE SA 342H GAZELLE

Role: Light utility and communications helicopter
Crew/Accommodation: One, plus up to four troops
Power Plant: One 590 shp Turbomeca Astazou IIIB turboshaft
Dimensions: Rotor diameter 10.5m (34.5 ft); length 11.97m (36.3 ft) rotors turning
Weights: Empty 908kg (2,002 lb); MTOW 1,800kg (3,970 lb)
Performance: Cruise speed 264km/h (164 mph) at sea-level; operational ceiling 5,000m (16,400 ft); range 670km (415 miles)
Load: Provision to carry four HOT anti-tank missiles

▲ AEROSPATIALE SA 360 DAUPHIN and SA 365 DAUPHIN 2
(France)

The Dauphin was designed as an Alouette III replacement, and first flew in 1972 with a single Astazou XVIIIA. Since then the type has been developed in a bewildering number of forms and variants. The SA 360 is a civil type, the similar SA 361H can carry 13 troops or operate in the anti-tank role with missile armament. The civil SA 365C introduced a retractable landing gear and a twin-engined power plant with two Arriel turboshafts. Military variants are the SA 365F anti-ship model with radar and four missiles, the SA 365M Panther battlefield helicopter with TM333 turboshafts, the SA 365N naval version of the SA 365C with anti-ship or anti-submarine sensors and weapons, and the SA 366G search-and-rescue type, which is used by the U.S. Coast Guard as the HH-65A Dolphin with Avco Lycoming LTS101 turboshafts. The SA 365N is also built in China at Harbin with the designation Z-9.

AEROSPATIALE SA 365N DAUPHIN 2

Role: Corporate/VIP transport/Air ambulance
Crew/Accommodation: One, plus up to 13 passengers
Power Plant: Two 724 shp Turbomeca Arriel 1C1 turboshafts
Dimensions: Rotor diameter 11.94m (39.17 ft); length 13.68m (44.9 ft) rotors turning
Weights: Empty 2,161kg (4,765 lb); MTOW 4,100kg (9,040 lb)
Performance: Maximum speed 295km/h (160 knots) at sea level; operational ceiling 3,600m (11,810 ft); range 855km (460 naut. miles)
Load: Up to 1,600kg (3,527 lb) external underslung loads

▼

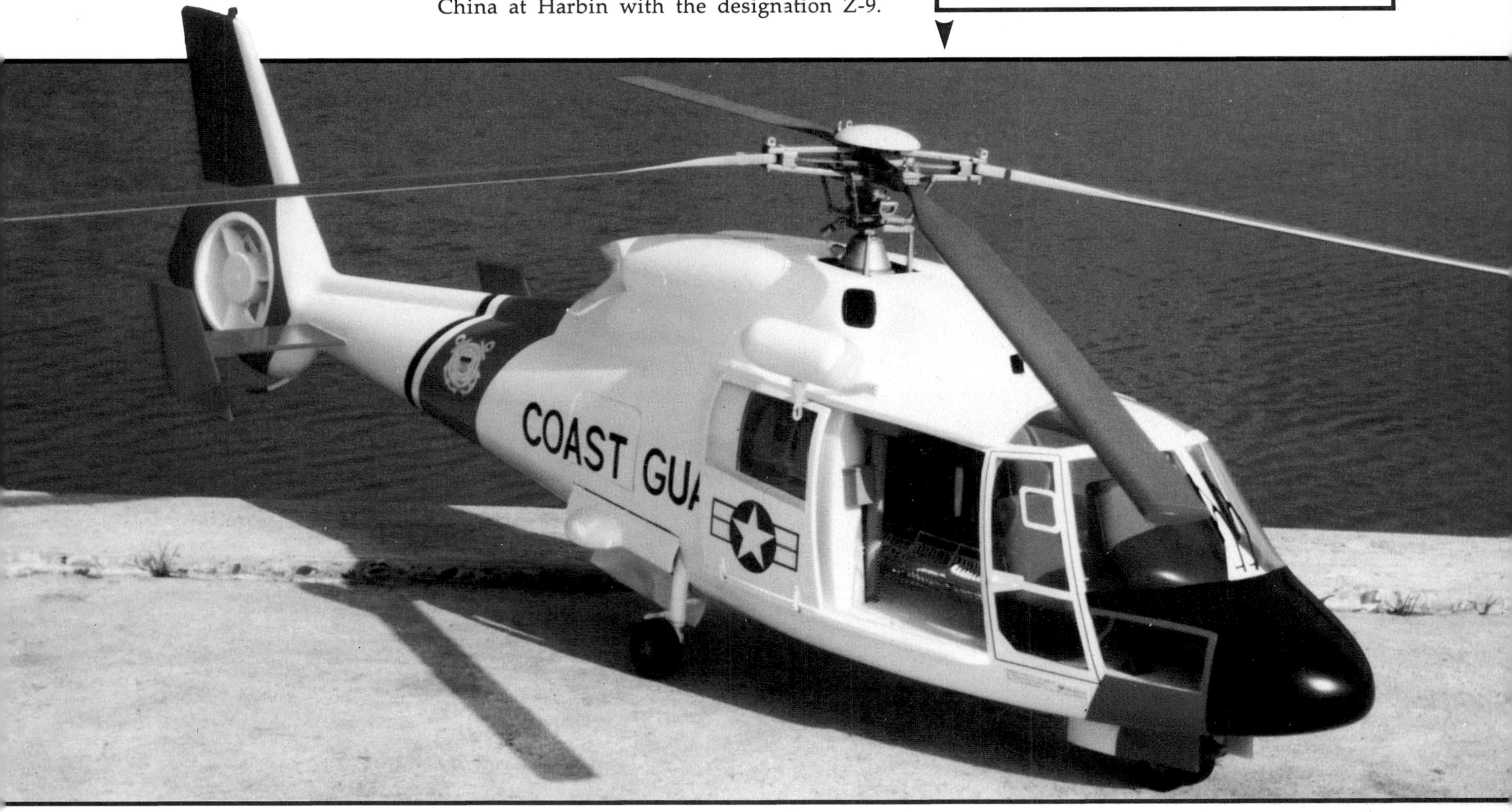

BRITISH AIRWAYS
G-BOAG

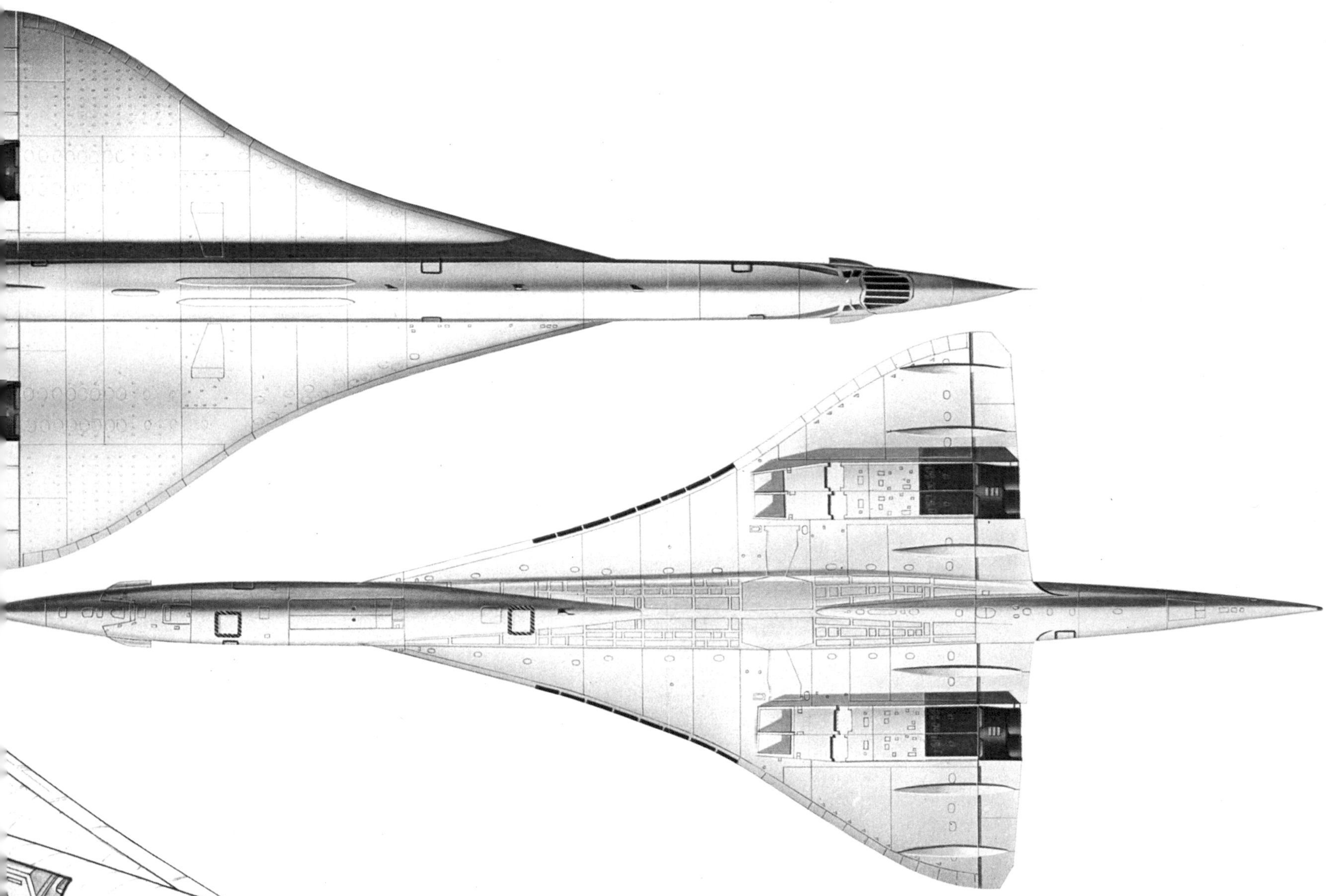

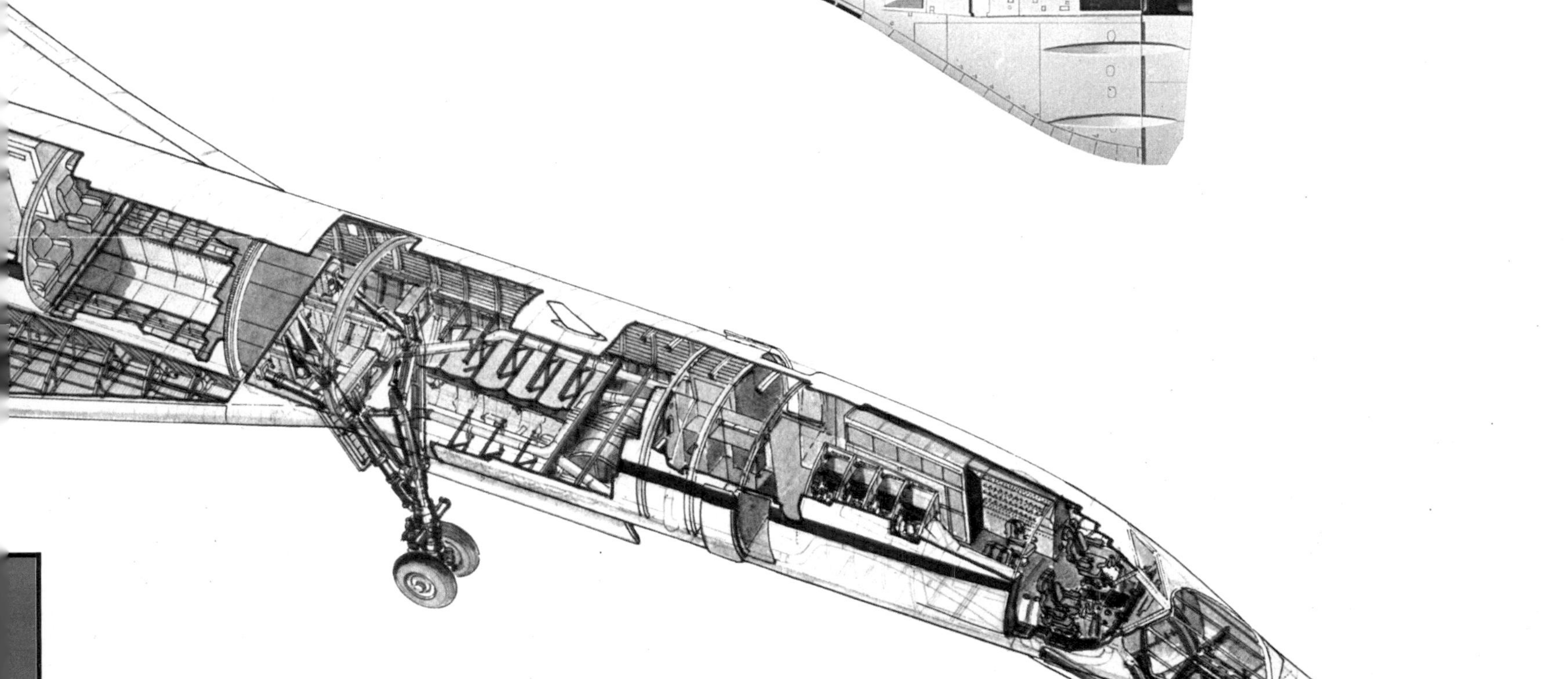

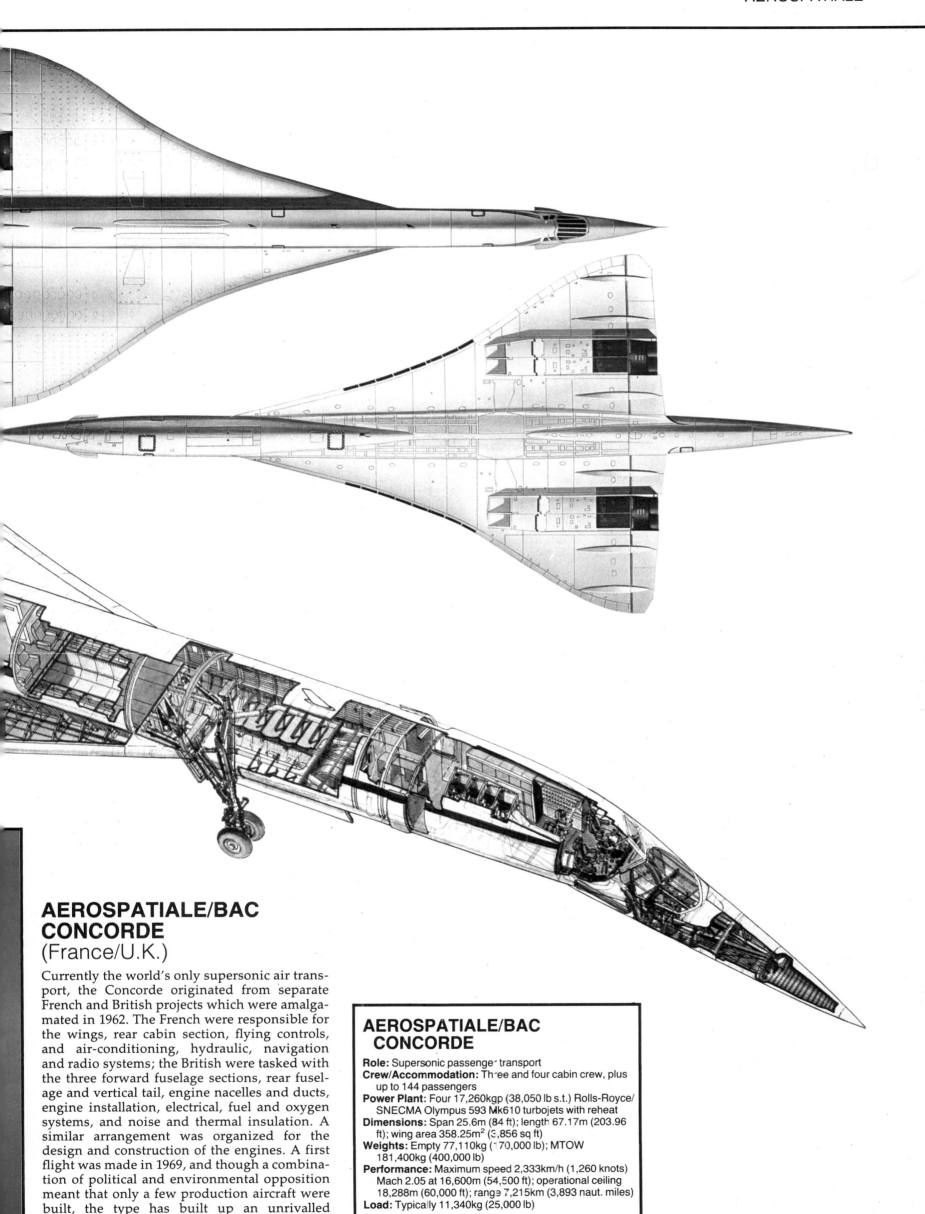

AEROSPATIALE/BAC CONCORDE
(France/U.K.)

Currently the world's only supersonic air transport, the Concorde originated from separate French and British projects which were amalgamated in 1962. The French were responsible for the wings, rear cabin section, flying controls, and air-conditioning, hydraulic, navigation and radio systems; the British were tasked with the three forward fuselage sections, rear fuselage and vertical tail, engine nacelles and ducts, engine installation, electrical, fuel and oxygen systems, and noise and thermal insulation. A similar arrangement was organized for the design and construction of the engines. A first flight was made in 1969, and though a combination of political and environmental opposition meant that only a few production aircraft were built, the type has built up an unrivalled reputation.

AEROSPATIALE/BAC CONCORDE

Role: Supersonic passenger transport
Crew/Accommodation: Three and four cabin crew, plus up to 144 passengers
Power Plant: Four 17,260kgp (38,050 lb s.t.) Rolls-Royce/SNECMA Olympus 593 Mk610 turbojets with reheat
Dimensions: Span 25.6m (84 ft); length 67.17m (203.96 ft); wing area 358.25m^2 (3,856 sq ft)
Weights: Empty 77,110kg (170,000 lb); MTOW 181,400kg (400,000 lb)
Performance: Maximum speed 2,333km/h (1,260 knots) Mach 2.05 at 16,600m (54,500 ft); operational ceiling 18,288m (60,000 ft); range 7,215km (3,893 naut. miles)
Load: Typically 11,340kg (25,000 lb)

AGUSTA A 109
(Italy)

After considerable experience in licence-production of helicopters, Agusta decided to try for independent commercial success with the A 109 originally planned round a single turboshaft, but then recast as a more reliable twin-engined machine with retractable landing gear and extremely clean design for maximum performance. The first example flew in 1971, and the A 109A initial production variant secured useful orders for civil and military tasks. This success has been continued by the updated A 109A Mk II version with a number of important improvements, and the manufacturer also offers a specialized A 109K model with Arriel 1K turboshafts for better performance in 'hot and high' conditions.

AGUSTA A 109

Role: Light utility/communications helicopter
Crew/Accommodation: One, plus up to seven troops
Power Plant: Two 420 shp Allison C250-C20B turboshafts
Dimensions: Rotor diameter 11m (36.08 ft); length 13.05m (42.95 ft) rotors turning
Weights: Empty 1,415kg (3,120 lb); MTOW 2,600kg (5,730 lb)
Performance: Maximum speed 266km/h (165 mph) at sea level; operational ceiling 4,968m (16,300 ft); range 615km (382 miles)
Load: Provision to carry two HOT or TOW anti-tank missiles

AICHI D3A 'VAL'
(Japan)

First flown in 1938, the D3A was Japan's most important naval dive-bomber at the beginning of World War II, and played a major part in the Pearl Harbor attack. The type's elliptical flying surfaces were aerodynamically elegant, and the combination of a lightweight structure and fixed but spatted landing gear provided good performance. Production totalled almost 1,500, the main variants being the D3A1 with the 798kW (1,070 hp) Kinsei 44 radial, the D3A2 with the 969kW (1,300 hp) Kinsei 54 plus greater fuel capacity for better performance and range, and the D3A2-K trainer conversion of the earlier models.

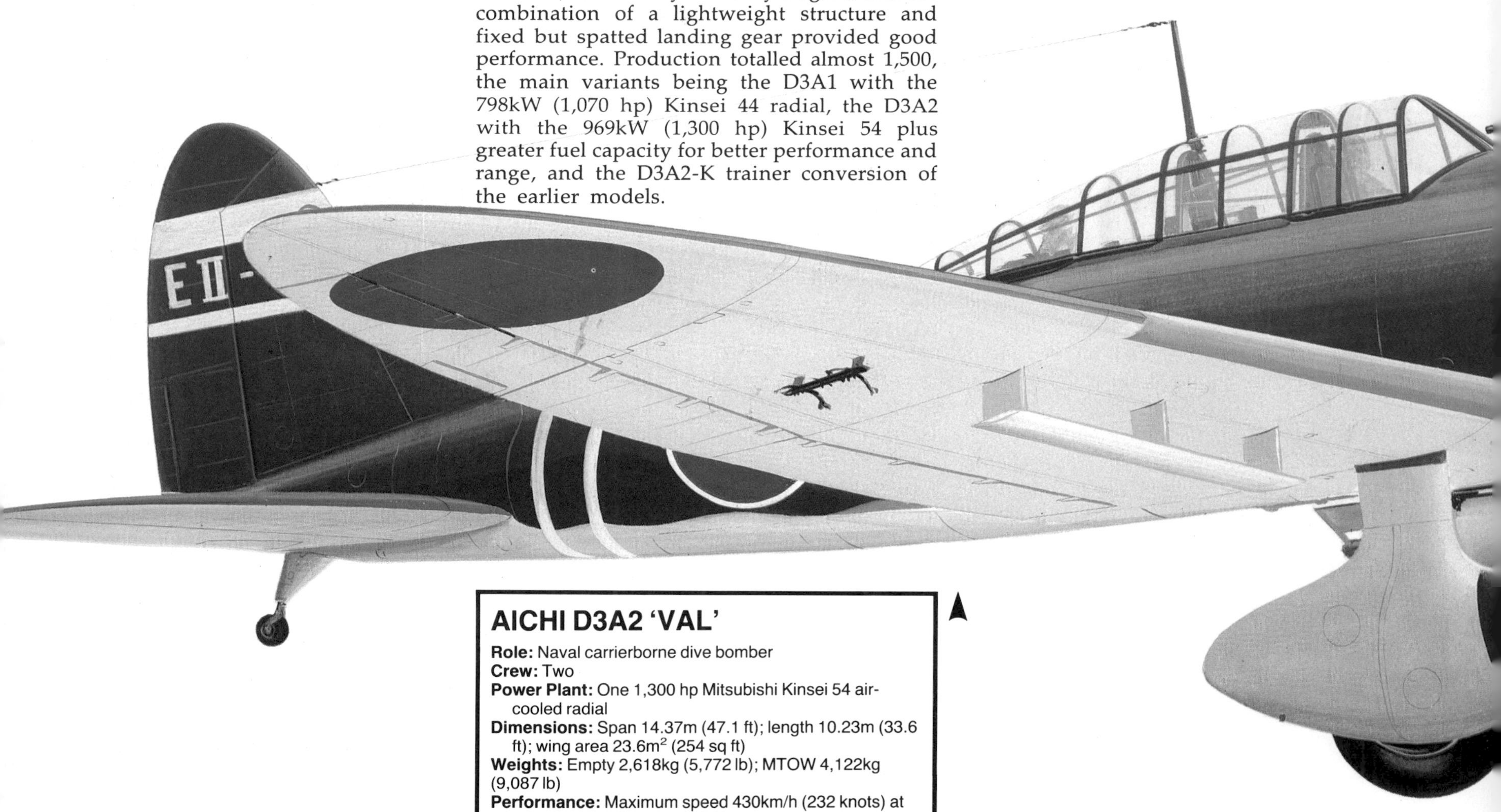

AICHI D3A2 'VAL'

Role: Naval carrierborne dive bomber
Crew: Two
Power Plant: One 1,300 hp Mitsubishi Kinsei 54 air-cooled radial
Dimensions: Span 14.37m (47.1 ft); length 10.23m (33.6 ft); wing area 23.6m^2 (254 sq ft)
Weights: Empty 2,618kg (5,772 lb); MTOW 4,122kg (9,087 lb)
Performance: Maximum speed 430km/h (232 knots) at 9,225m (20,340 ft); operational ceiling 10,888m (35,720 ft); range 1,561km (842 naut. miles) maximum
Load: Three 7.7mm machine guns, plus up to 370kg (816 lb) of bombs

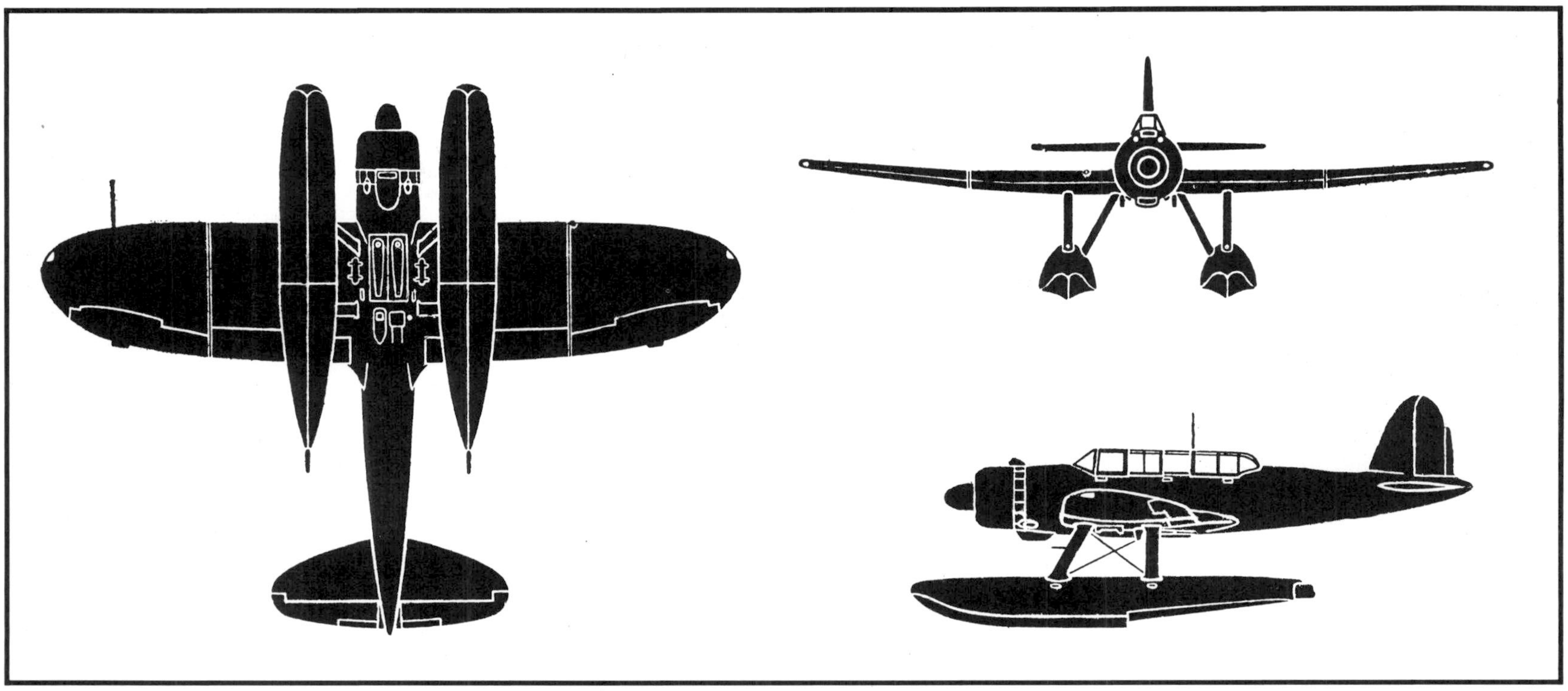

AICHI E13A 'JAKE'
(Japan)

First flown in 1938, the E13A was an important reconnaissance floatplane in World War II and operated from shore bases as well as warship catapults. Production of this twin-float seaplane totalled about 1,150, and the main variants were the E13A1 with the 805kW (1,080 hp) Kinsei 43 radial, the E13A1a with improved bracing of the floats and better radio equipment, and the E13A1b version of the E13A1a with anti-ship radar. The type's endurance of up to 15 hours made it an important patrol and SAR type later in the war, when it was also used in the *kamikaze* role.

AICHI E13A1 'JAKE'

Role: Naval shipboard reconnaissance floatplane
Crew/Accommodation: Three
Power Plant: One 1,080 hp Mitsubishi Kinsei 43 air-cooled radial
Dimensions: Span 14.5m (47.6 ft); length 11.27m (36.98 ft); wing area 39.7m^2 (427 sq ft)
Weights: Empty 2,642kg (5,825 lb); MTOW 4,000kg (8,818 lb)
Performance: Maximum speed 385km/h (208 knots) at 2,180m (7,152 ft); operational ceiling 3,962m (13,000 ft); range 1,535km (828 naut. miles) normal
Load: One 7.9mm machine gun, plus up to 240kg (530 lb) of bombs

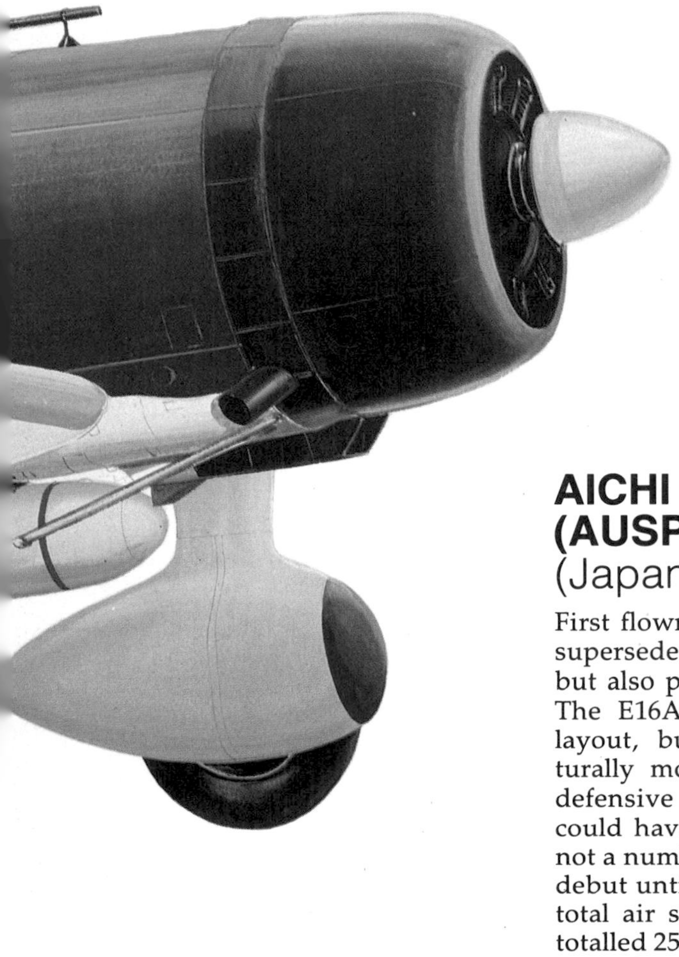

AICHI E16A ZUIUN (AUSPICIOUS CLOUD)'PAUL'
(Japan)

First flown in 1942, the E16A was designed to supersede the E13A in the reconnaissance role, but also possessed a dive-bombing capability. The E16A was similar to its predecessor in layout, but was aerodynamically and structurally more advanced. Useful offensive and defensive armament was carried, and the type could have given a good account of itself had not a number of problems delayed its in service debut until a time when the Allies had secured total air superiority. Production of the E16A1 totalled 252, and at the end of the war the E16A2 with the 1,163kW (1,560 hp) Kinsei 62 engine was under evaluation.

AICHI E16A1 'PAUL'

Role: Naval shipboard reconnaissance floatplane
Crew/Accommodation: Three
Power Plant: One 1,300 hp Mitsubishi Kinsei 54 air-cooled radial
Dimensions: Span 12.8m (42 ft); length 10.84m (35.6 ft); wing area 28m^2 (301 sq ft)
Weights: Empty 2,713kg (5,981 lb); MTOW 4,230kg (9,326 lb)
Performance: Maximum speed 448km/h (242 knots) at 5,580m (18,307 ft); operational ceiling 10,280m (33,727 ft); range 956km (516 naut. miles) normal
Load: One 20mm cannon, one 12.7mm machine gun, plus up to 500kg (1,102 lb) of bombs

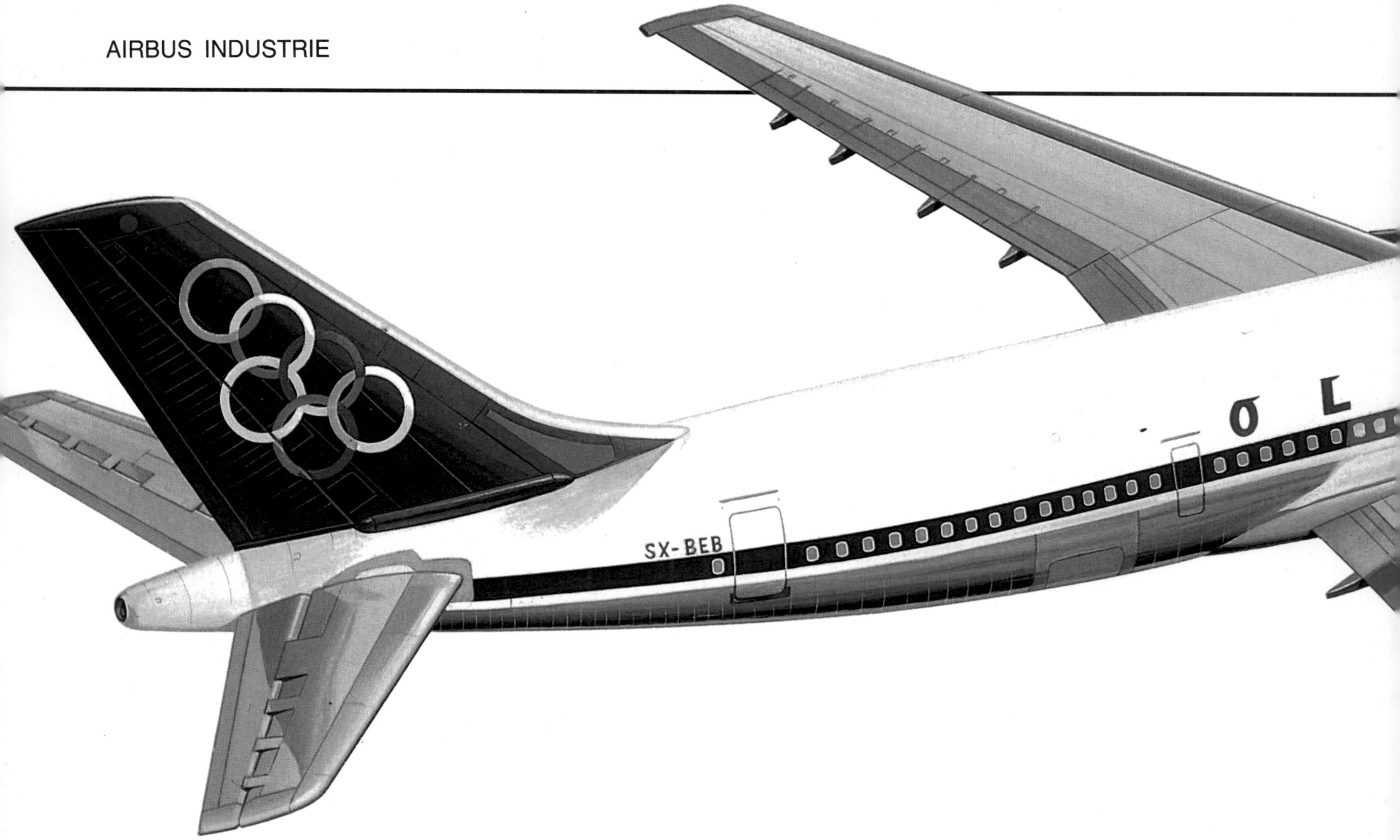

AIRBUS INDUSTRIE A300

(France/Spain/U.K./ West Germany)

The Airbus consortium was established in 1967 to co-ordinate a European challenge to the American 'big three' of airliner production: Boeing, Lockheed and Douglas (later McDonnell Douglas). A number of national designs had already been studied before the consortium was created to produce a 250-seater powered by two British or American turbofans. The first prototype A300B1 flew in 1972, and this was lengthened by 2.65m (8 ft 8 inches) to create the initial-production A300B2-100 with General Electric CF6-50 engines; variants are the A300B2-200 with wing root leading-edge flaps, the A300B2-220 with Pratt & Whitney JT9D-59A turbofans, and the A300B2-320 with higher take-off and landing weights. Then came the A300B4-100 long-range version with CF6 engines, the A300B4-200 with higher weights and the A300B4-200FF with a two-crew cockpit. The A300C4 is a convertible freighter based on the A300B4, and in 1980 Airbus launched the A300-600 as an advanced version of the A300B4 including a rear fuselage with the profile pioneered in the A310 and a choice of General Electric CF6, Pratt & Whitney PW4158 or Rolls-Royce RB.211 turbofans. The A300-600R improved version has wingtip fences and other modifications.

AIRBUS A300-600R

Role: Intermediate-range passenger transport
Crew/Accommodation: Three and six cabin crew, plus up to 344 passengers
Power Plant: Two 27,307kgp (60,200 lb s.t.) General Electric CF6-80C2A3 or 26,310kgp (58,000 lb s.t.) Pratt & Whitney PW4158 turbofans
Dimensions: Span 44.84m (147.1 ft); length 54.08m (177.4 ft); wing area 260m^2 (2,799 sq ft)
Weights: Empty 87,728kg (193,410 lb); MTOW 165,000kg (363,760 lb)
Performance: Maximum speed 891km/h (554 mph) at 9,450m (31,000 ft); operational ceiling 13,000+m (42,650+ ft); range 5,200km (3,430 miles) with maximum payload
Load: Up to 41,504kg (91,500 lb) including belly cargo

AIRBUS A310-300

Role: Intermediate-range passenger transport
Crew/Accommodation: Three and six cabin crew, plus up to 280 passengers
Power Plant: Two 22,680kgp (50,000 lb s.t.) General Electric CF6-80C2-A2 or Pratt & Whitney JT9D-7R4E turbofans
Dimensions: Span 43.9m (144.0 ft); length 44.66m (153.08 ft); wing area 219m^2 (2,357 sq ft)
Weights: Empty 77,040kg (169,840 lb); MTOW 150,000kg (330,693 lb)
Performance: Maximum speed 903km/h (561 mph) at 10,670m (35,000 ft); operational ceiling 13,000+m (42,650+ ft); range 6,950km (4,318 miles) with maximum payload
Load: Up to 35,108kg (77,400 lb) including belly cargo

AIRBUS INDUSTRIE A310
(France/Spain/U.K./West Germany)

The A310 resulted from an Airbus programme that at one time encompassed 11 proposals for a large-capacity airliner intended for short-haul routes. The final A310 was designed for as much commonality to the A300 as possible: the fuselage is essentially a shortened version of the A300's, and the power plant uses lower-rated versions of the CF6 and JT9D, with the RB.211 available as an option. Different features are the engine pylons, landing gear and low-drag wing. The type was first flown in April 1983 and though initially proposed in A310-100 and A310-200 short- and medium-range version, the former was then dropped in favour of different-weight versions of the A310-200. The A300-300 is a longer-range version with tail-plane trim tanks, and is available in the weight options for the A310-200. Convertible and freight versions are A310-200C and A310-200F.

AIRBUS INDUSTRIE A320
(France/Spain/U.K./West Germany)

The A320 was conceived as a 150-seat short/medium-range partner to the A310 and despite its slimmer fuselage bears a striking external resemblance to its larger brother. The type was initially marketed in 154- and 172-seat A320-100 and A320-200 versions, but the designations were soon altered to low- and high-weight versions each seating 162 passengers. In detail the A320 is a wholly new design using a large percentage of composite materials and advanced electronic features such as a digital fly-by-wire flight control system with sidestick controllers, an electronic flight instrument system and an electronic centralized aircraft monitor. The first A320 flew in February 1987, and the type is now one of the fastest-selling airliners in the market with CFM56 or International Aero Engines V2500 turbofans.

AIRBUS A320-200

Role: Intermediate/short-range passenger transport
Crew/Accommodation: Twoand four/five cabin crew, plus up to 164 passengers (single class cabin)
Power Plant: Two 11,340kgp (25,000 lb s.t.) General Electric/SNECMA CFM56-5 or IAE V2500 turbofans
Dimensions: Span 33.91m (111.25 ft); length 37.58m (123.25 ft); wing area 122.4m^2 (1,318 sq ft)
Weights: Empty 39,268kg (86,570 lb); MTOW 71,986kg (158,700 lb)
Performance: Maximum speed 903km/h (560 mph) at 8,535m (28,000 ft); operational ceiling 13,000+m (42,650+ ft); range 4,730km (2,940 miles) with maximum payload
Load: Up to 19,142kg (42,200 lb) including belly cargo

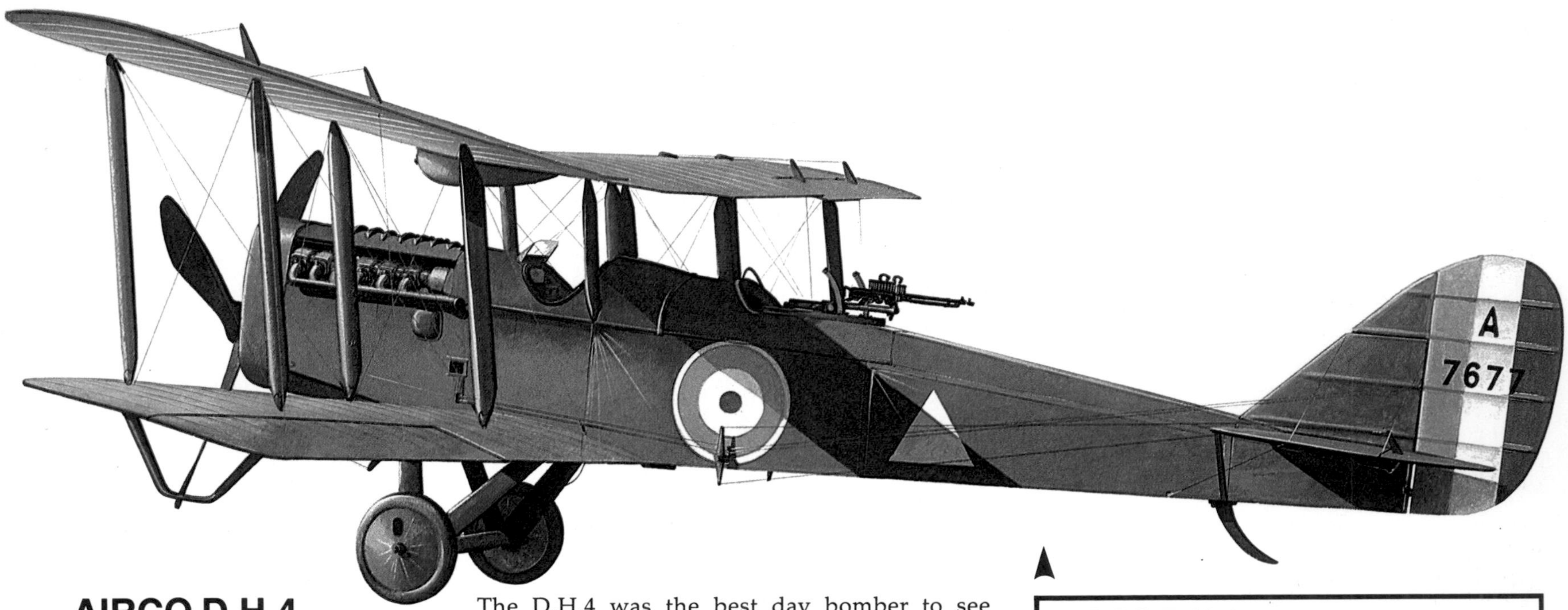

AIRCO D.H.4
(U.K.)

The D.H.4 was the best day bomber to see service in World War I. The type's primary limitation was the physical separation of the cockpits, making communication between pilot and gunner difficult, but this was more than offset by the type's excellent agility and performance on the Eagle engine substituted when the proposed B.H.P. engine suffered development delays. The type flew in August 1916, and production in Britain totalled 1,449 with a variety of engine types rated between 149 and 298kW (200 and 400 hp). The type was also built under licence in the U.S.A., where 4,846 D.H.4s were produced with the Liberty engine.

AIRCO D.H.4

Role: Light day bomber
Crew: Two
Power Plant: One 375 hp Rolls-Royce Eagle VIII water-cooled inline
Dimensions: Span 12.92m (42.4 ft); length 9.35m (30.66 ft); wing area 40.3m^2 (434 sq ft)
Weights: Empty 1,083kg (2,387 lb); MTOW 1,575kg (3,472 lb)
Performance: Maximum speed 230km/h (143 mph) at 1,830m (6,000 ft); operational ceiling 6,706m (22,000 ft); endurance 3.75 hours
Warload: Two .303 inch machine guns, plus up to 412kg (908 lb) of bombs

AIRCO D.H.9
(U.K.)

The D.H.9 was planned as a longer-range successor to the D.H.4, the same flying surfaces being combined with a new fuselage that located the pilot and gunner close together and provided better streamlining for the engine, in this instance a 171.5kW (230 hp) Galloway-built B.H.P. engine. The type flew in July 1917, and was ordered into large-scale production with the 224kW (300 hp) version of the B.H.P. developed by Siddeley-Deasy. The engine was unreliable and generally derated to 171.5kW (230 hp) resulting in performance inferior to that of the D.H.4. Some 3,200 were built, but far better were the 2,300 D.H.9As equipped with the 298kW (400 hp) Liberty 12 or 280kW (375 hp) Eagle VIII engines.

AIRCO D.H.9

Role: Light day bomber
Crew/Accommodation: Two
Power Plant: One 230 hp Siddeley-Deasy B.H.P. water-cooled inline
Dimensions: Span 12.92m (42.4 ft); length 9.28m (30.46 ft); wing area 40.32m^2 (430 sq ft)
Weights: Empty 1,012kg (2,230 lb); MTOW 1,508kg (3,325 lb)
Performance: Maximum speed 177km/h (110 mph) at 3,048m (10,000 ft); operational ceiling 4,724m (15,500 ft); endurance 4.5 hours
Load: Two .303 inch machine guns, plus up to 412kg (908 lb) bombload

AIRSPEED AMBASSADOR
(U.K.)

The Ambassador was one of the most elegant aircraft ever built, and resulted from the 1943 Brabazon Committee recommendation for a post-war 30-seat short/medium-range airliner. The first Ambassador flew in July 1947 and, with two Bristol Centaurus radials, had very promising performance. An order for 20 aircraft was received from BEA. The programme was then beset by a number of problems that so delayed its service entry up to 1952 that the initial 20-aircraft order was the only one fulfilled, as this piston-engined type had been overtaken in performance and operating economics by the turboprop-powered Vickers Viscount.

AIRSPEED AMBASSADOR

Role: Short range passenger transport
Crew/Accommodation: Three, plus three cabin crew and 47/49 passengers
Power Plant: Two 2,700 hp Bristol Centaurus 661 air-cooled radials
Dimensions: Span 35.05m (115 ft); length 24.69m (81 ft); wing area 111.48m^2 (1,200 sq ft)
Weights: Empty 16,277kg (35,884 lb); MTOW 23,590kg (52,000 lb)
Performance: Cruise speed 483km/h (300 mph) at 6,096m (20,000 ft); range 1,159km (720 miles) with maximum payload
Load: Up to 5,285kg (11,650 lb)

AIRSPEED ENVOY and OXFORD
(U.K.)

The Envoy was designed in 1933 as a larger twin-engined version of the single-engined Courier, and the prototype flew in June 1934. The type was mainly of wooden construction and carried, in addition to the pilot, eight passengers in an extensively glazed cabin. The 50 production aircraft were built in three types, the Series I aircraft lacking flaps, the Series II machines having split flaps and the Series III machines having a number of improvements. The aircraft were powered by a wide assortment of air-cooled radial piston engines and proved very popular. Developed from the Envoy for military training was the immensely successful Oxford, of which 8,586 were built.

AIRSPEED ENVOY III

Role: Short range passenger transport
Crew/Accommodation: One, plus up to eight passengers
Power Plant: Two 375 hp Armstrong Siddeley Cheetah AS9 air-cooled radials
Dimensions: Span 15.94m (52.33 ft); length 10.53m (34.5 ft); wing area 31.5m^2 (339 sq ft)
Weights: Empty 1,842kg (4,057 lb); MTOW 2,860kg (6,300 lb)
Performance: Cruise speed 309km/h (192 mph) at 2,226m (7,300 ft); operational ceiling 6,860m (22,500 ft); range 1,045km (650 miles)
Load: Up to 941kg (2,073 lb) including fuel

AIRSPEED HORSA
(U.K.).

This was the U.K.'s primary assault glider of World War II, and is best remembered for its part in the Arnhem operation. The type was designed to a 1940 specification, which required that the glider must be capable of easy assembly from a number of subcontracted assemblies. The Horsa was thus created from 30 subassemblies produced mainly by woodworking companies. The prototype flew in September 1941, and production totalled some 1,469 Horsa Mk Is, and 1,270 Horsa Mk IIs with twin nosewheels and a hinged nose for direct loading/unloading of light land vehicles.

AIRSPEED HORSA Mk I
Role: Military assault glider
Crew/Accommodation: Two, plus up to 25 troops
Power Plant: None
Dimensions: Span 26.8m (88 ft); length 20.4m (67 ft); wing area 102.5m^2 (1,104 sq ft)
Weights: Empty 3,800kg (8,370 lb); MTOW 7,030kg (15,500 lb)
Performance: Normal gliding speed 161km/h (100 mph) operational ceiling dependent upon tug aircraft
Load: Up to 2,835kg (6,250 lb)

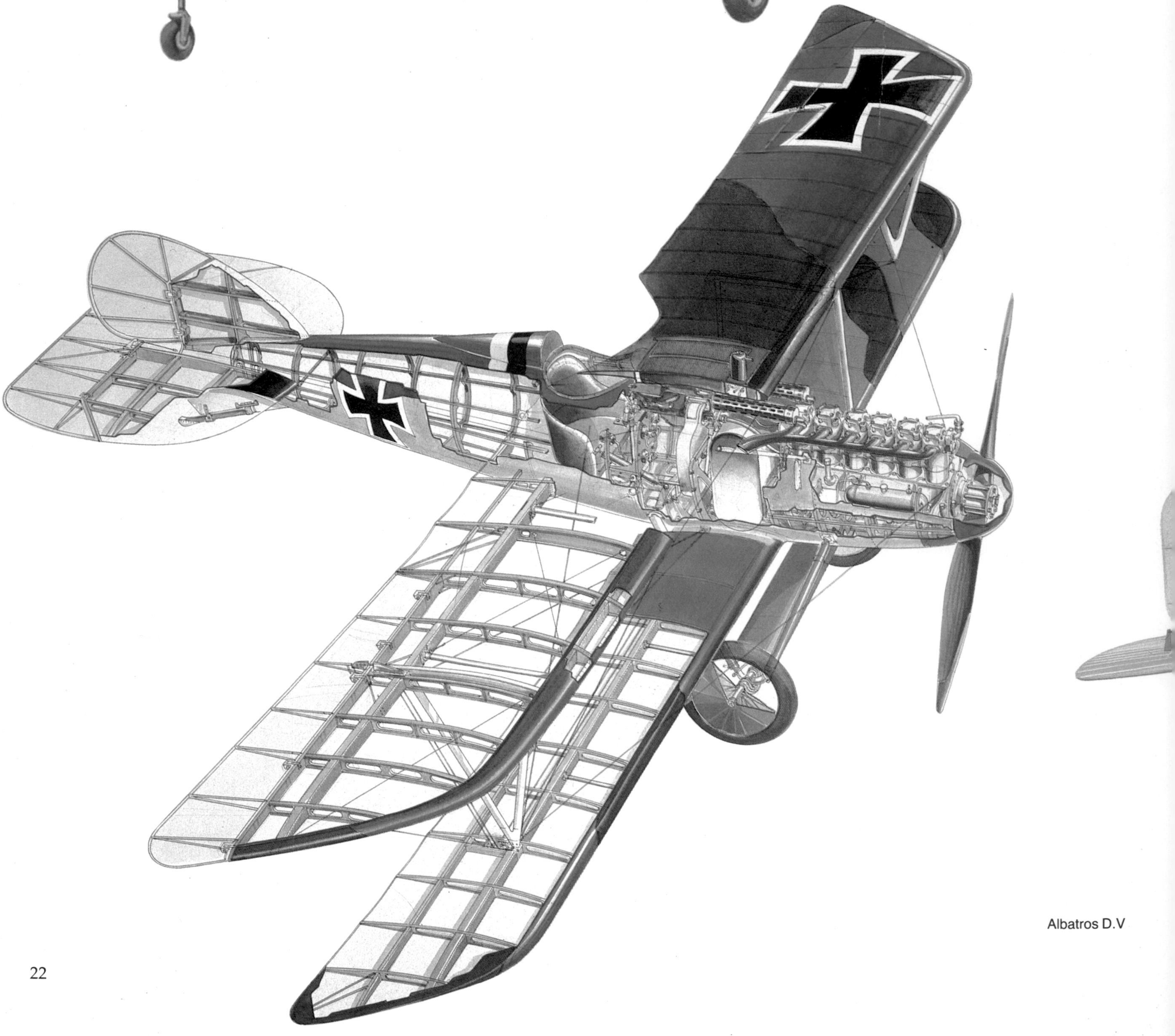

Albatros D.V

ALBATROS C VII
(Germany)

Typical of the general-purpose two-seaters of the 1917 period, the C VII was derived from the C V, which offered high performance but suffered from heavy controls and the overheating of its V-8 engine. The C VII was produced as an interim type pending the arrival of the definitive C X, and required considerable modification of the fuselage to accommodate the short V-6 power plant and its fuselage-mounted 'ear' radiators. The type proved both successful and popular in service, between 300 and 400 being produced.

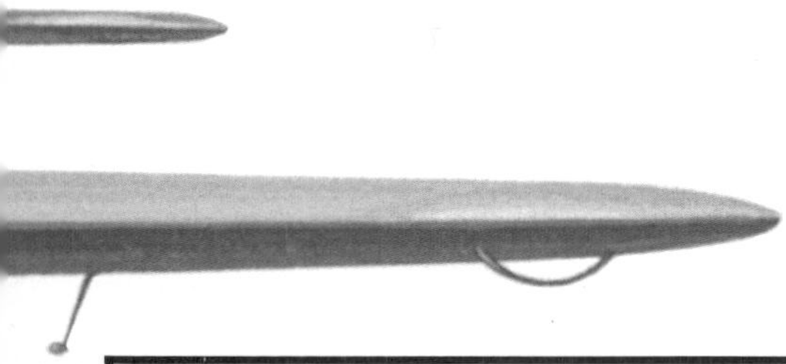

ALBATROS C VII

Role: Reconnaissance
Crew/Accommodation: Two
Power Plant: One 200 hp Benz Bz IV water-cooled inline
Dimensions: Span 12.78m (41.93 ft); length 8.70m (28.54 ft); wing area 43.4m² (468.7 sq ft)
Weights: Empty 989kg (2,176 lb); MTOW 1,550kg (3,410 lb)
Performance: Maximum speed 170km/h (106 mph) at sea level; operational ceiling 5,000m (16,404 ft); endurance 3.33 hours
Load: Two 7.9mm machine guns

ALBATROS D III and D V
(Germany)

The first Albatros fighters were the excellent D I and improved D II with virtually identical upper and lower wings. In an effort to improve manoeuvrability the designers then produced the D III with an increased-span upper wing connected to the smaller lower wing by V-section interplane struts. The D III entered service in spring 1917 and proved most successful until the Allies introduced types such as the Royal Aircraft Factory S.E.5, Sopwith Camel and Spad S.13. Albatros had introduced the D V in May 1917 with features such as a larger spinner and elliptical-section fuselage to reduce drag and so boost performance, and greater emphasis was then placed on this model and its D Va derivative with the upper wing and aileron control system of the D III. In fact the D V and D Va were outclassed by Allied fighters, and their lower wings were structurally deficient in the dive.

ALBATROS D III

Role: Fighter
Crew/Accommodation: One
Power Plant: One 175 hp Mercedes D.IIIa water-cooled inline
Dimensions: Span 9.05m (29.76 ft); length 7.33m (24 ft); wing area 20.5m² (220.7 sq ft)
Weights: Empty 680kg (1,499 lb); MTOW 886kg (1,953 lb)
Performance: Maximum speed 175km/h (109 mph) at 1,000m (3,280 ft); operational ceiling 5,500m (18,045 ft); endurance 2 hours
Load: Two 7.92mm machine guns

AMIOT 143
(France)

The Amiot 143 was surely one of the most ungainly aircraft ever built, and resulted from a 1928 French requirement for a day/night bomber also able to function in the reconnaissance and long-range escort roles. The prototype first flew in April 1931 with a narrow, very deep and angular fuselage, supporting wings big enough to allow internal access to the engines in flight, as well as fixed landing gear whose main wheels had huge speed fairings. Nose and dorsal turrets were complemented by a third gun position at the rear of the massive under-fuselage gondola that accommodated the navigator/bomb-aimer and bomb bay.

AMIOT 143M

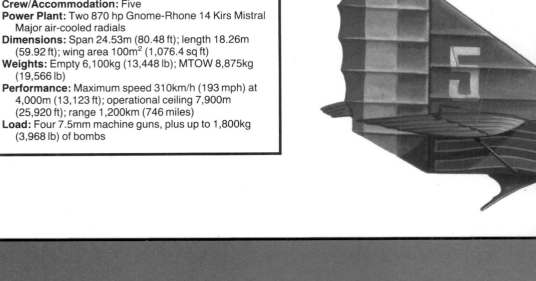

Role: Night bomber/reconnaissance
Crew/Accommodation: Five
Power Plant: Two 870 hp Gnome-Rhone 14 Kirs Mistral Major air-cooled radials
Dimensions: Span 24.53m (80.48 ft); length 18.26m (59.92 ft); wing area 100m^2 (1,076.4 sq ft)
Weights: Empty 6,100kg (13,448 lb); MTOW 8,875kg (19,566 lb)
Performance: Maximum speed 310km/h (193 mph) at 4,000m (13,123 ft); operational ceiling 7,900m (25,920 ft); range 1,200km (746 miles)
Load: Four 7.5mm machine guns, plus up to 1,800kg (3,968 lb) of bombs

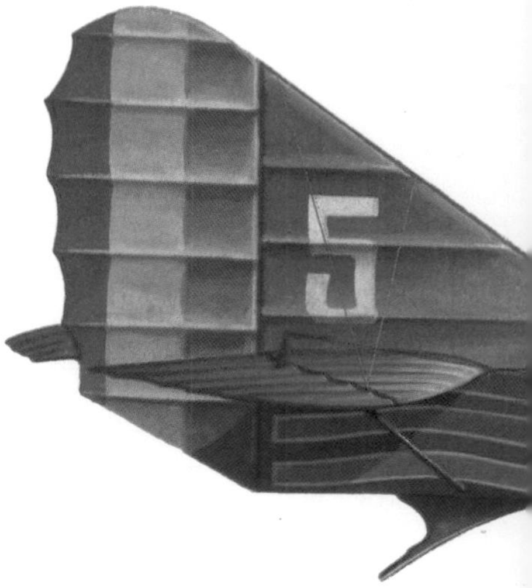

AMX INTERNATIONAL AMX
(Italy/Brazil)

Entering service in 1988, the AMX is a light attack fighter developed on a 70/30% basis by Italian and Brazilian interests. The type was designed for adequate performance but modest cost, and the limited but capable avionics fit was planned without high-performance radar for this reason. A wide assortment of weapons can be carried, and plans call for a two-seat version for training and a number of tactical roles, as well as a radar-equipped single-seater for the dedicated anti-ship role with missiles such as the West German Kormoran or Italian Sea Killer Mk 2.

AMX INTERNATIONAL AMX

Role: Light strike
Crew/Accommodation: One
Power Plant: One 5,000kgp (11,030 lb s.t.) Rolls-Royce Spey Mk 807 turbofan
Dimensions: Span 8.87m (29.1 ft); length 13.23m (47.38 ft); wing area 21m^2 (226.5 sq ft)
Weights: Empty 6,700kg (14,770 lb); MTOW 12,500kg (27,557 lb)
Performance: Maximum speed 913km/h (493 knots) at 10,975m (36,000 ft); radius 520km (322 miles) with 2,719kg (5,996 lb) of warload
Load: One 20mm rotary cannon or two 30mm cannon, plus 3,800kg (8,377 lb) of externally carried weapons, including two short-range air-to-air missiles

ANSALDO SVA
(Italy)

This aeroplane was designed in 1916 as a fighter with the 164kW (220 hp) SPA 6A engine, and first flew in March 1917. The type's most distinctive features were the triangular-section rear fuselage and Warren truss type of interplane strutting. The initial-production SVA 4s lacked the manoeuvrability to serve as fighters and were therefore used as fast reconnaissance aircraft. From this variant evolved the reduced-span SVA 3 interceptor, the SVA 5 long-range reconnaissance-bomber, the unsuccessful Idro-SVA twin-float fighter, the SVA 9 two-seat trainer and the SVA 10 two-seat reconnaissance-bomber.

ANSALDO SVA 1

Role: Fighter
Crew/Accommodation: One
Power Plant: One 220 hp SPA 6A water-cooled inline
Dimensions: Span 7.68m (25.2 ft); length 6.84m (22.44 ft)
Weights: MTOW 885kg (1,951 lb)
Performance: Maximum speed 220km/h (137 mph) at 2,000m (6,562 ft); operational ceiling 5,000m (16,404 ft); endurance 1.5 hours
Load: Two .303 inch machine guns

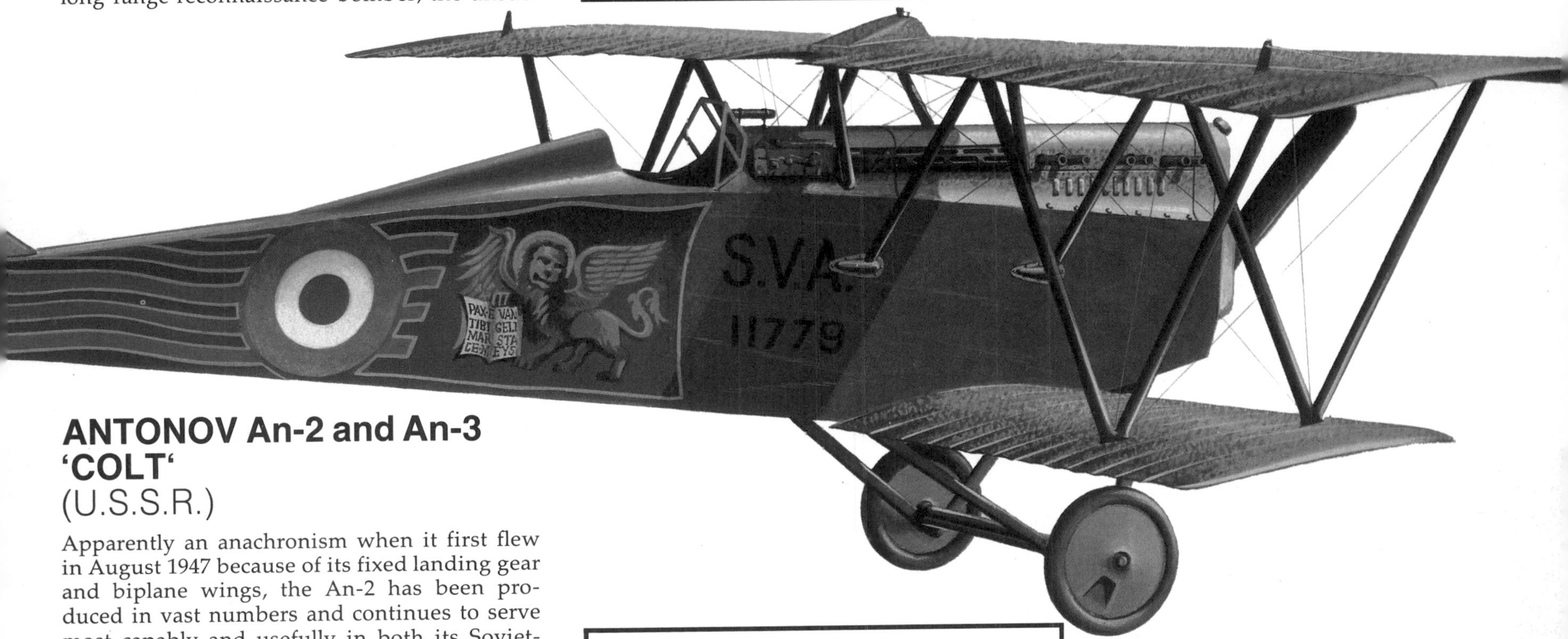

ANTONOV An-2 and An-3 'COLT'
(U.S.S.R.)

Apparently an anachronism when it first flew in August 1947 because of its fixed landing gear and biplane wings, the An-2 has been produced in vast numbers and continues to serve most capably and usefully in both its Soviet- and Polish-built variants. A large all-metal type with fabric covering on the tailplane and rear portions of the wings, the An-2 was designed for agricultural use and is exceptionally rugged. Other roles are transport (12 passengers or 1,240kg/2,733 lb of freight), float-equipped transport, ambulance work, fire-fighting, meteorological research, geophysical research, photogrametric survey and TV relay. The An-3 is the latest agricultural spraying version with a 701kW (940 shp) Glushenkov TVD-10B turboprop giving 40% more payload.

ANTONOV An-2 'COLT'

Role: Utility transport/agricultural spraying
Crew/Accommodation: Two, plus up to 12 passengers
Power Plant: One 1,000 hp Shvetsov ASh-62M air-cooled radial
Dimensions: Span 18.18m (59.65 ft); length 12.74m (41.8 ft); wing area 71.1m² (765 sq ft)
Weights: Empty 3,450kg (7,605 lb); MTOW 5,500kg (12,125 lb)
Performance: Cruise speed 185km/h (100 knots) at 1,750m (5,741 ft); range 900km (485 naut. miles) with 500kg (1,102 lb) payload
Load: Up to 1,500kg (3,306 lb)

ANTONOV An-12 'CUB'
(U.S.S.R.)

The An-12 was developed in parallel with the An-10 airliner as a tactical transport with an upswept rear fuselage/tail unit (complete with twin-cannon tail turret) to allow the incorporation of a rear ramp/door for the straight-in loading/unloading of bulky items and the dropping of paratroops. Some 850 were built up to 1973, and almost all of these serve with the Soviet air force. The type is now being phased out but is still important in its basic 'Cub-A' form. Variants are the 'Cub-B' electronic intelligence, and 'Cub-C' and 'Cub-D' electronic countermeasures models.

ANTONOV An-12 'CUB'

Role: Military troop/freight transport
Crew/Accommodation: Five, plus up to 90 troops
Power Plant: Four 4,000 shp Ivchenko AI-20K turboprops
Dimensions: Span 38m (124.67 ft); length 37m (121.39 ft); wing area 119.5m^2 (1,286 sq ft)
Weights: Empty 28,000kg (61,729 lb); MTOW 61,000kg (134,480 lb)
Performance: Maximum speed 715km/h (385 knots) at 5,100m (16,732 ft); operational ceiling 10,200m (33,464 ft); range 3,600km (1,942 naut. miles) with maximum payload
Load: Up to 20,000kg (44,092 lb)

ANTONOV An-22 'COCK'
(U.S.S.R.)

The An-22 was in its time the world's largest aeroplane, and was designed for the twin tasks of military heavy transport and of support for the resources exploitation industry in Siberia. The type first flew in February 1965, and only about 80 were built. Keynotes of the design are four potent turboprops driving immense contra-rotating propeller units, an upswept tail unit with endplate vertical surfaces above the rear-fuselage ramp/door allowing straight-in loading of items as large as tanks or complete missiles, and the 14-wheel landing gear that allows operations into and out of semi-prepared airstrips.

ANTONOV An-22 ANTHEUS 'COCK'

Role: Long-range freight transport
Crew/Accommodation: Five, plus up to 29 passengers/troops in upper cabin
Power Plant: Four 15,000 shp Kuznetsov NK 12MA turboprops
Dimensions: Span 64.4m (211.29 ft); length 57.8m (189.63 ft); wing area 345m^2 (3,713.6 sq ft)
Weights: Empty 114,000kg (251,327 lb); MTOW 250,000kg (551,156 lb)
Performance: Cruise speed 679km/h (366 knots) at 8,000m (26,247 ft); operational ceiling 10,000m (32,808 ft); range 5,000km (2,698 naut. miles) with maximum payload
Load: Up to 80,000kg (176,370 lb)

ANTONOV An-26 'CURL' and An-32 'CLINE'
(U.S.S.R.)

The An-26 was developed from the An-24 airliner as a military transport that began to enter service in 1970. The primary modifications were more powerful engines and a revised rear fuselage to permit the installation of the rear ramp/door typical of modern military transports. As in larger transports there are inbuilt conveyors and winches for the movement of freight in the hold. The An-26B is an improved version, and a specialized survey version is the An-30 with a raised flightdeck and a glazed nose, while the An-32 is an An-26 derivative for 'hot and high' operations with larger diameter propellers that require the radically more powerful engines (3,862kW/5,180 ehp Ivchyenko AI-20M turboprops) to be located above rather than on the leading edges of the wings.

ANTONOV An-26 'CURL'

Role: Short-range freight/troop transport
Crew/Accommodation: Five, plus up to 40 troops
Power Plant: Two 2,820 shp Ivchenko AI-24T turboprops, plus one 900kgp (1,984 lb s.t.) Tumansky RU-19-300 turbojet
Dimensions: Span 29.20m (95.8 ft); length 23.8m (78.08 ft); wing area 74.98m^2 (807 sq ft)
Weights: Empty 15,020kg (33,113 lb); MTOW 24,000kg (52,910 lb)
Performance: Cruise speed 440km/h (237 knots) at 7,000m (22,965 ft); operational ceiling 7,600m (24,934 ft); range 1,100km (594 naut. miles) with maximum payload
Load: Up to 5,500kg (12,125 lb)

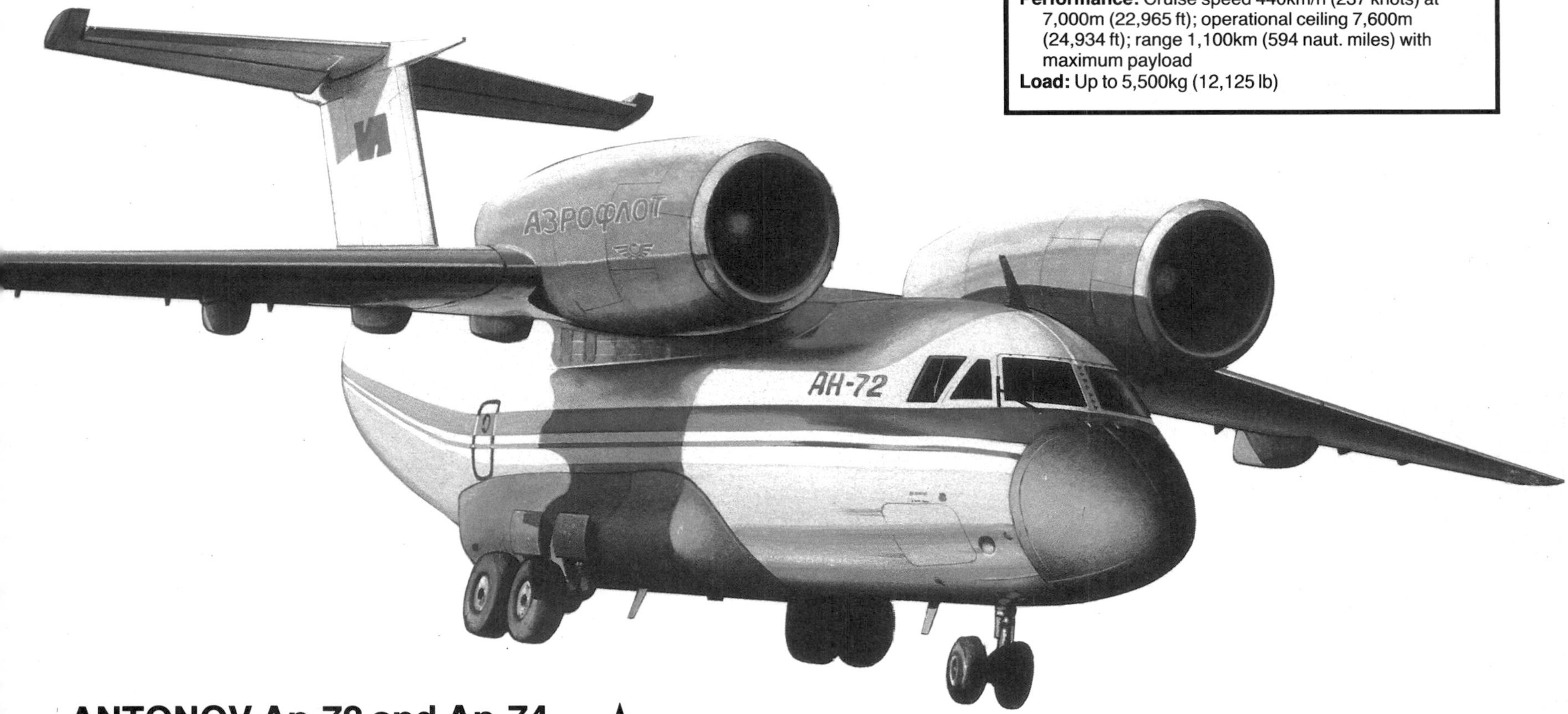

ANTONOV An-72 and An-74 'COALER'
(U.S.S.R.)

First flown in December 1977, the An-72 is the bureau's first jet-powered transport, and uses the advantages of upper-surface blowing, in which the twin turbofans exhaust over the upper surface of the wing and flaps to secure admirable STOL performance. The type is designed for utility operations and has a rear ramp/door under the T-tail. For arctic use the bureau has developed the An-74 variant with a revised wing, a longer fuselage with weather radar, and other modifications, and this has been further adapted as the basis of the 'Madcap' airborne early warning aeroplane, which has a radar rotodome on top of the fin.

ANTONOV An-72 'COALER'

Role: Military short/rough field-capable transport
Crew/Accommodation: Three, plus up to 32 troops
Power Plant: Two 6,500kgp (14,330 lb s.t.) Lotarev D-36 turbofans
Dimensions: Span 25.83m (84.74 ft); length 26.58m (87.2 ft)
Weights: MTOW 30,500kg (67,240 lb)
Performance: Cruise speed 720km/h (389 knots) at 10,000m (32,808 ft); operational ceiling 11,000m (36,089 ft); range 1,000km (539 naut. miles) with maximum payload
Load: Up to 7,500kg (16,534 lb)

ANTONOV An-124 'CONDOR'
(U.S.S.R.)

When introduced in December 1982 this was the world's largest aeroplane, and was designed as successor to the An-22 with the ability to carry larger payloads. The type is cast in the same basic mould as the An-22 apart from its power plant, which comprises four large turbofans rather than four turboprops and permits the carriage of a 150-tonne maximum payload over a range of 4,500km (2,800 miles). Mulit-wheel landing gear allows operation into and out of poor airstrips, and the type is equipped to obviate the need for ground support in remote areas.

ANTONOV An-124 'CONDOR'

Role: Long-range, heavy-lift transport
Crew/Accommodation: Six plus up to 88 seats in upper cabin
Power Plant: Four 23,430kgp (51,650 lb s.t.) Lotarev D-18T turbofans
Dimensions: Span 73.3m (240.49 ft); length 69.5m (228.02 ft); wing area 628m^2 (6,760 sq ft)
Weights: MTOW 450,000kg (892,863 lb)
Performance: Cruise speed 850km/h (459 knots) at 12,000m (39,370 ft); range 4,500km (2,428 naut. miles) with maximum payload
Load: Up to 150,000kg (330,688 lb)

ANTONOV An-225 MRIYA (DREAM)
(U.S.S.R.)

This first flew in December 1988 and succeeded the An-124 as the world's largest aeroplane. The type appears to be based on the outer wings and four-turbofan power plant of the An-124, these being married to a new centre section (with an additional pair of engines) and massive fuselage ending in an empennage with endplate vertical surfaces. This allows the carriage above the fuselage of the Soviet space shuttle or other loads too large to fit into the cavernous hold, which is accessed by a rear ramp/door arrangement. Multi-wheel landing gear bestows the ability to use semi-prepared airstrips.

ANTONOV An-225 MRIYA

Role: Heavy-lift cargo transport
Crew/Accommodation: Three
Power Plant: Six 24,400kgp (53,800 lb s.t.) Lotarev D-18 turbofans
Dimensions: Span 84.4m (290 ft); length 84m (275.6 ft)
Weights: MTOW 600,000kg (1,322,772 lb)
Performance: Cruise speed 700km/h (378 knots) range 4,500km (2,428 naut. miles) with 200,000kg (440,920 lb) payload
Load: Up to 250,000kg (551,145 lb)

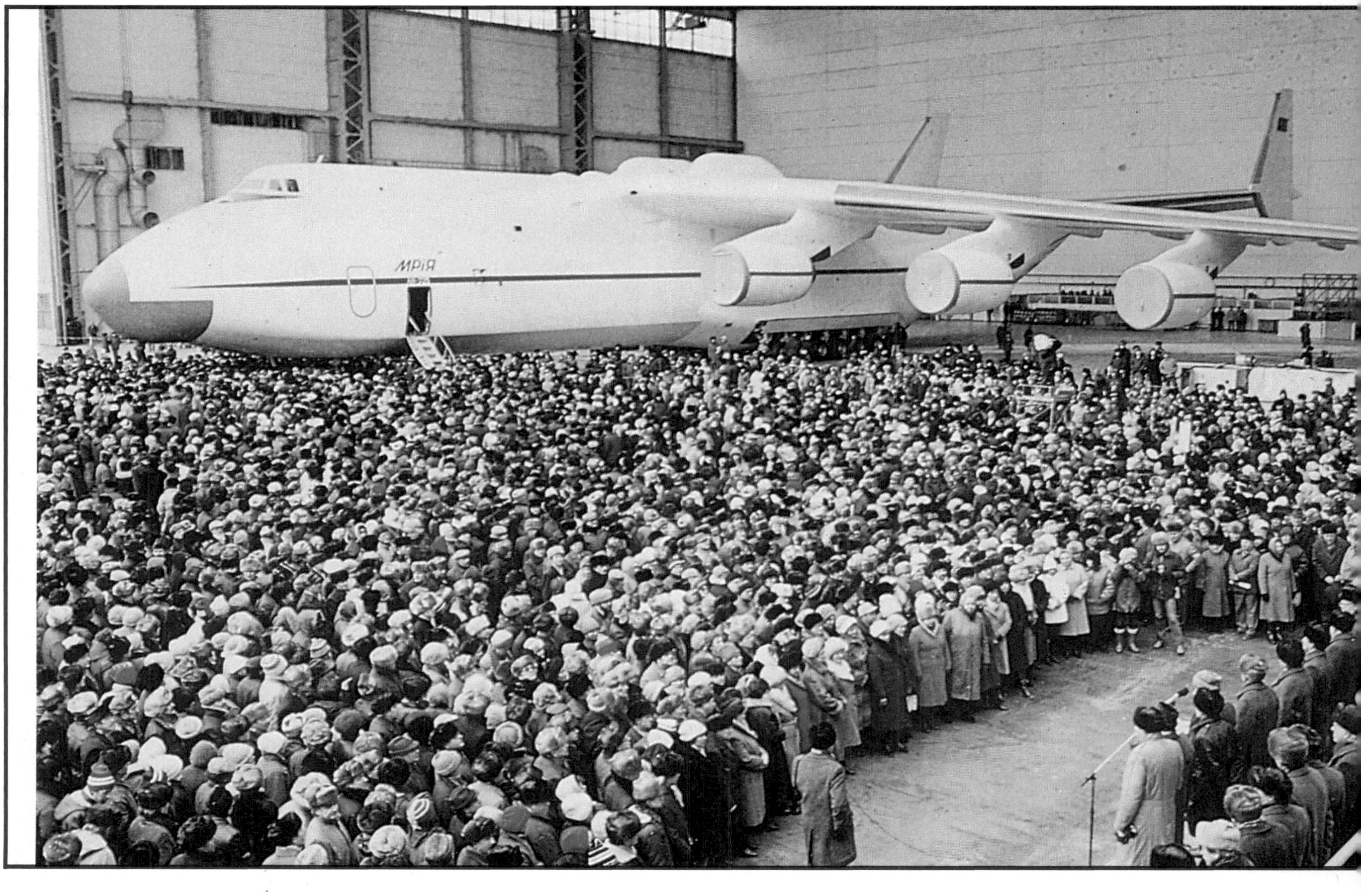

ARADO Ar 64 and Ar 65
(Germany)

The Ar 64 was designed to replace the Fokker D XIII fighters currently equipping the German fighter school at Lipetsk in the U.S.S.R., and first flew in 1930 as the Ar 64a. The small number of Ar 64c and d production aircraft had four- and two-bladed propellers respectively. The Ar 65 was essentially an inline-engined version of the radial-engined Ar 64, and was developed as the Ar 65a and improved Ar 65d prototypes leading to the Ar 65E and better-equipped Ar 65F production models. The type was replaced as a fighter after a few months by the Heinkel He 51 and was then used as a fighter trainer.

ARADO Ar 65E

Role: Fighter
Crew/Accommodation: One
Power Plant: One 600 hp BMW VI U water-cooled inline
Dimensions: Span 11.2m (36.75 ft); length 8.4m (27.53 ft); wing area 23m^2 (322.9 sq ft)
Weights: Empty 1,510kg (3,329 lb); MTOW 1,930kg (4,255 lb)
Performance: Maximum speed 300km/h (186 mph) at 1,650m (5,415 ft); operational ceiling 7,600m (24,935 ft); endurance 1.75 hours
Load: Two 7.92mm machine guns

Not illustrated

ARADO Ar 196
(Germany)

The Ar 196 was designed to meet a 1936 requirement for a floatplane reconnaissance aircraft suitable for the coastal and catapult-launched warship roles. The type clearly had kinship with the Ar 95 but was a monoplane and, after evaluation in the twin-float and single central/twin outrigger float layouts, was ordered as the twin-float Ar 196A. Production was more than 500 aircraft, these being a few Ar 196A-1s and strengthened Ar 196A-4s for shipboard operation plus, for the coastal role, the Ar 196A-2 with two 20mm fixed cannon, the strengthened Ar 196A-3 with a variable-pitch propeller, and the Ar 196A-5 with better radio and a twin instead of a single machine gun installation for the radio-operator/gunner.

▲

ARADO Ar 196A-3

Role: Shipborne reconnaissance floatplane
Crew/Accommodation: Two
Power Plant: One 960 hp BMW 132K air-cooled radial
Dimensions: Span 12.4m (40.68 ft); length 11m (36.09 ft); wing area 28.4m^2 (305.6 sq ft)
Weights: Empty 2,990kg (6,593 lb); MTOW 3,730kg (8,225 lb)
Performance: Maximum speed 310km/h (193 mph) at 4,000m (13,120 ft); operational ceiling 7,000m (22,960 ft); range 1,070km (665 miles)
Load: Two 20mm cannon and two 7.9mm machine guns, plus 100kg (210 lb) of bombs

Arado Ar 234C

Arado Ar 234B

ARADO Ar 234 BLITZ (LIGHTNING)
(Germany)

This was the world's first purpose-designed jet bomber. As first flown in June 1943 the Ar 234 took off from a trolley and landed on retractable skids, and the various prototypes trialled twin- or four-engined power plants. Pre-production aircraft featured a wider fuselage to make possible the installation of retractable tricycle landing gear. The Ar 234B series had two engines but no pressurization or ejector seat, and comprised the Ar 234B-1 reconnaissance and Ar 234B-2 bomber variants. The Ar 234C had four engines, cabin pressurization and an ejector seat, and comprised the Ar 234C-1 reconnaissance, Ar 234C-2 bomber, Ar 234C-3 multi-role and Ar 234C-4 armed reconnaissance variants. There were also a large number of development and projected models.

ARADO Ar 234B-2 BLITZ

Role: High speed bomber
Crew/Accommodation: One
Power Plant: Two 900kgp (1,980 lb s.t.) Junkers 004B Orkan turbojets
Dimensions: Span 14.44m (47.38 ft); length 12.64m (41.47 ft); wing area 27.3m^2 (284.2 sq ft)
Weights: Empty 5,200kg (11,464 lb); MTOW 9,800kg (21,715 lb)
Performance: Maximum speed 742km/h (461 mph) at 6,000m (19,685 ft); operational ceiling 10,000m (32,810 ft); range 1,556km (967 miles) with 500kg (1,102 lb) payload
Load: Two rear-firing 20mm cannon, plus up to 2,000kg (4,410 lb) of bombs

ARMSTRONG WHITWORTH SISKIN
(U.K.)

The Siskin was the mainstay of the RAF's fighter arm in the mid-1920s, and originated from the Siddeley-Deasy S.R.2 of 1919. The Siskin first flew with its definitive Jaguar radial in 1921, but had then to be recast as the all-metal Siskin Mk III of 1923. The 64 Siskin Mk IIIs began to enter service in 1924, and were supplemented by 348 Siskin Mk IIIAs with the supercharged Jaguar IV, and 53 Siskin Mk IIIDC dual-control trainers. The Siskin Mk IIIB, Mk IV and Mk V were experimental and racing types.

ARMSTRONG WHITWORTH SISKIN Mk IIIA

Role: Fighter
Crew/Accommodation: One
Power Plant: One 400 hp Armstrong Siddeley Jaguar IVS
Dimensions: Span 10.11m (33.16 ft); length 7.72m (25.33 ft); wing area 27.22m^2 (293 sq ft)
Weights: Empty 997kg (2,198 lb); MTOW 1,260kg (2,777 lb)
Performance: Maximum speed 227km/h (141 mph) at sea level; operational ceiling 6,401m (21,600 ft); endurance 2.75 hours
Load: Two .303 inch machine guns

ARMSTRONG WHITWORTH WHITLEY
(U.K.)

The Whitley was one of the U.K.'s most important bombers at the beginning of World War II, and was built to the extent of 1,814 examples. The prototype flew in February 1937, and production amounted to 34 Whitley Mk Is with Armstrong Siddeley Tiger X engines, 46 Whitley Mk IIs with Tiger VIIIs, 80 Whitley Mk IIIs with a powered nose turret and an enlarged bomb bay, 33 Whitley Mk IVs with Rolls-Royce Merlin IVs and a powered tail turret, seven Whitley Mk IVAs with Merlin Xs, and 1,446 improved Whitley Mk Vs. The 146 Whitley Mk VIIs were essentially Mk Vs modified for maritime reconnaissance, and the Whitley Mk VI was a proposed model equipped with Pratt & Whitney radial engines.

ARMSTRONG WHITWORTH WHITLEY Mk V

Role: Heavy bomber
Crew/Accommodation: Five
Power Plant: Two 1,145 hp Rolls-Royce Merlin X water-cooled inlines
Dimensions: Span 25.6m (84 ft); length 21.11m (69.25 ft); wing area 105.63m^2 (1,137 sq ft)
Weights: Empty 8,777kg (19,350 lb); MTOW 15,196kg (33,500 lb)
Performance: Maximum speed 370km/h (230 mph) at 5,000m (16,400 ft); operational ceiling 7,925m (26,000 ft)
Load: Five .303 inch machine guns, plus up to 3,175kg (7,000 lb) bombload

ATLANTIC F.9 FOKKER F.VII-3m) (Netherlands/U.S.A.)

The F.9 and military C-2 were versions of the F.VII-3m by Fokker's U.S. subsidiary, while the F.10 and military C-5 were larger versions. The F.VII first flew in 1924, and led to the F.VIIA with the 298 kW (400 hp) Liberty engine. Greater promise was offered by three lower-powered radials leading to the F.VIIA-3m and longer-span F.VIIB-3m that were of great importance in the development of European and third-world air transport. The type was also used extensively for route-prroving and record-breaking flights. The thick cantilever wings were of plywood-covered wooden construction, and the fuselage of welded steel tubes covered in fabric.

ATLANTIC-FOKKER C-2A

Role: Military transport
Crew/Accommodation: Two and up to 10 personnel
Power Plant: Three 220 hp Wright R-790 Whirlwind air-cooled radials
Dimensions: Span 22.61m (74.19 ft); length 14.73m (48.33 ft); wing area 66.71m^2 (718 sq ft)
Weights: Empty 2,952kg (6,507 lb); MTOW 4,715kg (10,394 lb)
Performance: Cruise speed 145km/h (90 mph) at sea level; operational ceiling 4,877m (16,000 ft); range 476km (296 miles)
Load: Up to 930kg (2,050 lb)

AVIA B.534 (Czechoslovakia)

The most important Czech aeroplane of the period between the world wars, the B.534 was built to the extent of 566 examples and was a classic biplane fighter. The B.534/1 prototype that flew in August 1933 was the unsuccessful B.34/2 re-engined with a Hispano-Suiza 12Ybrs, while the B.534/2 had an enclosed cockpit, larger rudder and revised landing gear. This led to the B.534-I initial production version with an open cockpit, the B.534-II with revised armament, the B.534-III with wheel spats, the B.534-IV with an enclosed cockpit, and the Bk.534 with an engine-mounted cannon or machine gun. There were also a number of experimental and development variants.

AVIA B.534-IV

Role: Fighter
Crew/Accommodation: One
Power Plant: One 830 hp Avia-built Hispano-Suiza 12Ybrs water-cooled inline
Dimensions: Span 9.4m (30.83 ft); length 8.2m (26.92 ft); wing area 23.56m^2 (253.6 sq ft)
Weights: Empty 1,460kg (3,219 lb); MTOW 1,985kg (4,376 lb)
Performance: Maximum speed 405km/h (252 mph) at 4,400m (14,435 ft); operational ceiling 10,600m (34,776 ft); range 580km (360 miles)
Load: Four 7.7mm machine guns

AVIATIK B and C SERIES
(Germany)

The Aviatik B I was one of Germany's first reconnaissance aircraft in World War I, and appeared in 1914 in two-bay as well as increased-span three-bay variants with the observer located forward of the pilot in a position where his fields of vision were seriously impaired by struts and rigging wires. The B II of 1915 retained the same layout, but had an 89kW (120 hp) Mercedes D.II rather than the B I's 75kW (100 hp) Mercedes D.I inline engine and a lighter but more refined structure for slightly better performance. Basically the same structure and layout were used in the C I armed two-seater that appeared in 1915. The C II was similar apart from the change from a Mercedes D.III to a Benz Bz.IV engine, and the main production model was the C III with reduced span and aerodynamic refinements.

AVIATIK CIII

Role: Reconnaissance
Crew/Accommodation: Two
Power Plant: One 160 hp Mercedes D.III water-cooled inline
Dimensions: Span 11.8m (38.71 ft); length 8.08m (26.51 ft); wing area 35m^2 (378 sq ft)
Weights: Empty 980kg (2,156 lb); MTOW 1,340kg (2,948 lb)
Performance: Maximum speed 160km/h (99.4 mph) at sea level; operational ceiling 4,500m (14,760 ft); endurance 3 hours
Load: Two 7.9mm machine guns

Aviatik B.I

AVIONS DE TRANSPORT REGIONAL ATR 42 and ATR 72
(France/Italy)

ATR was formed by Aerospatiale and Aeritalia to create the ATR 42 turboprop-powered regional airliner, of which the first example flew in August 1984. The designations ATR 42-100 and ATR 42-200 were at first planned as lighter and heavier-weight variants, but the initial production model was the ATR 42-200, with the ATR 42-300 as an optional higher-weight variant. In 1985 the consortium launched the ATR 72 with passenger capacity increased from 50 to 74, greater span and 1,790kW (2,400 shp) Pratt & Whitney PW124 turboprops. Other variants are the ATR 42F freighter and proposed ATM 42 military model.

AVIONS DE TRANSPORT REGIONAL ATR 42

Role: Short range passenger transport
Crew/Accommodation: Two and one cabin crew, plus up to 50 passengers
Power Plant: Two 2,277 shp Pratt & Whitney of Canada PW120 turboprops
Dimensions: Span 24.57m (80.58 ft); length 22.67m (74.5 ft); wing area 54.5m^2 (586 sq ft)
Weights: Empty 9,973kg (21,986 lb); MTOW 16,150kg (35,605 lb)
Performance: Cruise speed 490km/h (265 knots) at sea level; operational ceiling 7,620m (25,000 ft); range 1,668km (900 naut. miles) with maximum payload
Load: Up to 4,827kg (10,644 lb)

AVRO 504
(U.K.)

One of the most remarkable aircraft of all time, the Avro 504 was developed from the Avro 500 and first flew in July 1913. The 504 was ordered as a general-purpose aeroplane by the British army and navy, and after some front-line service (including the bombing of the Zeppelin sheds at Friedrichshafen in November 1914) the type found its metier as a trainer. The main first-line types were the 504, the 504A with smaller ailerons, the 504B with a larger fin, the 504C anti-Zeppelin single-seater and the 504D, all powered by a 60kW (80 hp) Gnome rotary engine. Trainer and civil variants were the 504E with less wing stagger, the 504J of 1916 with engines varying in power between 60 and 97 kW (80 and 130 hp), the 504J with a universal engine mounting, the 504L floatplane, the 504M cabin transport, and the 504N with a number of structural revisions and a radial engine. Well over 8,000 504s were built in World War I.

AVRO 504K

Role: Reconnaissance/trainer
Crew/Accommodation: Two
Power Plant: One 130 hp Clerget air-cooled rotary
Dimensions: Span 10.97m (36 ft); length 7.75m (25.42 ft); wing area 30.66m^2 (330 sq ft)
Weights: Empty 558kg (1,231 lb); MTOW 830kg (1,829 lb)
Performance: Maximum speed 153km/h (95 mph) at sea level; operational ceiling 5,486m (18,000 ft); endurance 3 hours
Load: A number of 504s were converted into home defence single-seat fighters with one .303 inch machine gun, plus stowage for hand-released small bombs

AVRO ANSON
(U.K.)

The Anson was built in very large numbers before, during and after World War II for a host of military and training tasks, but originated from a 1933 Imperial Airways' requirement for a four-passenger transport, which first flew as the Avro 652 in January 1935. Only two 652s were built, the basic design then being adapted as the Avro 652A to meet an RAF coastal reconnaissance requirement. The prototype flew in March 1935, and the type was accepted as the Anson Mk I, leaving first-line service in 1942. The Anson's greatest success was as a trainer and transport/communications aeroplane in a number of forms with different engines, and as such it remained in production up to May 1952.

AVRO ANSON Mk I

Role: General purpose reconnaissance
Crew/Accommodation: Three
Power Plant: Two 335 hp Armstrong Siddeley Cheetah IX air-cooled radials
Dimensions: Span 17.22m (56.5 ft); length 12.87m (42.25 ft); wing area 43.01m^2 (463 sq ft)
Weights: Empty 2,438kg (5,375 lb); MTOW 3,629kg (8,000 lb)
Performance: Maximum speed 302km/h (188 mph) at sea level; operational ceiling 5,791m (19,000 ft); range 1,062km (660 miles)
Load: Two .303 inch machine guns, plus up to 163kg (360 lb) of bombs

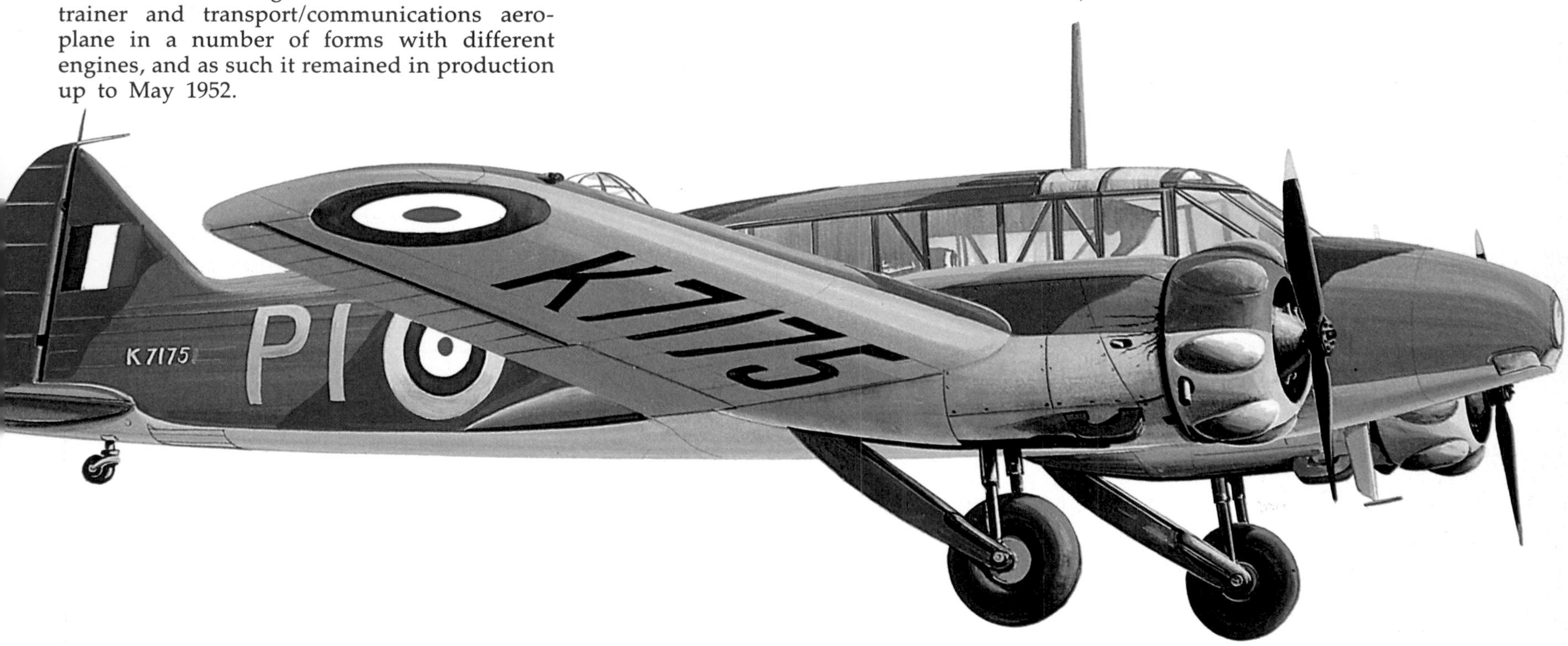

AVRO LANCASTER
(U.K.)

Certainly the best night bomber of World War II, the Lancaster was conceived as a four-engined development of the twin-engined Manchester, which failed because of the unreliability of its Rolls-Royce Vulture engines. The first Lancaster flew in January 1941 and the type was ordered into large-scale production as the Lancaster Mk I (later B.Mk I). A feared shortage of Merlins led to the development of the Lancaster Mk II with Bristol Hercules VI or XVI radial engines, but only 301 were built as performance was degraded and Merlins in abundant supply. The Mk Is were soon supplemented by B.Mk IIIs and Canadian-built B.Mk Xs with Packard-built Merlins. The final production version was the B.Mk VIII with an American dorsal turret, and total production was 7,378. After the war Lancasters were converted for a number of other roles.

AVRO LANCASTER Mk I

Role: Heavy night bomber
Crew/Accommodation: Seven
Power Plant: Four 1,640 hp Rolls-Royce Merlin 24 water-cooled inlines
Dimensions: Span 31.09m (102 ft); length 21.18m (69.5 ft); wing area 120.49m^2 (1,297 sq ft)
Weights: Empty 16,780kg (37,000 lb); MTOW 29,480kg (65,000 lb)
Performance: Maximum speed 394km/h (245 mph) at sea level; operational ceiling 6,706m (22,000 ft); range 3,589km (2,230 miles) with 3,182kg (7,000 lb) bombload
Load: Eight .303 inch machine guns, plus up to 8,165kg (18,000 lb) of bombs

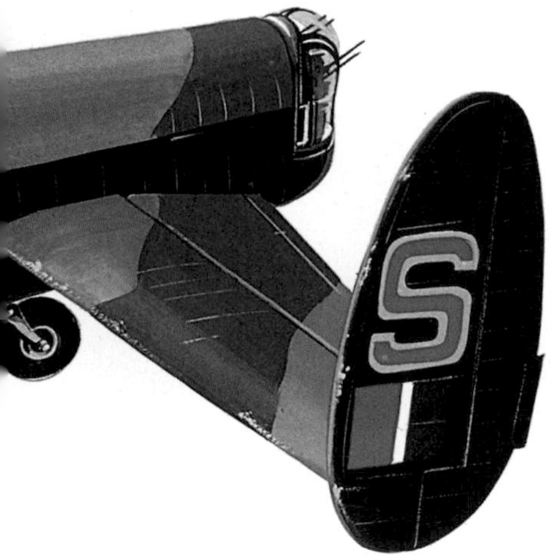

AVRO LINCOLN
(U.K.)

The Lincoln was planned as the Lancaster Mk IV long-range high-altitude bomber, but emerged as so different an aeroplane that it was given the name Lincoln. The prototype flew in June 1944, and though plans were laid for 2,254 aircraft, British post-war production amounted to only 72 Lincoln B.Mk 1s and 465 Lincoln B.Mk 2s. Canada completed one Lincoln B.Mk XV, but Australia completed 43 Lincoln B.Mk 30s and 30 Lincoln B.Mk 30As, and 20 of these were later modified to Lincoln B.Mk 31 standard with a longer nose accommodating search radar and two operators.

AVRO LINCOLN B.Mk 2

Role: Heavy bomber
Crew/Accommodation: Seven
Power Plant: Four 1,760 hp Rolls-Royce Merlin 68A water-cooled inlines
Dimensions: Span 36.57m (120 ft); length 23.85m (78.3 ft); wing area 132m^2 (1,421 sq ft)
Weights: Empty 20,044kg (44,188 lb); MTOW 37,194kg (82,000 lb)
Performance: Maximum speed 467km/h (290 mph) at 6,100m (20,000 ft); operational ceiling 6,710m (22,000 ft); range 3,620km (2,250 miles) with 6,364kg (14,000 lb) bombload
Load: Two .5 inch machine guns (nose turret), two 20mm cannon (mid upper turret), two .5 inch machine guns (tail turret), plus up to 9,979kg (22,000 lb) of bombs carried internally

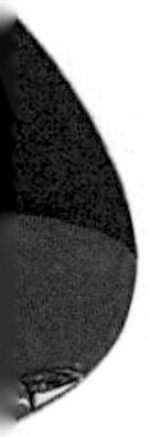

AVRO SHACKLETON
(U.K.)

After experience with the Lancaster in the maritime role after World War II, it was decided to develop a specialized aeroplane with the wing and landing gear of the Lincoln married to a new fuselage and revised empennage. The first Shackleton flew in March 1949, leading to production Shackleton GR.Mk 1 (later MR.Mk 1) and MR.Mk 1A aircraft. The Shackleton MR.Mk 2 had revised armament and search radar with its antenna in a retractable 'dustbin' rather than a chin radome, while the definitive Shackleton MR.Mk 3 was significantly updated, lost the dorsal turret but gained underwing hardpoints and a tricycle landing gear. Some 12 surplus MR.Mk 2s were converted in the 1970s to Shackleton AEW.Mk 2 airborne early warning aircraft with specialist radar.

AVRO SHACKLETON AEW.Mk 2

Role: Airborne early warning
Crew/Accommodation: Ten
Power Plant: Four 2,456 hp Rolls-Royce Griffon 57A water-cooled inlines
Dimensions: Span 36.52m (119.83 ft); length 28.19m (92.5 ft); wing area 135.45m^2 (1,458 sqft)
Weights: Empty 25,583kg (56,400 lb); MTOW 44,452kg (98,000 lb)
Performance: Cruise speed 322km/h (200 mph) at 3,050m (10,000 ft); operational ceiling 5,852m (19,200 ft); endurance 16 hours
Load: None, other than APS 20 long-range search radar

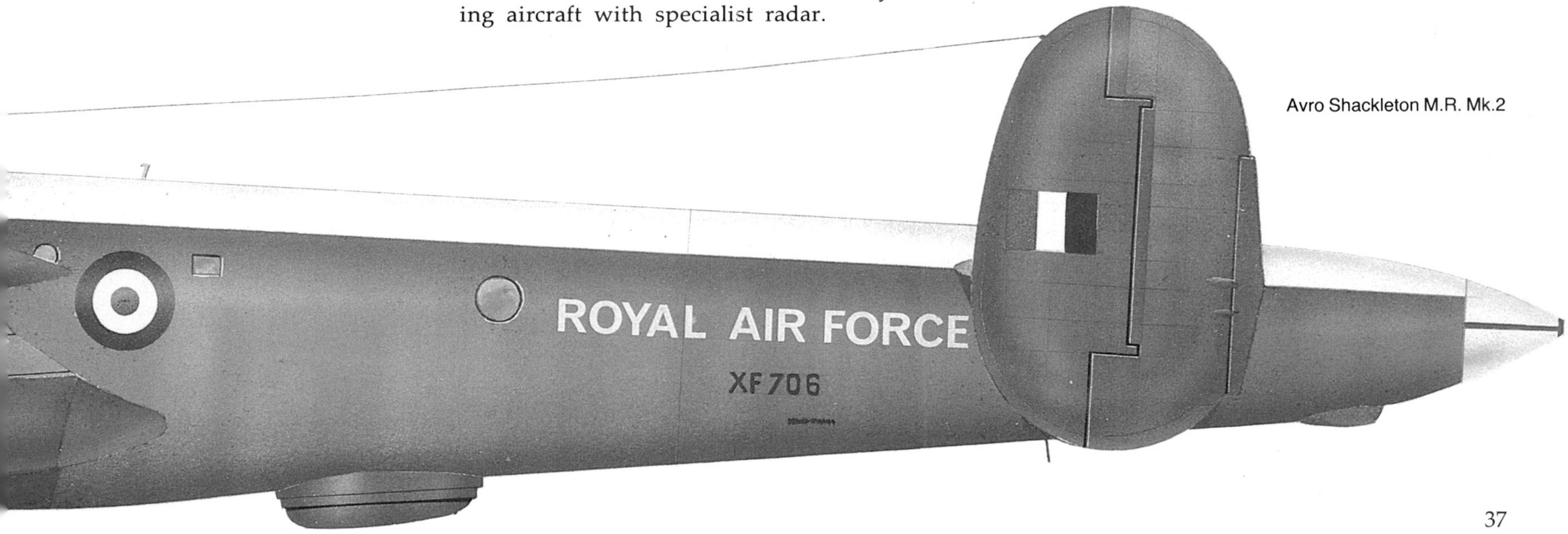

Avro Shackleton M.R. Mk.2

AVRO VULCAN
(U.K.)

This massive delta-winged bomber was an extraordinary aerodynamic feat, and with the Handley Page Victor and Vickers Valiant was one of the the U.K.'s nuclear 'V-bombers' in the 1950s and 1960s. The type was planned as a high-altitude bomber and first flew in August 1952, paving the way for the Vulcan B.Mk 1 initial production version with Olympus turbojets increased in power from 4,990 to 6,123kg (11,000 to 13,500 lb) thrust. In 1961 in-service aircraft were modified to B.Mk 1A standard with a bulged tail containing electronic countermeasures gear. The definitive model was the Vulcan B.Mk 2 with more powerful turbofans and a much-modified wing of greater area. The type was later modified for the low-level role with conventional bombs, and the Vulcan SR.Mk 2 was a strategic recce derivative.

AVRO VULCAN B.Mk 2

Role: Long-range bomber
Crew/Accommodation: Five
Power Plant: Four 9,072 kgp (20,000 lb s.t.) Bristol Siddeley Olympus 301 turbojets
Dimensions: Span 33.83m (111 ft); length 30.45m (99.92 ft); wing area 368.29m^2 (3,964 sq ft)
Weights: Empty 48,081kg (106,000 lb); MTOW 98,800kg (200,180 lb)
Performance: Maximum speed 1,041km/h (562 knots) Mach 0.98 at 12,192m (40,000 ft); operational ceiling 19,812m (65,000 ft); radius 3,701km (2,300 miles) at altitude with missile
Load: Up to 9,525kg (21,000 lb) of bombs, or one Blue Steel Mk 1 stand-off missile

AVRO CANADA CF-100 CANUCK
(Canada)

The CF-100 was an ambitious fighter tailored to an exacting Royal Canadian Air Force requirement of countering a trans-polar bomber attack. Powered by Rolls-Royce Avon turbojets, the prototype CF-100 Mk 1 flew in January 1950. Later CF-100 Mk 2 development aircraft had the more powerful Orenda 2 turbojet. The Orenda 8 was used in the CF-100 Mk 3 production aircraft with radar and eight heavy machine guns in the nose, and the Orenda 9 in the CF-100 Mk 4 with an armament of 106 unguided rockets in the wingtip pods and a ventral pack. The Mk 4 was redesignated Mk 4A after the introduction of the CF-100 Mk 4B with more powerful Orenda 11 engines. The final model was the CF-100 Mk 5 with Orenda 14 engines, a larger-span wing and rockets only in bigger wingtip pods.

AVRO CANADA CF-100 CANUCK Mk 5

Role: All-weather/night interceptor
Crew/Accommodation: Two
Power Plant: Two 3,314 kgp (7,300 lb s.t.) Avro Canada Orenda 14 turbojets
Dimensions: Span 17.44m (57.22 ft); length 19.75m (64.80 ft); wing area 55.14^2 (591 sq ft)
Weights: Empty 10,487kg (23,100 lb); MTOW 15,222kg (33,528 lb)
Performance: Maximum speed 891km/h (482 knots) at 9,143m (30,000 ft); operational ceiling 13,715m (45,000 ft); range 3,219km (2,000 miles)
Load: Twin wingtip pods, each containing a twenty-nine 2.75 inch folding fin anti-aircraft rockets

BEECH 17
(U.S.A.)

Generally known as the 'Staggerwing' because of the reverse stagger of its wings, the Beech 17 is unusual among biplanes in having retractable landing gear. The B17R prototype flew in November 1932 with landing gear in which the wheels were retracted into fixed spats, and demonstrated the remarkable speed range of 97 to 322km/h (60 to 200 mph) on its 313kW (420 hp) Wright R-975 radial engine. The B17L of 1934 introduced a deeper-section lower wing that could accommodate a conventional retracted gear, and from this time production accelerated rapidly. Production finally ended in 1949, by which time the 'Staggerwing' had been produced in a large number of variants including the UC-47 U.S. military liaison model.

BEECH D17S

Role: Executive transport
Crew/Accommodation: One, plus three passengers
Power Plant: One 450 hp Pratt & Whitney R-985-17 Wasp Junior air-cooled radial
Dimensions: Span 9.75m (32 ft); length 7.97m (26.16 ft); wing area 27.5m^2 (296 sq ft)
Weights: Empty 1,399kg (3,085 lb); MTOW 2,132kg (4,700 lb)
Performance: Maximum speed 322km/h (200 mph) at sea level; operational ceiling 6,096m (20,000 ft); range 805km (500 miles)
Load: Up to 272kg (600 lb)

BEECH 18
(U.S.A.)

In 1935 Beech began the development of an advanced monoplane light transport to supersede the 'Staggerwing'. The Beech 18A first flew in January 1937, and the type then remained in production for 32 years. The initial civil models were powered by a number of different types of radial engines. During World War II the type was produced as a staff transport with the U.S. designations C-45 (army) and JRB (navy) and the British name Expediter; there were also AT-7 and AT-11 (or SNB) navigation and bombing/gunnery trainers. After the war improved civil models appeared, and from 1963 retractable tricycle landing gear was an option. Substantial numbers were converted to turboprop power by several specialist companies.

BEECH D18S

Role: Light passenger/executive transport
Crew/Accommodation: Two/one, plus up to seven passengers
Power Plant: Two 450 hp Pratt & Whitney R.985-AN14B Wasp Junior air-cooled radials
Dimensions: Span 14.50m (47.58 ft); length 10.35m (33.96 ft); wing area 32.4m^2 (349 sq ft)
Weights: Empty 2,558kg (5,635 lb); MTOW 3,980kg (8,750 lb)
Performance: Cruise speed 338km/h (211 mph) at 3,050m (10,000 ft); operational ceiling 6,250m (20,500 ft); range 1,200km (750 miles)
Load: Up to 587kg (1,295 lb)

BEECH BONANZA
(U.S.A.)

Another long-lived Beech design, the four/five-seat Bonanza first flew in December 1945 as the V-tailed Beech 35. Large-scale production followed in a number of forms, but in 1959 the company introduced the Beech 33 Debonair with a conventional tail and lower-powered engine for those worried about the 'gimmickry' of the V-tail; in 1967 this became the Beech E33 Bonanza. The third type is the V-tailed Beech 36 Bonanza introduced in 1968 as a six-seater with double doors so that the type can double as a light freight transport. Many variants are available.

▲

BEECH BONANZA

Role: Tourer
Crew/Accommodation: One, plus up to three passengers
Power Plant: One 196 hp Continental E185-8 air-cooled flat-opposed
Dimensions: Span 10.01m (32.83 ft); length 7.67m (25.16 ft); wing area 16.49m^2 (177.6 sq ft)
Weights: Empty 715kg (1,575 lb); MTOW 1,203kg (2,650 lb)
Performance: Cruise speed 272km/h (170 mph) at 2,440m (8,000 ft); operational ceiling 5,485m (17,100 ft); range 1,207km (750 miles)
Load: Up to 373kg (822 lb)

BEECH BARON
(U.S.A.)

►

The Baron was developed from the Beech 95 Travel Air, and first flew as the Beech 55 in February 1960 as a four/five-seater with a swept vertical tail and other modern features. Models up to the Beech E55 have been produced, but there are also two other Baron variants. The Beech 56 appeared in September 1967 and offered turbocharged power, and the Beech 58 of late 1969 introduced a more spacious cabin, a longer wheelbase and double entry doors. The Beech 58 has since appeared in pressurized form as the Beech 58P which has a number of updated features, and the generally similar but unpressurized Beech 58TC.

BEECH BARON

Role: Light executive transport
Crew/Accommodation: One, plus up to four passengers
Power Plant: Two 260 hp Continental 10-470-L air-cooled flat-opposed
Dimensions: Span 11.53m (35.83 ft); length 8.53m (28 ft); wing area 18.5m^2 (199.2 sq ft)
Weights: Empty 1,466kg (3,233 lb); MTOW 2,313kg (5,100 lb)
Performance: Maximum speed 380km/h (236 mph) at sea level; operational ceiling 5,880m (19,300 ft); range 1,505km (935 miles) with full payload
Load: Up to 589kg (1,298 lb)

BEECH QUEEN AIR
(U.S.A.)

Like others of the post-war Beech range, the Queen Air has been developed in a number of related forms after its first flight in August 1958. The initial model was the Beech 65 developed to meet the market for a well-equipped seven/nine seater with retractable landing gear. In 1961 the company added the Beech 80 with more power and, in later variants, longer-span wings. The Beech 88 of August 1965 added pressurization. The Beech 70 of 1968 is the Beech 65 with the wing of the Beech 80. A bewildering series of variants is still in service, and the series has also been sold to the U.S. military as the U-8 Seminole.

BEECH QUEEN AIR

Role: Executive/passenger transport
Crew/Accommodation: One, plus up to nine passengers
Power Plant: Two 380 hp Lycoming IGSO-540-A1A air-cooled flat-opposed
Dimensions: Span 15.3m (50.25 ft); length 10.74m (35.25 ft); wing area 27.3m^2 (294 sq ft)
Weights: Empty 2,266kg (4,996 lb); MTOW 3,855kg (8,500 lb)
Performance: Maximum speed 370km/h (230 mph) at 4,572m (15,000 ft); operational ceiling 8,839m (29,000 ft); range 2,440km (1,517 miles) with 105kg (230 lb) payload
Load: Up to 1,325kg (2,921 lb)

BEECH KING AIR
(U.S.A.)

The King Air was developed as a turboprop-powered derivative of the Beech 65 Queen Air, and first flew in 1963 as a converted Queen Air. Initial orders were from the military for the unpressurized U-21 series, this being followed by the pressurized Beech A90 for civil operators. The type has developed through many variants to culminate in the Beech F90 Super King Air, with the fuselage of the Beech 90, the wings and power plant of the Beech 100, and the tail unit of the Beech 200. The Beech 100 of 1969 introduced a reduced-span wing and a fuselage lengthened to a capacity of 15 passengers, the Beech 200 of 1972 introduced more power, greater-span wing and a T-tail.

Beech King Air 200

Beech C-12

BEECH SUPER KING AIR 300

Role: Executive transport
Crew/Accommodation: Two, plus up to 13 passengers (normally 6)
Power Plant: Two 1,030 shp Pratt & Whitney of Canada PT6A-60A turboprops
Dimensions: Span 16.61m (54.5 ft); length 13.34m (43.75 ft); wing area 28.15m^2 (303 sq ft)
Weights: Empty 3,715kg (8,190 lb); MTOW 6,350kg (14,000 lb)
Performance: Maximum speed 583km/h (315 knots) at 7,315m (24,000 ft); operational ceiling 10,670m (35,000 ft); range 2,593km (1,400 naut. miles) with 8 passengers and baggage
Load: Up to 1,660kg (3,600 lb)

BEECH STARSHIP
(U.S.A.)

The Starship marks a departure from previous aircraft in the Beech line, being a futuristic swept canard with twin pusher turboprops. The type is also notable for the high percentage of composite materials used in the airframe. The Starship was first flown in August 1983 in the form of an 85% scale version developed by Burt Rutan's Scaled Composites Inc, and full-scale flight trials began in February 1986 with the first of six pre-production Starships, and the type gained F.A.A. certification in 1989.

BEECH STARSHIP 1

Role: Executive transport
Crew/Accommodation: Two, plus up to ten passengers
Power Plant: Two 1,100 shp Pratt & Whitney Canada PT6A-67 turboprops
Dimensions: Span 16.46m (54 ft); length 14.05m (46.08 ft); wing area 26.09m^2 (280.9 sq ft)
Weights: Empty 4,044kg (8,916 lb); MTOW 6,350kg (14,000 lb)
Performance: Cruise speed 652km/h (405 mph) at 7,620m (25,000 ft); operational ceiling 12,495m (41,000 ft); range 2,089km (1,298 miles) with maximum payload
Load: Up to 1,264kg (2,884 lb)

◄

BEECH SUNDOWNER
(U.S.A.)

The Beech 23 was developed as a low-cost business aeroplane with fixed tricycle landing gear, and is generally similar to the company's Debonair. First flown in October 1961 as the Musketeer, the type proved highly popular and was developed in a number of forms, including aerobatic two-seaters and the Musketeer Super R with retractable landing gear. In 1971 the series was redesignated as the Sundowner (standard four-seater), Sport (four- and aerobatic two-seater) and Sierra (four/six-seater with retractable landing gear), a later modification adding a suffix indicating engine horsepower.

BEECH SUNDOWNER 180

Role: Primary trainer
Crew/Accommodation: Two
Power Plant: One 180 hp Avco Lycoming 0.360-A4K air-cooled flat-opposed
Dimensions: Span 9.98m (39.75 ft); length 7.62m (25.75 ft); wing area 13.57m^2 (146 sq ft)
Weights: Empty 681kg (1,502 lb); MTOW 1,111kg (2,449 lb)
Performance: Maximum speed 228km/h (141 mph) at sea level; operational ceiling 3,840m (12,600 ft); range 1,420km (883 miles)
Load: Up to 252kg (555 lb) including student pilot

BELL P-39 AIRACOBRA
(U.S.A.)

This fighter was an ingenious attempt to allow powerful nose-mounted armament and to improve manoeuvrability by locating the engine on the centre of gravity aft of the cockpit, from where it drove the propeller by means of an extension shaft. The type flew in October 1937 and was ordered in large numbers. However, it was never more than adequate as a fighter and found its real milieu in the low-level attack role, where its firepower and ruggedness were very useful. Large numbers were supplied to the U.S.S.R. in World War II.

BELL P-39D AIRACOBRA

Role: Fighter
Crew/Accommodation: One
Power Plant: One, 1,150 hp Allison V-1710-35 water-cooled inline
Dimensions: Span 10.36m (34 ft); length 9.19m (30.16 ft); wing area 19.79m^2 (213 sq ft)
Weights: Empty 2,478kg (5,462 lb); MTOW 3,720kg (8,200 lb)
Performance: Maximum speed 592km/h (368 mph) at 4,206m (13,800 ft); operational ceiling 9,784m (32,100 ft); range 1,287km (800 miles) with 227kg (500 lb) of bombs
Load: One 37mm cannon, plus two .5 inch and four .303 inch machine guns, along with 227kg (500 lb) of bombs

BELL MODEL 47 and H-13
(U.S.A.)

First flown during December 1945, the Bell 47 in 1947 became the first helicopter in the world to secure civil certification. The type was built in a number of civil forms including the classic Bell 47D with the 'goldfish bowl' cockpit, and the definitive Bell 47G. It was also built under licence in Italy, Japan and the U.K. The Bell 47 was adopted for the military as the R-13, later changed to H-13, in the training, observation and casevac (casualty evacuation) roles. The U.S. Navy operated the HTL trainer model.

BELL MODEL 47/UH-13H

Role: Civil and military light utility helicopter
Crew/Accommodation: One, plus up to two passengers
Power Plant: One 250 hp Lycoming 0-435.23 air-cooled, flat-opposed
Dimensions: Overall length rotors turning 13.11m (43 ft); rotor diameter 10.69m (35.08 ft)
Weights: Empty 710kg (1,564 lb); MTOW 1,225kg (2,700 lb)
Performance: Cruise speed 137km/h (85 mph) at sea level; operational ceiling 4,023m (13,200 ft); range 383km (238 miles) with 2 passengers
Load: Up to 454kg (1,000 lb) of externally slung load

Bell UH-1D

BELL MODELS 204, 205 and 212 (UH-1 'HUEY')

(U.S.A.)

The 'Huey' series is more formally known as the Iroquois, and was initially developed for the U.S. Army as the Bell 204, which first flew in October 1956 and entered service as the HU-1, altered in 1962 to UH-1. The Bell 204 went through several variants with increased power and payload, and was then complemented by the Bell 205 with a larger cabin for up to 14 troops. This entered U.S. military service as the UH-1D and later UH-1H series, which was in turn partnered by the generally similar Bell 212 (UH-1N) with a twin-turboshaft power plant. Civil variants of the three basic models, but especially the Bell 205 and 212, have also been successfully produced.

BELL MODEL 212/UH-1N

Role: Military utility transport helicopter
Crew/Accommodation: One, plus up to 14 troops
Power Plant: One 1,800 shp Pratt & Whitney Canada PT6T-3B Turbo Twin Pac coupled turboshaft
Dimensions: Overall length rotors turning 17.45m (57.25 ft); rotor diameter 14m (48.2 ft)
Weights: Empty 2,517kg (5,549 lb); MTOW 5,085kg (11,200 lb)
Performance: Maximum speed 203km/h (110 knots) at sea level; operational ceiling 5,305m (17,400 ft); range 439km (237 naut. miles)
Load: Up to 2,268kg (5,000 lb) of external load

BELL MODEL 206 JETRANGER and OH-58 KIOWA

(U.S.A.)

Though now known as one of the world's most successful civil helicopters, the Bell 206 was developed as a scout for the U.S. Army and first flew in January 1966. The type entered service as the OH-58 Kiowa, and in the late 1980s reached the level of the OH-58D armed scout with a mast-mounted sight and a varied armament capability. The civil JetRangers have been produced in three basic variants with steadily increased power, and also as the LongRanger, with a lengthened cabin for greater capacity and passenger comfort.

BELL MODEL 206 JETRANGER/ TH-57/OH-58 KIOWA

Role: Civil and military light utility transport helicopter
Crew/Accommodation: One, plus up to four passengers
Power Plant: One 420 shp Allison 250-C20 turboshaft
Dimensions: Overall length rotors turning 11.82m (38.8 ft); rotor diameter 10.16m (33.33 ft)
Weights: Empty 680kg (1,500 lb); MTOW 1,451kg (3,200 lb)
Performance: Maximum speed 225km/h (121 knots) at sea level; operational ceiling 5,760m (18,900 ft); range 360km (194 naut. miles) with 227kg (500 lb) payload
Load: Up to 449kg (990 lb) internal, or 545kg (1,200 lb) externally slung

BELL MODEL 209 (AH-1 HUEYCOBRA)
(U.S.A.)

This is a gunship developed as an escort for UH-1 troop helicopters in Vietnam with Huey's dynamic system and a slim fuselage mounting a chin turret and stub wings for disposable armament. The type has been developed as the single-engined HueyCobra for the U.S. Army and the twin-engined SeaCobra for the U.S. Marine Corps with steadily improving capabilities in terms of avionics and weapons, the latter including wire-guided anti-tank missiles and multi-barrelled cannon.

BELL MODEL 209/AH-1T SEACOBRA

Role: Armed and armoured attack helicopter
Crew/Accommodation: Two
Power Plant: One 2,050 shp Pratt & Whitney Canada T400-WV-402 coupled turboshaft
Dimensions: Overall length rotors turning 18.2m (59.7 ft); rotor diameter 14.6m (48 ft)
Weights: Empty 3,635kg (8,014 lb); MTOW 6,350kg (14,000 lb)
Performance: Maximum speed 291km/h (157 knots) at sea level; operational ceiling 3,795m (12,450 ft); radius 209km (113 naut. miles) with full warload
Load: One 20mm multi-barrel cannon, plus 4 store stations to carry anti-armour missiles, minigun pods or rocket launchers

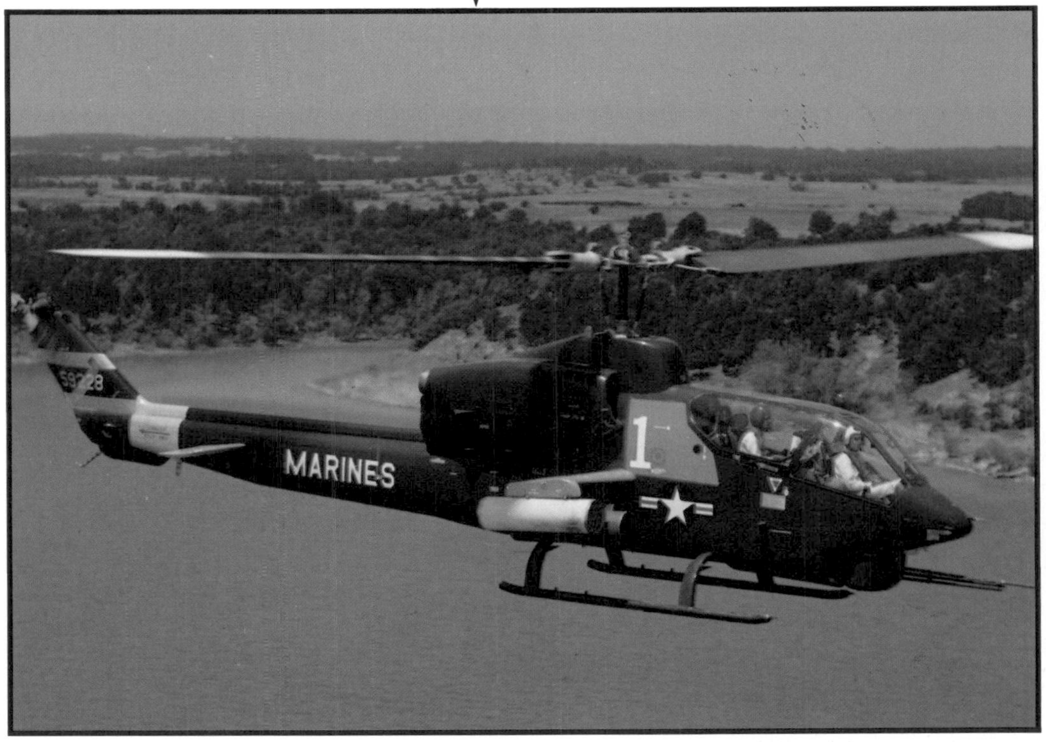

BELL MODEL 222
(U.S.A.)

The Bell 222 first flew in August 1977 as a high-performance helicopter with retractable landing and twin-engined power plant. The type is offered with a number of options in standard, executive and offshore versions, and has secured useful sales because of its performance combined with advanced features such as nodal suspension, elastomeric main rotor bearing and glassfibre/stainless steel rotor blades. It can carry up to eight passengers.

BELL MODEL 222B

Role: Light commercial/executive transport helicopter
Crew/Accommodation: One/two, plus up to eight passengers (with one pilot)
Power Plant: Two 684 shp Lycoming LTS101-750C-1 turboshafts
Dimensions: Overall length rotors turning 15.36m (50.40 ft); rotor diameter 12.8m (42 ft)
Weights: Empty 2,223kg (4,900 lb); MTOW 3,742kg (8,250 lb)
Performance: Cruise speed 259km/h (140 knots) at 1,220m (4,000 ft); operational ceiling 4,815m (15,800 ft); range 532km (287 naut. miles) with seven passengers
Load: Up to 540kg (1,190 lb)

BELL/BOEING V-22 OSPREY
(U.S.A.)

When it reaches operators during the 1990s the Osprey will be the world's first tilt-rotor convertiplane to enter service. Making its first flight in April 1989, the Osprey is derived from Bell's long experience with such aircraft at the experimental level, and has its two engines at the tips of the wing in nacelles that can be turned to the vertical so that the two large-diameter propellers/rotors provide direct lift, and then to the horizontal for translation into wingborne forward flight. The Osprey's main role will be as an assault transport aircraft for the U.S. Marines.

BELL/BOEING MV-22 OSPREY

Role: Military/naval vertical take-off and landing assault transport
Crew/Accommodation: Three, plus up to 24 troops
Power Plant: Two 6,150 shp Allison T406-AD-400 turboshafts
Dimensions: Overall length rotors turning 17.65m (57.90 ft); rotor diameter 11.58mg (38 ft)
Weights: Empty 14,412kg (31,772 lb); MTOW 21,546kg (47,500 lb) with vertical take-off
Performance: Maximum speed 571km/h (308 knots) at sea level; operational ceiling 9,144m (30,000 ft); radius 278km (150 naut. miles) with 4,545kg (10,000 lb) payload
Load: Up to 6,804kg (15,000 lb).
Note: the overall span, rotors turning, is 25.75m (84.5ft) and conventional take-off MTOW is 27,443kg (60,500 lb)

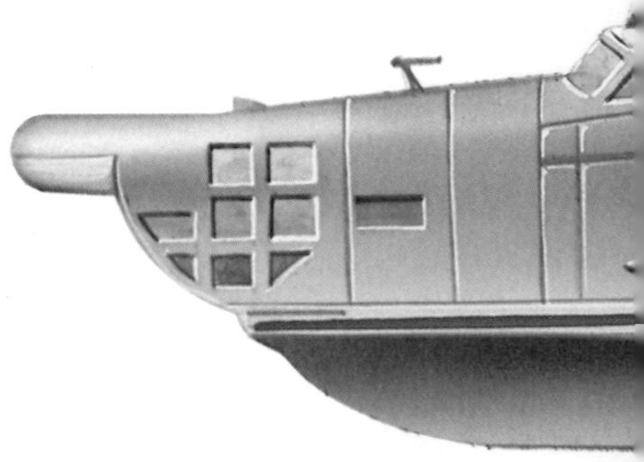

BERIEV Be-12 'MAIL'
(U.S.S.R.)

The Be-12 is one of the few amphibians still in military service, and was developed in the early 1960s as a turboprop-powered maritime patrol type modelled conceptually in the obsolescent Be-6. The most distinctive features are the gull wing to keep the propellers clear of the water, the sturdy retractable landing gear, nose radome for the search radar, and the tail 'sting' for the magnetic anomaly detector used in the search for submerged submarines.

BERIEV Be-12 'MAIL'

Role: Maritime patrol and anti-submarine
Crew/Accommodation: Five
Power Plant: Two 4,250 shp Ivchenko AI-20M turboprops
Dimensions: Span 29.67m (97.3 ft); length 32.22m (105.7 ft); wing area 95.7m^2 (1,030 sq ft)
Weights: Empty 19,500kg (42,990 lb); MTOW 25,599kg (65,255 lb)
Performance: Maximum speed 426km/h (230 knots) at 3,048m (10,000 ft); operational ceiling 8,200m (26,900 ft); radius 998km (538 naut. miles)
Load: Up to 2,000kg (4,409 lb)

BLACKBURN SKUA
(U.K.)

The Skua was the Royal Navy's first monoplane dive-bomber (and later an emergency fighter) and was also a new departure as that service's first all-metal carrierborne aeroplane with retractable landing gear, flaps and a variable-pitch propeller. The first prototype flew in February 1937 with a Mercury IX radial, but as this engine was required for other aircraft the 190 Skua Mk II production aircraft had the slightly more powerful Perseus XII. The type was obsolescent as it entered service, but did score a few successes right at the beginning of World War II.

BLACKBURN SKUA Mk II

Role: Naval carrierborne dive bomber
Crew/Accommodation: Two
Power Plant: One 890 hp Bristol Perseus XII air-cooled radial
Dimensions: Span 14.07m (46.16 ft); length 10.84m (35.56 ft); wing area 29.64m^2 (319 sq ft)
Weights: Empty 2,493kg (5,496 lb); MTOW 3,732kg (8,228 lb)
Performance: Maximum speed 369km/h (229 mph) at 1,980m (6,500 ft); operational ceiling 5,822m (19,100 ft); range 890km (553 miles)
Load: Five .303 inch machine guns, plus up to 227kg (500 lb) of bombs

BLACKBURN BEVERLEY (U.K.)

The Beverley heavy tactical freighter was developed by General Aircraft before its purchase by Blackburn, and was an ungainly machine of the pod-and-boom type with fixed tricycle landing gear and Bristol Hercules radial engines. The type first flew in June 1950 and in re-engined production form proved itself able to carry a 22.4-tonne freight payload, or alternatively up to 94 troops or 70 paratroops, of whom 36 or 30 were accommodated in the large boom. A total of 47 was produced for the RAF, entering service in 1955.

BLACKBURN BEVERLEY C.Mk 1

Role: Military freight/troop transport
Crew/Accommodation: Four, plus up to 94 troops
Power Plant: Four 2,850 hp Bristol Centaurus 173 air-cooled radials
Dimensions: Span 49.38m (162 ft); length 30.3m (99.42 ft); wing area 270.92m^2 (2,916 sq ft)
Weights: Empty 37,240kg (82,100 lb); MTOW 64,864kg (143,000 lb)
Performance: Cruise speed 359km/h (223 mph) at 1,524m (5,000 ft);range 322km (200 miles) with 19,958kg (44,000 lb) payload
Load: Up to 22,400kg (49,380 lb)

BLACKBURN BUCCANEER (U.K.)

This superb aeroplane was planned as a low-level transonic carrierborne anti-ship strike platform, and first flew in April 1958. The Buccaneer S.Mk 1 was powered by 3,221kg (7,100 lb) thrust de Havilland Gyron Junior 101 turbojets and entered Royal Navy service despite being underpowered. The Buccaneer S.Mk 2 switched to the Spey turbofan and displayed an all-round improvement in performance. With the demise of the navy's large carriers the aircraft were reallocated to the RAF, which also ordered additional aircraft which have been steadily upgraded for service into the mid-1990s.

BLACKBURN BUCCANEER S.Mk 2

Role: Low-level strike
Crew/Accommodation: Two
Power Plant: Two 5,035 kgp (11,100 lb s.t.) Rolls-Royce Spey Mk 101 turbofans
Dimensions: Span 13.41m (44 ft); length 19.33m (63.42 ft); wing area 47.82m^2 (514.7 sq ft)
Weights: Empty 13,517kg (29,800 lb); MTOW 28,123kg (62,000 lb)
Performance: Maximum speed 1,040km/h (561 knots) Mach 0.85 at 76m (250 ft); operational ceiling 12,192m (40,000+ ft); radius 1,738km (938 naut. miles) with full warload
Load: Up to 3,175kg (7,000 lb) of ordnance, including up to 1,815kg (4,000 lb) internally, the remainder, typically Martel or Sea Eagle anti-ship missiles, being carried externally under the wings

BLERIOT MONOPLANES
(France)

Louis Bleriot was one of the true pioneers of aviation, and secured his place in history during 1911 as the first man to fly a heavier-than-air craft across the English Channel. This machine was a Bleriot XI with a 18.7kW (25 hp) Anzani engine, the culmination of a series of monoplanes that had started with the unsuccessful Bleriot V of 1906. Bleriot's cross-Channel triumph secured a comparative flood of orders for his aeroplane, which was steadily upgraded with more powerful engines. The type was also developed for the military as a reconnaissance machine in Bleriot XI-2 and -3 two- and three-seat forms.

BLERIOT XI-2

Role: Reconnaissance/training
Crew/Accommodation: Two
Power Plant: One 80 hp Gnome air-cooled rotary
Dimensions: Span 10.35m (33.96 ft); length 8.4m (27.56 ft); wing area 19m^2 (205 sq ft)
Weights: Empty 335kg (786 lb); MTOW 585kg (1,290 lb)
Performance: Maximum speed 120km/h (75 mph) at sea level; endurance 3.5 hours
Load: None other than crew

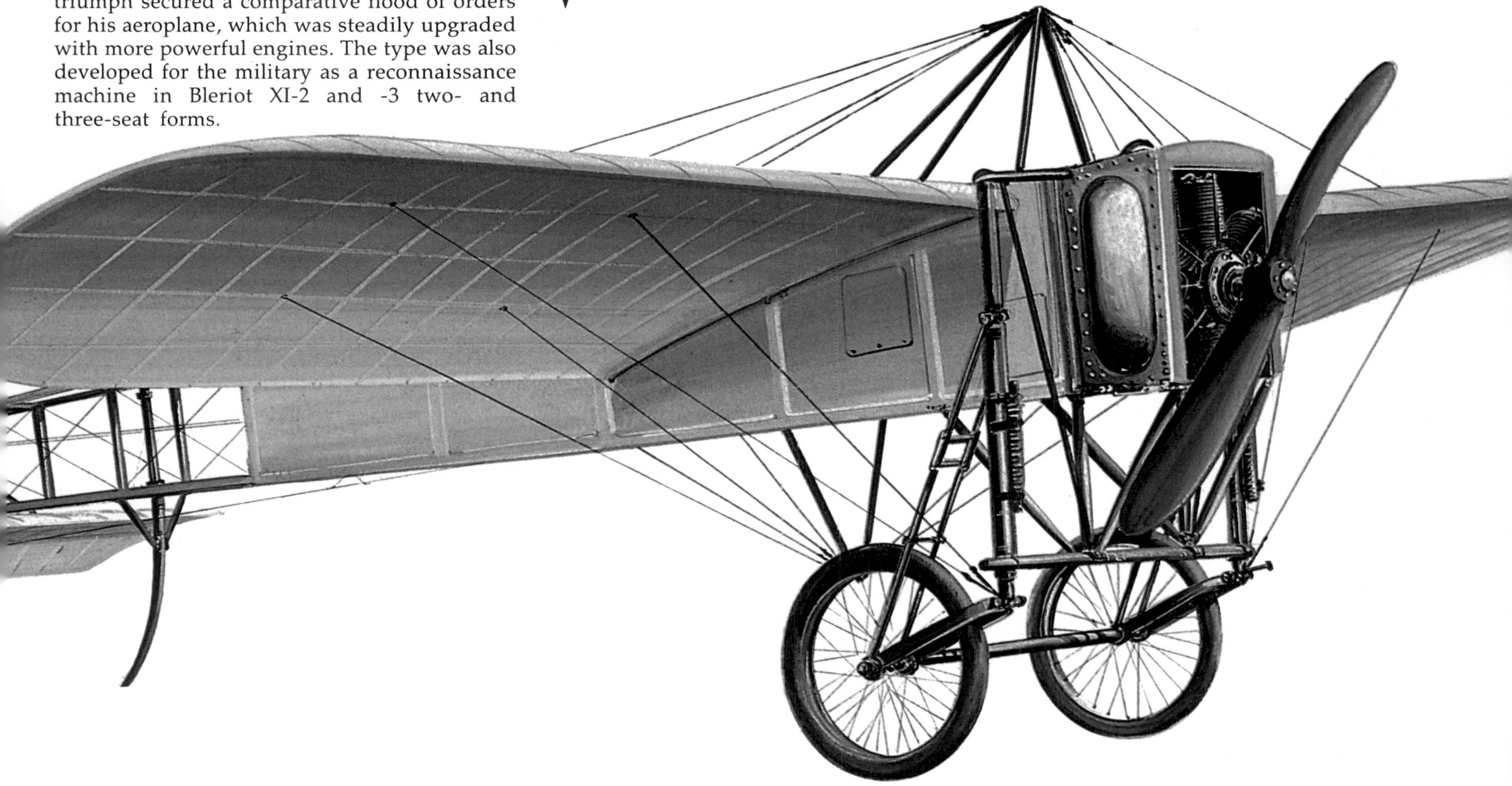

BLOCH M.B.151 and M.B.152
(France)

The M.B.151 was one of France's first 'modern' monoplane fighters, and resulted from the unsuccessful M.B.150 prototype produced to meet a 1934 requirement. The type first flew in October 1937 after being fitted with a larger wing and more powerful engine, and was ordered into production with the Gnome-Rhone 14N-11 radial. An improved version was developed as the M.B.152 with the more powerful Gnome-Rhone 14N-25 or 14N-49. Production was slow, and only a handful were combat ready by the time of the German invasion of France in May 1940.

BLOCH M.B.152

Role: Fighter
Crew/Accommodation: One
Power Plant: One 1,000 hp Gnome-Rhone 14N-25 air-cooled radial
Dimensions: Span 10.54m (34.58 ft); length 9.1m (29.86 ft); wing area 17.32m^2 (186.4 sq ft)
Weights: Empty 2,158kg (4,758 lb); MTOW 2,800kg (6,173 lb)
Performance: Maximum speed 509km/h (316 mph) at 4,500m (14,765 ft); operational ceiling 10,000m (32,808 ft); range 540km (335 miles)
Load: Two 20mm cannon and two 7.5mm machine guns

BLOCH M.B.174
(France)

This multi-role bomber and reconnaissance aeroplane was one of the best aircraft possessed by the French in the opening stages of World War II, but in common with other types had been developed at too leisurely a pace to be available in significant numbers. The type first flew as the M.B.170 prototype in February 1938 with Gnome-Rhone 14N-6/7 engines, but time was then lost in the exploration of several production variants before the M.B.174 was finalized.

BLOCH M.B.174A3

Role: Light bomber/reconnaissance
Crew/Accommodation: Three
Power Plant: Two 1,140 hp Gnome-Rhone 14N 48/49 air-cooled radials
Dimensions: Span 17.92m (58.79 ft); length 12.23m (40.12 ft); wing area 41.13m^2 (442.7 sq ft)
Weights: Empty 5,600kg (12,346 lb); MTOW 7,160kg (15,784 lb)
Performance: Maximum speed 530km/h (329 mph) at 5,200m (17,060 ft); operational ceiling 11,000m (36,090 ft); range 1,650km (1,025 miles) in reconnaissance role
Load: Seven 7.5mm machine guns, plus up to 500kg (1,102 lb) of bombs

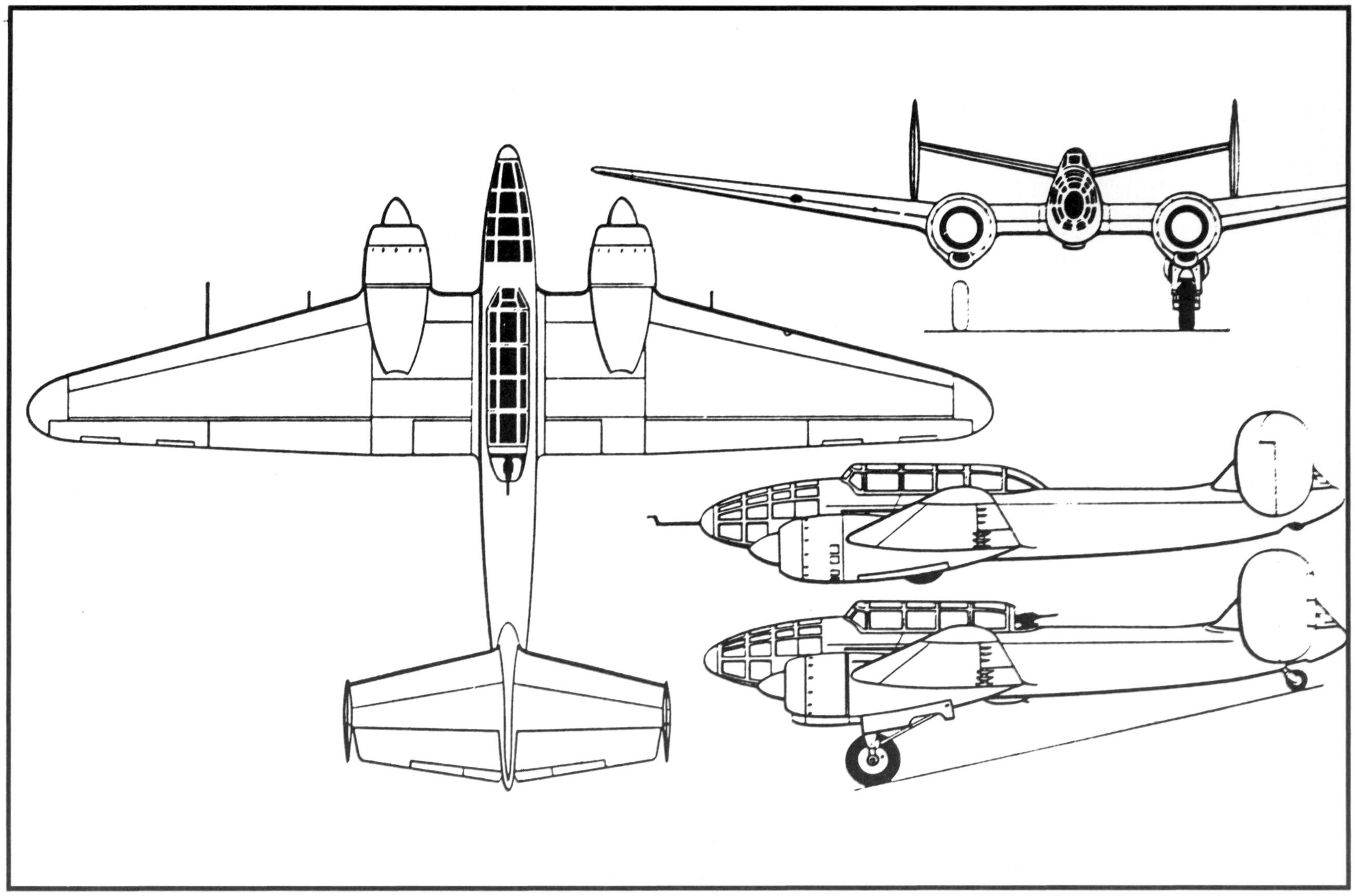

BLOCH M.B.220
(France)

The M.B.220 may be regarded as the French counterpart to the Douglas DC-3 and Junkers Ju 52/3m, but was not as successful as either of its rivals. The first example flew in December 1935 and was followed by 16 production aircraft with Gnome-Rhone 14N-16/17 radials. The type was used in World War II, and at least five were modified after the war to M.B.221 standard with Wright R-1820-97 Cyclone radials, being retired in 1950.

BLOCH M.B.220

Role: Short-range passenger transport
Crew/Accommodation: Four, plus up to 16 passengers
Power Plant: Two 985 hp Gnome-Rhone 14N 16/17 air-cooled radials
Dimensions: Span 22.82m (74.87 ft); length 19.25m (63.16 ft); wing area 75m^2 (807.3 sq ft)
Weights: Empty 6,713kg (14,799 lb); MTOW 9,500kg (20,943 lb)
Performance: Maximum speed 330km/h (205 mph) at 3,800m (12,467 ft); operational ceiling 7,000m (22,965 ft); range 1,400km (869 miles)
Load: Up to 1,860kg (4,101 lb)

BLOHM UND VOSS Ha 139 and Bv 139
(Germany)

This twin-float type resulted from a 1935 Lufthansa requirement for a seaplane offering capability on the airline's newly established transatlantic air mail route. The aeroplane had to be stressed for catapult launches and also to operate from rough water for the carriage of 500kg (1,102 lb) of mail over a range of 5,000km (3,107 miles). The Ha 139 prototype flew in the autumn of 1936 on a quartet of Junkers Jumo 205 diesels selected for their reliability and low specific fuel consumption. By March 1937 the first two aircraft were in service, staging via catapult-equipped ships in the middle of the Atlantic. The slightly larger and hevier Ha 139B joined these initial aircraft in mid-1938. In 1939 all three were impressed for service as reconnaissance and transport aircraft with the Luftwaffe, and were subsequently redesignated Bv 139.

BLOHM UND VOSS Ha 139A

Role: Shipborne long-range mailplane
Crew/Accommodation: Four
Power Plant: Four 605 hp Junkers Jumo 205C water-cooled diesels
Dimensions: Span 27m (88.58 ft); length 19.5m (63.98 ft); wing area 117m² (1,259 sq ft)
Weights: Empty 10,360kg (22,840 lb); MTOW 17,500kg (38,581 lb) for catapult launch
Performance: Cruise speed 260km/h (161.5 mph) at sea level; operational ceiling 3,500m (11,482 ft); range 5,300km (3,293 miles)
Load: Up to 480kg (1,058 lb)

BOEING PW-9 and FB
(U.S.A.)

After learning the craft from the manufacture of other company's designs, Boeing entered the fighter market with the Model 15 that first flew in June 1923 as an unequal-span biplane with a massive Curtiss D-12 inline engine. The performance was impressive, and orders were placed by the U.S. Army (PW-9 series) and U.S. Marines (FB series). The 30 PW-9s were followed by 25 PW-9As with the D-12C, 15 PW-9Cs with the D-12D and 16 PW-9Ds with a balanced rudder. A total of 14 FB-1s was ordered for the U.S. Marines, the last four being used for experimental purposes with designations from FB-2 to FB-6 except FB-5, which was reserved for 27 aircraft with the Packard 2A-1500 engine.

BOEING PW-9

Role: Fighter
Crew/Accommodation: One
Power Plant: One 435 hp Curtiss D-12 water-cooled inline
Dimensions: Span 9.75m (32 ft); length 7.14m (23.42 ft); wing area 24.15m² (260 sq ft)
Weights: Empty 878kg (1,936 lb); MTOW 1,415kg (3,120 lb)
Performance: Maximum speed 256km/h (159.1 mph) at sea level; operational ceiling 5,768m (18,925 ft); range 628km (390 miles)
Load: One .5 inch and one .303 inch machine guns

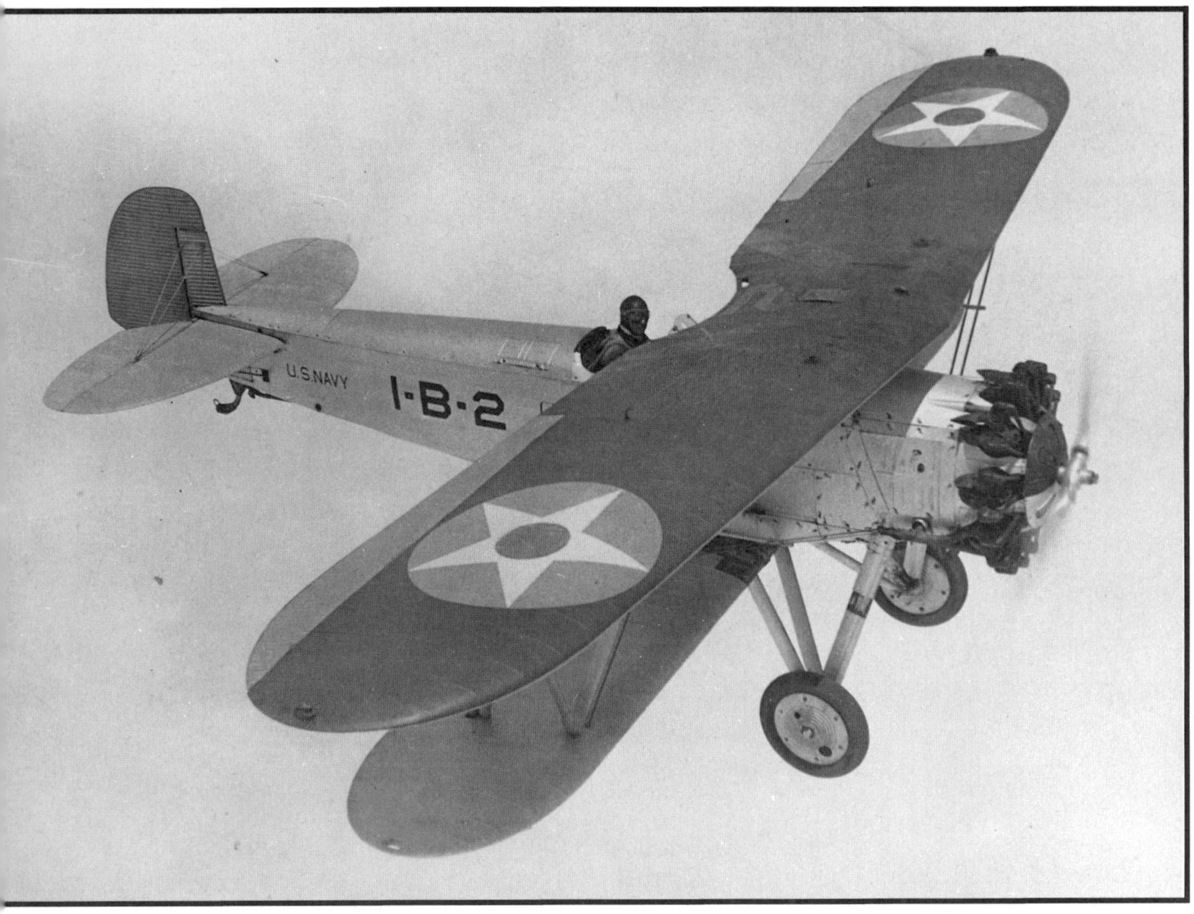

BOEING F2B and F3B
(U.S.A.)

The F2B carrierborne fighter was developed from the unsuccessful XP-8 landplane, the main difference in the naval version being the use of an R-1340 Wasp radial in place of the original Packard inline. The XF2B-1 first flew in November 1926 and production F2B-1 aircraft totalled 32. From this basic design Boeing evolved the F3B with a longer fuselage, completely new and bigger wings and more power. The first example flew in February 1928, and production totalled 74 F3B-1 fighters including the prototype.

BOEING F3B-1

Role: Naval carrierborne fighter
Crew/Accommodation: One
Power Plant: One 425 hp Pratt & Whitney R.1340B Wasp air-cooled radial
Dimensions: Span 9.17m (30.08 ft); length 7.57m (24.83 ft); wing area 25.55m^2 (275 sq ft)
Weights: Empty 988kg (2,179 lb); MTOW 1,336kg (2,945 lb)
Performance: Maximum speed 253km/h (157 mph) at sea level; operational ceiling 6,553m (21,500 ft); range 547km (340 miles)
Load: One .5 inch and one .303 inch machine guns

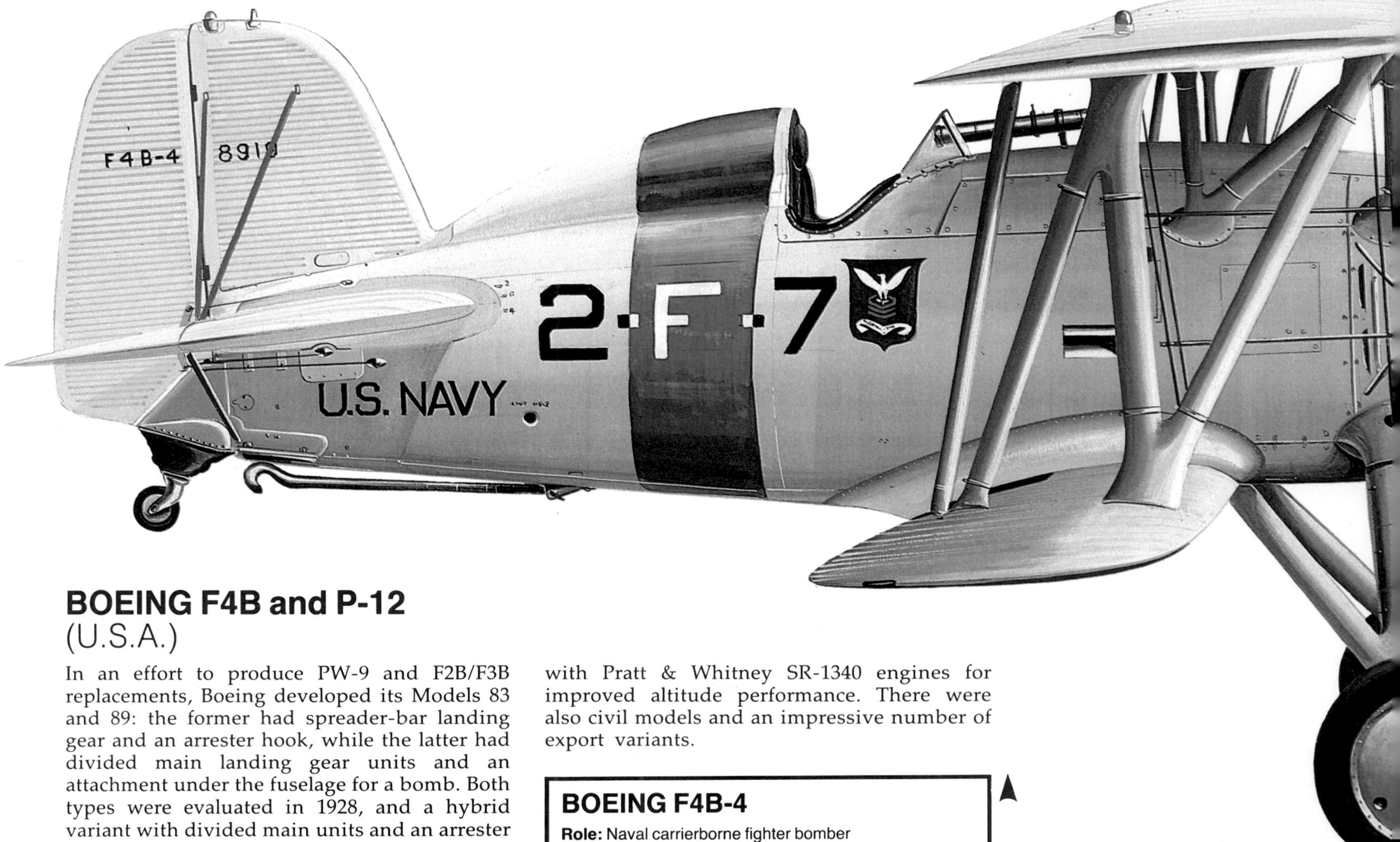

BOEING F4B and P-12
(U.S.A.)

In an effort to produce PW-9 and F2B/F3B replacements, Boeing developed its Models 83 and 89: the former had spreader-bar landing gear and an arrester hook, while the latter had divided main landing gear units and an attachment under the fuselage for a bomb. Both types were evaluated in 1928, and a hybrid variant with divided main units and an arrester hook was ordered for the U.S. Navy as the F4B-1. These 27 aircraft were followed by 46 F4B-2s with a drag-reducing cowling ring and spreader-bar landing gear, 21 F4B-3s with a semi-monocoque fuselage and 92 F4B-4s with a larger fin. The U.S. Army ordered the type as the P-12, the first 10 aircraft being generally similar to the Model 89; later aircraft were 90 P-12Bs with revised ailerons, 95 P-12Cs similar to the F4B-2, 36 improved P-12Ds, 110 P-12Es with a semi-monocoque fuselage, and 25 P-12Fs with Pratt & Whitney SR-1340 engines for improved altitude performance. There were also civil models and an impressive number of export variants.

BOEING F4B-4

Role: Naval carrierborne fighter bomber
Crew/Accommodation: One
Power Plant: One 500 hp Pratt & Whitney R-1340-D Wasp air-cooled radial
Dimensions: Span 9.14m (30 ft); length 7.75m (25.42 ft); wing area 21.18m^2 (228 sq ft)
Weights: Empty 1,049kg (2,312 lb); MTOW 1,596kg (3,519 lb)
Performance: Maximum speed 301km/h (187 mph) at sea level; operational ceiling 8,382m (27,500 ft); range 941km (585 miles)
Load: One .5 inch and one .303 inch machine guns, plus one 227kg (500 lb) bomb

BOEING/STEARMAN MODEL 75 (PT-13 and PT-17)
(U.S.A.)

In 1939 Boeing bought Stearman Aircraft, and as a result acquired the excellent Model 75 developed by Stearman from the X-70 first flown in December 1933. The U.S. Navy had taken 61 of these with the designation NS-1, and development had then led to the Model 75 accepted by the U.S. Army as the PT-13 with the 160kW (215 hp) Lycoming R-680-5 radial engine. Further evolution led to the PT-13A with the R-680-7 engine and improved instrumentation, the PT-13B with the R-680-11 engine and the night-flying PT-13C. A change was then made to the Continental R-670-5 radial engine for the PT-17, of which 3,510 were built. Specialist versions were the blind-flying PT-17A and pest-control PT-17B. The Continental engine was also used in the naval N2S, which went through -1 to -4 variants before a common army/navy model was developed as the PT-13D and N2S-5. Variants with Jacobs R-755-7 radials were designated PT-18 and, in the blind-flying role, PT-18A. Aircraft supplied to Canada were designated PT-27 by the Americans, but were called Kaydet in the receiving country, and this name is generally applied today to all Model 75 variants.

BOEING/STEARMAN PT-17A

Role: Basic trainer
Crew/Accommodation: Two
Power Plant: One 220 hp Continental R-670-5 air-cooled radial
Dimensions: Span 9.8m (32.16 ft); length 7.32m (24.02 ft); wing area 27.63m^2 (297.4 sq ft)
Weights: Empty 878kg (1,936 lb); MTOW 1,232kg (2,717 lb)
Performance: Maximum speed 200km/h (124 mph) at sea level; operational ceiling 3,414m (11,200 ft); range 813km (505 miles)
Load: None

BOEING P-26
(U.S.A.)

This was a step towards the 'modern' monoplane fighter of the mid-1930s, but as first flown in March 1932 had a monoplane wing wire-braced to the fuselage and the well-faired main units of the fixed landing gear. The U.S. Army ordered 136 examples of the P-26A, generally dubbed 'Peashooter' in service. In-service aircraft were retrofitted with flaps to reduce landing speed, and other variants were two P-26Bs with fuel-injected R-1340-33 radials, and 23 P-26Cs with modified fuel systems. A few aircraft were also exported to China.

BOEING P-26C

Role: Fighter
Crew/Accommodation: One
Power Plant: One 600 hp Pratt & Whitney R-1340-33 Wasp air-cooled radial
Dimensions: Span 8.52m (27.96 ft); length 7.24m (23.75 ft); wing area 13.89m^2 (149.5 sq ft)
Weights: Empty 1,058kg (2,333 lb); MTOW 1,395kg (3,075 lb)
Performance: Maximum speed 378km/h (235 mph) at sea level; operational ceiling 8,230m (27,000 ft); range 1,022km (635 miles)
Load: Two .5 inch machine guns, plus 90.8kg (200 lb) of bombs

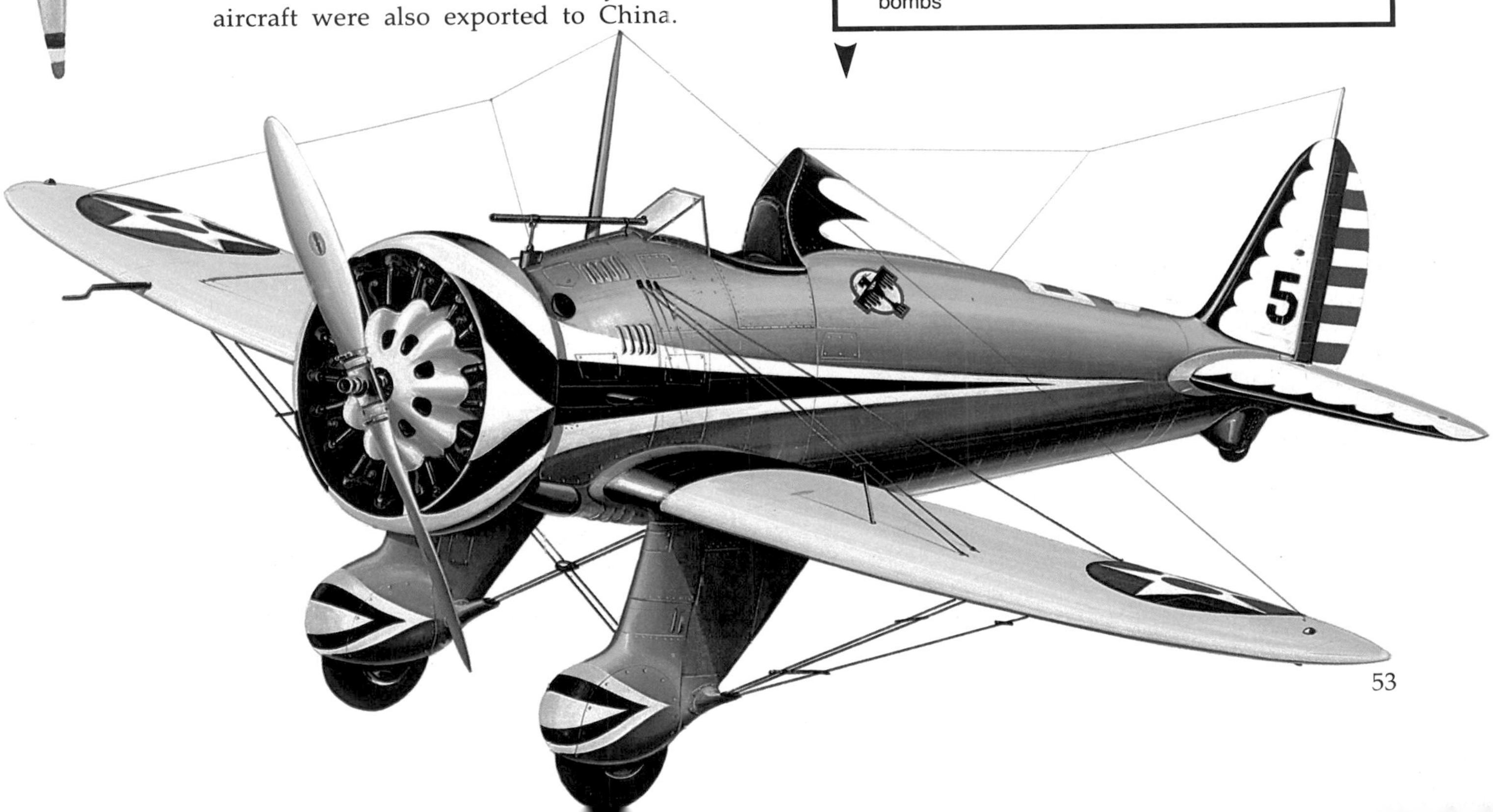

BOEING MODEL 40
(U.S.A.)

The Model 40 first flew in July 1925, and was beaten for a production contract by a Douglas machine. However, in 1927 the Model 40 was revised as the Model 40A with a Pratt & Whitney Wasp radial, a fabric-covered steel-tube rather than composite fuselage, and the internal payload volume was revised for two passengers separated from the cockpit and engine by two mail compartments. Boeing Air Transport ordered 24 of the type, which were redesignated Model 40B when retrofitted with a more powerful Pratt & Whitney Hornet radial engine, and then Model 40B-2 when the 38 four-passenger Model 40B-4s began to enter service. There were also 10 Model 40Cs with the Wasp engine and seating for four passengers.

BOEING MODEL 40B-4

Role: Short-range mail/passenger transport
Crew/Accommodation: One, plus up to four passengers
Power Plant: One 525 hp Pratt & Whitney R-1690 Hornet air-cooled radial
Dimensions: Span 13.47m (44.19 ft); length 10.14m (33.27 ft); wing area 50.82m^2 (547 sq ft)
Weights: Empty 1,688kg (3,722 lb); MTOW 2,756kg (6,075 lb)
Performance: Cruise speed 201km/h (125 mph) at sea level; operational ceiling 4,907.28m (16,100 ft); range 861km (535 miles)
Load: Up to 517kg (1,140 lb), including 227kg (500 lb) of mail

BOEING MODEL 80
(U.S.A.)

Flown in August 1928, this was Boeing's first custom-designed passenger transport for the burgeoning routes of Boeing Air Transport. The type was a large unequal-span biplane with an enclosed cockpit, three Pratt & Whitney Wasp radials, and provision for 12 passengers and one stewardess. Production of four such aircraft was followed by that of 10 Model 80As with many improvements including accommodation for 18 passengers. The aircraft were later modified to Model 80A-1 standard, giving them an increased fin area.

BOEING MODEL 80A

Role: Passenger transport
Crew/Accommodation: Two, one cabin crew, plus up to 18 passengers
Power Plant: Three 525 hp Pratt & Whitney R-1690 Hornet air-cooled radials
Dimensions: Span 24.38m (80 ft); length 17.22m (56.5 ft); wing area 113.34m^2 (1,220 sq ft)
Weights: Empty 4,800kg (10,582 lb); MTOW 7,938kg (17,500 lb)
Performance: Cruise speed 201km/h (125 mph) at sea level; operational ceiling 4,267m (14,000 ft); range 740km (460 miles)
Load: Up to 1,877kg (4,138 lb)

BOEING MODEL 247
(U.S.A.)

This has many claims to the title of the world's first 'modern' air transport with features such as all-metal construction, cantilever wings, de-icing of the flying surfaces, a semi-monocoque fuselage, retractable landing gear and fully enclosed accommodation. The type first flew in February 1933 but, being sized to Boeing Air Transport's exact requirement for 10 passengers, it proved to lack the accommodation required for other airlines. Thus BAT's 60 aircraft were complemented by only 15 more aircraft for companies or individuals. Most were later brought to Model 247D standard.

BOEING MODEL 247D

Role: Passenger transport
Crew/Accommodation: Two crew, one cabin crew, plus up to ten passengers
Power Plant: Two 550hp Pratt & Whitney Wasp S1H1G air-cooled radials
Dimensions: Span 22.56m (74 ft); length 15.72m (51.58 ft); wing area 77.67m^2 (836 sq ft)
Weights: Empty 4,148kg (9,144 lb); MTOW 6,192kg (13,650 lb)
Performance: Cruise speed 304km/h (189 mph); operational ceiling 7,742m (25,400 ft); range 1,199km (745 miles)
Load: Up to 998kg (2,200 lb).

BOEING MODEL 299 (B-17 FORTRESS)
(U.S.A.)

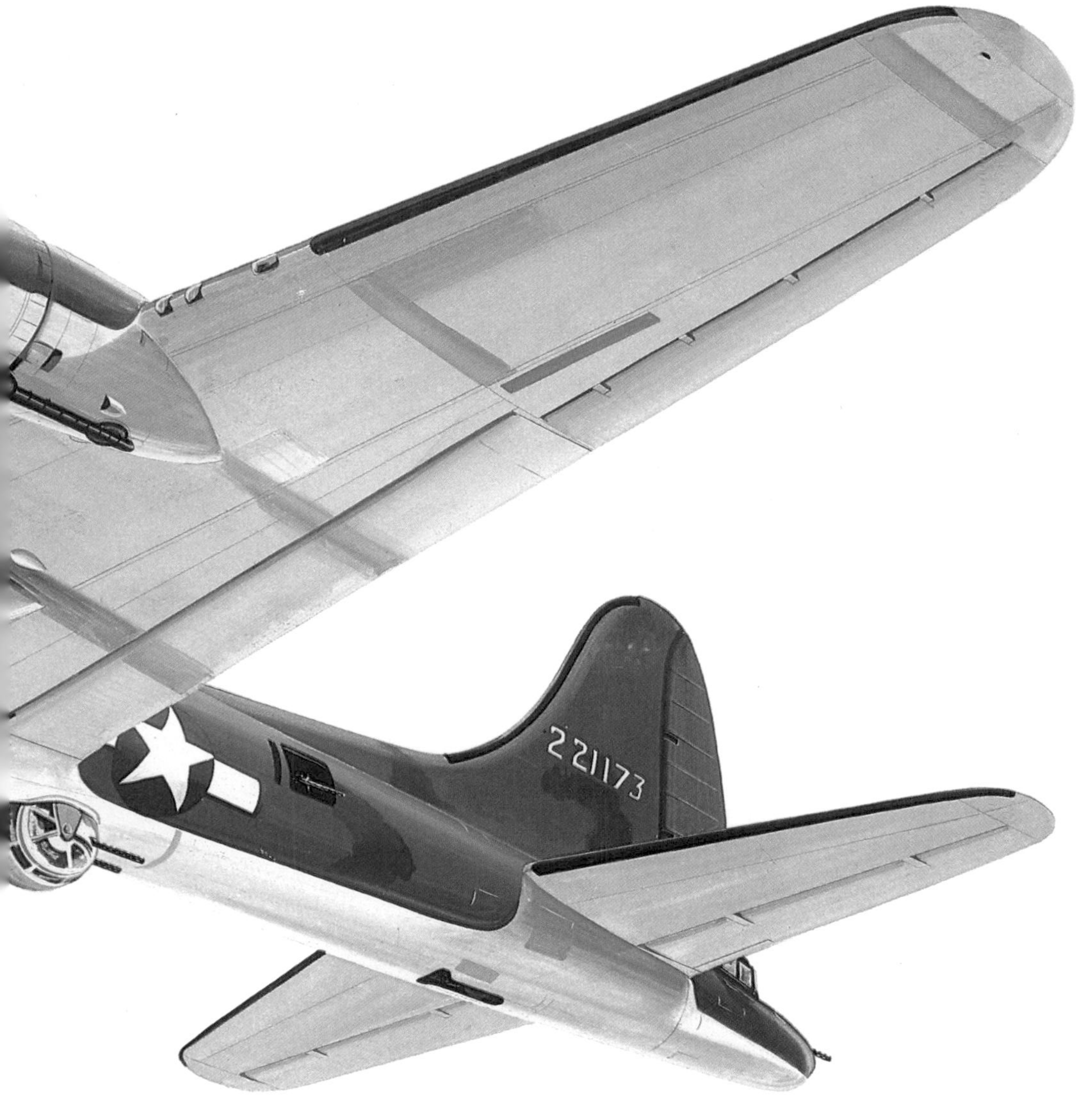

The Fortress was one of the U.S.A.'s most important warplanes of World War II, and resulted from a 1934 requirement for a multi-engined bomber with a 907 kg (2,000 lb) bomb load. The prototype flew in July 1935, and though it crashed this was followed by 14 pre-production aircraft including the static test airframe brought up to flight standard. The early production models were really development variants and included 39 B-17Bs, 38 upengined B-17Cs and 42 B-17Ds with self-sealing tanks and better armour. At this stage the tail was redesigned with a large dorsal fillet, and this led to the first of the definitive Fortresses, the B-17E of which 512 were built and B-17F with improved defensive armament of which 3,405 were built. The ultimate bomber variant was the B-17G with a chin turret and improved turbochargers for better ceiling, and this accounted for 8,680 of the almost 13,000 Fortresses. The type was discarded immediately after World War II, only a few special-purpose variants remaining. There were also a number of experimental and navy models.

BOEING B-17G FORTRESS

Role: Long-range day bomber
Crew/Accommodation: Ten
Power Plant: Four 1,200 hp Wright R-1820-97 Cyclone air-cooled radials
Dimensions: Span 31.62m (103.75 ft); length 22.66m (74.33 ft); wing area 131.92^2 (1,420 sq ft)
Weights: Empty 16,391kg (36,135 lb); MTOW 29,484kg (65,000 lb)
Performance: Maximum speed 462km/h (287 mph) at 7,620m (25,000 ft); operational ceiling 10,851m (35,600 ft); range 5,472km (3,400 miles)
Load: Twelve .5 inch machine guns, plus up to 2,722kg (6,000 lb) of bombs

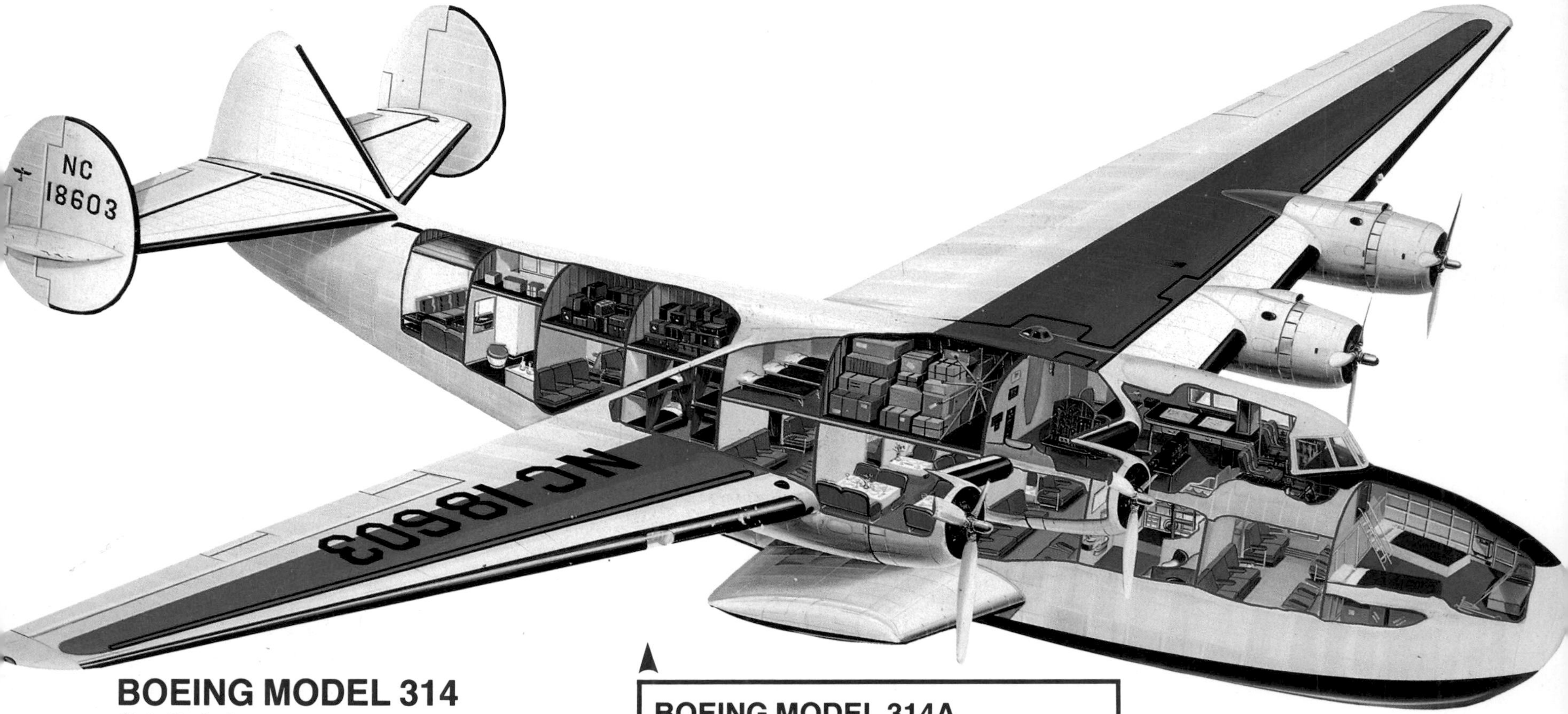

BOEING MODEL 314
(U.S.A.)

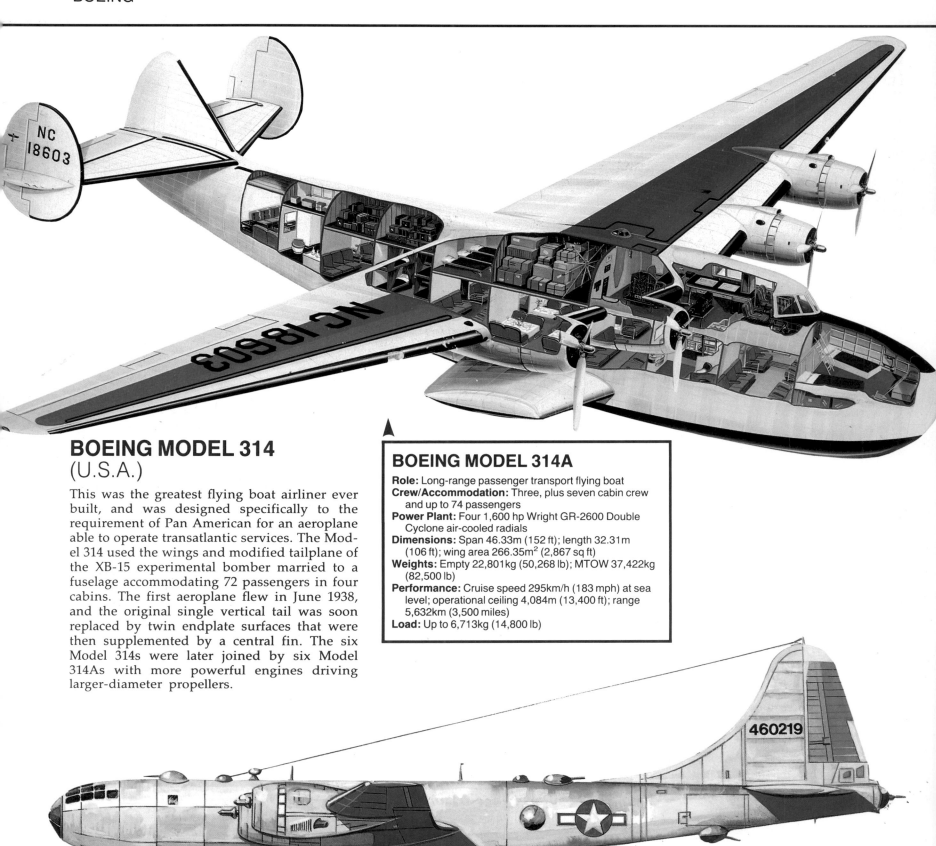

This was the greatest flying boat airliner ever built, and was designed specifically to the requirement of Pan American for an aeroplane able to operate transatlantic services. The Model 314 used the wings and modified tailplane of the XB-15 experimental bomber married to a fuselage accommodating 72 passengers in four cabins. The first aeroplane flew in June 1938, and the original single vertical tail was soon replaced by twin endplate surfaces that were then supplemented by a central fin. The six Model 314s were later joined by six Model 314As with more powerful engines driving larger-diameter propellers.

BOEING MODEL 314A

Role: Long-range passenger transport flying boat
Crew/Accommodation: Three, plus seven cabin crew and up to 74 passengers
Power Plant: Four 1,600 hp Wright GR-2600 Double Cyclone air-cooled radials
Dimensions: Span 46.33m (152 ft); length 32.31m (106 ft); wing area 266.35m^2 (2,867 sq ft)
Weights: Empty 22,801kg (50,268 lb); MTOW 37,422kg (82,500 lb)
Performance: Cruise speed 295km/h (183 mph) at sea level; operational ceiling 4,084m (13,400 ft); range 5,632km (3,500 miles)
Load: Up to 6,713kg (14,800 lb)

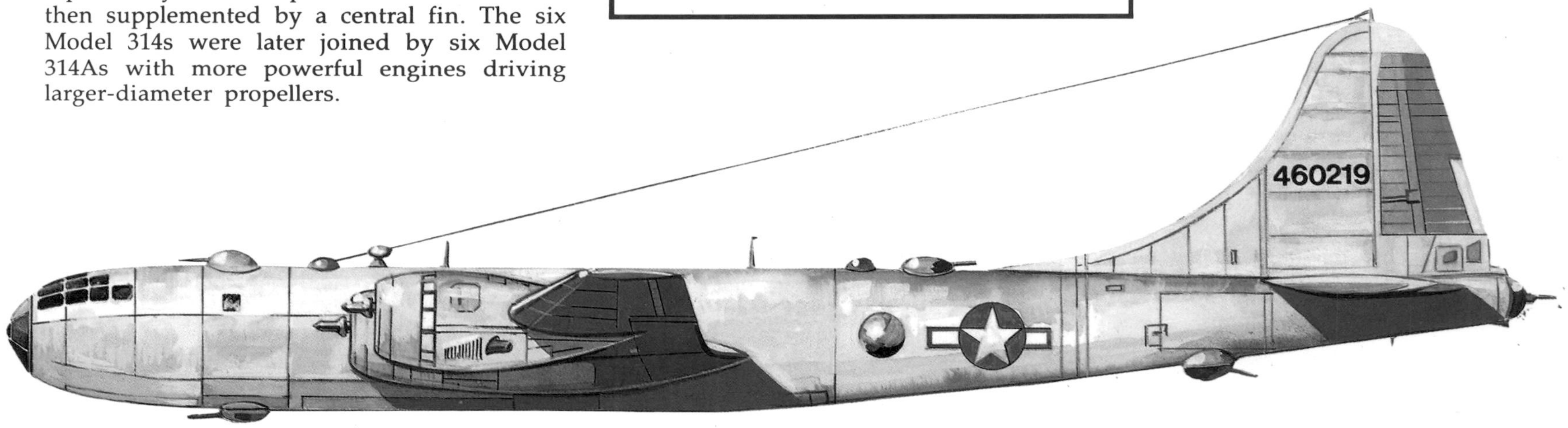

BOEING MODEL 345 (B-29 and B-50 SUPERFORTRESS)
(U.S.A.)

The B-29 was the world's first genuinely effective long-range strategic bomber, and was an extremely advanced design with pressurized accommodation, remotely controlled defensive armament, very high performance including great ceiling and range, and a formidable offensive load. The first example flew in September 1942, and a prodigious effort was made to bring the Superfortress into the bombing campaign against Japan during World War II, which it ended with the A-bombings of Hiroshima and Nagasaki. Some 2,848 B-29s were followed by 1,122 B-29As with slightly greater span and revised defensive armament, and by 311 B-29Bs with no defensive armament other than a radar-directed tail barbette. The type was also developed for reconnaissance and experimental roles, and was then revised with Wright R-3350 engines as the B-29D, which entered production as the B-50A. This was followed by its own series of bomber, reconnaissance and tanker aircraft.

BOEING B-29A SUPERFORTRESS

Role: Long-range, high altitude day bomber
Crew/Accommodation: Ten
Power Plant: Four 2,200 hp Wright R-3350-23 Cyclone Eighteen air-cooled radials
Dimensions: Span 43.05m (141.25 ft); length 30.18m (99 ft); wing area 161.56m^2 (1,739 sq ft)
Weights: Empty 32,369kg (71,360 lb); MTOW 62,823kg (138,500 lb)
Performance: Maximum speed 576km/h (358 mph) at 7,620m (25,000 ft); operational ceiling 9,708m (31,850 ft); range 6,598km (4,100 miles) with 7,258kg (16,000 lb) bombload
Load: One 20mm cannon and twelve .5 inch machine guns, plus up to 9,072kg (20,000 lb) of bombs

BOEING MODEL 450 (B-47 STRATOJET)
(U.S.A.)

The B-47 was a phenomenal technical achievement, and as a swept-wing strategic bomber in the medium-range bracket, formed the main strength of the U.S. Strategic Air Command in the early 1950s. The first aeroplane flew in December 1947, and was notable for many features including the 'bicycle' type landing gear whose twin main units retracted into the fuselage. The B-47A was essentially a development model, and the first true service variant was the B-47B, followed by the B-47E with a host of operational improvements. The B-47B and B-47E were both strengthened structurally later in their lives, leading to the designation B-47B-II and B-47E-II. There were also reconnaissance plus several special-purpose and experimental variants.

BOEING B-47E STRATOJET

Role: Heavy bomber
Crew/Accommodation: Three
Power Plant: Six 3,266 kgp (7,200 lb s.t.) General Electric J47-GE-25 turbojets, plus a 16,329 kgp (36,000 lb s.t.) rocket pack for Jet Assisted Take-Off (JATO)
Dimensions: Span 35.36m (116 ft); length 32.92m (108 ft); wing area 132.67m^2 (1,428 sq ft)
Weights: Empty 36,631kg (80,756 lb); MTOW 93,759kg (206,700 lb) with JATO rocket assistance
Performance: Maximum speed 975km/h (606 mph) at 4,968m (16,300 ft); operational ceiling 12,344m (40,500 ft); range 6,228km (3,870 miles) with 4,536kg (10,000 lb) bombload
Load: Two rear-firing 20mm cannon, plus up to 9,979kg (22,000 lb) of bombs

BOEING MODEL 464 (B-52 STRATOFORTRESS) (U.S.A.)

Numerically the B-52 is still the most important bomber in the U.S. Strategic Air Command inventory, offering great range and very large payload. The type was planned as a turboprop-powered successor to the B-50, but was then recast as a turbojet-powered type using eight Pratt & Whitney J57s podded in four pairs under the swept wings. The B-52 employs the same type of 'bicycle' landing gear as the B-47, and first flew in April 1952. The B-52A, B and C were in reality development models, and the first true service version was the B-52D, of which 170 were built. This model was followed by 100 B-52Es with improved navigation and weapon systems, 89 B-52Fs with greater power, 193 B-52Gs with a shorter fin and integral fuel tankage, and 102 B-52Hs with Pratt & Whitney TF33 turbofans and structural strengthening for the low-altitude role. The main types are the B-52G and H, which have carried Hound Dog and SRAM supersonic nuclear missiles and are now configured for the AGM-86B air-launched cruise missile.

BOEING B-52H STRATOFORTRESS

Role: Long-range bomber
Crew/Accommodation: Six
Power Plant: Eight 7,718 kgp (17,000 lb s.t.) Pratt & Whitney TF33-P-3 turbofans
Dimensions: Span 56.42m (185 ft); length 48.03m (157.6 ft); wing area 371.6m^2 (4,000 sq ft)
Weights: Empty 78,355kg (172,740 lb); MTOW 221,350kg (488,000 lb)
Performance: Maximum speed 1,013km/h (630 mph) at 7,254m (23,800 ft); operational ceiling 14,540m (47,700 ft); radius 7,000km (4,350 miles) with maximum bombload
Load: Up to 16,330kg (36,000 lb) of bombs or missiles

▼

▲ BOEING MODEL 717 (KC-135 STRATOTANKER) (U.S.A.)

From its Model 367-80 prototype Boeing developed the KC-135 inflight-refuelling tanker. The type was ordered on the basis of trials with the 'Dash 80' prototype fitted with Boeing's patented flying boom refuelling system, and the first KC-135A flew in August 1956. Such was the priority allocated to this essential support for the U.S.A.'s strategic bombers that production built up very rapidly, and eventually more than 800 KC-135 series aircraft were produced. Some were converted for special-purposes and experimental platforms, but the type remains so important that most are being re-engined and strengthened as KC-135REs.

BOEING KC-135A STRATOTANKER

Role: Military tanker-transport
Crew/Accommodation: Four, including fuel boom operator
Power Plant: Four 6,237 kgp (13,750 lb s.t.) Pratt & Whitney J57P-59W turbojets
Dimensions: Span 39.88m (130.83 ft); length 41.53m (136.25 ft); wing area 226.03m^2 (2,433 sq ft)
Weights: Empty 44,664kg (98,466 lb); MTOW 134,718kg (297,000 lb)
Performance: Cruise speed 888km/h (552 mph) at 9,144m (30,000 ft); operational ceiling 15,240m (50,000 ft); range 1,850m (1,150 miles) with maximum payload
Load: Up to 54,432kg (120,000 lb)

BOEING MODELS 707 and 720
(U.S.A.)

Though beaten into service by the de Havilland Comet, the Boeing 707 is regarded as the world's first effective long-range jet transport. The type was derived from the Model 367-80 prototype that first flew in July 1954, and once the U.S. Air Force had given clearance, the company started marketing it as an important civil transport. Some 844 of the type were built, and the major variants were the Model 707-120 transcontinental airliner with Pratt & Whitney JT3C turbojets, the Model 707-120B with JT3D turbofans, the Model 707-220 with JT4A turbojets, the Model 707-320 intercontinental airliner with longer wing and fuselage plus more powerful JT4A turbojets, the Model 707-320B with aerodynamic refinements and turbofans, the Model 707-320C convertible or freighter variants with turbofans, and the Model 707-420 with Rolls-Royce Conway turbofans. The Model 720 is aerodynamically similar to the Model 707-120 but has a shorter fuselage plus a new and lighter structure optimized for the intermediate-range role. There is also a Model 720B turbofan-powered variant of which 154 were built.

BOEING MODEL 707-320C

Role: Long-range passenger/cargo transport
Crew/Accommodation: Three, plus five/six cabin crew, plus up to 189 passengers
Power Plant: Four 8,618 kgp (19,000 lb s.t.) Pratt & Whitney JT3D-7 turbofans
Dimensions: Span 44.42m (145.71 ft); length 45.6m (152.92 ft); wing area 283.4m^2 (3,050 sq ft)
Weights: Empty 66,224kg (146,000 lb); MTOW 151,315kg (333,600 lb)
Performance: Cruise speed 886km/h (550 mph) at 8,534m (28,000 ft); operational ceiling 11,885m (39,000 ft); range 6,920km (4,300 miles) with maximum payload
Load: Up to 41,453kg (91,390 lb)

BOEING MODEL 727
(U.S.A.)

The Model 727 was conceived as a short/medium-range partner to the Model 707, and uses the same fuselage section. Otherwise the Model 727 has three rear-mounted engines, a T-tail and a clean wing with triple-slotted trailing-edge flaps. Independence of airport services is ensured by an auxiliary power unit and a ventral airstair/door. The Model 727 first flew in February 1963, and production totalled 1,932 in two main variants. The basic variant was the Model 727-100 with its convertible and quick-change convertible derivatives. Then came the Model 727-200 lengthened by 6.1m (20 ft) and with structural modification for operation at higher weights; the latest version was the Advanced 727-200 with a performance data computer system to improve operating economy and safety.

BOEING MODEL 727-200

Role: Intermediate-range passenger transport
Crew/Accommodation: Three and four cabin crew, plus up to 189 passengers
Power Plant: Three 7,257 kgp (16,000 lb s.t.) Pratt & Whitney JT8D-17 turbofans
Dimensions: Span 32.9m (108 ft); length 46.7m (153.17 ft); wing area 153.3m^2 (1,650 sq ft)
Weights: Empty 46,164kg (101,773 lb); MTOW 95,028kg (209,500 lb)
Performance: Cruise speed 982km/h (530 knots) at 7,620m (25,000 ft); operational ceiling 11,582+m (38,000+ ft); range 5,371km (2,900 naut. miles)
Load: Up to 18,597kg (41,000 lb)

BOEING MODEL 737
(U.S.A.)

The short-range Model 737 is the small brother of the Models 707 and 727, and is the world's best-selling airliner with more than 2,500 ordered. Intended for short sectors, the Model 737 first flew in April 1967 and is distinguished by a fuselage section similar to that of the Models 707 and 727, JT8D turbofans pod-mounted onto the under surfaces of the wings, main wheel units that are left uncovered when retracted, and a tail unit reminiscent of that of the Model 707. The initial variant was the Model 737-100 for 100 passengers, but only a few were built before there appeared the larger Model 727-200 for 130 passengers; there are also convertible, quick-change convertible, and advanced derivatives. Later came the Model 737-300 with CFM56-3 turbofans and further lengthening for 148 passengers, the further stretched Model 737-400 for 156 passengers and the updated Model 737-500 with more power.

BOEING MODEL 737-300

Role: Intermediate/short-range passenger transport
Crew/Accommodation: Two and four cabin crew, plus up to 149 passengers
Power Plant: Two 9,072 kgp (20,000 lb s.t.) General Electric/SNECMA CFM56-3-BI turbofans
Dimensions: Span 28.9m (94.75 ft); length 33.4m (109.58 ft); wing area 91m^2 (980 sq ft)
Weights: Empty 31,630kg (69,730 lb); MTOW 56,470kg (124,500 lb)
Performance: Cruise speed 908km/h (491 knots) at 7,925m (26,000 ft); operational ceiling 00,000m (00,000 ft); range 2,923km (1,580 miles) with maximum payload
Load: Up to 16,030kg (35,270 lb)

Boeing 737-400

Boeing 737-200

Boeing 747-400

BOEING MODEL 747
(U.S.A.)

Known universally as the 'jumbo', this is the world's largest airliner, and the mainstay of the Western world's long-range high-capacity routes. The type first flew in February 1969 and introduced the modern concept of wide-body airliners. With 900 aircraft ordered it is still in production with a choice of General Electric, Pratt & Whitney and Rolls-Royce turbofans. The main variants have been the initial Model 747-100 and strengthened Model 747-100B, the Model 747-200 with uprated engines, increased fuel capacity and structural strengthening (also available in convertible and freighter versions), the Model 747SP lighter-weight long-range version, the Model 747SR short-range version of the Model 747-100B, the Model 747-300 with a stretched upper deck increasing this area's accommodation from 16 first-class to 69 economy-class passengers, and the Model 747-400 with a revised and modernized structure, a two-crew flightdeck with the latest cockpit electronic displays and instrumentation, extended wings with drag-reducing winglets, and the latest lean-burn turbofans.

BOEING MODEL 747-400

Role: Long-range passenger/cargo transport
Crew/Accommodation: Two, plus up to 412 passengers and cabin crew in 3-class configuration
Power Plant: Four 26,263kg (57,900 lb s.t.) General Electric CF6-80C2 or 25,741kgp (56,750 lb s.t.) Pratt & Whitney PW4000 or 26, 308kg (58,000 lb s.t.) Rolls-Royce RB211-524G turbofans
Dimensions: Span 64.8m (213 ft); length 70.7m (231.87 ft); wing area 525m^2 (5,650 sq ft)
Weights: Empty 178,262kg (393,000 lb); MTOW 324,625kg (870,000 lb)
Performance: Cruise speed 939km/h (507 knots) at 10,670m (35,000 ft); operational ceiling 13,000+m (42,650+ ft); range 13,658km (7,370 naut. miles) with maximum fuel
Load: Up to 65,230kg (143,800 lb) including belly cargo

BOEING MODEL 757
(U.S.A.)

This may be regarded as replacement for the Model 727, and is a narrow-body type (with the same fuselage diameter as the Models 707, 727 and 737) offered with Rolls-Royce RB.211-535 or Pratt & Whitney PW2037 turbofans. The type is optimized for short/medium-range routes, and was originally planned in short- and long-fuselage variants: launch customers opted for the latter, and the shorter variant was dropped. The type first flew in February 1982, and is now offered in versions at three different weights.

BOEING MODEL 757-200

Role: Intermediate/short-range passenger transport
Crew/Accommodation: Two, four cabin crew, plus up to 239 passengers (charter)
Power Plant: Two 18,189 kgp (40,100 lb s.t.) Rolls-Royce RB211-535E4 or 17,327 kgp (38,200 lb s.t.) Pratt & Whitney PW2037 turbofans
Dimensions: Span 38.05m (124.83 ft); length 47.32m (155.25 ft); wing area 181.3m^2 (1,951 sq ft)
Weights: Empty 58,264kg (128,450 lb); MTOW 108,862kg (240,000 lb)
Performance: Cruise speed 850km/h (459 knots) at 11,887m (39,000 ft); operational ceiling 13,000+m (42,650+ ft); range 6,150km (3,320 naut. miles) with maximum payload
Load: Up to 26,349kg (58,090 lb)

BOEING MODEL 767
(U.S.A.)

This was planned in concert with the Model 757, and though a wide-body type, it has so much in common with its half-brother that pilots can secure a single rating for both types. The type was schemed as a high-capacity airliner for medium-range routes, and is offered with a choice of General Electric CF6 or Pratt & Whitney turbofans. The first Model 767 flew in September 1981 and, with the cancellation of the planned shorter-fuselage Model 76-100, the Model 767-200 became the standard variant. This is also available in Model 767-200ER form with additional fuel and higher weights for greater range. Greater capacity is provided by the Model 767-300, which has its fuselage stretched by 6.42m (21 ft 1 inch): this also is available in an extended-range variant.

Boeing 767-200

BOEING MODEL 767-300

Role: Intermediate-range passenger transport
Crew/Accommodation: Two/three, six cabin crew, plus up to 290 passengers (charter)
Power Plant: Two 23,814kgp (52,500 lb s.t.) General Electric CF6-80C2B2 or 22,770kgp (50,200 lb s.t.) Pratt & Whitney PW4050 turbofans
Dimensions: Span 47.57m (156.08 ft); length 54.94m (180.25 ft); wing area 283.4m^2 (3,050 sq ft)
Weights: Empty 86,954kg (191,700 lb); MTOW 159,211kg (351,000 lb)
Performance: Maximum speed 906km/h (489 knots) at 11,887m (39,000 ft); operational ceiling 13,000+m (42,650+ ft); range 5,965km (3,220 naut. miles) with maximum payload
Load: Up to 39,145kg (86,300 lb)

BOEING CH-46 SEA KNIGHT
(U.S.A.)

This twin-rotor medium-lift helicopter was developed as the Vertol (later Boeing Vertol) Model 107 with two turboshafts driving mechanically linked three-blade rotors. The type first flew in April 1958 and was adopted as the CH-46 assault transport for the U.S. Marines and UH-46 replenishment helicopter for the U.S. Navy, and these successively improved and upengined variants with heavier payloads were complemented by Model 107-II civil helicopters which are now licence-produced in Japan by Kawasaki.

BOEING VERTOL CH-46E SEA KNIGHT

Role: Naval medium-lift assault helicopter
Crew/Accommodation: Three, plus up to 26 troops
Power Plant: Two 1,400 shp General Electric T58-GE-16 turboshafts
Dimensions: Overall length rotors turning 25.69m (84.3 ft); rotor diameter 15.54m (51 ft)
Weights: Empty 6,057kg (13,342 lb); MTOW 10,569kg (23,300 lb)
Performance: Cruise speed 248km/h (133 knots) at sea level; operational ceiling 4,625m (14,000 ft); range 382km (206 naut. miles) with full payload
Load: Up to 1,905kg (4,200 lb) internally-carried, including ramp-loaded vehicles

BOEING CH-47 CHINOOK
(U.S.A.)

This is essentially an enlarged and more powerful version of the Model 107, and was developed as the Model 114 for a first flight in September 1961. As in the Model 107, the hold is accessed by a rear ramp/door, but quadricycle rather than tricycle landing gear is fitted. The Chinook has gone through a number of variants culminating in the current CH-47D with well over twice the power of the original CH-47A, and a civil version is manufactured as the Model 234. Licence production is undertaken by Meridionali in Italy.

BOEING VERTOL CH-47D CHINOOK

Role: Military medium-lift assault helicopter
Crew/Accommodation: Three, plus up to 44 troops
Power Plant: Two 3,750 shp Lycoming T55-L-712 turboshafts
Dimensions: Overall length rotors turning 30.15m (98.92 ft); rotor diameter 18.29m (60 ft)
Weights: Empty 10,247kg (22,591 lb); MTOW 22,680kg (50,000 lb)
Performance: Cruise speed 211km/h (114 knots) at sea level; operational ceiling 2,560m (8,400 ft); range 370km (200 miles) with 44 troops
Load: Up to 12,712kg (28,000 lb) of externally-slung cargo

▲ Boulton Paul Sidestrand

BOULTON PAUL SIDESTRAND and OVERSTRAND
(U.K.)

The Sidestrand was designed to a 1924 medium day bomber requirement, and first flew in 1926 as the Sidestrand Mk I. Only 18 were ordered, these being delivered as nine examples each of the Sidestrand Mks II and III, the former with ungeared Jupiter VI and the latter with geared Jupiter VIIIF radial engines. The Overstrand was a 1933 development of the Sidestrand with Pegasus I radials, and was a monumentally ungainly type with an enclosed cockpit, a large nose turret and a servo-operated rudder. The first aircraft were conversions originally designated Sidestrand Mk V, and these were followed by 24 production aircraft.

BOULTON PAUL OVERSTRAND Mk 1

Role: Day bomber
Crew/Accommodation: Four/five
Power Plant: Two 635 hp Bristol Pegasus air-cooled radials
Dimensions: Span 21.95m (72 ft); length 14.02m (46 ft); wing area 91.05m² (980 sq ft)
Weights: Empty 3,600kg (7,936 lb); MTOW 5,307kg (11,700 lb)
Performance: Maximum speed 232km/h (144 mph) at 1,981m (6,500 ft); operational ceiling 6,888m (22,600 ft); range 805km (500 miles) with full warload
Load: Three .303 inch machine guns, plus up to 499kg (1,100 lb) of bombs

BOULTON PAUL DEFIANT ▲
(U.K.)

Conceived to a 1935 requirement for a day fighter in which all armament would be concentrated in a four-gun turret, the Defiant first flew in August 1937 and began to enter service as the Defiant Mk I in 1939. Early operations revealed the inadequacy of the complete concept, and existing aircraft were converted to Defiant NF.Mk IA night-fighters with primitive radar. Upengined aircraft were Defiant NF.Mk IIs, of which many were converted to Defiant TT.Mk I target-tugs; similarly converted Mk Is became Defiant TT.Mk IIIs. About 50 were also modified for air/sea rescue, and total production was 1,065.

BOULTON PAUL DEFIANT Mk II

Role: Night fighter
Crew/Accommodation: Two
Power Plant: One 1,280 hp Rolls-Royce Merlin XX water-cooled inline
Dimensions: Span 11.99m (39.33 ft); length 10.77m (35.33 ft); wing area 23.23m² (250 sq ft)
Weights: Empty 2,850kg (6,282 lb); MTOW 3,773kg (8,318 lb)
Performance: Maximum speed 507km/h (315 mph) at 5,029m (16,500 ft); operational ceiling 9,251m (30,350 ft); range 748km (465 miles)
Load: Four .303 inch machine guns in power-operated turret.
Note: the Defiant Mk III was retrofitted to embody AIMk4 radar.

BREDA Ba 65
(Italy)

Designed as a multi-role type able to fulfil the interceptor, light bomber, attack and reconnaissance roles, the Ba 65 failed in all of these tasks despite the faith pinned on the type by the Italian air force. The type first flew in September 1935, and with the single-seat aircraft of the initial batch being powered by the Gnome-Rhone 14K radial engine, and later aircraft, of two-seat configuration with a dorsal gunner, by the Fiat A.80 radial engine. Production totalled about 275 including export aircraft.

BREDA Ba 65
Role: Ground attack
Crew/Accommodation: One
Power Plant: One 900 hp Isotta-Fraschini K.14 air-cooled radial
Dimensions: Span 11.9m (39.04 ft); length 9.6m (31.5 ft); wing area 23.5m^2 (253 sq ft)
Weights: Empty 1,950kg (4,290 lb); MTOW 2,500kg (5,512 lb)
Performance: Maximum speed 415km/h (258 mph) at 5,000m (16,400 ft); operational ceiling 7,800m (25,590 ft); range 550km (342 miles) with full warload
Load: Two 12.7mm and two 7.7mm machine guns, plus up to 1,000kg (2,205 lb) of bombs

BREGUET Bre.14
(France)

This was France's single most important warplane of World War I, first flown in November 1916 and developed during that war as a two-seat reconnaissance and single/two-seat bomber aeroplane. Production of more than 7,800 aircraft continued to 1928, and the type was phased out of French service only in 1932. The type was sturdy, and with a variety of engines possessed good performance. Post-war variants were developed for the colonial, ambulance, training and civil transport roles.

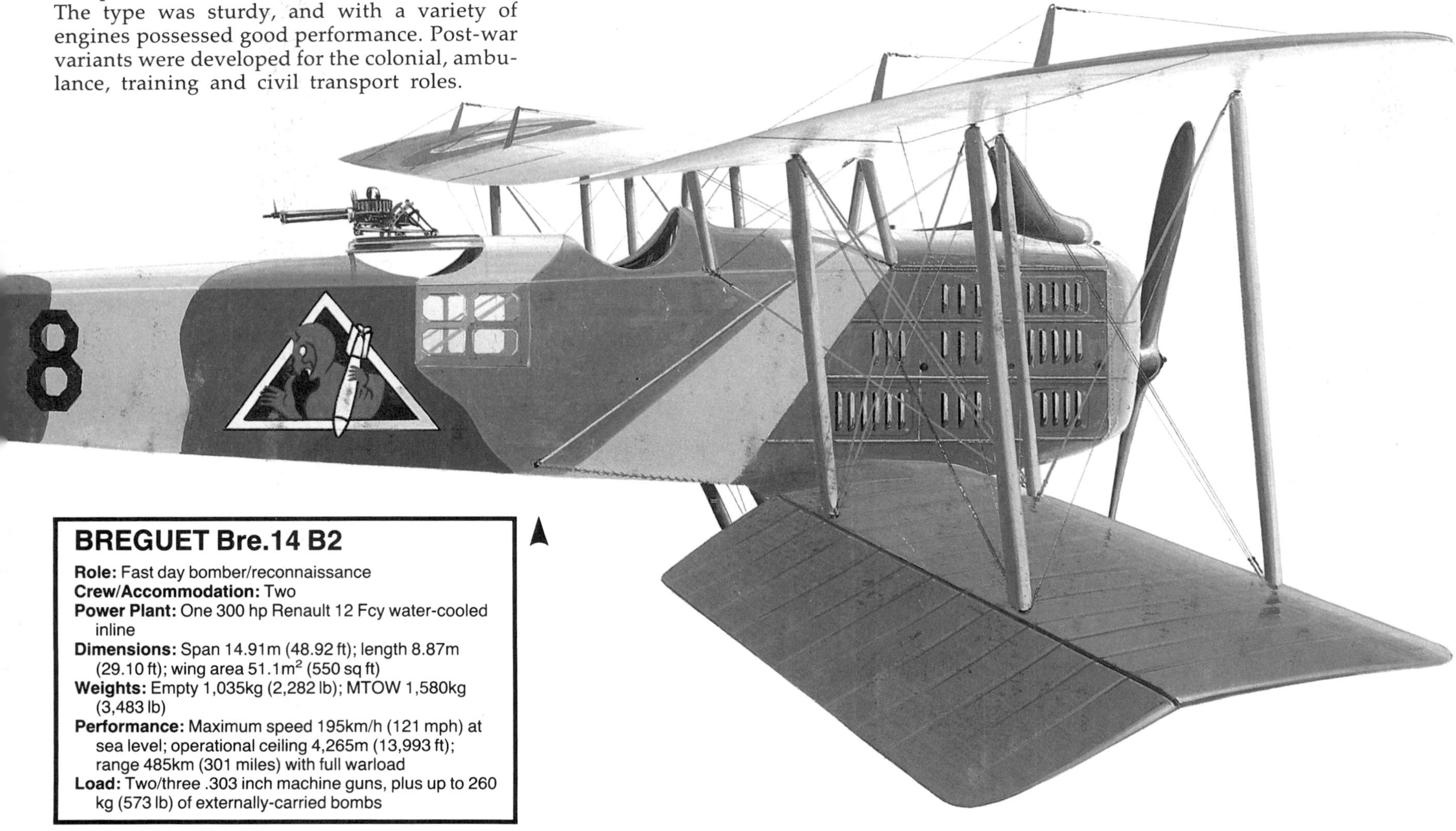

BREGUET Bre.14 B2
Role: Fast day bomber/reconnaissance
Crew/Accommodation: Two
Power Plant: One 300 hp Renault 12 Fcy water-cooled inline
Dimensions: Span 14.91m (48.92 ft); length 8.87m (29.10 ft); wing area 51.1m^2 (550 sq ft)
Weights: Empty 1,035kg (2,282 lb); MTOW 1,580kg (3,483 lb)
Performance: Maximum speed 195km/h (121 mph) at sea level; operational ceiling 4,265m (13,993 ft); range 485km (301 miles) with full warload
Load: Two/three .303 inch machine guns, plus up to 260 kg (573 lb) of externally-carried bombs

BREGUET Bre.19
(France)

First flown in March 1922, the Bre.19 was planned as a Bre.14 successor, but was produced in parallel to its predecessor for service with units based in metropolitan France. The type was used mainly in the two-seat reconnaissance and light bomber roles, and considerable export sales were secured mainly of the Bre.19GR long-range variant. The Bre.19 was also developed in Bidon (petrol can) and Super Bidon variants for a succession of record-breaking distance flights.

BREGUET Bre.19 B2

Role: Light day bomber
Crew/Accommodation: Two
Power Plant: One 550 hp Renault 12Kc water-cooled inline
Dimensions: Span 14.8m (48.5 ft); length 8.89m (29.16 ft); wing area 50m^2 (538.4 sq ft)
Weights: Empty 1,485kg (3,273 lb); MTOW 2,310kg (5,093 lb)
Performance: Maximum speed 240km/h (149 mph) at sea level; operational ceiling 7,800m (25,590 ft); range 800km (497 miles) with full warload
Load: Three/four .303 inch machine guns, plus up to 700kg (1,543 lb) of bombs

BREGUET Br.1050 ALIZE (TRADEWIND)
(France)

This anti-submarine type was developed from the abortive Br.960 Vultur strike aeroplane that had featured a Mamba turboprop for cruise and a tail-mounted turbojet for combat performance. Removal of the turbojet and revision of the fuselage allowed the incorporation of a weapon bay, a two-man mission crew and search radar with its antenna in a retractable 'dustbin' radome. The prototype flew in October 1956, and 87 were built for the French and Indian navies. French aircraft have been extensively updated.

BREGUET ALIZE

Role: Naval carrierborne anti-submarine
Crew/Accommodation: Three
Power Plant: One 2,100 shp Rolls-Royce Dart R.Da21 turboprop
Dimensions: Span 15.6m (51.2 ft); length 13.9m (45.5 ft); wing area 36m^2 (388 sq ft)
Weights: Empty 5,700kg (12,566 lb); MTOW 8,200kg (18,100 lb)
Performance: Maximum speed 460km/h (247 knots) at sea level; operational ceiling 6,096m (20,000 ft); endurance 7.66 hours
Load: Up to 880kg (1,940 lb) of weaponry

BREGUET Br.1150 ATLANTIC 1 and DASSAULT-BREGUET ATLANTIQUE 2
(France)

The Atlantic was planned as a NATO-wide maritime patroller, but was adopted only by France, Italy, the Netherlands and West Germany, with second-hand aircraft later going to Pakistan. The 87 aircraft were built by a European consortium, another such organization being responsible for the engines. The prototype flew in October 1961, and so impressive has been the type's reliability and range that the basic airframe has been retained for the exclusively French Atlantique, of which the first flew in May 1981. Some 42 are planned with a completely modernized suite of mission equipment.

DASSAULT-BREGUET ATLANTIC 2

Role: Land-based, long-range anti-submarine and maritime patrol
Crew/Accommodation: Twelve
Power Plant: Two 5,750 shp Rolls-Royce Tyne Mk.21 turboprops
Dimensions: Span 37.36m (122.6 ft); length 31.75m (104.2 ft); wing area 120m^2 (1,292 sq ft)
Weights: Empty 25,600kg (56,320 lb); MTOW 46,200kg (111,640 lb)
Performance: Maximum speed 658km/h (355 knots) at 5,486m (18,000 ft); operational ceiling 9,144m (30,000 ft); endurance 18 hours
Load: Up to 2,000 kg (4,409 lb), including up to four Exocet anti-ship cruise missiles

BREWSTER F2A BUFFALO
(U.S.A.)

This was the U.S. Navy's first production monoplane fighter, but generally a poor type. The prototype flew in June 1936, and orders were placed for 54 F2A-1 production aircraft, most of these being diverted to Finland. There followed 43 upengined F2A-2s and 108 better-protected F2A-3s for the U.S. Navy, together with B-339 variants for Belgium, the Netherlands and the U.K., of which the latter two used their aircraft against the Japanese in the Far East in World War II.

BREWSTER F2A-1

Role: Naval carrierborne fighter
Crew/Accommodation: One
Power Plant: One 940 hp Wright R-1820-34 Cyclone air-cooled radial
Dimensions: Span 11.48m (35 ft); length 7.92m (26 ft); wing area 19.42m^2 (209 sq ft)
Weights: Empty 1,717kg (3,785 lb); MTOW 2,436kg (5,370 lb)
Performance: Maximum speed 484km/h (301 mph) at 5,182m (17,000 ft); operational ceiling 9,906m (32,500 ft); range 2,486km (1,545 miles)
Load: Four .5 inch machine guns, plus up to 91kg (200 lb) of bombs

BRISTOL BOXKITE
(U.K.)

Otherwise known as the 1910 Biplane, this was one of the first series-produced aircraft of British design and originated from a Gabriel Voisin design further developed by Henry Farman. The type first flew in July 1910, and was produced in a number of forms with different engine types. The type was important in developing British civil and military aviation in its formative years.

BRISTOL BOXKITE

Role: Trainer
Crew/Accommodation: Two
Power Plant: One 50 hp Gnome air-cooled rotary or one 60 hp Renault air-cooled inline
Dimensions: Span 14.17m (46.5 ft); length 11.73m (38.5 ft); wing area 48.96m^2 (527 sq ft)
Weights: Empty 408kg (900 lb); MTOW 522kg (1,500 lb)
Performance: Maximum speed 64.4km/h (40 mph) at sea level
Load: None

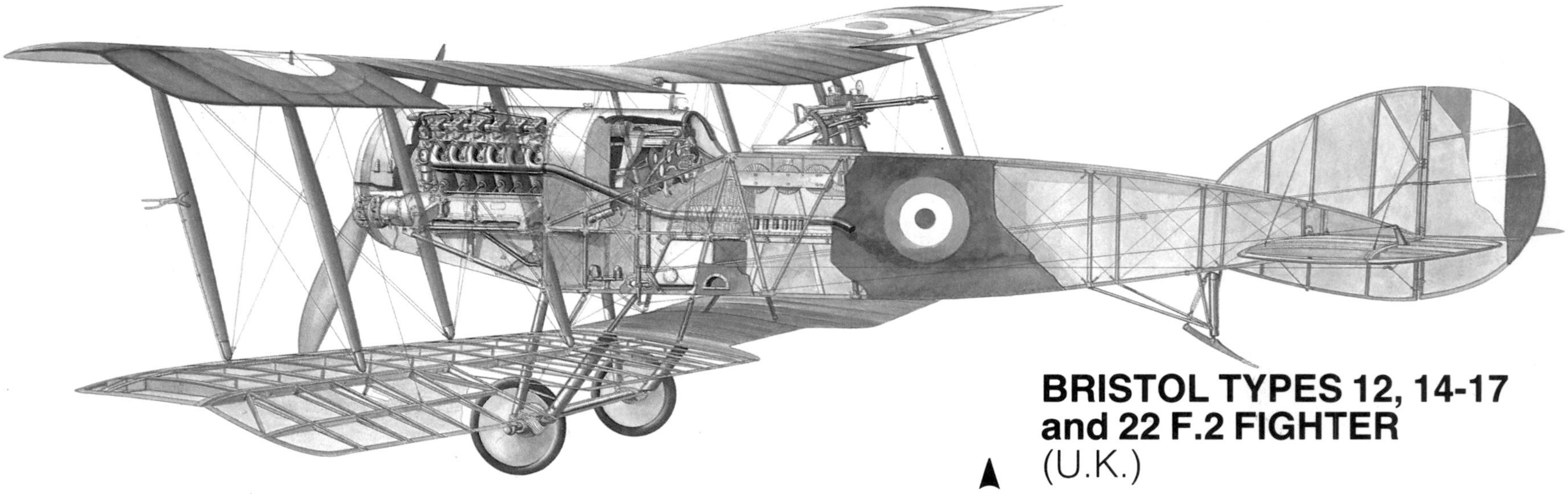

BRISTOL TYPES 12, 14-17 and 22 F.2 FIGHTER
(U.K.)

The F.2 Fighter was undoubtedly the best two-seat combat aeroplane of World War I, and was designed in 1916 as a reconnaissance type. The type was ordered as the R.2A, but such was the promised performance that even before the first prototype flew in September 1916 the type had been redesignated F.2A with a 142kW (190 hp) Rolls-Royce Falcon engine. F.2A production was 50 the 5,308-aircraft production run being of the F.2B variant with successively more powerful engines and modifications to improve fields of vision and combat-worthiness.

BRISTOL F.2B

Role: Fighter
Crew/Accommodation: Two
Power Plant: One 275 hp Rolls-Royce Falcon III water-cooled inline
Dimensions: Span 11.96m (39.25 ft); length 7.87m (25.83 ft); wing area 37.6m^2 (405 sq ft)
Weights: Empty 875kg (1,930 lb); MTOW 1,270kg (2,800 lb)
Performance: Maximum speed 201km/h (125 mph) at sea level; operational ceiling 6,096m (20,000 ft); endurance 3 hours
Load: Two or three .303 inch machine guns, plus up to 54.4kg (120 lb) of bombs

BRISTOL TYPES 10, 11, 20 and 77 M.1 MONOPLANE SCOUT
(U.K.)

This excellent fighter was distrusted by the British authorities because of its 'structurally suspect' monoplane layout, so despite their high flight performance the 125 production aircraft were relegated to secondary theatres. The type originated in the M.1A prototype that flew in July 1916 and led to four improved M.1Bs before the advent of the definitive M.1C. One M.1B was re-engined with a 76kW (100 hp) Bristol Lucifer as the post-war M.1D racer.

BRISTOL M.1C
Role: Fighter
Crew/Accommodation: One
Power Plant: One 110 hp Le Rhone 9J air-cooled rotary
Dimensions: Span 9.37m (30.75 ft); length 6.24m (20.46 ft); wing area 13.47m^2 (145 sq ft)
Weights: Empty 394kg (896 lb); MTOW 611kg (1,348 lb)
Performance: Maximum speed 204km/h (127 mph) at 1,524m (5,000 ft); operational ceiling 6,096m (20,000 ft); endurance 1.75 hours
Load: One .303 inch machine gun

BRISTOL TYPE 105 BULLDOG
(U.K.)

The U.K.'s primary fighter of the late 1920s and early 1930s, the Bulldog Mk I first flew in May 1927. There followed 312 production aircraft, of which the major variants were the basic Bulldog Mk II with a lengthened fuselage and a Jupiter VII radial, the Bulldog Mk IIA with a strengthened structure and Jupiter VIIF engine, the Bulldog Mk IVA with strengthened ailerons and a Mercury VIS.2 radial, and the Bulldog TM trainer with a second cockpit in a rear fuselage section that could be replaced by that of the standard fighters in times of crisis.

BRISTOL BULLDOG Mk IVA
Role: Fighter
Crew/Accommodation: One
Power Plant: One 640 hp Bristol Mercury VIS2 air-cooled radial
Dimensions: Span 10.26m (33.66 ft); length 7.72m (25.33 ft); wing area 27.31m^2 (294 sq ft)
Weights: Empty 1,220kg (2,690 lb); MTOW 1,820kg (4,010 lb)
Performance: Maximum speed 360km/h (224 mph) at sea level; operational ceiling 10,180m (33,400 ft); endurance 2.25 hours
Load: Two .303 inch machine guns, plus up to 36kg (80 lb) of bombs

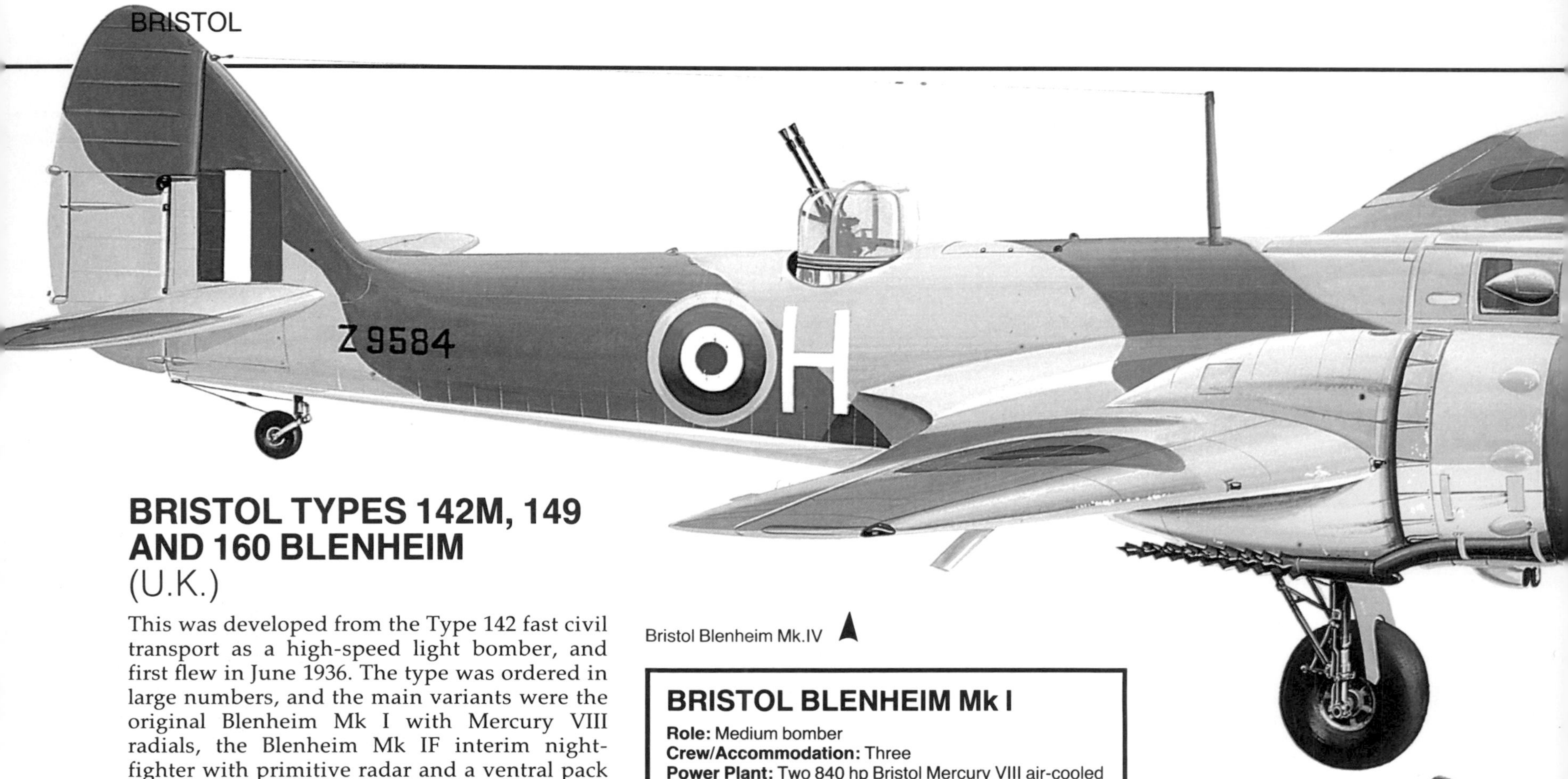

BRISTOL TYPES 142M, 149 AND 160 BLENHEIM
(U.K.)

This was developed from the Type 142 fast civil transport as a high-speed light bomber, and first flew in June 1936. The type was ordered in large numbers, and the main variants were the original Blenheim Mk I with Mercury VIII radials, the Blenheim Mk IF interim night-fighter with primitive radar and a ventral pack of four machine guns, the Blenheim Mk IV with more fuel and a lengthened nose, the Blenheim Mk IVF extemporized night-fighter, and the Blenheim Mk V with a solid nose housing four machine guns. The same basic type was built in Canada as the Bolingbroke coastal reconnaissance and light bomber aeroplane, variants being the Mk I with Mercury VIIIs, the Mk IV with Mercury XVs and the Mk IV W with Pratt & Whitney R-1830 Wasps.

Bristol Blenheim Mk.IV ▲

BRISTOL BLENHEIM Mk I

Role: Medium bomber
Crew/Accommodation: Three
Power Plant: Two 840 hp Bristol Mercury VIII air-cooled radials
Dimensions: Span 17.17m (56.33 ft); length 12.11m (39.75 ft); wing area 43.57m^2 (469 sq ft)
Weights: Empty 3,674kg (8,100 lb); MTOW 5,670kg (12,500 lb)
Performance: Maximum speed 459km/h (285 mph) at 4,572m (15,000 ft); operational ceiling 8,315m (27,280 ft); range 1,810km (1,125 miles) with full bombload
Load: Two .303 inch machine guns, plus up to 454kg (1,000 lb) of bombs

BRISTOL TYPE 152 BEAUFORT
(U.K.)

▲

The Beaufort was a development of the Blenheim Mk IV for the torpedo-bomber role. The type first flew in October 1938, and differed from the Blenheim in having more powerful Taurus radials and a revised fuselage providing internal accommodation for one torpedo as well as a dorsal turret and an undernose gun blister. The only other British production version was the Beaufort Mk II with Pratt & Whitney Twin Wasp radial engines for markedly improved performance. Australian production encompassed the Beaufort Mks V, VA, VI, VII and VIII with Twin Wasp engines and a number of detail modifications.

BRISTOL BEAUFORT Mk I

Role: Torpedo bomber/reconnaissance
Crew/Accommodation: Four
Power Plant: Two 1,130 hp Bristol Taurus VI/XII or XVI air-cooled radials
Dimensions: Span 17.63m (57.83 ft); length 13.49m (44.25 ft); wing area 46.73m^2 (503 sq ft)
Weights: Empty 5,942kg (13,000 lb); MTOW 9,630kg (21,230 lb)
Performance: Maximum speed 418km/h (260 mph) at sea level; operational ceiling 5,029m (16,500 ft); range 2,574km (1,600 miles) with full bombload
Load: Six .303 inch machine guns, plus up to 728kg (1,605 lb) in the shape of a single torpedo, or 680kg (1,500 lb) of bombs

BRISTOL TYPE 156 BEAUFIGHTER
(U.K.)

This robust twin-engined machine was born of the need for a heavy fighter, and was planned round the wings, tail unit and landing gear of the Beaufort married to a new fuselage and two Hercules radial engines. The first prototype flew in July 1939, and production then separated into two streams. First of these was the night-fighter type as exemplified by the Beaufighter Mk IF with Hercules XI radials, nose radar and an armament of four 20mm nose cannon plus six wing machine guns; further evolution led to the Beaufighter Mk IIF with Merlin XX inlines and finally to the Beaufighter Mk VIF with Hercules VI or XVI radials. More significant was the anti-ship version first developed as the Beaufighter Mk IC and then evolved via the torpedo-carrying Beaufighter Mk VIC to the classic Beaufighter TF.Mk X with search radar and an armament of one torpedo plus light bombs or eight rockets. The Beaufighter TF.Mk XI was similar but lacked torpedo capability and the Mk 21 was the Australian-built version of the TF.Mk X.

Bristol Beaufighter TF Mk.X

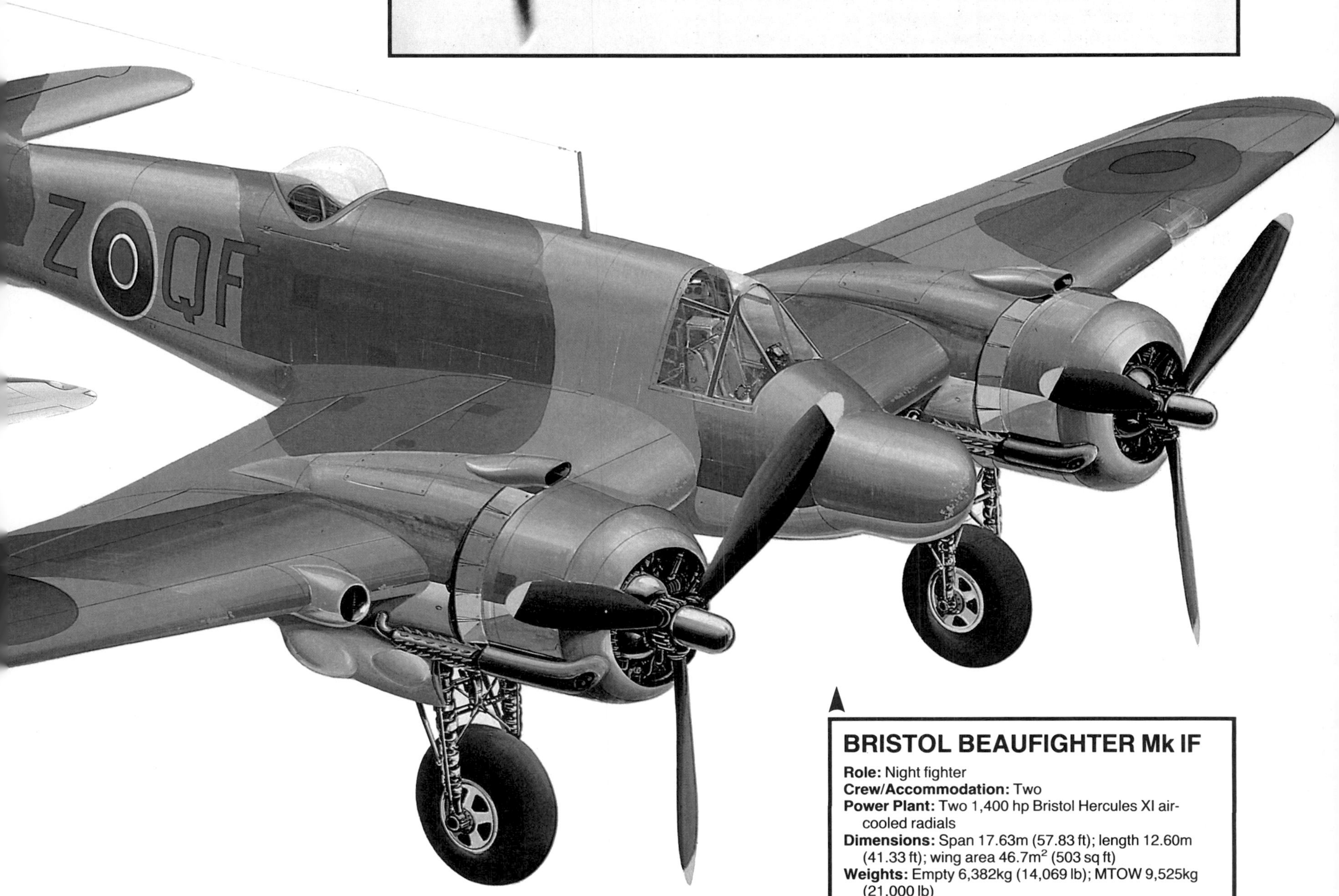

BRISTOL BEAUFIGHTER Mk IF

Role: Night fighter
Crew/Accommodation: Two
Power Plant: Two 1,400 hp Bristol Hercules XI air-cooled radials
Dimensions: Span 17.63m (57.83 ft); length 12.60m (41.33 ft); wing area 46.7m^2 (503 sq ft)
Weights: Empty 6,382kg (14,069 lb); MTOW 9,525kg (21,000 lb)
Performance: Maximum speed 520km/h (323 mph) at 4,572m (15,000 ft); operational ceiling 8,839m (29,000 ft); range 2,413km (1,500 miles) internal fuel only
Load: Four 20mm cannon and six .303 machine guns (interception guided by AI Mk IV radar)

BRISTOL TYPE 163 BUCKINGHAM and TYPE 164 BRIGAND
(U.K.)

The Buckingham was planned as a Blenheim replacement and first flew in February 1944, when the superlative Mosquito was already filling the type's niche. The 119 production aircraft were thus completed as 54 Buckingham B.Mk 1 tactical day bombers and 65 Buckingham C.Mk 1 fast courier transports. The generally similar Brigand first flew in December 1944 and was schemed as a Beaufort replacement, but with the Beaufighter fulfilling this role the type was developed as the Brigand B.Mk 1 light attack bomber for Far Eastern operations. Production totalled 144, a few as Brigand Met.Mk 3 meteorological reconnaissance and Brigand T.Mk 4 radar training aircraft.

BRISTOL BRIGAND B.Mk 1

Role: Long-range strike
Crew/Accommodation: Three
Power Plant: Two 2,470 hp Bristol Centaurus 57 air-cooled radials
Dimensions: Span 22m (72.33 ft); length 14.2m (46.43 ft); wing area 66.7m^2 (718 sq ft)
Weights: Empty 11,611kg (25,598 lb); MTOW 17,690kg (39,000 lb)
Performance: Maximum speed 576km/h (358 mph) at 4,880m (16,000 ft); operational ceiling 7,930m (26,000 ft); range 4,506km (2,800 miles)
Load: Four 20mm cannon, plus 908kg (2,000 lb) of bombs carried internally

▼

BRISTOL TYPE 175 BRITANNIA
(U.K.)

▲ Probably the finest turboprop airliner ever built, the Britannia was so delayed by engine problems that it was overtaken by turbojet-powered airliners and failed to fulfil its great commercial promise. The type was planned as a 36-passenger medium-range airliner for British Imperial routes, but as the proposed Bristol Centaurus radials were more than powerful enough for this load the type was enlarged to 48-passenger capacity, and then revised for 90 passengers on the power of four 2,088kW (2,800 ehp) Bristol Proteus turboprops. The first prototype flew in August 1952, and the Britannia Series 100 entered service in 1957 with 2,819kW (3,780 ehp) Proteus 705s. There followed the Britannia Series 300 with its fuselage lengthened by 3.12m (10 ft 3 inches) for a maximum of 133 passengers carried over transatlantic routes, and the longer-range Britannia Series 310 with 3,072kW (4,120 ehp) Proteus 755s and greater fuel capacity. Only two were built of the final Britannia Series 320 with 3,318kW (4,450 ehp) Proteus 765s, and the last variant was the Britannia Series 250 modelled on the Series 310 but intended for RAF use as the Britannia C.Mk 1 and C.Mk 2. The same basic airframe was used by Canadair as the core of two aircraft, the CL-28 Argus maritime patroller with Wright R-3350-EA1 Turbo-Compound piston engines, and the CL-44 long-range transport with Rolls-Royce Tyne 515 Mk 10 turboprops.

BRISTOL BRITANNIA 310 SERIES

Role: Long-range passenger transport
Crew/Accommodation: Four, four cabin crew and up to 139 passengers
Power Plant: Four 4,120 ehp Bristol Siddeley Proteus 755 turboprops
Dimensions: Span 43.37m (142.29 ft); length 37.87m (124.25 ft); wing area 192.78m^2 (2,075 sq ft)
Weights: Empty 37,438kg (82,537 lb); MTOW 83,915kg (185,000 lb)
Performance: Cruise speed 660km/h (410 mph) at 6,401m (21,000 ft); operational ceiling 9,200+m (30,184 ft); range 6,869km (4,268 miles) with maximum payload
Load: Up to 15,830kg (34,900 lb)

BRITISH AEROSPACE (HAWKER/HAWKER SIDDELEY) HARRIER and SEA HARRIER

(U.K.)

The Harrier was the world's first operational VTOL combat aircraft, and at its core is the remarkable Rolls-Royce (Bristol Siddeley) Pegasus vectored-thrust turbofan. The type was pioneered in the P.1127 prototype that first flew in October 1960, and such was the potential of this experimental type that Kestrel F(GA).Mk 1 evaluation aircraft were built for a combined British, U.S. and West German trials squadron. The Harrier is the true operational version, and with successively more powerful Pegasus engines the main types have been the Harrier GR.Mk 1, the GR.Mk 1A and the GR.Mk 3 with a revised nose accommodating a laser ranger and marked-target seeker. Combat-capable two-seat trainers are the Harrier T.Mk 2, T.Mk 2A and T.Mk 4. The U.S. Marine Corps uses the Harrier as the AV-8A single-seater (of which many have been upgraded to AV-8C standard) and TAV-8A two-seater. A much improved variant has been developed by McDonnell Douglas and BAe as the Harrier II with a larger wing and a number of advanced features. This serves with the U.S. Marine Corps as the AV-8B and with the RAF as the Harrier GR.Mk 5 or, with night-vision capability, the Harrier GR.Mk 7. The companies are also working on a radar-carrying model. BAe has developed the specialized Sea Harrier FRS.Mk 1 with a fighter-type cockpit, radar and other equipment for the Fleet Air Arm, and these aircraft are being upgraded to Sea Harrier FRS.Mk 2 standard with more advanced radar and the capability for advanced air-to-air and anti-ship missiles.

BRITISH AEROSPACE SEA HARRIER FRS Mk 1

Role: Strike fighter/reconnaissance
Crew/Accommodation: One
Power Plant: One 9,752kgp (21,500 lb s.t.) Rolls-Royce Pegasus 104 vectored thrust turbofan
Dimensions: Span 7.7m (25.25 ft); length 14.52m (47.58 ft); wing area 18.68m^2 (201.1 sq ft)
Weights: Empty 6,374kg (14,052 lb); MTOW 11,880kg (26,200 lb) with short take-off roll
Performance: Maximum speed 1,185km/h (640 knots) at sea level; operational ceiling 15,240+m (50,000+ ft); radius 520km (280 naut. miles) with two anti-ship missiles
Load: Two 30mm cannon, plus up to 3,630kg (8,000 lb) of externally carried ordnance

BRITISH AEROSPACE (HAWKER SIDDELEY) HAWK and MCDONNELL DOUGLAS T-45A GOSHAWK
(U.K.)

The Hawk was developed to replace the Hawker Siddeley (Folland) Gnat in the training role, and first flew in August 1974. The type is operated by the RAF as the Hawk T.Mk 1 and secondary air-defence Hawk T.Mk 1A, and variously upengined models serve with several other air arms. Other variants are the Hawk 100 two-seat ground-attack aeroplane and the Hawk 200 single-seat attack model with radar capability and advanced weapons. The basic design has been adapted by McDonnell Douglas as the T-45A Goshawk carrier-capable trainer for the U.S. Navy.

BRITISH AEROSPACE (SIDDELEY) HAWK 60 SERIES

Role: Light strike/trainer
Crew/Accommodation: Two
Power Plant: One 2,585kgp (5,700 lb s.t.) Rolls-Royce Turbomeca Adour 861 turbofan
Dimensions: Span 9.4m (30.83 ft); length 11.85m (38.92 ft); wing area 16.69m^2 (179.64 sq ft)
Weights: Empty 3,750kg (8,270 lb); MTOW 8,500kg (18,740 lb)
Performance: Maximum speed 1,037km/h (560 knots) Mach 0.81 at seal level; operational ceiling 15,240m (50,000 ft); radius 1,093km (590 naut. miles) with 998kg (2,000 lb) warload
Load: Up to 3,084kg (6,800 lb) of weapons/fuel carried externally.

McDonnell Douglas T-45A Goshawk ▲

BAe. Hawk T. Mk.1 ▼

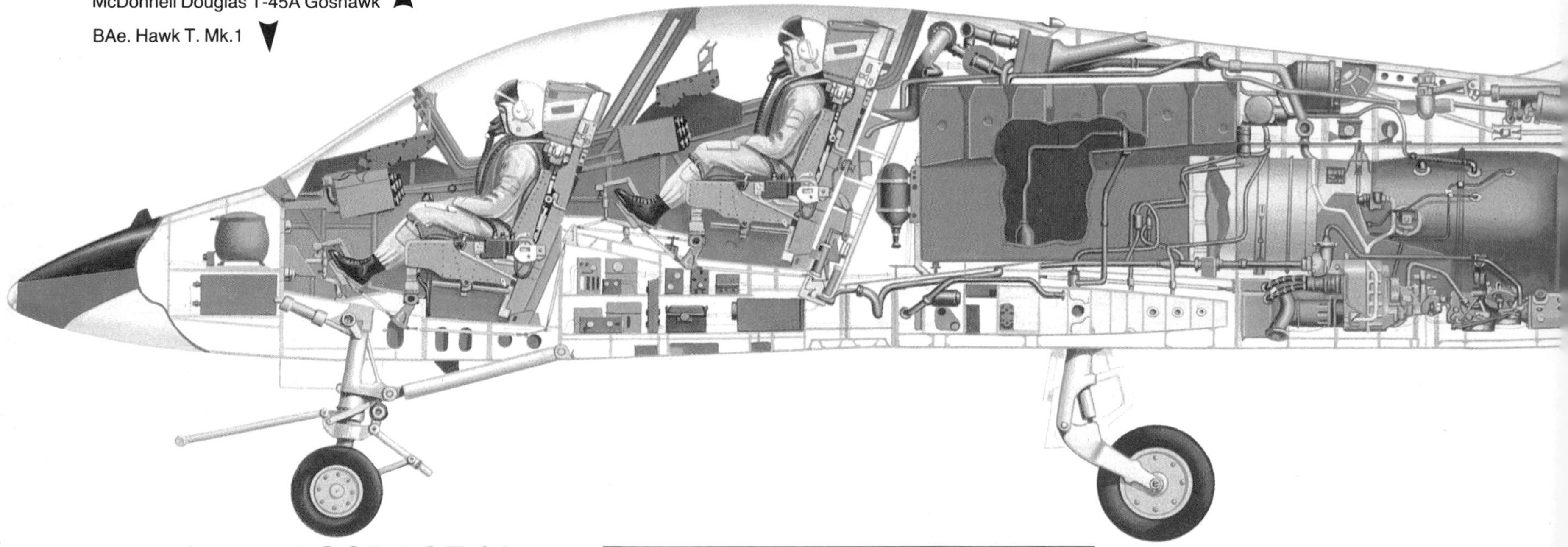

BRITISH AEROSPACE (de HAVILLAND/HAWKER SIDDELEY) H.S.125 and DOMINIE
(U.K.)

Certainly the most successful post-World War II civil aircraft of British origins, the H.S.125 was initially schemed by de Havilland as a long-range executive transport, and first flew in August 1962 with two 1,361kg (3,000 lb) thrust Armstrong Siddeley Viper 20 turbojets, replaced in the enlarged D.H.125 Series 1 production aircraft by Viper 520s. Successive variants have been the D.H.125 Series 1A/1B with higher weights and 1,406kg (3,100 lb) thrust Viper 521 and 522 engines, the H.S.125 Series 2 operated by the RAF as the Dominie navigation trainer and communications aircraft, the H.S.125 Series 3 and H.S.125 Series 400 with more powerful Viper 522s, the H.S.125 Series 600 with the fuselage lengthened by 0.61 m (24 inches) for 14 passengers, the H.S.125 Series 700 with 1,678kg (3,700 lb) thrust Garrett TFE731-3-1H turbofans for markedly improved range with a considerable decrease in noise, and the improved BAe 125 Series 800.

BRITISH AEROSPACE BAe 125 SERIES 800

Role: Executive Jet transport
Crew/Accommodation: Two and one cabin crew, plus up to 14 passengers
Power Plant: Two 1,950kgp (4,300 lb s.t.) Garrett AiResearch TFE731-5R-1H turbofans
Dimensions: Span 15.66m (51.38 ft); length 15.6m (51.17 ft); wing area 34.75m^2 (374 sq ft)
Weights: Empty 6,858kg (15,120 lb); MTOW 12,430kg (27,400 lb)
Performance: Cruise speed 741km/h (400 knots) at 11,887m (39,000 ft); operational ceiling 13,100m (49,000 ft); range 5,318km (2,870 naut. miles) with maximum payload
Load: Up to 1,088kg (2,400 lb)

▼

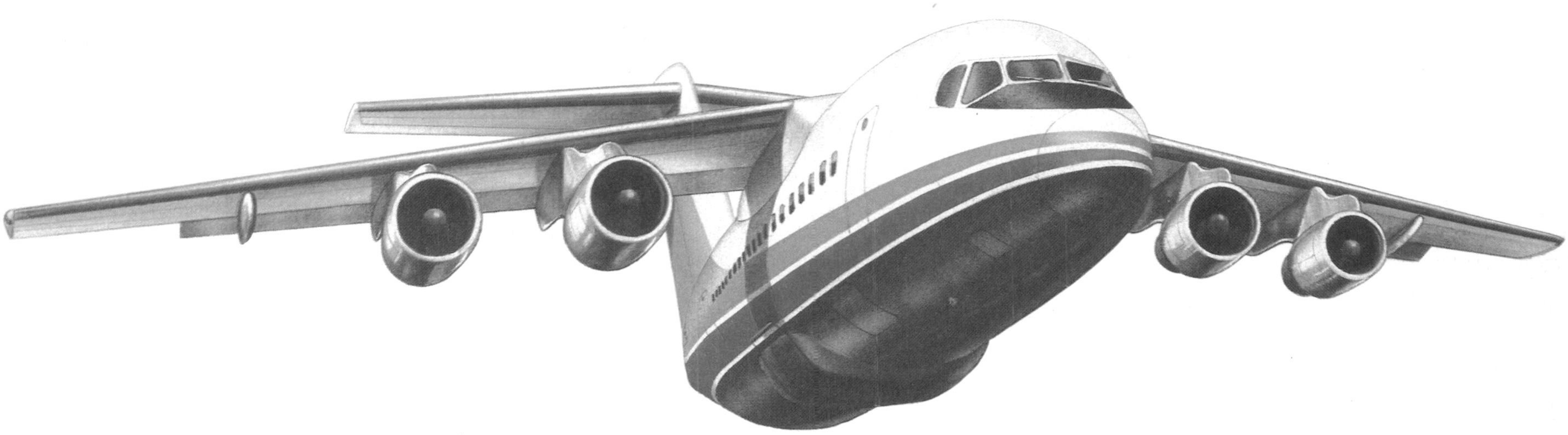

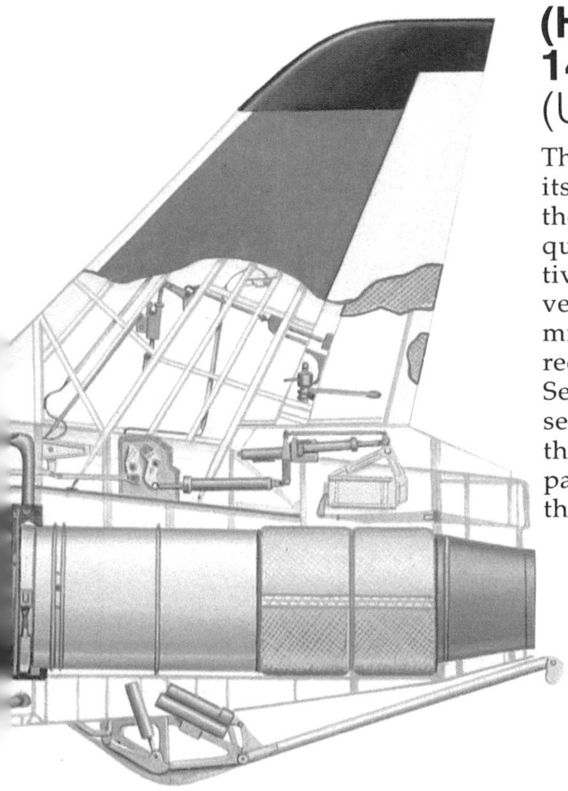

▲ BRITISH AEROSPACE (HAWKER SIDDELEY) BAe 146
(U.K.)

The BAe. 146 short-range airliner is unusual for its type in having four rather than two engines, though these are exceptionally economical and quiet turbofans that have proved highly effective in service. The type's development was very protracted because of a suspension in the mid-1970s as a result of the world economic recession, and the first example flew only in September 1981. Since that time the airliner has secured a useful niche in the marketplace, and the main variants are the baseline 146-100 for 93 passengers, the 146-200 with its fuselage lengthened by 2.41m (7 ft 11 inches) for 109 passengers, and the 146-300 with its fuselage lengthened by another 2.38m (7 ft 10 inches) for 130 passengers. There are also freighter and military versions in service or planned.

BRITISH AEROSPACE BAe 146-300

Role: Short-range passenger/cargo transport
Crew/Accommodation: Two, three cabin crew, plus up to 128 passengers
Power Plant: Four 3,162kgp (6,970 lb s.t.) Avco Lycoming ALF502R-5 turbofans
Dimensions: Span 26.34m (86.42 ft); length 30.99m (101.67 ft); wing area 77.3m^2 (832 sq ft)
Weights: Empty 24,471kg (53,950 lb); MTOW 44,225kg (97,500 lb)
Performance: Maximum speed 784km/h (423 knots) at 7,315m (24,000 ft); operational ceiling 9,449m (31,000 ft); range 1,761km (950 naut. miles) with maximum payload
Load: Up to 11,135kg (24,549 lb)

BRITISH AEROSPACE (HANDLEY PAGE/ SCOTTISH AVIATION) JETSTREAM ►
(U.K.)

The Jetstream was designed as the Handley Page H.P.137 executive transport and feederliner, but the development cost broke the parent company and further work was transferred to Scottish Aviation, which became a division of BAe in 1978. The original Jetstream 1 (Jetstream Series 200) was underpowered with two 626kW (840 ehp) Turbomeca Astazou XIV turboprops, though better performance was attained by U.S. completed aircraft with 743kW (996 ehp) Astazou XVIs. British military examples are generally similar to this latter standard. However, the engine that turned the Jetstream into a winner is the Garrett TPE331 turboprop as used in the definitive Jetstream 31. The latest development is the Jetstream 41 series with a 4.87m (16 ft) longer fuselage seating up to 29 passengers.

BRITISH AEROSPACE JETSTREAM 31

Role: Short-range passenger transport
Crew/Accommodation: Two, plus one cabin crew and up to 19 passengers
Power Plant: Two 900 shp Garrett AiResearch TPE 331-10 turboprops
Dimensions: Span 15.85m (52 ft); length 14.37m (47.15 ft); wing area 25.08m^2 (270 sq ft)
Weights: Empty 6,000kg (13,228 lb); MTOW 6,600kg (14,550 lb)
Performance: Cruise speed 482km/h (260 knots) at 6,100m (20,000 ft); operational ceiling 7,620m (25,000 ft); range 1,185km (640 naut. miles) with full payload
Load: Up to 1,897kg (4,182 lb)

BRITISH AEROSPACE (HAWKER SIDDELEY) NIMROD

(U.K.)

The Nimrod was developed on the aerodynamic basis of the Comet airliner as a maritime patroller to replace the Avro Shackleton, and features a deepened fuselage to accommodate a large weapons bay, a turbofan power plant, and highly advanced mission electronics including radar, MAD and an acoustic data-processing system using dropped sonobuoys. The first prototype flew in May 1967, and successful trials led to production of 43 Nimrod MR.Mk 1s. A variant of this baseline version is the Nimrod R.Mk 1, a special electronic intelligence variant of which three were produced. Further development in the electronic field led to the improved Nimrod MR. Mk2, of which 32 were produced by conversion of MR. Mk 1 airframes. The project to produce a Nimrod AEW.Mk 3 airborne early warning version was cancelled in 1987.

BRITISH AEROSPACE (HAWKER SIDDELEY) NIMROD MR. Mk 2

Role: Long-range maritime reconnaissance/anti submarine
Crew/Accommodation: Twelve
Power Plant: Four, 5,507kgp (12,140 lb s.t.) Rolls-Royce Spey Mk 250 turbofans
Dimensions: Span 35m (114.83 ft); length 38.63m (126.75 ft); wing area 197m^2 (2,121 sq ft)
Weights: Empty 39,000kg (86,000 lb); MTOW 87,090kg (192,000 lb)
Performance: Maximum speed 926km/h (500 knots) at 610m (2,000 ft); operational ceiling 12,802m (42,000 ft); endurance 12 hours
Load: Up to 6,120kg (13,500 lb) including up to nine lightweight anti-submarine torpedoes

BRITISH AIRCRAFT CORPORATION ONE-ELEVEN

(U.K.)

This pioneering airliner was developed initially as the Hunting H.107 with aft-mounted engines and a T-tail. Hunting was bought by BAC and the H.107 became the BAC 107 and, when enlarged from 59 to 80 passengers, BAC 111 (later One-Eleven). The prototype flew in August 1963, and useful sales were secured for the basic One-Eleven Series 200 with Spey Mk 506 turbofans, the One-Eleven Mk 300 with Spey Mk 511s, the generally similar One-Eleven Mk 400 for U.S. airlines, the stretched One-Eleven Series 500 for 119 passengers, and the 'hot and high' One-Eleven Series 475 with fuselage of the Series 400 plus the wings and power plant of the Series 500. The One-Eleven production line was later bought by Rombac in Romania, where One-Elevens are still in production.

BRITISH AIRCRAFT CORPORATION ONE-ELEVEN 500

Role: Short-range passenger transport
Crew/Accommodation: Two and three/four cabin crew, plus up to 119 passengers
Power Plant: Two 5,692kgp (12,550 lb s.t.) Rolls-Royce Spey 512-DW turbofans
Dimensions: Span 28.5m (92.5 ft); length 32.61m (107 ft); wing area 95.78m^2 (1,031 sq ft)
Weights: Empty 24,758kg (54,582 lb); MTOW 47,000kg (104,500 lb)
Performance: Maximum speed 871km/h (470 knots) at 6,400m (21,000 ft); range 2,380km (1,480 miles) with full passenger load
Load: Up to 11,983kg (26,418 lb) including belly cargo

BRITTEN-NORMAN ISLANDER and DEFENDER
(U.K.)

The Islander was designed as a simple feederliner for operators in remoter areas, and first flew in September 1964. The type was then modified with greater span wing and 0-540 piston engines in place of the original Rolls-Royce 10-360 units, and entered production as the nine-passenger BN-2. Later variants have been the improved BN-2A and the heavier BN-2B Islander II. The type has also been produced in Defender and Maritime Defender militarized models, and the BN-2T has 239kW (320 shp) Allison 250 turboprops. The largest derivative is the BN-2A Mk III Trislander with a third piston engine added at the junction of the enlarged vertical tail and the mid-set tailplane; the Trislander has a lengthened fuselage for 17 passengers.

BRITTEN-NORMAN BN-2B-26 ISLANDER

Role: Light short-field utility/passenger transport
Crew/Accommodation: One, plus up to nine passengers
Power Plant: Two 260hp Lycoming 0-540-E4C5 air-cooled flat-opposed
Dimensions: Span 14.94m (49 ft); length 12.02m (35.67 ft); wing area 30.2m^2 (325 sq ft)
Weights: Empty 1,866kg (4,114 lb); MTOW 2,993kg (6,600 lb)
Performance: Cruise speed 257km/h (160 mph) at 2,140m (7,000 ft); operational ceiling 3,597m (11,800 ft); range 1,251km (675 naut. miles) with full payload
Load: Up to 755kg (1,665 lb)

BUCKER Bu 131 JUNGMANN AND Bu 133 JUNGMEISTER
(Germany)

The Jungmann first flew in April 1934 as a compact biplane trainer with a fabric-covered airframe comprising wooden wings, and a steel-tube fuselage and empennage. The Bu 131A had the 60kW (980 hp) HM 60R engine, and the improved Bu 131B had the 78kW (105 hp) HM 504A-2. The type was built in Japan as the Watanabe Ki-86 and K9W. For advanced training Bucker developed the basically similar Jungmeister, of which the major variants were the Ju 133A with the 101kW (135 hp) HM 6 inline engine, the Bu 133B with a 119kW (160 hp) HM 506 inline engine and the Bu 133C with the 119kW Siemens Sh 14 radial engine.

BUCKER Bu 131B JUNGMANN

Role: Trainer
Crew/Accommodation: Two
Power Plant: One 105 hp Hirth HM 504A-2 air-cooled inline
Dimensions: Span 7.4m (24.28 ft); length 6.62m (21.72 ft); wing area 13.5m^2 (145.3 sq ft)
Weights: Empty 390kg (860 lb); MTOW 680kg (1,500 lb)
Performance: Maximum speed 183km/h (114 mph) at sea level; operational ceiling 3,000m (9,843 ft); range 650km (404 miles)
Load: None

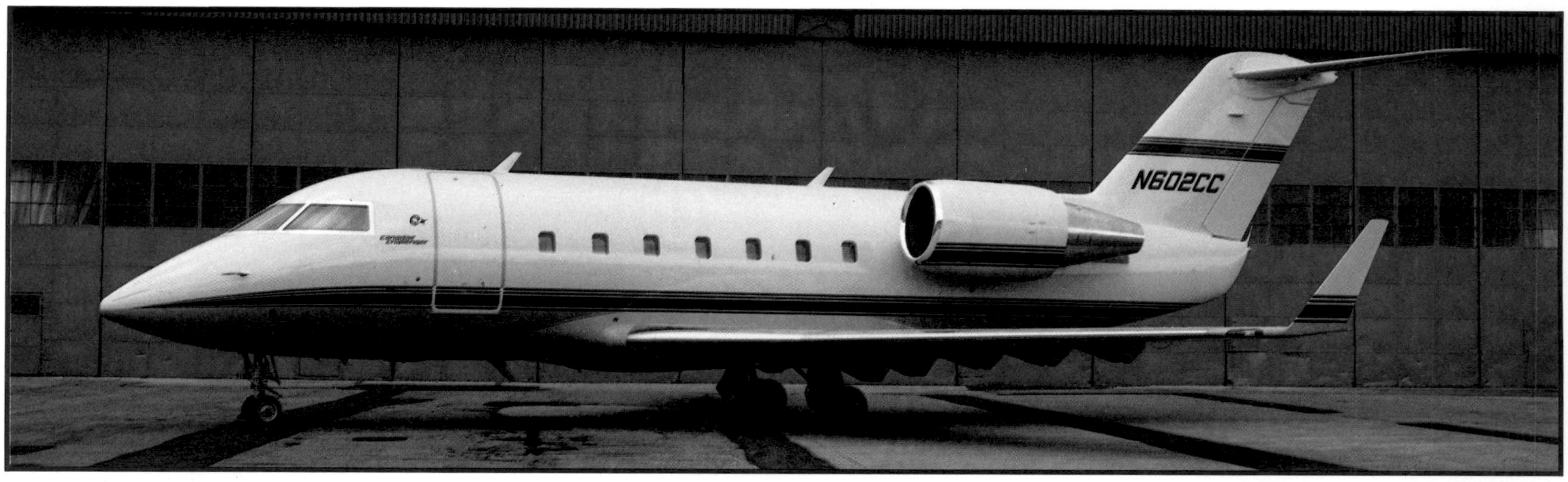

CANADAIR 601 CHALLENGER
(Canada)

This important 'bizjet' was designed by Bill Lear of Learjet fame, but was sold to Canadair in 1976. The type was first flown in November 1978, and important selling points have been the type's long range and comfort, the latter provided by a capacious cabin with good headroom. The CL-600 initial production version was powered by 3,402kg (7,500 lb) thrust Avco Lycoming ALF 502L turbofans, while the improved CL-601 has drag-reducing winglets and 3,924kg (8,650lb) thrust General Electric CF34-1A turbofans, soon to be replaced by 4,146kg (9,140 lb) thrust CF34-3As in the CL-601-3A version.

◄

CANADAIR 601 CHALLENGER

Role: Executive jet transport
Crew/Accommodation: Two, plus, typically, ten passengers
Power Plant: Two 3,924kgp (8,650 lb s.t.) General Electric CF34-1A turbofans
Dimensions: Span 19.61m (64.33 ft); length 20.85m (68 ft); wing area 450m² (41.81 sq ft)
Weights: Empty 14,062kg (31,000 lb); MTOW 19,550kg (43,100 lb)
Performance: Cruise speed 819km/h (442 knots) at 10,973m (36,000 ft); operational ceiling 12,497m (41,000 ft); range 4,725km (2,550 naut. miles)
Load: Up to 2,910kg (6,415 lb)

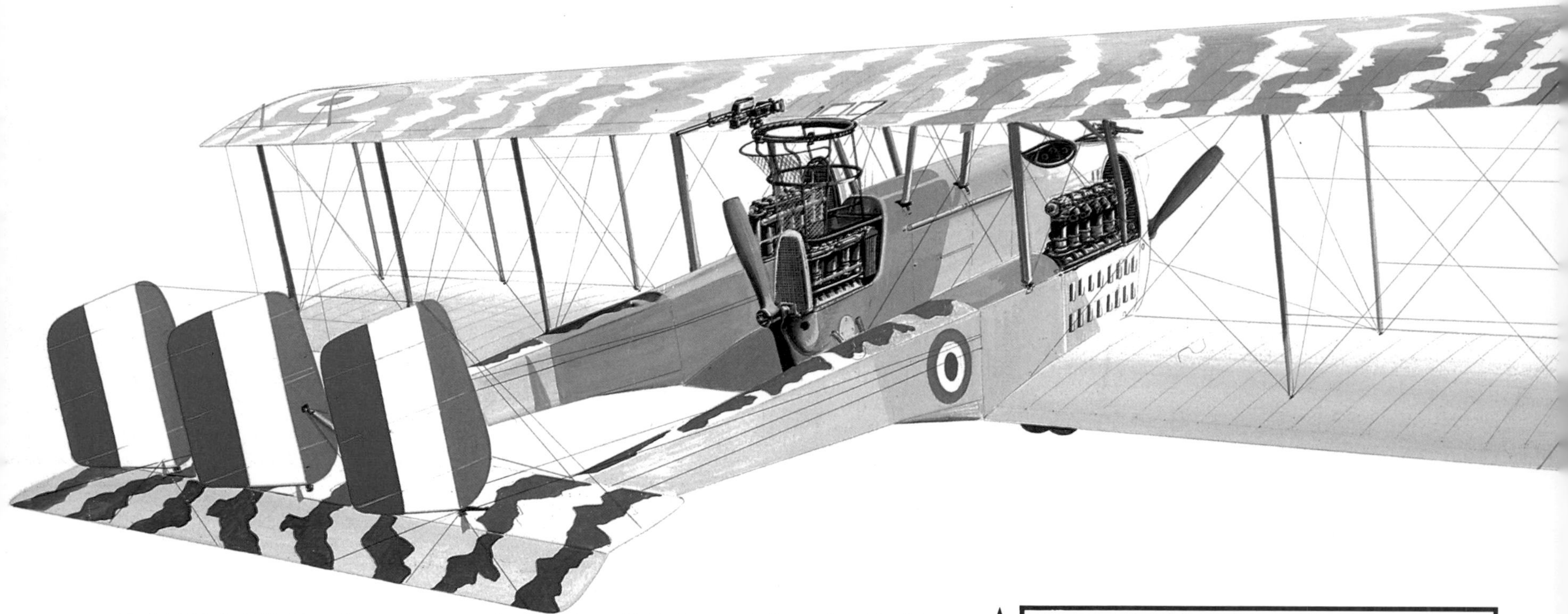

CAPRONI Ca 3
(Italy)

In October 1914 Caproni flew the Ca 30 prototype of a three-engined bomber with a central nacelle and twin booms supporting the tail unit. This was put in production as the Ca 1, and 162 were built with 74kW (100 hp) Fiat A.10 engines. Developments were the nine Ca 23s with the central engine replaced by a 112kW (150 hp) Isotta-Faschini V.4B engine, 299 Ca 3s (plus 83 French-built aircraft) with three V.4B engines, and 153 improved but structurally simplified Ca 3Ms. Ca 1s and Ca 3s converted as primitive airliners were designated Ca56 and Ca 56A respectively.

▲

CAPRONI Ca 33

Role: Heavy bomber
Crew/Accommodation: Four
Power Plant: Three 150 hp Isotta-Fraschini V-4B water-cooled inlines
Dimensions: Span 22.74m (74.61 ft); length 11.05m (36.25 ft); wing area 95.64m² (1,029.4 sq ft)
Weights: Empty 2,300kg (5,071 lb); MTOW 3,890kg (8,576 lb)
Performance: Maximum speed 140km/h (87 mph) at sea level; operational ceiling 4,100m (13,451 ft); range 450km (280 miles) with full warload
Load: Two to four 7.7mm machine guns, plus up to 450kg (992 lb) bombload

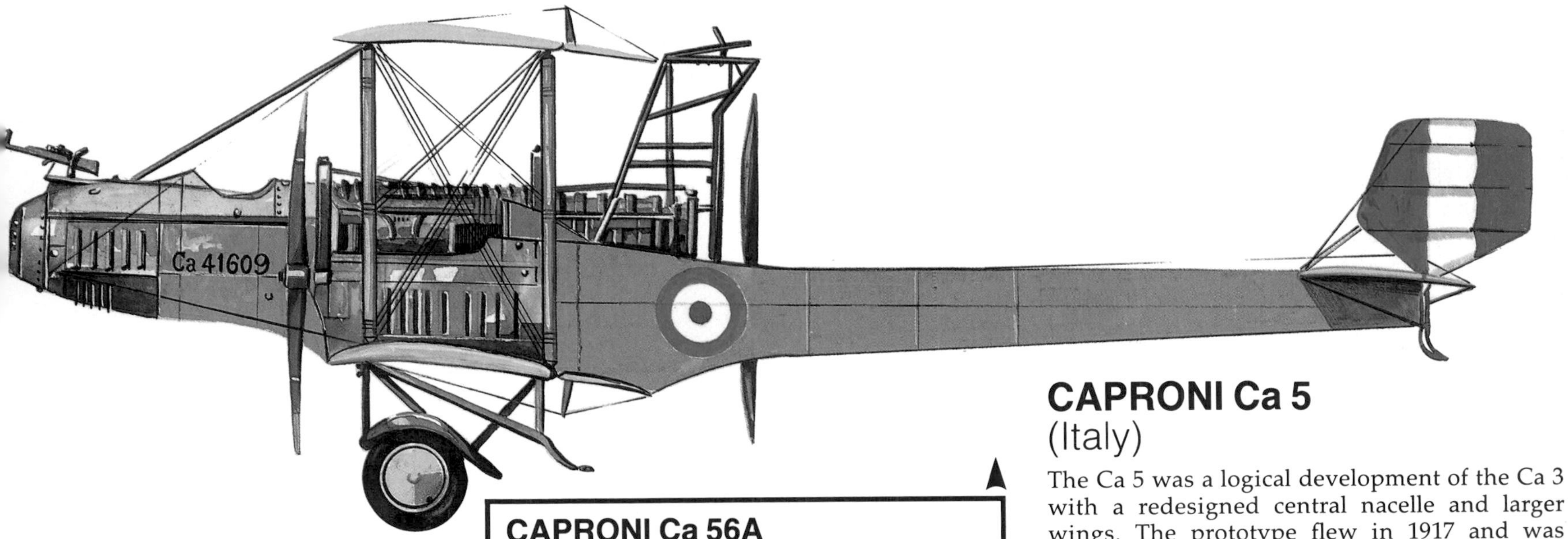

CAPRONI Ca 5
(Italy)

The Ca 5 was a logical development of the Ca 3 with a redesigned central nacelle and larger wings. The prototype flew in 1917 and was followed by 659 production bombers with 186/261kW (250/350 hp) inline engines. The Americans planned to produce 1,500 such bombers, but the programme was cancelled after just three had been produced. After World War I the type was redesignated Ca 44 with Fiat A.12bis engines, Ca 45 with Isotta-Fraschini V.6 engines and Ca 46 with Liberty engines. The designation Ca 57 was used for eight passenger conversions of Ca 45s and Ca 46s.

CAPRONI Ca 56A

Role: Passenger transport
Crew/Accommodation: Two, plus up to six passengers
Power Plant: Three 150 hp Isotta-Fraschini V-4B water-cooled inlines
Dimensions: Span 22.74m (74.61 ft); length 11.05m (36.25 ft); wing area 95.64m^2 (1,029.4 sq ft)
Weights: Empty 2,400kg (5,291 lb); MTOW 3,900kg (8,598 lb)
Performance: Maximum speed 130km/h (8- mph) at sea level
Load: Up to 450kg (993 lb)

CASA C-212 AVIOCAR
(Spain)

This simple yet effective aeroplane was developed to replace the Spanish Air Force's miscellany of obsolete transports, and first flew in March 1971. The type's ruggedness and low costs appealed to third-world civil operators as well as air forces, and the type has been widely built by Nurtanio in Indonesia as well as in Spain. The basic variants are the C-212A military transport, the C-212-5 (later C-212-100) civil type with 579kW (776 ehp) Garrett TPE5331-5-251 turboprops, the heavier C-212-10 (C-212-200) with TPE331-10-501Cs, and the still heavier C-212-300 with 671kW (900 shp) TPE331-10R-512Cs driving Dowty propellers.

CASA C-212 AVIOCAR SERIES 200

Role: Short, rough field-going utility transport
Crew/Accommodation: Two, plus up to 24 troops
Power Plant: Two 900 shp Garrett AiResearch TPE 331-10-501C turboprops
Dimensions: Span 19m (62.33 ft); length 15.2m (49.75 ft); wing area 40m^2 (430.6 sq ft)
Weights: Empty 6,550kg (14,440 lb); MTOW 7,300kg (16,096 lb)
Performance: Cruise speed 353km/h (190 knots) at 3,048m (10,000 ft); operational ceiling 8,534m (28,000 ft); range 760km (410 naut. miles) with maximum payload
Load: Up to 2,250kg (4,960 lb)

CESSNA MODELS 150 and 152
(U.S.A.)

In September 1957 Cessna flew the first Model 150 and thus re-entered the high-wing two-seater market in which its earlier competitors had been the highly successful Models 120 and 140. The Model 150 was similar to the 140 but had more power, optional dual controls and fixed tricycle rather than tailwheel landing gear. Production ended in 1977 after 23,836 aircraft had been built, including 1,754 French-built Reims F150 aircraft. The Model 150s' successor is the Model 152 with an 82kW (110 hp) Avco Lycoming 0-235-L2C engine in place of the earlier model's 75kW (100 hp) Continental 0-200. ➤

CESSNA 150K

Role: Light tourer/trainer
Crew/Accommodation: Two
Power Plant: One 100 hp Continental O-200A air-cooled flat-opposed
Dimensions: Span 9.97m (32.71 ft); length 7.24m (23.75 ft); wing area 14.57m^2 (156.86 sq ft)
Weights: Empty 456kg (1,005 lb); MTOW 726kg (1,600 lb)
Performance: Cruise speed 188km/h (117 mph) at 2,134m (7,000 ft); operational ceiling 3,856m (12,650 ft); range 909km (565 miles) with no reserves
Load: Up to 111kg (245 lb)

Cessna 172 ▲

CESSNA MODELS 170, 172, 175 and 182
(U.S.A.)

This series enjoys the distinction of being the most successful lightplane of all time. The Model 170 first flew in 1948 as a Model 120 re-engineered as a four-seater, and good initial sales were later boosted by the advent of the Model 170B with slotted flaps. The 1955 Model 172 was basically the Model 170B with fixed tricycle rather than tailwheel landing gear. In 1958 there appeared the Model 175 with a number of refinements and more power, and this short-lived variant also appeared in Model 175A and de luxe Skylark forms. The de luxe version of the Model 172 was the Skyhawk, which was later revised with a swept vertical tail. The Model 182 introduced more power, and was also produced in Skylane upgraded versions. Further development of the Models 172 and 182 has produced a host of versions with improved furnishing, better instrumentation, retractable landing gear and turbocharged engines. The Model 172 has also been produced in T-41 Mescalero trainer form.

CESSNA 172 SKYHAWK (T-41A)

Role: Light touring (and military basic trainer)
Crew/Accommodation: One, plus up to three passengers
Power Plant: One 160 hp Lycoming 0-320 air-cooled flat-opposed
Dimensions: Span 10.92m (35.83 ft); length 8.20m (26 ft); wing area 16.16m^2 (174 sq ft)
Weights: Empty 636kg (1,402 lb); MTOW 1,043kg (2,300 lb)
Performance: Cruise speed 226km/h (122 knots) at 2,438m (8,000 ft); operational ceiling 4,328m (14,200 ft); range 1,065km (575 naut. miles) with full payload
Load: Up to 299kg (660 lb)

CESSNA MODEL 208 CARAVAN I
(U.S.A.)

The Caravan can be considered as Cessna's replacement for the elderly Model 185 Skywagon with considerably greater capacity and performance, combined with more advanced features such as tricycle landing gear and a turboprop power plant. The first Model 208 flew in December 1982 with deliveries following in 1985. The type can operate on sturdy wheeled or float landing gear, and current versions are the Model 208 Caravan, the Model 208A Cargomaster freighter with an under-fuselage cargo pannier and no fuselage windows, and the Model 208B Super Cargomaster with a 1.22m (4ft) fuselage stretch for greater capacity. The type has proved highly successful in several roles, including the bush task. There is also a U-27 militarized variant.

CESSNA 208 CARAVAN I

Role: Commercial and military short field-capable utility transport
Crew/Accommodation: One, plus up to 14 passengers
Power Plant: One 600 shp Pratt & Whitney Canada PT6A-114 turboprop
Dimensions: Span 15.88m (52.08 ft); length 11.46m (37.58 ft); wing area 25.96m^2 (279.4 sq ft)
Weights: Empty 1,724kg (3,800 lb); MTOW 3,311kg (7,300 lb)
Performance: Cruise speed 341km/h (184 knots) at 3,048m (10,000 ft); operational ceiling 8,410m (27,600 ft); range 2,362km (1,275 naut. miles) with full payload
Load: Up to 1,360kg (3,000 lb)

CESSNA MODEL 305 (L-19/ O-1 BIRD DOG)
(U.S.A.)

This aeroplane, designed to meet a U.S. Army requirement for an observation type, was based on the Model 170 with a new power plant, cut-down rear fuselage to offer better fields of vision (enhanced by transparencies in the wing centre section), and a flapped wing to provide better field performance. Some 3,431 Bird Dogs were produced between 1950 and later in the decade in variants up to L-19E. In 1962 the series was redesignated as the O-1.

CESSNA L-19E/O-1E BIRD DOG

Role: Battlefield observation/artillery direction
Crew/Accommodation: Two
Power Plant: One 213 hp Continental 0-470-11 air-cooled flat-opposed
Dimensions: Span 10.97m (36 ft); length 7.87m (25.83 ft); wing area 16.17m^2 (174 sq ft)
Weights: Empty 732kg (1,614 lb); MTOW 1,089kg (2,400 lb)
Performance: Cruise speed 167km/h (104 mph) at 1,524m (5,000 ft); operational ceiling 5,639m (18,500 ft); range 853km (530 miles)
Load: None, other than special-to-task communications equipment

CESSNA MODELS 310, 320, 335 and 340
(U.S.A.)

The Model 310 was designed in 1952 and the prototype flew in January 1953. In 1966 the company introduced an improved version with turbocharged engines as the Model 320 and then Turbo T310. In 1971 there appeared the pressurized Model 340 with the wing and landing gear of the Model 414. An unpressurized, and therefore lighter, variant is the Model 335. Military aircraft were originally designated L-27, but were redesignated in the U-3 series from 1962.

CESSNA 310L (USAF U-3)

Role: Light twin passenger/utility transport
Crew/Accommodation: Two 260 hp Continental 10-47 0-V0 air-cooled flat-opposed
Dimensions: Span 11.25m (36.92 ft); length 8.99m (29.5 ft); wing area 16.63m^2 (179 sq ft)
Weights: Empty 1,418kg (3,125 lb); MTOW 2,360kg (5,200 lb)
Performance: Maximum speed 357km/h (222 mph) at 1,981m (6,500 ft); operational ceiling 6,065m (19,900 ft); range 1,554km (966 miles) with full payload
Load: Up to 420kg (925 lb)

CESSNA MODEL 318 (T-37) and MODEL 318E (A-37 DRAGONFLY)
(U.S.A.)

The Model 318 was designed to a U.S. Air Force specification of the early 1950s for a jet-powered trainer, and first flew in October 1954 with U.S. licence-built Turbomeca Marbore turbojets. The type was ordered into production as the T-37A, and these 534 aircraft were followed by T-37Bs with more powerful engines and by T-37Cs with light armament and wingtip fuel tanks. Production of the T-37 for U.S. and export services was 1,268. A special counter-insurgency and light attack version was developed as the YAT-37D with more powerful General Electric J85-GE-5 turbojets for the carriage of a substantial warload. Some 25 of the type were evaluated from 1967 as A-37As, and their success led to development of the beefed-up model 318E, which was ordered into production as the A-37B.

CESSNA A-37B DRAGONFLY

Role: Light strike
Crew/Accommodation: Two
Power Plant: Two, 1,293kgp (2,850 lb s.t.) General Electric J85-GE-17A turbojets
Dimensions: Span 11.71m (38.42 ft); length 9.69m (31.83 ft); wing area 17.09m^2 (183.9 sq ft)
Weights: Empty 1,845kg (4,067 lb); MTOW 6,350kg (14,000 lb)
Performance: Maximum speed 771km/h (479 mph) at 4,724m (15,500 ft); operational ceiling 7,620m (25,000 ft); radius 380km (236 miles) with 843kg (1,858 lb) bombload
Load: One 7.62mm multi-barrel machine gun, plus up to 2,576kg (5,680 lb) of bombs or air-to-ground rockets carried on underwing pylons

CESSNA MODELS 336 SKYMASTER and 337 SUPER SKYMASTER
(U.S.A.)

Cessna developed the Model 336 with its unusual push/pull engine arrangement on the centreline to avoid asymmetric thrust problems in the event of an engine failure. The prototype flew in February 1961, and entered production with seating for six on the power of two 157kW (210 hp) Continental IO-360-A engines. The company soon realized, however, that there was considerable customer resistance to the fixed tricycle landing gear and the design was amended to produce the Model 337 with retractable gear and provision for an underfuselage baggage pannier. Additional construction was undertaken in France of a version designated Reims F337, which had a Milirole utility military version. In 1970 Cessna introduced the Model 337 Turbo-System Super Skymaster with turbocharged engines, and in the following year the pressurized T337 Skymaster. The Model 337 was also developed by Cessna for the military as the O-2.

CESSNA O-2A

Role: Battlefield observation/artillery direction
Crew/Accommodation: Two, plus up to two observers
Power Plant: Two 210 hp Continental IO-360 C/D air-cooled flat-opposed
Dimensions: Span 11.63m (38.16 ft); length 9.07m (29.75 ft); wing area 18.81m^2 (202.5 sq ft)
Weights: Empty 1,520kg (3,350 lb); MTOW 2,449kg (5,400 lb)
Performance: Maximum speed 286km/h (178 mph) at sea level; operational ceiling 4,602m (15,100 ft); endurance 5.5 hours
Load: Four underwing pylons to carry 7.62mm minigun pods, flare packs or other special-to-mission equipment

CESSNA MODELS 401, 402 and 411
(U.S.A.)

First flown in July 1962, the Model 411 was at the time the company's largest business aeroplane and, by comparison with the Model 310, had more power, slightly greater span and a longer fuselage. In August 1965 the company flew the first example of a generally similar, but lower-powered and therefore cheaper, type that was built in two variants as the Model 401 for six passengers and the Model 402 with a quick-change interior for nine commuter passengers or freight. The Model 401 was phased out of production in 1972 to allow concentration on the Model 402 which then became the original Utililiner, the six/eight-seat Businessliner, and upgraded Utililiner II and Businessliner II and III variants.

CESSNA 401

Role: Executive transport
Crew/Accommodation: Two, plus up to six pasengers
Power Plant: Two 300 hp Continental TS10-520-E air-cooled flat-opposed
Dimensions: Span 12.15m (39.86 ft); length 10.27m (33.75 ft); wing area 18.18m^2 (195.7 sq ft)
Weights: Empty 1,651kg (3,640 lb); MTOW 2,858kg (6,300 lb)
Performance: Cruise speed 386km/h (240 mph) at 6,096m (20,000 ft); operational ceiling 7,980m (26,180 ft); range 1,300km (808 miles) with full payload
Load: Up to 925kg (2,040 lb)

CESSNA MODEL 500 CITATION
(U.S.A.)

With the Citation Cessna moved into the market for high-performance 'bizjets', and the Fanjet 500 prototype first flew in September 1969. The type was then renamed Citation, and this straight-winged type entered service in 1971 with Pratt & Whitney Canada JT15D-1 turbofans pod-mounted on the rear fuselage. Later developments have been the Citation I of 1976 with greater span, the Citation II of 1978 with greater span, a lengthened fuselage and 1,134kg (2,500 lb) thrust JT15D-4 engines, the completely revised Citation III of 1982 with swept wings of supercritical section, a lengthened fuselage for two crew and 13 passengers, a T-tail and 1,656kg (3,650 lb) thrust Garrett TFE731-3B-100S turbofans, and the most recent Citation V. This combines the short-field performance of the Citation II with a larger cabin and the speed and cruising altitude of the Citation III.

CESSNA CITATION S/II (T-47A)

Role: Executive transport and military trainer
Crew/Accommodation: Two, plus up to eight passengers
Power Plant: Two, 1,134kgp (2,500 lb s.t.) Pratt & Whitney JT15D-4B turbofans
Dimensions: Span 15.90m (52.21 ft); length 14.39m (47.21 ft); wing area 31.83m^2 (342.6 sq ft)
Weights: Empty 3,655kg (8,059 lb); MTOW 6,849kg (15,100 lb)
Performance: Cruise speed 746km/h (403 knots) Mach 0.70 at 10,670m (35,000 ft); operational ceiling 13,105m (43,000 ft); range 3,223km (1,739 naut. miles) with four passengers
Load: Up to 871kg (1,920 lb)

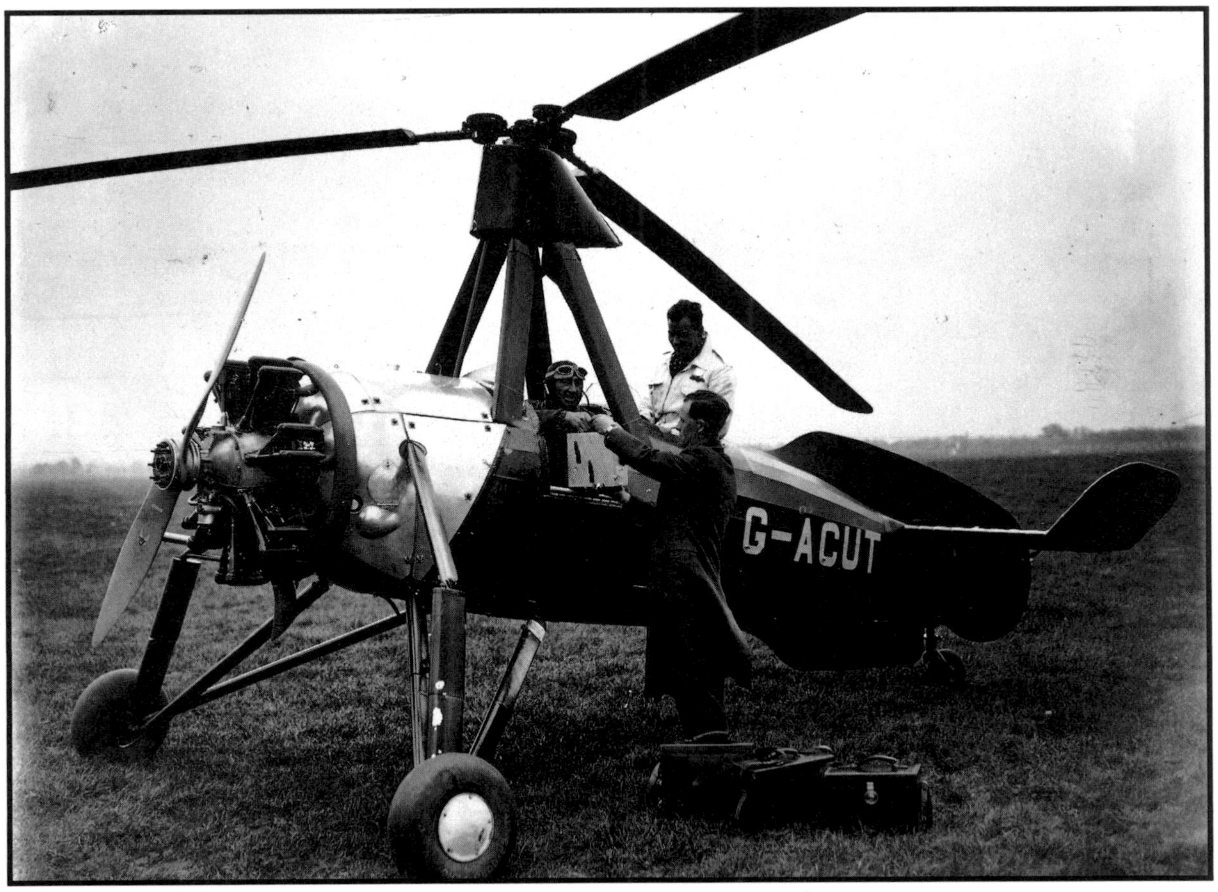

CIERVA AUTOGIROS
(Spain)

Appalled by the loss of life when one of his early fixed-wing aircraft stalled and crashed, Juan de la Cierva turned his attention to unstallable rotary-wing aircraft of the Autogiro type with an unpowered rotor. The first successful model was the C.4 that flew in January 1923, and there followed a stream of successively improved developments, many of them based on the fuselage/engine combination of established fixed-wing aircraft. Models went up to the C.40, and production was undertaken mainly by Avro in the U.K., Liore-et-Olivier in France and Pitcairn in the U.S.A. Possibly the most successful models were the C.19 and C.30, of which the latter was used by the RAF as the Rota.

CIERVA C-30A (as built by AVRO)

Role: Light short take-off and landing utility/communications autogiro
Crew/Accommodation: One/two
Power Plant: One 140 hp Armstrong Siddeley Genet Major Ia air-cooled radial
Dimensions: Overall length rotors turning 11.28m (37 ft); rotor diameter 11.28m (37 ft)
Weights: Empty 553kg (1,220 lb); MTOW 816kg (1,800 lb)
Performance: Maximum speed 177km/h (110 mph) at sea level; operational ceiling 2,438m (8,000 ft); range 465km (285 miles)
Load: Up to 169kg (372 lb) including second person

CONSOLIDATED MODEL 28 (PBY and CATALINA)
(U.S.A.)

Designed as a patrol flying boat with greater range and payload than the Consolidated P2Y and Martin P3M, the XP3Y-1 prototype first flew in March 1935. It was a large machine with a strut-braced parasol wing mounted on top of a massive pylon that accommodated the flight engineer, and drag was reduced by the use of stabilizing floats that retracted in flight to become the wingtips. The type clearly possessed considerable potential, and after being reworked as a patrol bomber was ordered into production as the PBY-1, the following PBY-2 having equipment improvements, and the PBY-3 and PBY-4 offering more powerful engines. The PBY-5 was generally improved but built in fairly small numbers as it was more than complemented by the PBY-5A amphibian. The Naval Aircraft Factory produced a PBN-1 Nomad version with aerodynamic and hydrodynamic improvements, and a comparable model was also produced by Consolidated as the PBY-6A. The type was also produced in Canada and the U.S.S.R. as the Boeing PB2B and GST respectively, while the U.K. adopted the aeroplane as several variants named Catalina.

CONSOLIDATED PBY-5 (RAF CATALINA Mk IV)

Role: Long-range maritime patrol bomber flying boat
Crew/Accommodation: Nine
Power Plant: Two, 1,200 hp Pratt & Whitney R-1830-92 Twin Wasp air-cooled radials
Dimensions: Span 37.10m (104 ft); length 19.47m (63.88 ft); wing area 130.1m^2 (1,400 sq ft)
Weights: Empty 7,809kg (17,200 lb); MTOW 15,436kg (34,000 lb)
Performance: Cruise speed 182km/h (113 mph) at sea level; operational ceiling 5,517m (18,100 ft); range 4,812km (2,990 miles) with full warload
Load: Two .5 inch and two .303 inch machine guns, plus up to 1,816kg (4,000 lb) of torpedoes, depth charges or bombs carried externally

Consolidated B-24H ▲

CONSOLIDATED MODEL 32 (B-24 LIBERATOR)
(U.S.A.)

The Liberator was a remarkably versatile aeroplane, and was built in greater numbers than any other U.S. warplane of the World War II period. The type was designed as successor to machines such as the Boeing XB-15 and Douglas XB-19, neither of which entered production, and was based on the exceptional wing of the Model 31 flying boat, whose high aspect ratio offered low drag for high speed and great range. The XB-24 prototype flew in December 1939, and YB-24 pre-production machines were followed by small numbers of B-24A, turbocharged B-24B and upgunned B-24C production aircraft before the advent of the first major production models, the B-24D and generally similar B-24E, the B-24G and B-24H with a longer nose terminating in a turret, the slightly modified B-24J which was the major variant (6,678 built), the B-24L based on the B-24D but with modified tail armament, and the B-24M based on the B-24J, but with a different tail turret. There were also a number of experimental bomber variants, while other roles included transport (LB-30, Air Force C-87 and Navy RY variants), fuel tanking (C109), photographic reconnaissance (F-7), patrol bombing (PB4Y-1 and specially developed PB4Y-2 with a single vertical tail surface instead of the standard machine's twin endplate surfaces) and maritime reconnaissance (British Liberator GR Models).

CONSOLIDATED B-24J LIBERATOR

Role: Long-range day bomber
Crew/Accommodation: Ten
Power Plant: Four 1,200 hp Pratt & Whitney R.1830-65 Twin Wasp air-cooled radials
Dimensions: Span 33.53m (110 ft); length 20.47m (67.16 ft); wing area 97.36m^2 (1,048 sq ft)
Weights: Empty 16,556kg (36,500 lb); MTOW 29,484kg (65,000 lb)
Performance: Maximum speed 467km/h (290 mph) at 7,620m (25,000 ft); operational ceiling 8,534m (28,000 ft); range 3,379km (2,100 miles) with full bombload
Load: Ten .5 inch machine guns, plus up to 3,992kg (8,800 lb) of internally carried bombs

CONVAIR MODEL 8 (F-102 DELTA DAGGER and F-106 DELTA DART)
(U.S.A.)

These two delta-winged fighters were designed specifically for air-defence of the continental U.S.A., and were among the first aircraft in the world designed as part of a complete weapon system integrating airframe, sensors and weapons. The YF-102 first flew in October 1952 and proved to have disappointing performance. The airframe was redesigned with Whitcomb area ruling to reduce drag, and this improved performance to the degree that made feasible the introduction of F-102A single-seat fighter and TF-102A two-seat trainer variants. Greater effort went into the development of the true Mach 2 version, which was developed as the F-102B but then ordered as the F-106 before the prototype flew in December 1956. The F-106A single-seat fighter and F-102B two-seat trainer versions served into the late 1980s.

CONVAIR F-106A DELTA DART

Role: All-weather interceptor
Crew/Accommodation: One
Power Plant: One 11,115kgp (24,500 lb s.t.) Pratt & Whitney J75-P-17 turbojet with reheat
Dimensions: Span 11.67m (38.29 ft); length 21.56m (70.73 ft); wing area 64.83m^2 (697.8 sq ft)
Weights: Empty 10,904kg (24,038 lb); MTOW 17,779kg (39,195 lb)
Performance: Maximum speed 2,135km/h (1,152 knots) at 10,668m (35,000 ft); operational ceiling 16,063m (52,700 ft); radius 789km (490 miles) on internal fuel only
Load: One 20mm multi-barrel cannon, plus one long-range and four medium-range air-to-air missiles

CONVAIR CV-240, 340 and 440 CONVAIRLINER
(U.S.A.)

The CV-240 series was developed in the hope of producing a successor to the legendary Douglas DC-3, and though the type was in every aspect good it failed to make a decisive impression on the vast numbers of C-47s released on to the civil market when they became surplus to military requirements. The CV-110 prototype first flew in July 1946 with pressurized accommodation for 30 passengers, but the type was then revised for 40 passengers as the CV-240, of which 179 were built as airliners. There followed the 44-passenger CV-340 with more power, and finally the similar CV-440 with high-density seating for 52 passengers. Variants of the series for the military were T-29 U.S. Air Force crew trainer, the C-131 air ambulance and transport for the U.S.A.F. and R4Y transport for the U.S. Navy.

CONVAIR 440 CONVAIRLINER

Role: Short-range passenger transport
Crew/Accommodation: Two, plus up to 52 passengers
Power Plant: Two 2,500 hp Pratt & Whitney R-2800-CB16/17 Double Wasp air-cooled radials
Dimensions: Span 31.10m (105.33 ft); length 24.84m (81.5 ft); wing area 85.47m^2 (920 sq ft)
Weights: Empty 15,111kg (33,314 lb); MTOW 22,544kg (49,700 lb)
Performance: Cruise speed 465km/h (289 mph) at 6,096m (20,000 ft); operational ceiling 7,590m (24,900 ft); range 459km (285 miles) with maximum payload
Load: Up to 5,820kg (12,836 lb)

CONVAIR CV-540, 580, 600 and 640
(U.S.A.)

This was the Convairliner series converted for turboprop power, and began with the CV-540, which was a version of the CV-340 modified by the British engine manufacturer Napier with two of its 2,282kW (3,060 ehp) NEL.1 Eland engines. A comparable but larger programme was undertaken in the U.S.A. by PacAero Engineering, which evolved the CV-580 from CV-340 and 440 aircraft with 2,796kW (3,750 shp) Allison 501-D13 engines. Convair also got into the act with a modification based on 2,256kW (3,025 ehp) Rolls-Royce Dart RDa.10/1 Dart Mk 542 engines: the aircraft were originally designated CV-240D, 340D and 440D, the first later becoming the CV-600 and the last two the CV-640.

CONVAIR CV-580

Role: Short-range passenger transport
Crew/Accommodation: Two and two cabin crew, plus up to 52 passengers
Power Plant: Two 3,750 shp Allison 501-D13H turboshafts
Dimensions: Span 32.1m (105.33 ft); length 24.84m (81.50 ft); wing area 85.47m² (920 sq ft)
Weights: Empty 13,732kg (30,275 lb); MTOW 26,371kg (58,140 lb)
Performance: Cruise speed 550km/h (342 mph) at 6,100m (20,000 ft); operational ceiling 7,590m (24,900 ft); range 2,980km (1,605 miles) with 2,268kg (5,000 lb) payload
Load: Up to 4,717kg (10,400 lb)

CONVAIR MODEL 4 (B-58 HUSTLER)
(U.S.A.)

This supersonic medium strategic bomber resulted from a 1949 requirement, and was for its time a stupendous technical achievement. The airframe was designed round the smallest possible delta-winged airframe and four podded turbojets, the outward-leg fuel load and all of the offensive weaponry being accommodated in a massive pod carried under the fuselage for release over the target. The prototype flew in November 1956 and proved tricky to fly. Production amounted to only 86 B-58As, training being carried out in eight TB-58As converted from YB-58A pre-production aircraft.

▲

CONVAIR B-58A HUSTLER

Role: Supersonic bomber
Crew/Accommodation: Three
Power Plant: Four 7,076kgp (15,600 lb s.t.) General Electric J79-GE-3B turbojets with reheat
Dimensions: Span 17.32m (56.83 ft); length 29.49m (96.75 ft); wing area 143.26m² (1,542 sq ft)
Weights: Empty 25,202kg (55,560 lb); MTOW 73,936kg (163,000 lb)
Performance: Maximum speed 2,126km/h (1,147 knots) Mach 2.1 at 12,192m (40,000 ft); operational ceiling 19,202m (63,000 ft); range 8,247km (4,450 naut. miles) unrefuelled
Load: One 20mm multi-barrel cannon, plus detachable mission pod containing nuclear weapon, plus fuel

CONSOLIDATED / CONVAIR MODEL 36 (B-36)
(U.S.A.)

This extraordinary machine was the world's first genuine intercontinental strategic bomber, and resulted from a 1941 requirement for a machine able to carry a maximum bombload of 32,659kg (72,000 lb) and able to deliver 4,536kg (10,000 lb) of bombs on European targets from bases in the U.S.A. The XB-36 first flew in August 1946 and featured a pressurized fuselage and six pusher engines buried in the trailing edges of wings sufficiently deep to afford inflight access to the engines. Subsequently multi-wheel main landing gear units and a raised cockpit were introduced, and these were features of the first production model, the B-36A unarmed crew trainer. The B-36B introduced armament, and most were later revised as B-36Ds with greater weights and performance through the addition of four General Electric J47-GE-19 turbojets in podded pairs under the outer wing. Later bombers with greater power and improved electronics were the B-36F, H and J. There were also RB-36D, E, F and H reconnaissance versions, and even the GRB-36F with an embarked fighter for protection over the target area.

CONSOLIDATED/CONVAIR B-36D

Role: Long-range heavy bomber
Crew/Accommodation: Fifteen, including four relief crew members
Power Plant: Six 3,500 hp Pratt & Whitney R-4360-41 air-cooled radials, plus four 2,359kgp (5,200 lb s.t.) General Electric J47-GE-19 turbojets
Dimensions: Span 70.1m (230 ft); length 49.4m (162.08 ft); wing area 443m² (4,772 sq ft)
Weights: Empty 72,051kg (158,843 lb); MTOW 162,161kg (357,500 lb)
Performance: Maximum speed 706km/h (439 mph) at 9,790m (32,120 ft); operational ceiling 13,777m (45,200 ft); range 12,070km (7,500 miles) with 4,535kg (10,000 lb) bombload
Load: Twelve 20mm cannon, plus up to 39,009kg (86,000 lb) of bombs

CRDA CANT Z.1007 ALCIONE
(Italy)

The Z.1007 was planned from 1935 as a high-performance medium bomber of the typically Italian three-engined configuration, and first flew in March 1937 with 615kW (825 hp) Isotta-Fraschini Asso XI engines. Only a few were built before the development of the main variant, the Z.1007bis with 746kW (1,000 hp) velopment was the Z.1007ter with 858kW (1,150 single vertical tail surface, but later machines had twin endplate surfaces. The ultimate development was the Z.1007ter with 858kW (1,150 hp) P.XIX radial engines and reduced bombload for high performance.

CANT Z.1007bis ALCIONE

Role: Medium bomber
Crew/Accommodation: Five
Power Plant: Three 1,000 hp Piaggio P.XIbis RC 40 air-cooled radials
Dimensions: Span 24.8m (81.36 ft); length 18.35m (60.20 ft); wing area 70m^2 (753.4 sq ft)
Weights: Empty 9,396kg (20,715 lb); MTOW 13,620kg (30,027 lb)
Performance: Maximum speed 455km/h (283 mph) at 6,000m (19,685 ft); operational ceiling 7,500m (24,600 ft); range 1,795km (1,115 miles) with full warload
Load: Two 12.7mm and two 7.7mm machine guns, plus up to 1,200kg (2,645 lb) of bombs, or two torpedoes

CURTISS TRIAD and A-SERIES
(U.S.A.)

Glenn Curtiss was one of the great enthusiasts of the waterplane concept, and though his early efforts to develop a 'hydroaeroplane' were only marginally successful he persevered and finally achieved the world's first effective amphibian with the Triad. This flew in February 1911, and Curtiss then moved forward to a series of pusher flying boats for the U.S. Navy. Eventually the service took 14 such aircraft based on the Model E, and these were of several types flown in various landplane, floatplane and amphibian configurations.

CURTISS TRIAD A-1

Role: General-purpose
Crew/Accommodation: Two
Power Plant: One 50 hp/later 75 hp Curtiss V-8 water-cooled inline
Dimensions: Span 11.28m (37 ft); length 8.72m (28.59 ft); wing area 26.57m^2 (286 sq ft)
Weights: Empty 419.6kg (926 lb); MTOW 714.4kg (1,575 lb)
Performance: Maximum speed 96.6km/h (60 mph) at sea level; operational ceiling 274.3m (900 ft); range 180km (112 miles) with full payload
Load: 81.6kg (180 lb)

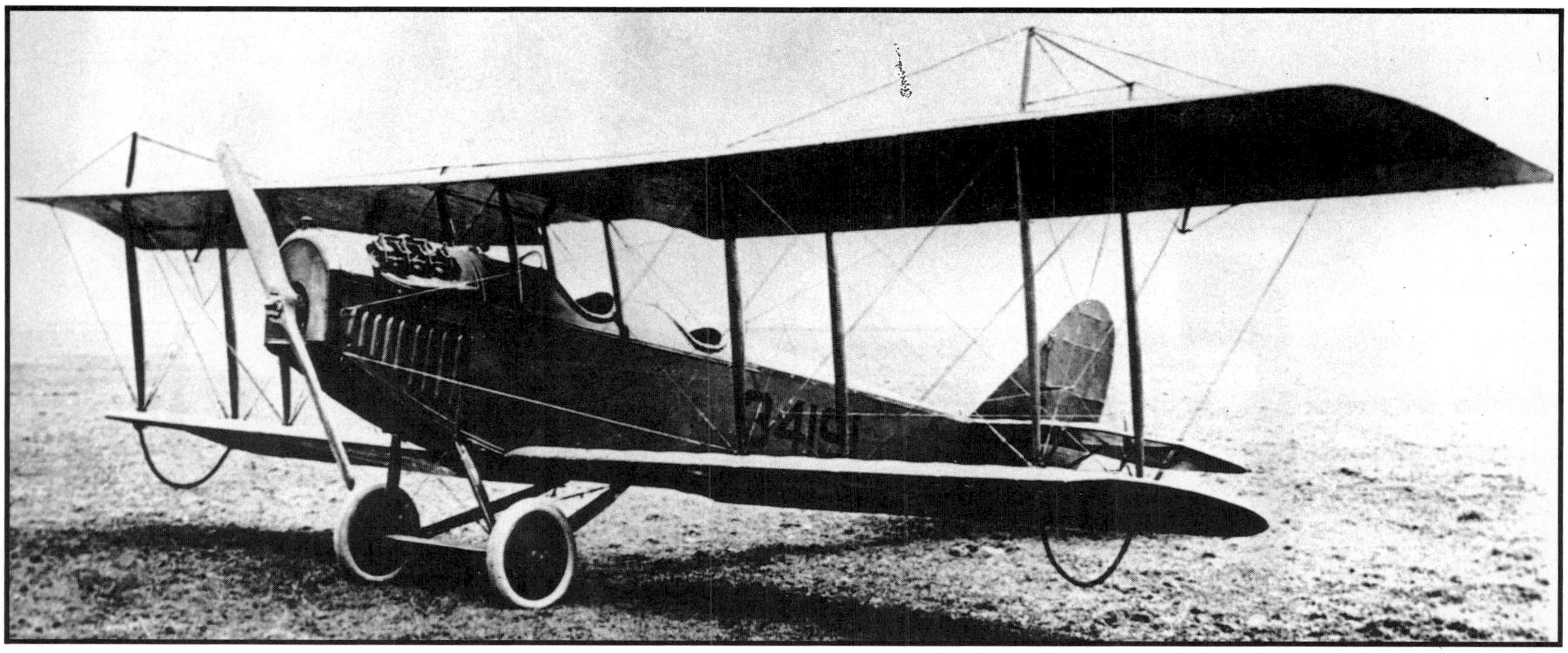

CURTISS MODEL J (JN 'JENNY') (U.S.A.)

▲ Generally known as the 'Jenny', this celebrated trainer resulted from the Model J. This had a 67kW (90 hp) Curtiss O inline engine and unequal-span biplane wings with upper-wing ailerons operated by the obsolete Curtiss shoulder-yoke system. The contemporary Model N had the 75kW (100 hp) Curtiss OXX inline and interplane ailerons. Features of both types were included in the JN-2 that appeared in 1915 with equal-span biplane wings each carrying an aileron, and powered by a 67kW (90 hp) Curtiss OX engine. The succeeding JN-3 had unequal-span wings with ailerons only on the upper surfaces, but these control surfaces were operated by a conventional joystick. In July 1916 there appeared the definitive JN-4, and this was built in large numbers for useful service well into the 1920s. Variants produced in smaller numbers were the JN-5 and JN-6 trainers.

CURTISS JN-P4D

Role: Primary trainer
Crew/Accommodation: Two
Power Plant: One 90 hp Curtiss OX-5 water-cooled inline
Dimensions: Span 13.29m (43.6 ft); length 8.33m (27.33 ft); wing area 32.7m^2 (352 sq ft)
Weights: Empty 630.5kg (1,390 lb); MTOW 871kg (1,920 lb)
Performance: Cruise speed 97km/h (60 mph) at sea level; operational ceiling 1,981m (6,500 ft); range 402km (250 miles)
Load: None

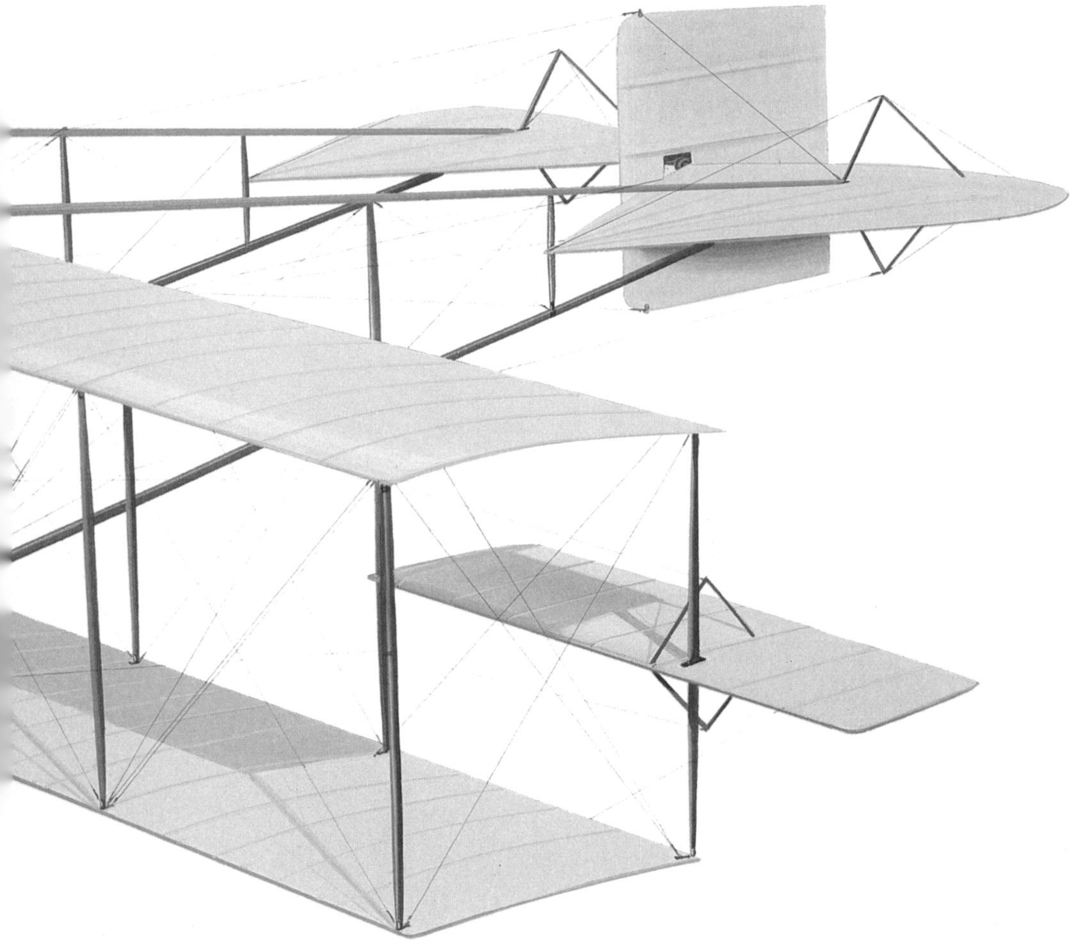

CURTISS MODEL 6 (H-12 SMALL AMERICA and H-16 LARGE AMERICA) (U.S.A.)

The H-12 when it appeared in autumn 1916 was essentially a larger version of the H-4 with two 119kW (160 hp) Curtiss V-X-X engines. 84 such reconnaissance bombers were delivered to the U.K., and the U.S. Navy then took 20 similar aircraft with 149kW (200 hp) Curtiss V-2-3 engines. Both variants were underpowered, the British turning their 'boats into H-12As with 205kW (275 hp) Rolls-Royce Eagle Is or H-12Bs with 280kW (375 hp) Eagle VIIIs, and the Americans the H-12L with 268kW (360 hp) Liberty engines. Further development led to the H-16 with a refined hull, larger-span wings of reduced area, and greater engine power: U.S. 'boats had Liberty engines, and British 'boats 257kW (345 hp) Eagle IV engines.

CURTISS H-16

Role: Maritime patrol bomber flying boat
Crew/Accommodation: Four
Power Plant: Two 400 hp Liberty 12A water-cooled inlines
Dimensions: Span 28.97m (95.05 ft); length 14.05m (46.1 ft); wing area 108.13m^2 (1,164 sq ft)
Weights: Empty 3,357kg (7,400 lb); MTOW 4,944kg (10,900 lb)
Performance: Maximum speed 153km/h (95 mph) at sea level; operational ceiling 3,033m (9,950 ft); range 608km (378 miles)
Load: Five or six .303 inch machine guns, plus up to 417kg (920 lb) of bombs

CURTISS MODEL 34 (P-1 and F6C HAWK SERIES)
(U.S.A.)

With the Model L-18-1 Curtiss began the private-venture development of an advanced fighter that was to prove one of the decisive designs of the 1920s. The type first flew late in 1922, but was followed by only 25 PW-8 production fighters for the U.S. Army. The XPW-8B experimental variant with the 328kW (440 hp) Curtiss D-12 engine introduced tapered wings and other alterations, resulting in an order for the P-1 production variant. This was then produced in a bewildering number of variants, of which the most significant were the P-1A with detail improvements, the P-1B with the 324kW (435 hp) Curtiss V-1150-3 and larger-diameter wheels, and the P-1C with wheel brakes. The army's P-1 series was also attractive to the U.S. Navy, which ordered the type with the designation F6C. The F6C-1 was intended for land-based use by the U.S. Marine Corps and was all but identical with the P-1, and the F6C-2 added carrier landing equipment including an arrester hook. The F6C-3 was a modified F6C-2, and the F6C-4 introduced the 313kW (420 hp) Pratt & Whitney R-1340 Wasp radial in place of the original D-12 inline.

CURTISS F6C-3

Role: Naval carrierborne fighter
Crew/Accommodation: One
Power Plant: One 400 hp Curtiss D.12 Conqueror water-cooled inline.
Dimensions: Span 9.63m (31.6 ft); length 6.96m (22.83 ft); wing area 23.41m^2 (252 sq ft)
Weights: Empty 980kg (2,161 lb); MTOW 1,519kg (3,349 lb)
Performance: Maximum speed 248km/h (154 mph) at sea level; operational ceiling 6,187m (20,300 ft); range 565km (351 miles)
Load: Two .303 inch machine guns

CURTISS MODEL 35 (P-6 HAWK)
(U.S.A.)

Further development of the P-1 led to the P-6 series with the Curtiss V-1570 Conqueror engine. The main variants were the original P-6, the P-6A with Prestone-cooled engine, and the P-6E with the 522kW (700 hp) V-1570C Conqueror. There were many experimental variants including the radial-engined P-3 and P-21, and the turbocharged P-5 and P-23. The type also secured comparatively large export orders under the generic designation Hawk I; the same basic type with a Wright Cyclone radial was sold with the name Hawk II. The U.S. Navy also ordered the latter variant with the designations F11C-2 and, with manually operated landing gear that retracted into a bulged lower fuselage, BF2C-1.

CURTISS P-6E

Role: Fighter
Crew/Accommodation: One
Power Plant: One 700 hp Curtiss V-1570C Conqueror water-cooled inline
Dimensions: Span 9.6m (31.5 ft); length 6.88m (22.58 ft); wing area 23.4m^2 (252 sq ft)
Weights: Empty 1,231kg (2,715 lb); MTOW 1,558kg (3,436 lb)
Performance: Maximum speed 311km/h (193 mph) at sea level; operational ceiling 7,285m (23,900 ft); range 393km (244 miles)
Load: Two .303 inch machine guns

Curtiss XF11C-3

CURTISS MODEL 84 (SB2C HELLDIVER)
(U.S.A.)

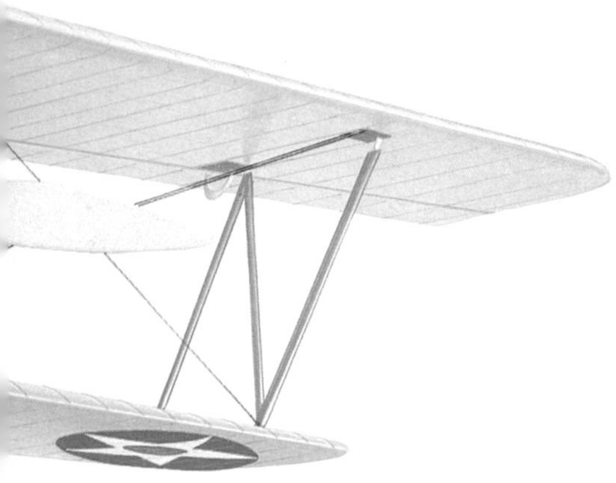

The SB2C was the third Curtiss design to bear the name Helldiver, the first two having been the F8C/O2C biplanes of the early 1930s and SBC biplane of the late 1930s. The SB2C monoplane was designated as successor to the SBC in scout bomber/dive-bomber role. The prototype flew in December 1940 but, because of the need to co-develop an A-25A Shrike version for the U.S. Army, the first SB2C-1 production aeroplane with the 1,268kW (1,700 hp) Wright R-2600-8 Cyclone 14 radial engine did not emerge until June 1942. The SB2C-1A was the A-25A reassigned to the U.S. Marine Corps, and other variants were the SB2C-1C with the four wing-mounted machine guns replaced by two 20mm cannon, the SB2C-3 with the 1,417kW (1,900 hp) R-2600-20, the SB2C-4 with underwing bomb/rocket racks, the radar-fitted SB2C-4E, and the SB2C-5 with greater fuel capacity. Similar versions were built by Fairchild and Canadian Car & Foundry with the basic designation SBF and SBW respectively.

CURTISS SB2C-5 HELLDIVER

Role: Naval carrierborne bomber/reconnaissance
Crew/Accommodation: Two
Power Plant: One 1,900 hp Wright R-2600-20 Double Cyclone air-cooled radial
Dimensions: Span 15.15m (49.75 ft); length 11.17m (36.66 ft); wing area 39.2m^2 (422 sq ft)
Weights: Empty 4,799kg (10,580 lb); MTOW 7,388kg (16,287 lb)
Performance: Maximum speed 418km/h (260 mph) at 4,907m (16,100 ft); operational ceiling 8,047m (26,400 ft); range 1,875km (1,165 miles) with 454 kg (1,000 lb) bombload
Load: Two 20mm cannon and two .303 inch machine guns, plus up to 907 kg (2,000 lb) of bombs

CURTISS MODEL 75 (P-36 and HAWK 75)
(U.S.A.)

Designed as a private-venture fighter from 1934, the Model 75 prototype first flew in May 1935 as a monoplane with retractable landing gear and a 671kW (900 hp) Wright XR-1670-5 radial engine. The type was evaluated as the Model 75B with the 634kW (750 hp) Wright R-1820 radial engine, and ordered into production as the P-36A. Later variants were the P-36B with the 746kW (1,000 hp) Pratt & Whitney R-1830-25 radial engine, and the P-36C with two additional guns and more power. The type was exported in fairly large numbers as the H75A, British aircraft being named Mohawk. Curtiss also developed a less advanced version as the Hawk 75, generally similar to the pre-production Y1P-36 but with a lower-powered 652kW (875 hp) Wright GR-1820 radial engine and fixed landing gear.

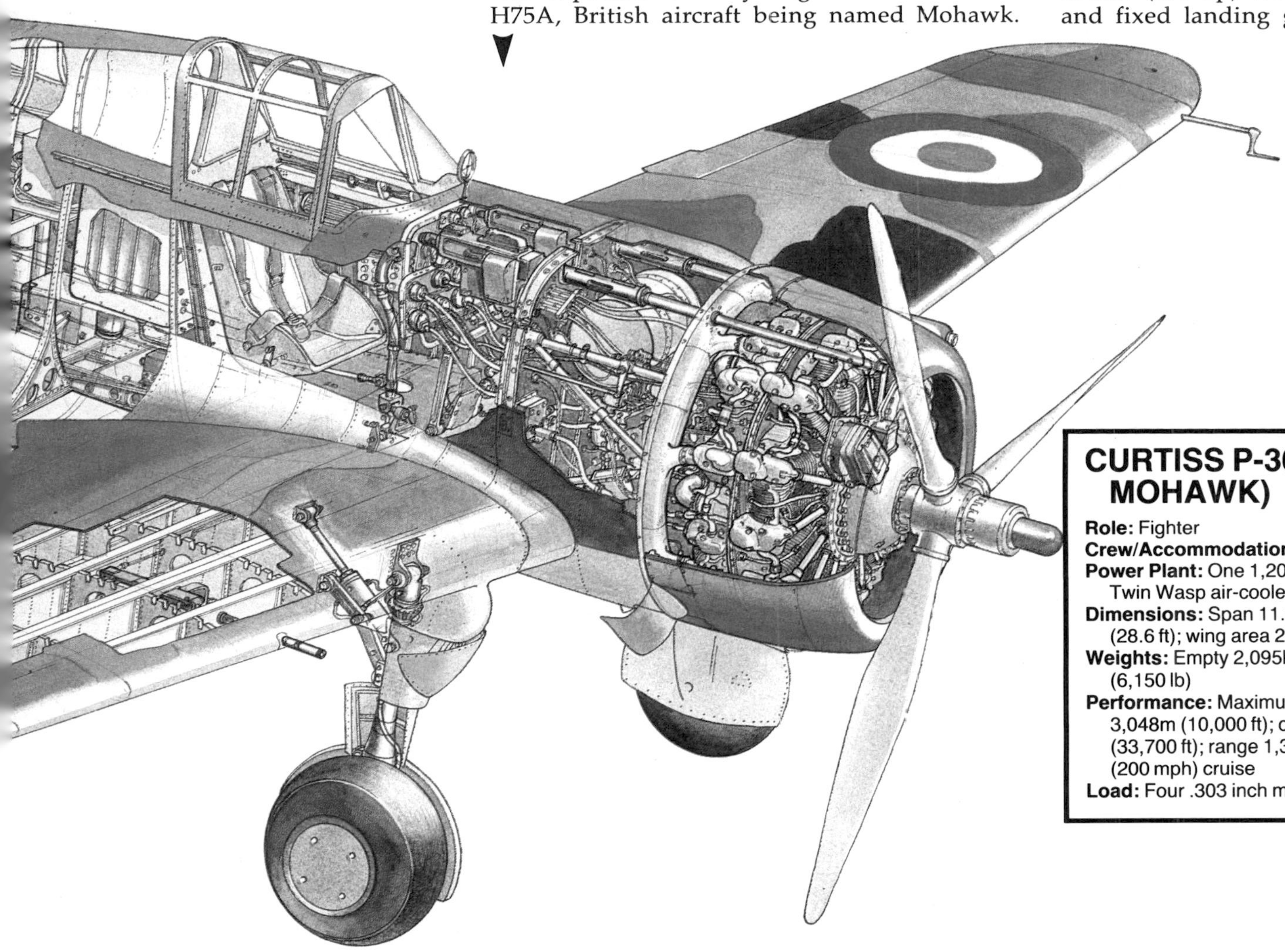

CURTISS P-36C (RAF MOHAWK)

Role: Fighter
Crew/Accommodation: One
Power Plant: One 1,200 hp Pratt & Whitney R-1830-17 Twin Wasp air-cooled radial
Dimensions: Span 11.35m (37.33 ft); length 8.72m (28.6 ft); wing area 21.92m^2 (236 sq ft)
Weights: Empty 2,095kg (4,619 lb); MTOW 2,790kg (6,150 lb)
Performance: Maximum speed 501km/h (311 mph) at 3,048m (10,000 ft); operational ceiling 10,272m (33,700 ft); range 1,320km (820 miles) at 322km/h (200 mph) cruise
Load: Four .303 inch machine guns

CURTISS MODELS 81 and 87 (P-40 WARHAWK/ TOMAHAWK and KITTYHAWK)
(U.S.A.)

The basis for the P-40 series was the Model 75I, a standard Model 75/XP-37A airframe modified to take the 858kW (1,150 hp) Allison V-1710-11 inline engine. This became the first U.S. fighter to exceed 300 mph (483km/h) in level flight, and the type was ordered by the U.S. Army in modified form with the designation P-40 and the less powerful V-1710-33; export versions were the Hawk 81-A1 for France and Tomahawk Mk I for the U.K. Improved models were the P-40B (Tomahawk Mk IIA) with self-sealing tanks, armour and better armament, the P-40C (Tomahawk Mk IIB) with improved self-sealing tanks and two more wing guns, the P-40D (Kittyhawk Mk I) with the 858kW (1,150 hp) V-1710-39 with better supercharging to maintain performance to a higher altitude, and the P-40E with four wing guns plus the similar Kittyhawk Mk IA with six wing guns. The P-40 series had all along been limited by the indifferent supercharging of the V-1710, and this situation was remedied in the P-40F and generally similar P-40L (Kittyhawk Mk II) by the adoption of the 969kW (1,300 hp) Packard V-1650-1 (licence-built Rolls-Royce Merlin). The type's forte was still the fighter-bomber role at low altitude, and further developments included the P-40K (Kittyhawk Mk III) version of the P-40E with the V-1710-33 engine, the P-40M with the V-1710-71 engine, and the definitive P-40N (Kittyhawk Mk IV) with the V-1710-81/99/115 engine and measures to reduce weight significantly as a means of improving performance.

▲ Curtiss Tomahawk Mk.IIb

CURTISS P-40F WARHAWK

Role: Fighter
Crew/Accommodation: One
Power Plant: One 1,300 hp Packard-built Rolls-Royce V-1650-1 Merlin water-cooled inline
Dimensions: Span 11.38m (37.33 ft); length 10.16m (33.33 ft); wing area 21.93m^2 (236 sq ft)
Weights: Empty 2,989kg (6,590 lb); MTOW 4,241kg (9,350 lb)
Performance: Maximum speed 586km/h (364 mph) at 6,096m (20,000 ft); operational ceiling 10,485m (34,400 ft); range 603km (375 miles)
Load: Six .5 inch machine guns, plus up to 227 kg (500 lb) of bombs

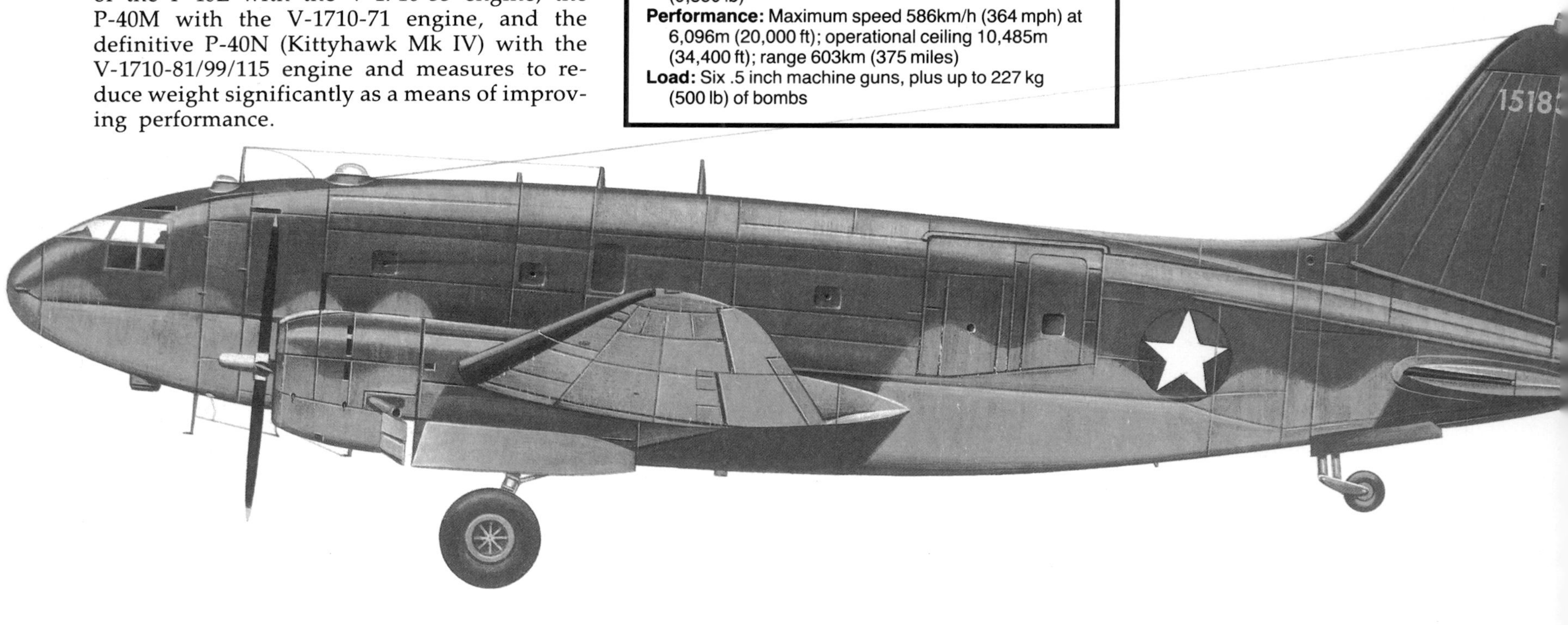

CURTISS-WRIGHT C-46 COMMANDO ▲
(U.S.A.)

The aeroplane that entered widespread production as the C-46 Commando troop and freight transport was conceived to pick up where the Douglas DC-3 left off by offering advantages such as cabin pressurization, larger capacity, longer range and higher cruising speed. The twin-finned CW-20T prototype flew in March 1940 with 1,268kW (1,700 hp) Wright R-2600 radial engines, but subsequent development was geared to the needs of the U.S. Army, which ordered the C-46A version with 1,491kW (2,000 hp) R-2800-51 radial engines and accommodation for 45 troops. The C-46 series was used almost exclusively in the Pacific theatre during World War II, and the several variants were all comparable to the C-46A apart from minor modifications and adaptations. The type was also used by the U.S. Navy with the designation R5C, and after the war many ex-military machines were released onto the civil market where a not inconsiderable number remains to the present.

CURTISS-WRIGHT C-46A COMMANDO

Role: Long-range transport
Crew/Accommodation: Four, plus up to 50 troops
Power Plant: Two 2,000 hp Pratt & Whitney R-2800-51 Double Wasp air-cooled radials
Dimensions: Span 32.91m (108 ft); length 23.26m (76.33 ft); wing area 126.34m^2 (1,360 sq ft)
Weights: Empty 13,608kg (30,000 lb); MTOW 25,401kg (56,000 lb)
Performance: Cruise speed 278km/h (173 mph) at 4,572m (15,000 ft); operational ceiling 7,468m (24,500 ft); range 1,931km (1,200 miles) with full payload
Load: Up to 6,804 kg (15,000 lb)

DASSAULT M.D.452 MYSTERE IV and SUPER MYSTERE B2
(France)

After World War II Marcel Bloch changed his name to Dassault and rebuilt his original company as Avions Marcel Dassault, the premier French manufacturer of warplanes. After gaining experience in the design and construction of jet-powered fighters with the straight-winged Ouragan fighter-bomber, Dassault turned his attention to a swept-wing design, the Mystere. This flew in Mystere I prototype form during February 1951, and was followed by eight more prototypes each with a Rolls-Royce turbojet. Then came 11 pre-production Mystere IICs with the 3,000kg (6,614 lb) thrust SNECMA Atar 101 turbojet, and these paved the way for the Mystere IVA full-production variant, which first flew in September 1952 with the 3,500kg (7,716 lb) thrust Hispano-Suiza Verdon 350 turbojet. Further development led to the Mystere IVB prototype with a Rolls-Royce Avon turbojet, a thinner and more highly swept wing, and a revised fuselage of lower drag. The resulting Super Mystere B1 production prototype flew in March 1955 with an afterburning turbojet as the first genuinely supersonic aeroplane of European design, and was followed by the Super Mystere B2 production model with the 4,460kg (9,833 lb) thrust Atar 101G-2/3 afterburning turbojet.

DASSAULT MYSTERE IVA

Role: Strike fighter
Crew/Accommodation: One
Power Plant: One 3,500 kgp (7,716 lb s.t.) Hispano-Suiza Verdun 350 turbojet
Dimensions: Span 11.13m (36.5 ft); length 12.83m (42.1 ft); wing area 32m² (344.5 sq ft)
Weights: Empty 5,875kg (12,950 lb); MTOW 9,095kg (20,050 lb)
Performance: Maximum speed 1,120km/h (604 knots) at sea level; operational ceiling 13,716m (45,000 ft); range 460km (248 naut. miles)
Load: Two 30mm DEFA cannon, plus to up 907kg (2,000 lb) of externally carried bombs

DASSAULT M.D.550 MIRAGE III and MIRAGE 5
(France)

The Mirage was designed to meet a 1954 requirement for a small all-weather supersonic interceptor, and emerged as the delta-winged Mirage I for a first flight in June 1955 with two Armstrong Siddeley Viper turbojets. The type was too small for any realistic military use, and a slightly larger Mirage II was planned: this was not built, both these initial concepts being abandoned in favour of the still larger Mirage III that first flew in November 1956 with an Atar 101G-1 afterburning turbojet. Further development led to the Mirage IIIA pre-production type with an Atar 9B of 6,000kg (13,228 lb) afterburning thrust boosting speed from Mach 1.65 to 2.2 at altitude. The type went into widespread production for the French forces and for export, the basic variants being the Mirage IIIB two-seat trainer, the Mirage IIIC single-seat interceptor, the Mirage IIIE single-seat strike fighter and the Mirage IIIR reconnaissance aeroplane. The Mirage 5 was produced as a clear-weather type, though the miniaturization of electronics in the 1970s and 1980s have allowed the installation or retrofit of avionics that make most Mirage 5 and more powerful Mirage 50 models superior to the baseline Mirage III models. Israel produced a Mirage 5 variant as the IAI Kfir with a General Electric J79 afterburning turbojet and advanced electronics, and this has spawned the very impressive Kfir-C2 variant with canard foreplanes for much improved field and combat performance. Many surviving Mirage III aircraft have been modernized to a similar electronic standard.

DASSAULT MIRAGE IIIE

Role: Strike fighter
Crew/Accommodation: One
Power Plant: One 6,200 kgp (13,670 lb s.t.) SNECMA Atar 9C turbojet, plus provision for one 1,500 kgp (3,307 lb s.t.) SEPR 844 rocket engine
Dimensions: Span 8.22m (27 ft); length 15.03m (49.26 ft); wing area 34.85m² (375 sq ft)
Weights: Empty 7,050kg (15,540 lb); MTOW 13,000kg (29,760 lb)
Performance: Maximum speed 2,350km/h (1,268 knots) Mach 2.21 at 12,000m (39,375 ft); operational ceiling 17,000m (55,775 ft); radius 1,200km (648 naut. miles)
Load: Two 30mm DEFA cannon, plus up to 1,362 kg (3,000 lb) of externally carried ordnance

DASSAULT-BREGUET MIRAGE F1
(France)

This was produced to succeed the Mirage III/5 family, and is a markedly different aeroplane with 'conventional' flying surfaces. The type first flew in December 1966, and was ordered into production with the Atar 9K-50 afterburning turbojet. The main variants have been the Mirage F1A clear-weather ground-attack fighter, the Mirage F1B two-seat trainer, the Mirage F1C (and Mirage F1C-200 long-range) single-seat multi-role attack fighter and the Mirage F1CR-200 long-range reconnaissance aeroplane.

DASSAULT-BREGUET MIRAGE F1C

Role: Strike fighter
Crew/Accommodation: One
Power Plant: One 7,200 kgp (15,873 lb s.t.) SNECMA Atar 9K-50 turbojet with reheat
Dimensions: Span 8.4m (27.55 ft); length 15m (49.24 ft); wing area 25m^2 (270 sq ft)
Weights: Empty 7,400kg (16,315 lb); MTOW 14,900kg (32,850 lb)
Performance: Maximum speed 2,335km/h (1,260 knots) Mach 2.2 at 12,000m (39,370 ft); operational ceiling 20,000m (65,600 ft); range 3,300km (1,781 miles)
Load: Two 30mm DEFA cannon, plus up to 4,000 kg (8,818 lb) of externally carried weapons

DASSAULT-BREGUET MIRAGE 2000
(France)

With the Mirage 2000 the manufacturer reverted to the delta-wing planform, but in this instance of the relaxed-stability type with an electronic fly-by-wire control system to avoid many of the aerodynamically similar Mirage III/5's low-level handling limitations. The type is powered by an afterburning turbofan, and first flew in prototype form during March 1978. The type remains in production, the primary variants being the Mirage 2000B two-seat trainer, the Mirage 2000C single-seat interceptor and multi-role fighter, the Mirage 2000N two-seat nuclear strike fighter, the Mirage 2000N-1 two-seat conventional strike fighter and the Mirage 2000R single-seat reconnaissance fighter.

DASSAULT-BREGUET MIRAGE 2000C

Role: Air superiority fighter
Crew/Accommodation: One
Power Plant: One 9,700 kgp (21,385 lb s.t.) SNECMA M53-P2 turbofan with reheat
Dimensions: Span 9.13m (29.95 ft); length 15m (49.2 ft); wing area 41m^2 (441.3 sq ft)
Weights: Empty 7,500kg (16,534 lb); MTOW 17,000kg (37,480 lb)
Performance: Maximum speed 2,335km/h (1,260 knots) Mach 2.2 at 11,000m (36,000 ft); operational ceiling 18,000m (60,000 ft); range 1,600+ km (850+ naut miles) with external fuel
Load: Two 30mm DEFA cannon, two Matra 550 Magic and two Matra Super 530 missiles

DASSAULT-BREGUET RAFALE
(France)

The Rafale is to become France's main combat aeroplane of the 1990s, and is of the modern compound delta canard configuration with relaxed stability, fly-by-wire control system and a large proportion of composite materials in the structure. The type first flew in ACX or Rafale-A prototype form during July 1984 with two General Electric F404 afterburning turbofans, and planned production variants are the Rafale-D for the French air force in single- and two-seat variants for the combat and operational training roles, and the Rafale-M for the French navy in a carrierborne single-seat form.

DASSAULT-BREGUET ACX

Role: Developmental advanced fighter
Crew/Accommodation: One
Power Plant: Two 7,258 kgp (16,000 lb) General Electric F404-GE-100 turbofans with reheat
Dimensions: Span 11.18m (36.75 ft); length 15.79m (51.85 ft); wing area 47m^2 (506 sq ft)
Weights: Empty 9,500kg (20,950 lb); MTOW 20,000kg (44,090 lb)
Performance: Maximum speed 2,124km/h (1,146 knots) Mach 2.2 at 11,000m (36,090 ft); operational ceiling (not available); range (not available)
Load: One 30mm DEFA 554 cannon, plus four Matra Mica and two Matra Magic missiles

DASSAULT-BREGUET/DORNIER ALPHA JET
(France/West Germany)

This Franco-West German aeroplane was designed as an advanced two-seat trainer and light attack type of high subsonic performance, and first flew in prototype form during October 1973 with two specially developed Turbomeca Larzac non-afterburning turbofans. The design has proved highly successful in the domestic and export markets, the main variants being the Alpha Jet-A close support/attack aeroplane used by West Germany, and the Alpha Jet-E advanced flying/weapons trainer used by France plus most export customers. Derivatives of the Alpha Jet-E are the Alpha Jet MS2 with a more advanced nav/attack system, the Alpha Jet NGEA (Alpha Jet 2) with the MS2's nav/attack system, more powerful engines and air-to-air missile capability, the Lancier (Alpha Jet 3) derived from the NGEA with radar and the capability for several advanced weapon types, and the Alpha Jet ATS based on the Alpha Jet 3 with state-of-the-art-cockpit displays.

DASSAULT-BREGUET/DORNIER ALPHA JET 2

Role: Light strike/advanced trainer
Crew/Accommodation: Two
Power Plant: Two 1,440 kgp (3,175 lb s.t.) SNECMA/Turbomeca Larzac 04-C20 turbofans
Dimensions: Span 9.11m (29.95 ft); length 12.29m (40.25 ft); wing area 17.5m^2 (188 sq ft)
Weights: Empty 3,515kg (7,749 lb); MTOW 8,000kg (17,637 lb)
Performance: Maximum speed 1,038km/h (560 knots) at sea level; operational ceiling 14,630m (48,000 ft); radius 1,075km (580 naut. miles)
Load: Up to 2,500 kg (5,510 lb) of externally carried armament

de HAVILLAND D.H.60 MOTH
(U.K.)

First flown in February 1925, the Moth was designed as a sport aeroplane for the 'man in the street', and was a conventional light biplane built in large numbers. The type was flown with a variety of engines, and the name of the individual engine was generally prefixed to the basic name. Perhaps the most celebrated variant was the D.H.60G Gipsy Moth with the 75kW (100 hp) de Havilland Gipsy I inline engine, and this was used in a number of celebrated long-distance flights.

de HAVILLAND D.H.60 MOTH

Role: Tourer/trainer
Crew/Accommodation: Two
Power Plant: One 60 hp ADC Cirrus I air-cooled inline
Dimensions: Span 8.84m (29 ft); length 7.16m (23.5 ft); wing area 20.9m^2 (225 sq ft)
Weights: Empty 388kg (855 lb); MTOW 612kg (1,350 lb)
Performance: Maximum speed 146km/h (91 mph) at sea level; operational ceiling 3,962m (13,000 ft); range 515km (320 miles)
Load: 81.6kg (180 lb)

de HAVILLAND D.H.82 TIGER MOTH
(U.K.)

From the Moth was developed the D.H.60T Moth Trainer as a two-seat basic trainer with a strengthened airframe for a crew of two. From this latter was derived the Tiger Moth. This retained the straight lower wing and dihedralled upper wing of the Moth Trainer, but in the D.H.82 prototype that first flew in October 1931 the lower wing was given dihedral for improved ground clearance. Large-scale production followed, all but a few of them of the Tiger Moth Mk II (D.H.82A) variant in which the ridged stringer/fabric rear decking of the Tiger Moth Mk I was replaced by smooth plywood decking. The D.H.82B was the Queen Bee remotely controlled target drone, and the D.H.82C was a winterized variant built by de Havilland Canada.

de HAVILLAND D.H.82A TIGER MOTH

Role: Trainer/tourer
Crew/Accommodation: Two
Power Plant: One 130 hp De Havilland Gipsy Major I air-cooled inline
Dimensions: Span 8.94m (29.3 ft); length 7.3m (23.95 ft); wing area 22.2m^2 (239 sq ft)
Weights: Empty 506kg (1,115 lb); MTOW 828kg (1,825 lb)
Performance: Maximum speed 167km/h (104 mph) at sea level; operational ceiling 4,267m (14,000 ft); range 483km (300 miles)
Load: 81.6kg (180 lb)

de HAVILLAND D.H.84 DRAGON and D.H.86
(U.K.)

The Dragon was designed for commercial operations between southern England and Paris, and emerged for its first flight in November 1932. It was an elegant equal-span biplane with a smoothly contoured nose and twin engines mounted on the lower wings with the main units of the fixed landing gear directly below them. The three variants were the Dragon I with a single long window along each side of the cabin, the Dragon 2 with an individual window by each seat, and the D.H.84M military variant with a dorsal gun ring and fin fillet. The same basic configuration was retained in the D.H.86, a larger aeroplane with four 149kW (200 hp) de Havilland Gipsy Six inline engines designed to a QANTAS requirement for an airliner able to fly the route across the Timor Sea between Singapore and Darwin. The D.H.86 first flew in January 1934, and variants were the basic D.H.86, the improved D.H.86A and the D.H.86B conversion with small fins at the tips of the tailplane.

de HAVILLAND D.H.84 DRAGON 1

Role: Short-range passenger transport
Crew/Accommodation: One, plus up to eight passengers
Power Plant: Two 130 hp De Havilland Gipsy Major 1 air-cooled inlines
Dimensions: Span 14.43m (47.3 ft); length 10.52m (34.5 ft); wing area 34.9m^2 (376 sq ft)
Weights: Empty 1,043kg (2,300 lb); MTOW 1,905kg (4,200 lb)
Performance: Maximum speed 206km/h (128 mph) at sea level; operational ceiling 3,810m (12,500 ft); range 740km (468 miles)
Load: 363 kg (800 lb)

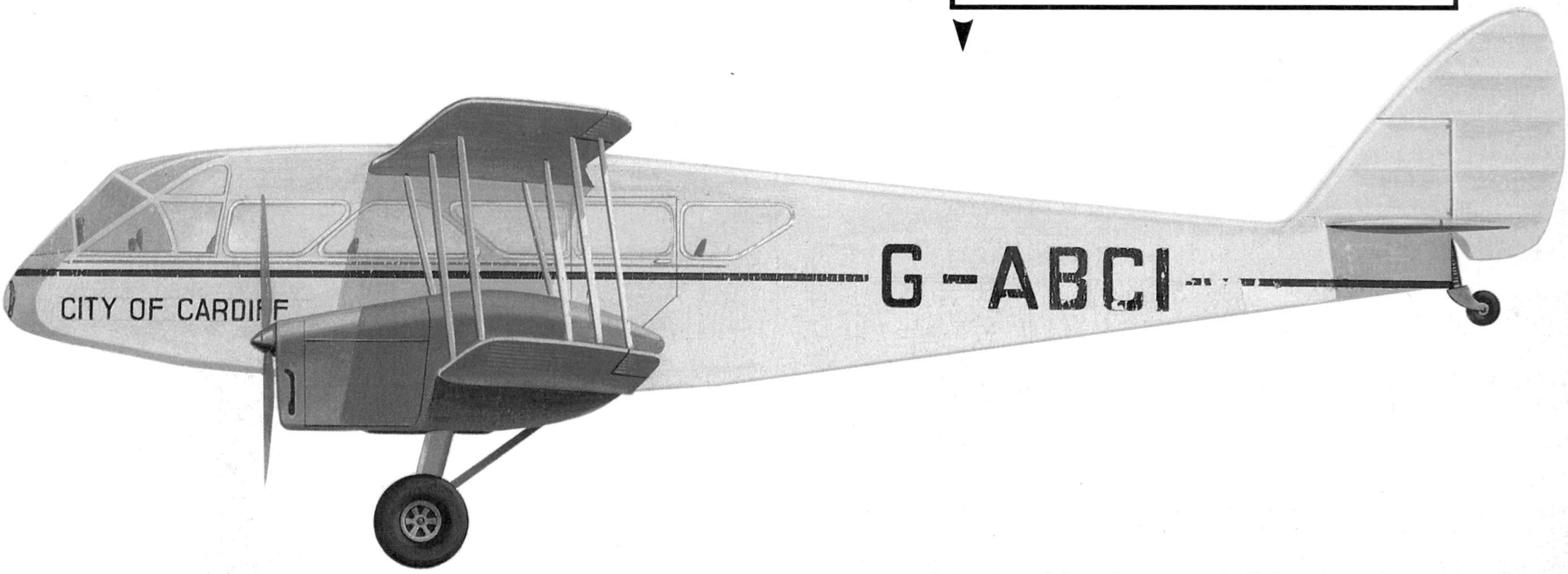

de HAVILLAND D.H.89 DRAGON RAPIDE and D.H.90 DRAGONFLY
(U.K.)

In the Dragon Rapide the manufacturer capitalized on its experience with the D.H.84 and D.H.86 to produce a supremely elegant light transport resembling the D.H.84 but with fully spatted main landing gear units. The Dragon Six prototype flew in April 1934, and variants were the basic D.H.89, the D.H.89A with trailing-edge flaps on the lower wings outboard of the engines, and the D.H.89M military version that served in Dominie Mk 1 radio trainer and Dominie Mk 2 communications models. The D.H.90 first flew in August 1935 and resembled a scaled-down D.H.89. It had a completely different fuselage structure, however, with a preformed plywood monocoque shell in place of the previous spruce and plywood box. The production version was the D.H.90A.

de HAVILLAND D.H.89A DRAGON RAPIDE

Role: Short-range passenger transport
Crew/Accommodation: One, plus up to eight passengers
Power Plant: Two 200 hp De Havilland Gipsy Queen 3 air-cooled inlines
Dimensions: Span 14.63m (48 ft); length 10.52m (34.5 ft); wing area 31.2m^2 (336 sq ft)
Weights: Empty 1,486kg (3,276 lb); MTOW 2,495kg (5,500 lb)
Performance: Maximum speed 252km/h (157 mph) at sea level; operational ceiling 5,944m (19,500 ft); range 930km (578 miles)
Load: 363 kg (800 lb)

de HAVILLAND D.H.98 MOSQUITO
(U.K.)

Perhaps the most versatile warplane of World War II, and certainly one of the classic warplanes of all time, the Mosquito began as a private venture and was based on the type of composite wooden construction raised to a fine art by de Havilland in aircraft such as the D.H.91 Albatross mailplane/airliner. The type was planned as a light bomber with performance so high that no defensive armament would be required, and the Mosquito Mk I prototype flew in November 1940. The type possessed fighter-type performance and considerable agility, and was immediately ordered into production in photographic reconnaissance, fighter, trainer and bomber forms. The PR versions were the Mosquito PR.Mk IV with four cameras, PR.Mk VIII with Merlins using two-stage superchargers, PR.Mk IX based on the Mk VIII but with greater fuel capacity, PR.Mk XVI based on the B.Mk XVI with a pressurized cockpit, PR.Mk 32 based on the NF. Mk XV, PR.Mk 34 with extra fuel in a 'bulged' bomb bay, PR.Mk 40 Australian development of the FB.Mk 40, and PR.Mk 41 version of the PR. Mk 40 with two-stage engines.

Fighter versions were the Mosquito NF.Mk II night-fighter with AI.Mk IV radar and a nose-mounted armament of four 20mm cannon and four 7.7mm (0.303 inch) machine guns, FB.Mk VI fighter-bomber with provision for internal and external bombs plus underwing rockets, NF.Mk XII conversion of the NF.Mk II with centimetric AI. MkVIII radar, NF.Mk XIII new-production equivalent to the NF.Mk XII, NF.Mk XV conversion of the B.Mk IV for high-altitude interception with a pressure cabin, extended-span wings and AI.Mk VIII radar, NF.Mk XVII conversion of the NF.Mk II with U.S. AI.Mk X radar, FB.Mk XVIII conversion of the FB.Mk VI for the anti-ship role with a 57mm gun and rockets, NF.Mk XIX development of the NF. Mk XIII with a 'universal' nose able to accept British or U.S. radar, FB.Mk 21 Canadian-built FB.Mk VI, FB.Mk 26 version of the FB.Mk 21 with Packard-built Merlin engines, NF.Mk 30 high-altitude model with two-stage Merlins and early ECM equipment, TR.Mk 33 carrier-capable naval torpedo fighter, NF.Mk 36 higher-altitude equivalent to the NF.Mk 30, TR.Mk 37 version of the TR.Mk 33 with British rather than U.S. radar, and FB.Mk 40 Australian-built equivalent of the FB.Mk VI.

Trainer versions were the Mosquito T.Mk III, T.Mk 22 Canadian-built equivalent to the T.Mk III, T.Mk 27 version of the T.Mk 22 with Packard-built engines, T.Mk 29 conversion of the FB.Mk 26, and T.Mk 43 Australian-built equivalent to the T.Mk III.

Bomber versions were the basic Mosquito B.Mk IV, B.Mk VII Canadian-built type with underwing hardpoints, B.Mk IX high-altitude type with the ability to carry a single 1,814kg (4,000 lb) 'blockbbuster' bomb, B.Mk XVI development of the B.Mk IX with pressurized cockpit, B.Mk 20 Canadian-built equivalent to the B.Mk IV, B.Mk 25 version of the B.Mk 20 with Packard-built engines, and B.Mk 35 long-range high-altitude model with a pressurized cockpit.

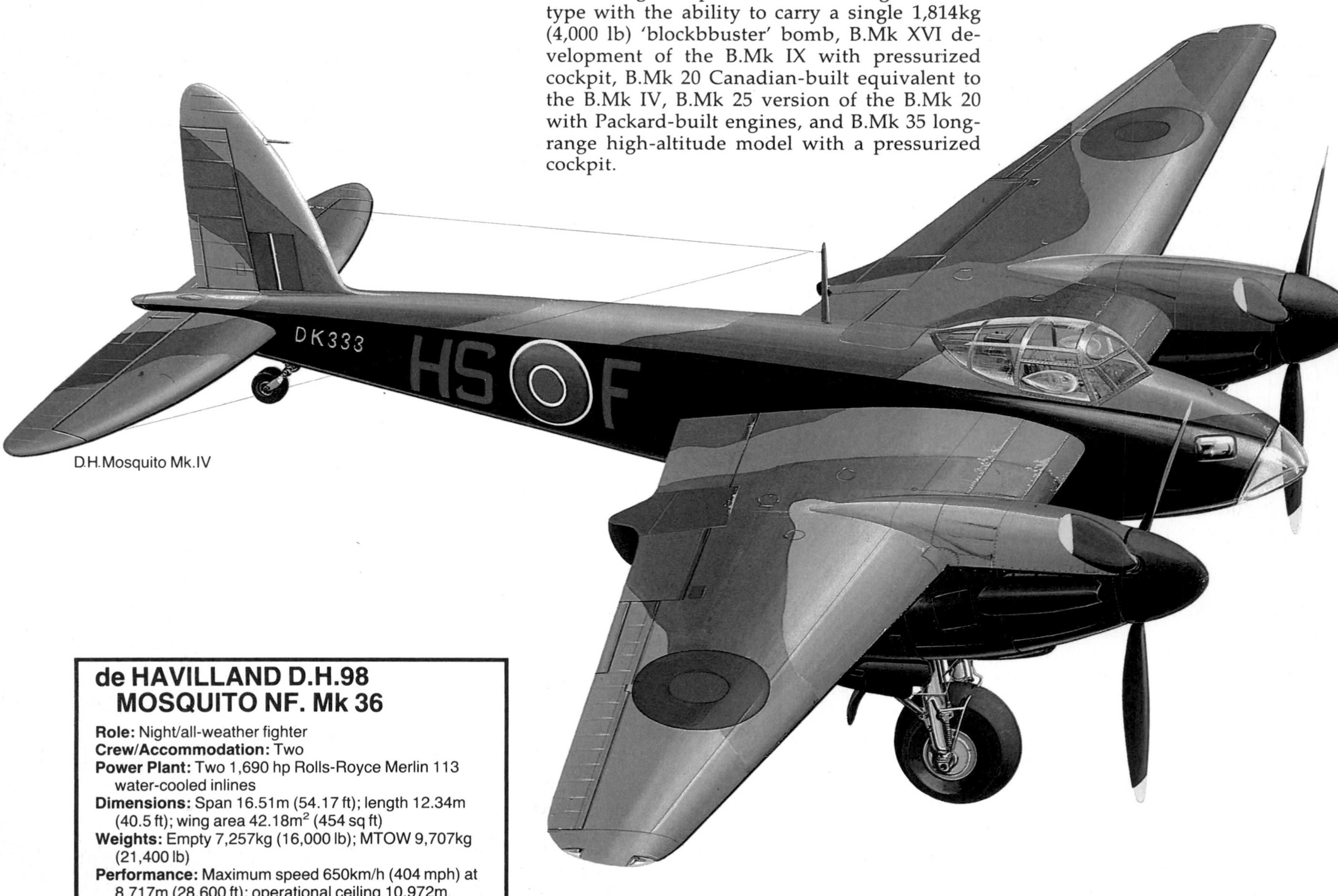

D.H.Mosquito Mk.IV

de HAVILLAND D.H.98 MOSQUITO NF. Mk 36

Role: Night/all-weather fighter
Crew/Accommodation: Two
Power Plant: Two 1,690 hp Rolls-Royce Merlin 113 water-cooled inlines
Dimensions: Span 16.51m (54.17 ft); length 12.34m (40.5 ft); wing area 42.18m^2 (454 sq ft)
Weights: Empty 7,257kg (16,000 lb); MTOW 9,707kg (21,400 lb)
Performance: Maximum speed 650km/h (404 mph) at 8,717m (28,600 ft); operational ceiling 10,972m (36,000 ft); range 2,704km (1,680 miles)
Load: Four 20mm cannon (interception guided by AI Mk 10 radar)

de HAVILLAND D.H.100 VAMPIRE, D.H.113 VAMPIRE NIGHT-FIGHTER and D.H.115 VAMPIRE TRAINER
(U.K.)

The Vampire was the U.K.'s second turbojet-powered fighter, but was just too late for World War II. The type was planned round a portly central nacelle and twin booms to allow the use of a short and therefore more efficient jetpipe for the large-diameter engine of the centrifugal-flow type, and first flew in September 1943. The Vampire F.Mk 1 entered service in 1946 with the 1,225kg (2,700 lb) de Havilland Goblin I turbojet, and was followed by the F.Mk 3 with provision for underwing stores and modifications to improve longitudinal stability. Next came the FB.Mk 5 fighter-bomber with a wing of reduced span but greater strength, and finally in the single-seat stream the FB.Mk 9 for tropical service with a cockpit air conditioner. British variants on the FB.Mk 5 theme were the Sea Vampire FB.Mks 20 and 21, while export variants include the generally similar FB.Mk 6 for Switzerland and a number of FB.Mk 50 variants with Goblin and Rolls-Royce Nene engines, the latter featuring in the licence-built French version, the Sud-Est S.E.535 Mistral. A side-by-side two-seater for night fighting was also produced as the Vampire NF.Mk 10 (export NF.Mk 54 for France), and a similar accommodation layout was retained in the Vampire T.Mk 11 and Sea Vampire T.Mk 22. Australia produced the trainer in Vampire T.Mks 33, 34 and 35 variants, and de Havilland exported the type as the Vampire T.Mk 5.

de HAVILLAND D.H.100 VAMPIRE FB Mk 5

Role: Strike fighter
Crew/Accommodation: One
Power Plant: One 1,420 kgp (3,100 lb s.t.) De Havilland Goblin 2 turbojet
Dimensions: Span 11.6m (38 ft); length 9.37m (30.75 ft); wing area 28.7m^2 (266 sq ft)
Weights: Empty 3,310kg (7,253 lb); MTOW 5,600kg (12,290 lb)
Performance: Maximum speed 861km/h (535 mph) at 5,791m (19,000 ft); operational ceiling 12,192m (40,000 ft); range 1,883km (1,170 miles) with maximum fuel
Load: Four 20mm cannon, plus up to 904 kg (2,000 lb) of ordnance

de HAVILLAND D.H.104 DOVE

Role: Short-range passenger transport
Crew/Accommodation: Two, plus up to 11 passengers
Power Plant: Two 345 hp De Havilland Gipsy Queen Series 70 air-cooled inlines
Dimensions: Span 17.37m (57 ft); length 11.98m (39.33 ft); wing area 31.12m^2 (335 sq ft)
Weights: Empty 2,551kg (5,625 lb); MTOW 3,855kg (8,500 lb)
Performance: Cruise speed 322km/h (200 mph) at 2,590m (8,500 ft); operational ceiling 6,100m (20,000 ft); range 805km (500 miles) with 771kg (1,700 lb) payload
Load: Up to 861 kg (1,893 lb)

de HAVILLAND D.H.104 DOVE
(U.K.)

The Dove was planned as post-World War II successor to the Dragon Rapide, and appeared for its first flight in September 1945 as an all-metal monoplane with graceful lines and retractable tricycle landing gear. The type entered service as the Dove 1 11-passenger light transport and Dove 2 six-passenger executive transport, later variants being the Dove 3 military transport (Devon C.Mk 1 and Sea Devon C.Mk 20), the Dove 5 and 6 feederliner and executive transport with uprated Gipsy Queen 70 Mk 2 engines, and the Dove 7 and 8 equivalent versions with Gipsy Queen 70 Mk 3 engines. In North America four of the type were re-engined with Garrett turboprops under the designation CJ-600A Jet Liner and carrried 18 passengers in a longer fuselage.

DH Comet 1

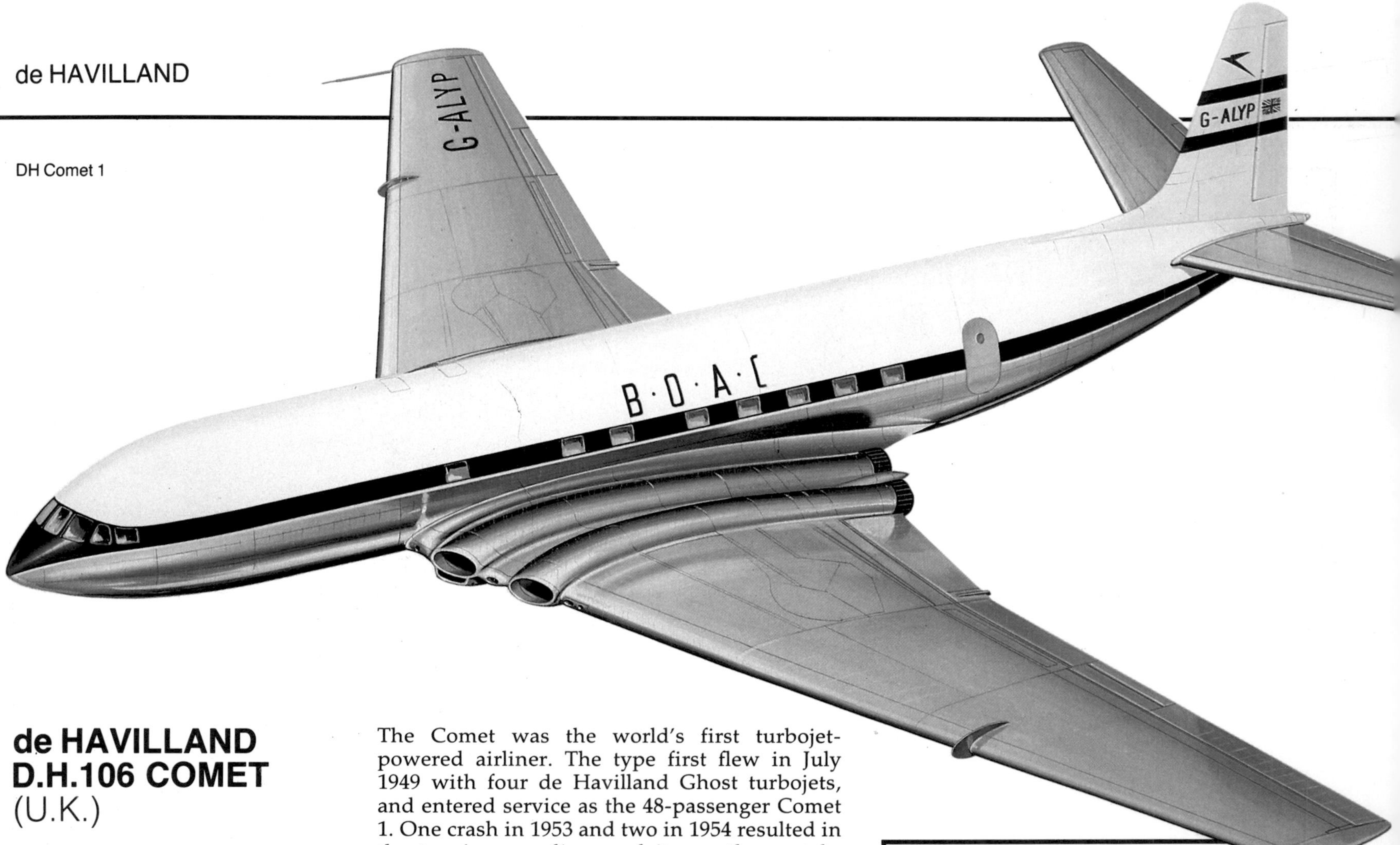

de HAVILLAND D.H.106 COMET
(U.K.)

The Comet was the world's first turbojet-powered airliner. The type first flew in July 1949 with four de Havilland Ghost turbojets, and entered service as the 48-passenger Comet 1. One crash in 1953 and two in 1954 resulted in the type's grounding, and it was then established that fatigue failures at the corners of the rectangular windows were to blame. Rounded windows were introduced on the Comet 2, a stretched 70-seat version with axial-flow Avon 503 engines, but these BOAC aircraft were diverted to the RAF as Comet C.Mk 2s. The Comet 3 was precursor to a transatlantic version that entered service as the 78-seat Comet 4 with Avon RA29 engines in May 1958. Derivatives of this last variant were the shorter-range Comet 4B with a shorter wing but longer fuselage for 99 passengers, and the Comet 4C combining the wing of the Comet 4 with the fuselage of the Comet 4B.

de HAVILLAND D.H.106 COMET 4

Role: Intermediate range passenger transport
Crew/Accommodation: Three, plus four cabin crew and up to 78 passengers
Power Plant: Four 4,649 kgp (10,250 lb s.t.) Rolls-Royce Avon RA29 turbojets
Dimensions: Span 35m (114.83 ft); length 33.99m (111.5 ft); wing area 197m² (2,121 sq ft)
Weights: Empty 34,200kg (75,400 lb); MTOW 72,575kg (160,000 lb)
Performance: Cruise speed 809km/h (503 mph) at 12,802m (42,000 ft); operational ceiling 13,411+m (44,000+ ft); range 5,190km (3,225 miles) with full load
Load: Up to 9,206kg (20,286 lb)

D.H.Sea Vixen Mk.1

de HAVILLAND D.H.110 SEA VIXEN
(U.K.)

First flown in September 1951, the D.H.110 was designed as an all-weather interceptor for the RAF, but with selection going to the Gloster Javelin the type was taken in hand for development as the Sea Vixen carrierborne strike fighter, flying in this form during June 1955. The first production model was the Sea Vixen FAW.Mk 1 with naval features such as folding wings and radome, an arrester hook and catapult point. The only other variant was the FAW.Mk 2 with greater fuel capacity and provision for Red Top rather than Firestreak air-to-air missiles.

HAWKER SIDDELEY (de HAVILLAND D.H. 110) SEA VIXEN FAW. Mk 2

Role: All-weather/night fighter
Crew/Accommodation: Two
Power Plant: Two 5,102kgp (11,250 lb s.t.) Rolls-Royce Avon Mk .208 turbojets
Dimensions: Span 15.24m (50 ft); length 16.94m (55.58 ft); wing area 62.2m² (648 sq ft)
Weights: MTOW 16,783kg (37,000 lb)
Performance: Maximum speed 1,070km/h (577 mph) Mach 0.94 at 6,096m (20,000 ft); operational ceiling 14,630m (48,000 ft); range 1,271km (686 naut miles) on internal fuel only
Load: Four short-range air-to-air missiles, plus two packs each with fourteen 2 inch folding fin anti-aircraft rockets

de HAVILLAND D.H.121 TRIDENT
(U.K.)

▲ The Trident was a considerable technical achievement, but suffered commercially from being sized to the specific requirements of BEA rather than an assortment of customers: the type had been planned with three Rolls-Royce Medway turbofans to carry 111 passengers over 2,900km (1,800 miles), but was then scaled down to three Spey turbofans to carry 100 passengers over 1,300km (810 miles). The first example flew in January 1962, and was a highly distinctive machine with its T-tail and three turbofans grouped in the tail to leave the wings uncluttered as high-performance lifting surfaces fitted with advanced high-lift devices. The original Trident 1 was followed by the Trident 1E with increased weights, Spey 511-5s and improved high-lift devices, the Trident 2E with uprated Spey 512-5Ws and greater range, the Trident 3B with a stretched fuselage for 180 passengers and a 2,381kg (5,250 lb) thrust Rolls-Royce RB.162 boost engine to improve 'hot-and-high' performance with three uprated Speys, and the Super Trident 3B with higher weights.

HAWKER SIDDELEY (de HAVILLAND D.H. 121) TRIDENT 3B

Role: Intermediate/short-range passenger transport
Crew/Accommodation: Three and up to six cabin crew, plus up to 164 passengers
Power Plant: Three 5,411kgp (11,930 lb s.t.) Rolls-Royce Spey Mk 512-5W turbofans, plus one 2,381 kgp (5,250 lb s.t.) Rolls-Royce RB, 162 turbojet for take-off boost
Dimensions: Span 29.87m (98 ft); length 36.55m (119.92 ft); wing area 138.7m^2 (1,493 sq ft)
Weights: Empty 37,695kg (83,104 lb); MTOW 68,040kg (150,000 lb)
Performance: Cruise speed 967km/h (522 knots) at 8,625m (28,300 ft); operational ceiling 13,000+m (42,650+ ft); range 1,761km (1,094 miles) with maximum payload
Load: Up to 14,695kg (32,396 lb)

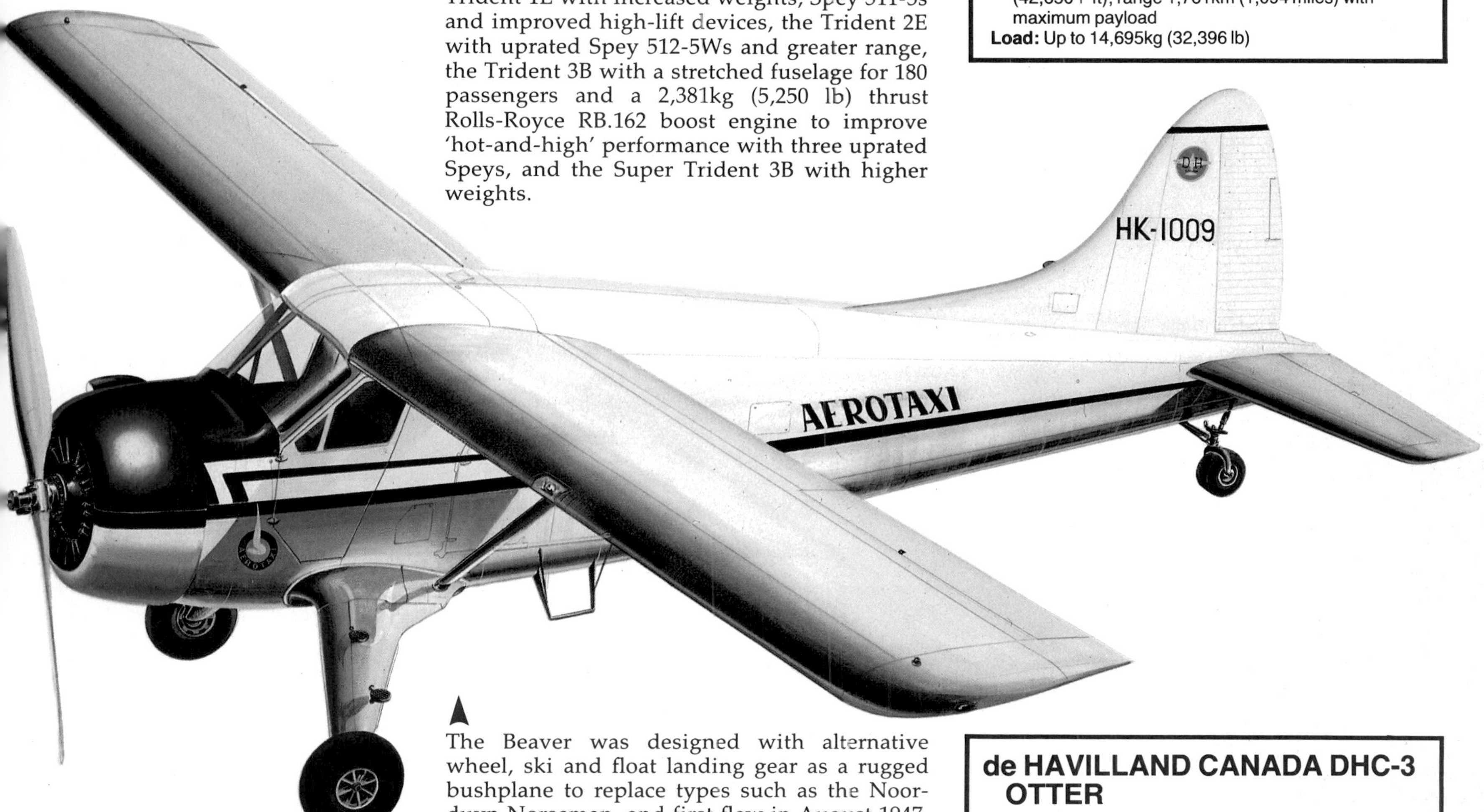

de HAVILLAND CANADA DHC-2 BEAVER and DHC-3 OTTER
(Canada)

▲ The Beaver was designed with alternative wheel, ski and float landing gear as a rugged bushplane to replace types such as the Noorduyn Norseman, and first flew in August 1947. The only model to achieve mass production was the Beaver I, which was also used by the U.S. Army with the designation L-20 (from 1962 U-6). One Beaver II was produced with a 410kW (550 hp) Alvis Leonides radial engine, and there were also a few Turbo-Beaver IIIs with the 431kW (578 ehp) Pratt & Whitney Canada PT6A-6/20 turboprop. The type had a payload of seven passengers or 680kg (1,500 lb) of freight, and was supplemented by the similar but larger and more capacious DHC-3 Otter with the 447kW (600 hp) Pratt & Whitney R-1340 Wasp radial engine. This first flew in December 1951 and was also used by the U.S. Army with the designation U-1.

de HAVILLAND CANADA DHC-3 OTTER

Role: Utility transport
Crew/Accommodation: Two, plus up to nine passengers
Power Plant: One 600 hp Pratt & Whitney R-1340 Wasp air-cooled radial
Dimensions: Span 17.68m (58 ft); length 12.75m (41.83 ft); wing area 34.83m^2 (375 sq ft)
Weights: Empty 2,398kg (5,287 lb); MTOW 3,629kg (8,000 lb)
Performance: Cruise speed 222km/h (138 mph) at 1,524m (5,000 ft); operational ceiling 4,998m (16,400 ft); range 1,419km (882 miles) with 1,055 kg (2,325 lb) payload
Load: Up to 1,430kg (3,153 lb)

de HAVILLAND CANADA DHC-4 CARIBOU and DHC-5 BUFFALO
(Canada)

The Caribou was developed as a STOL transport combining the load capability of the Douglas DC-3/C-47 Dakota and field performance of the DHC-2 and -3. The type emerged for its first flight in July 1958 as an ungainly machine with two 1,081kW (1,450 hp) Pratt & Whitney R-2000 Twin Wasp radial engines and an upswept tail allowing access to a hold able to carry 32 troops or 3,048kg (6,720 lb) of freight. The only production variant was the DHC-4A, which was used by the U.S. services as the C-7. The Bufffalo is a turboprop-powered equivalent, and first flew in April 1964. The initial variant was the DHC-5A, but the definitive model is the DHC-5D with uprated engines.

de HAVILLAND CANADA DHC-5D BUFFALO

Role: Short/rough field-capable bulk cargo transport
Crew/Accommodation: Three, plus up to 41 troops
Power Plant: Two 3,133 shp General Electric CT64-820-4 turboprops
Dimensions: Span 29.26m (96 ft); length 24.08m (79 ft); wing area 87.8m^2 (945 sq ft)
Weights: Empty 11,340kg (25,000 lb); MTOW 18,598kg (41,000 lb)
Performance: Cruise speed 421km/h (262 mph) at sea level; operational ceiling 8,382m (27,500 ft); range 1,112km (691 miles) with maximum payload
Load: Up to 8,164kg (18,000 lb)

de HAVILLAND CANADA DHC-6 TWIN OTTER
(Canada)

The Twin Otter was designed to offer Otter-type payload with the reliability and performance advantages of a twin-turboprop power plant, and the prototype of this attractive type first flew in May 1965. The type entered production as the Twin Otter Series 100 with 432kW (579 ehp) PT6A-20 turboprops, and skis or floats can be used instead of the standard wheels of the fixed tricycle landing gear. The Twin Otter Series 200 featured increased baggage capacity and higher weights, and was followed by the Twin Otter Series 300 with 486kW (652 ehp) PT6A-27 engines for higher payload.

de HAVILLAND CANADA DHC-6 TWIN OTTER SERIES 300

Role: Short take-off and landing passenger transport
Crew/Accommodation: Two, plus up to 20 passengers
Power Plant: Two 652 ehp Pratt & Whitney Canada PT6A-27 turboprops
Dimensions: Span 19.81m (65 ft); length 15.77m (51.75 ft); wing area 39.02m^2 (420 sq ft)
Weights: Empty 3,121kg (6,881 lb); MTOW 5,670kg (12,500 lb)
Performance: Maximum speed 315km/h (170 mph) at sea level; operational ceiling 8,138m (26,700 ft); range 1,297km (806 miles) with full payload
Load: Up to 1,135 kg (2,500 lb)

de HAVILLAND CANADA DHC-7 DASH 7
(Canada)

Seeking ways to capitalize on its STOL experience with larger aircraft offering greater financial reward, de Havilland Canada decided in the early 1970s that there was scope for a rugged STOL airliner with 50 seats and four turboprops driving slow-turning propellers. These would boost the STOL qualities of the wings but generate very little noise. The first DHC-7 flew in March 1975, and the type secured a major slice of the specialized market for which it was designed. The baseline passenger version is the Dash 7 Series 100, and other variants are the Series 101 freight or convertible freight/passenger model, the Series 150 passenger model with greater weights and fuel, and the Series 151 freight or convertible freight/passenger model.

de HAVILLAND CANADA DHC-7 DASH 7

Role: Short take-off and landing passenger transport
Crew/Accommodation: Two, plus up to 50 passengers
Power Plant: Four 1,120 shp Pratt & Whitney Canada PT6A-50 turboprops
Dimensions: Span 28.35m (93 ft); length 24.58m (80.66 ft); wing area 79.9m^2 (860 sq ft)
Weights: Empty 12,406kg (27,350 lb); MTOW 19,958kg (44,000 lb)
Performance: Cruise speed 428km/h (231 knots) at 2,438m (8,000 ft); operational ceiling 6,222m (20,400 ft); range 1,300km (700 naut. miles) with full payload
Load: Up to 5,284kg (11,650 lb)

de HAVILLAND CANADA DHC-8 DASH 8
(Canada)

The DHC-8 was developed to the same basic operating philosophy as the DHC-7, but is sized for 40 passengers in the commuterliner role. The type first flew in June 1983, and the baseline variant is the Dash 8 Series 100 with 1,432kW (1,800 shp) Pratt & Whitney Canada PW120A turboprops. The only other variant to have entered production (after Boeing's purchase of the company) is the Dash 8 Series 300 with greater span and length to allow the carriage of 56 passengers on two 1,776kW (2,380 shp) turboprops.

de HAVILLAND CANADA DHC-8 DASH 8 SERIES 100

Role: Short field-capable passenger transport
Crew/Accommodation: Two, plus up to 39 passengers
Power Plant: Two 1,800 shp Pratt & Whitney Canada PW 120A turboprops
Dimensions: Span 25.91m (85 ft); length 22.25m (73 ft); wing area 54.4m^2 (585 sq ft)
Weights: Empty 9,978kg (21,998 lb); MTOW 15,649kg (34,500 lb)
Performance: Maximum speed 554km/h (265 knots) at 7,620m (25,000 ft); operational ceiling 7,620m (25,000 ft); range 1,019km (633 miles) with full payload
Load: Up to 4.128kg (9,100 lb)

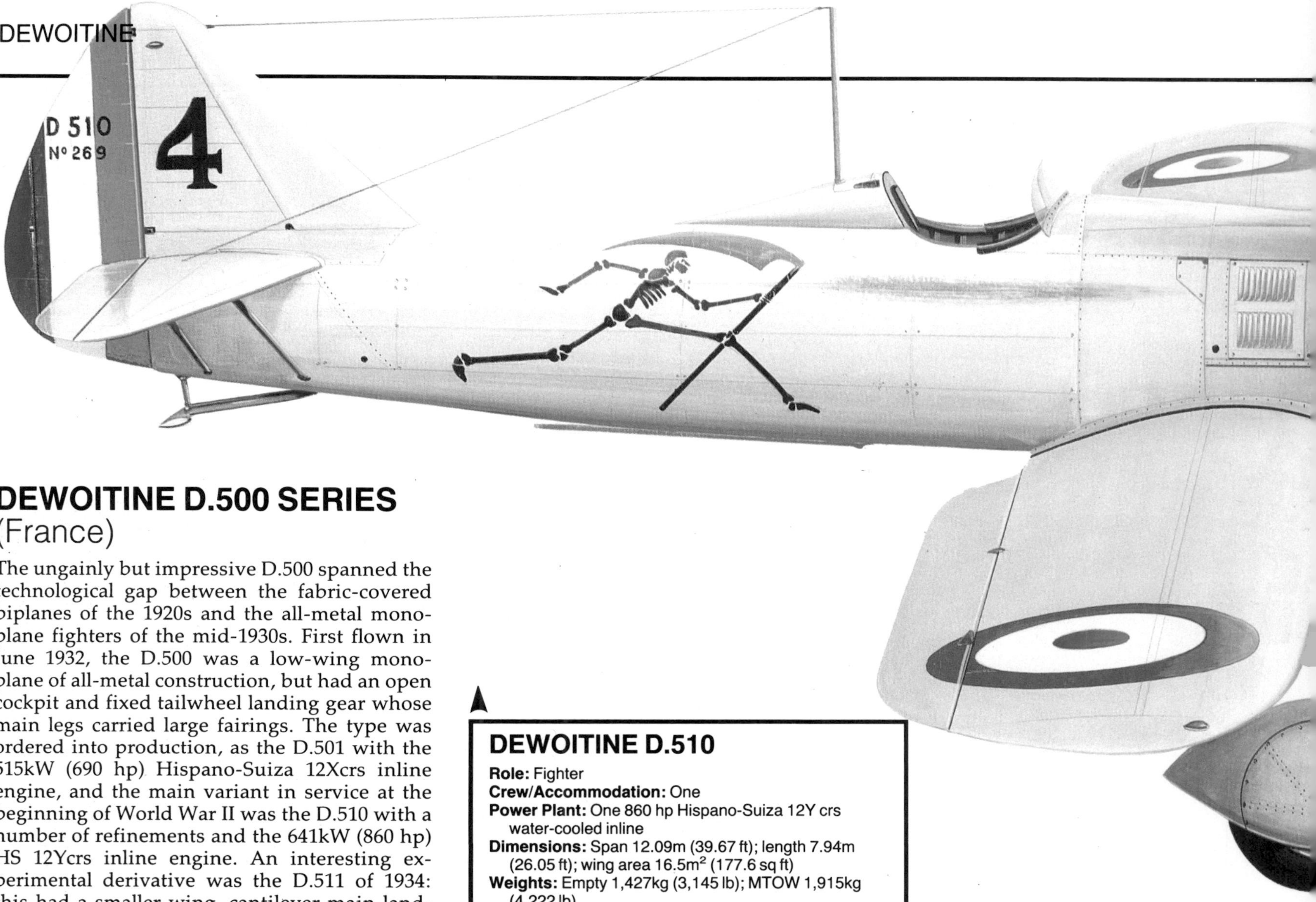

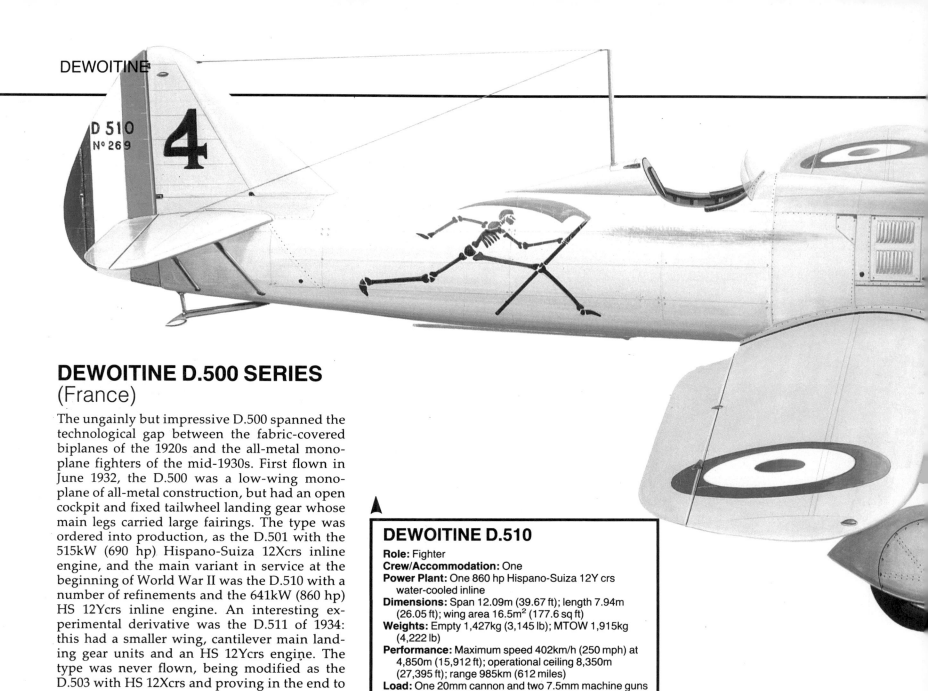

DEWOITINE D.500 SERIES
(France)

The ungainly but impressive D.500 spanned the technological gap between the fabric-covered biplanes of the 1920s and the all-metal monoplane fighters of the mid-1930s. First flown in June 1932, the D.500 was a low-wing monoplane of all-metal construction, but had an open cockpit and fixed tailwheel landing gear whose main legs carried large fairings. The type was ordered into production, as the D.501 with the 515kW (690 hp) Hispano-Suiza 12Xcrs inline engine, and the main variant in service at the beginning of World War II was the D.510 with a number of refinements and the 641kW (860 hp) HS 12Ycrs inline engine. An interesting experimental derivative was the D.511 of 1934: this had a smaller wing, cantilever main landing gear units and an HS 12Ycrs engine. The type was never flown, being modified as the D.503 with HS 12Xcrs and proving in the end to be inferior to the D.501.

DEWOITINE D.510

Role: Fighter
Crew/Accommodation: One
Power Plant: One 860 hp Hispano-Suiza 12Y crs water-cooled inline
Dimensions: Span 12.09m (39.67 ft); length 7.94m (26.05 ft); wing area 16.5m^2 (177.6 sq ft)
Weights: Empty 1,427kg (3,145 lb); MTOW 1,915kg (4,222 lb)
Performance: Maximum speed 402km/h (250 mph) at 4,850m (15,912 ft); operational ceiling 8,350m (27,395 ft); range 985km (612 miles)
Load: One 20mm cannon and two 7.5mm machine guns

DEWOITINE D.520
(France)

One of the most advanced fighters to serve with the French Air Force in the disastrous early campaign in 1940, the D.520 was a modern fighter of trim lines embodying an enclosed cockpit, flaps, retractable landing gear and a variable-pitch propeller. The type first flew in October 1938, but only comparatively small numbers had been delivered before the fall of France in June 1940. The type was used in other French theatres, and generally performed well.

DEWOITINE D.520

Role: Fighter
Crew/Accommodation: One
Power Plant: One 920 hp Hispano-Suiza 12Y45 water-cooled inline
Dimensions: Span 10.2m (33 ft); length 8.76m (28 ft); wing area 15.95m^2 (171.7 sq ft)
Weights: Empty 2,092kg (4,612 lb); MTOW 2,783kg (6,134 lb)
Performance: Maximum speed 535km/h (332 mph) at 6,000m (19,685 ft); operational ceiling 11,000m (36,090 ft); range 900km (553 miles)
Load: One 20mm cannon and four 7.5mm machine guns

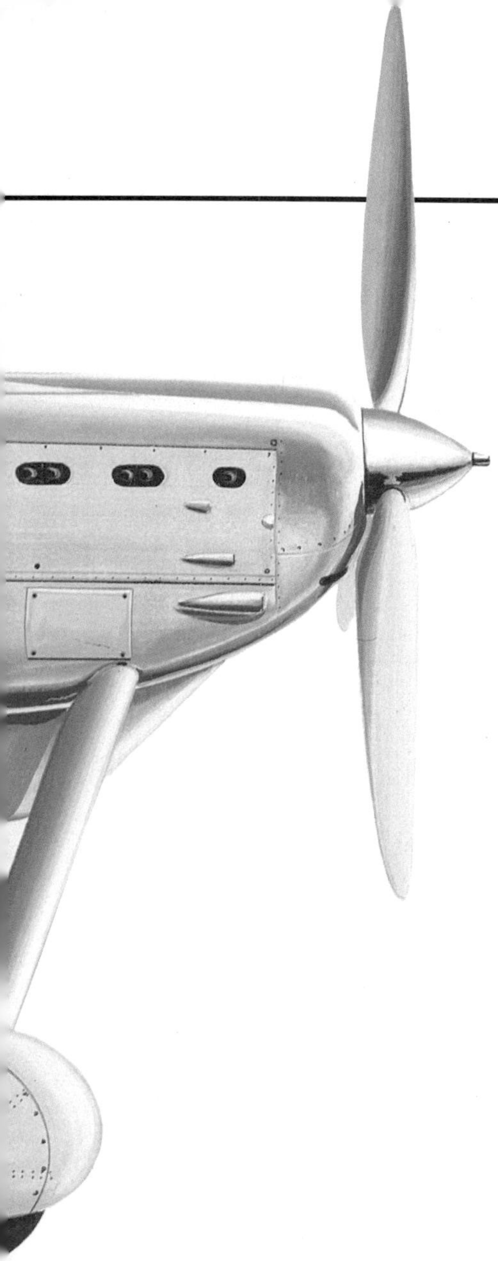

DORNIER D I
(Germany)

Also known as a Zeppelin-Landau type, as this was the concern for which Claudius Dornier was working during World War I, the D I was an interesting experimental fighter intended to validate a cantilever biplane layout with aluminium alloy wings covered as far aft as the rear spar with alloy sheet. The rest of the wings and the tailplane were fabric-covered, and the fuselage was an all-metal stressed-skin structure with integral vertical surfaces. The type was flown in the summer of 1918, and proved to have only indifferent performance.

DORNIER D 1
Role: Fighter
Crew/Accommodation: One
Power Plant: One 185 hp BMW IIIa
Dimensions: Span 7.8m (25.59 ft); length 6.37m (20.9 ft); wing area 18.7m^2 (2020 sq ft)
Weights: Empty 710kg (1,562 lb); MTOW 890kg (1,958 lb)
Performance: No reliable performance data survives
Load: Two 7.92mm machine guns

DORNIER Do 17, Do 215 and Do 217
(Germany)

The origins of this important German bomber lie with a 1933 Lufthansa requirement for a six-passenger mailplane, though this requirement also produced the small-capacity 'pencil' fuselage that was one of the main hindrances to the type's later development as a warplane. The type first flew in the autumn of 1934, and its performance suggested a military development with single vertical tail surface replaced by endplate surfaces to increase the dorsal gunner's field of fire. Six military prototypes were followed by two pre-production types, the Do 17E-1 bomber with a shortened but glazed nose and 500kg (1,102 lb) bomb load, and the Do 17F-1 photographic reconnaissance type, both powered by 559kW (750 hp) BMW VI inline engines. There followed a number of experimental and limited-production variants before the advent of the definitive bomber, the Do 17Z built to the extent of 1,700 aircraft of several subvariants in 1939 and 1940. From this was developed the Do 215 with Daimler-Benz DB 601A inline engines: the two main variants were the Do 215B-4 reconnaissance and Do 215B-5 night-fighter aircraft. The Do 217 that first flew in September was essentially a Do 17 with 802kW (1,075 hp) DB 601A engines, a larger fuselage and a revised empennage. Production amounted to 1,730 aircraft in a number of bomber, reconnaissance, fighter-bomber, night-fighter and missile launcher versions up to Do 217R indesignation.

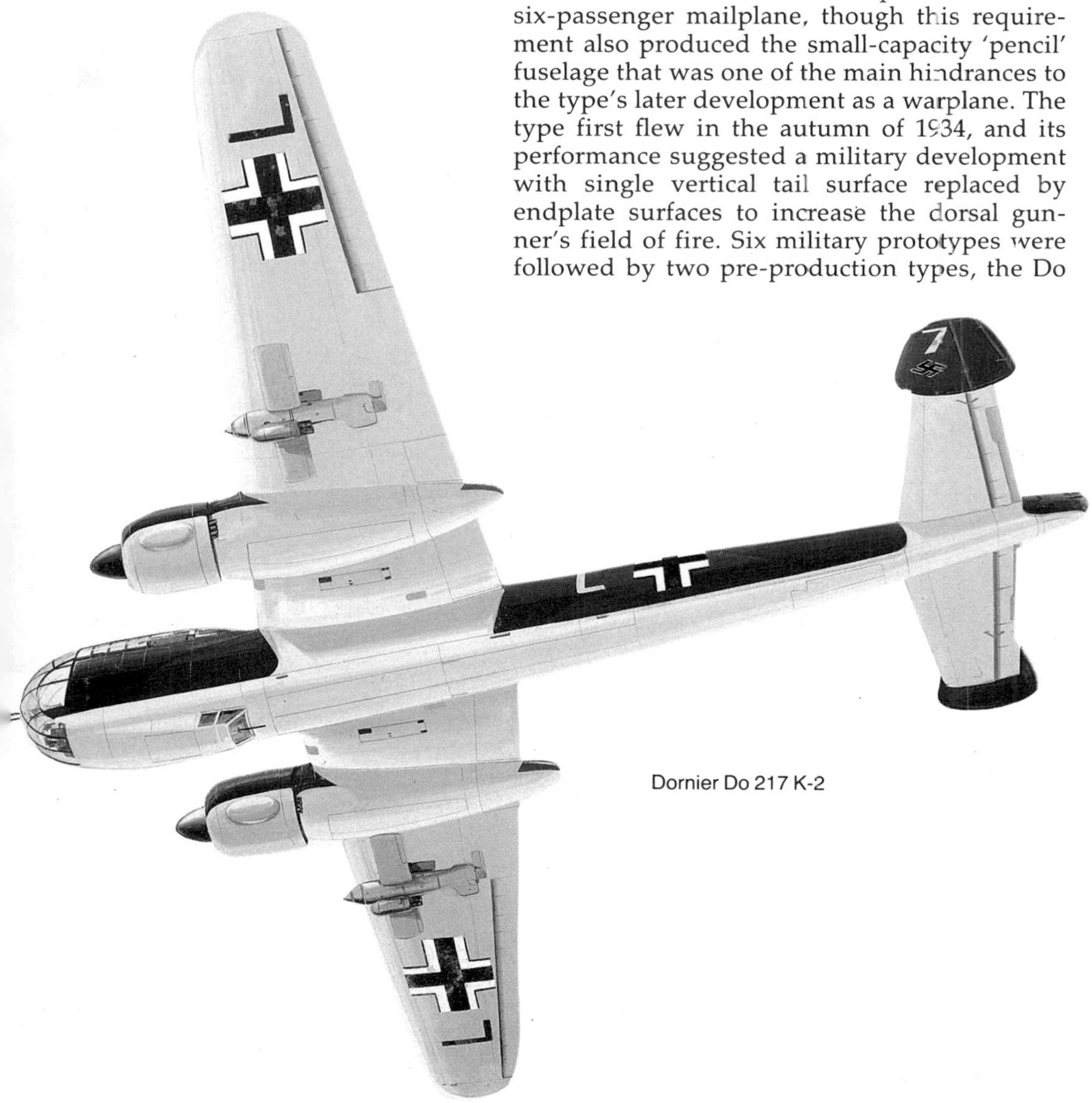

Dornier Do 217 K-2

DORNIER Do 17Z-2
Role: Medium bomber
Crew/Accommodation: Five
Power Plant: Two 1,000 hp BMW Bramo 323P Fafnir air-cooled radials
Dimensions: Span 18m (59.06 ft); length 15.8m (51.84 ft); wing area 55m^2 (592 sq ft)
Weights: Empty 5,210kg (11,488 lb); MTOW 8,590kg (18,940 lb)
Performance: Maximum speed 410km/h (255 mph) at 4,000m (13,124 ft); operational ceiling 8,200m (26,904 ft); radius 330km (205 miles) with full bombload
Load: Eight 7.9mm machine guns, plus up to 1,000kg (2,205 lb) of bombs

DORNIER Do 18
(Germany)

The Do 18 was designed as a transoceanic mailplane successor to the highly successful series of Wal flying boats. The type first flew in March 1935, and used a modernized version of the same basic design concept with stabilizing fuselage sponsons and a parasol wing with two diesel engines mounted as a tandem push/pull pair on the centreline. The type was later adapted for the coastal reconnaissance role as the Do 18D with Jumo 205C engines. Later models were the Do 18G with Jumo 205D engines, the Do 18H dual-control trainer and the Do 18N air/sea rescue variant.

DORNIER Do 18G-1

Role: Maritime patrol flying boat
Crew/Accommodation: Four
Power Plant: Two 880 hp Junkers Jumo 205D water-cooled inline diesels
Dimensions: Span 23.7m (77.75 ft); length 19.25m (63.16 ft); wing area 97.5m^2 (1,049.1 sq ft)
Weights: Empty 5,850kg (12,900 lb); MTOW 10,000kg (22,046 lb)
Performance: Cruise speed 220km/h (137 mph) at sea level, operational ceiling 4,200m (13,776 ft); range 3,500km (2,174 miles)
Load: One 20mm cannon and one 13mm machine gun, plus up to 100kg (220 lb) of bomb/depth charges

DORNIER Do 24
(Germany)

Designed to a Dutch requirement and first flown in July 1937, the Do 24 adhered to Dornier's well established flying boat formula, though in this instance with three tractor engines and a tailplane carrying endplate vertical surfaces. The Dutch version was the Do 24K with Wright R-1820 Cyclone radial engines, while versions for the German air arms were the Do 24N air/sea rescue model with R-1820 radial engines and the Do 24T maritime patroller with BMW-Bramo 323 Fafnir radial engines.

DORNIER Do 24T-1

Role: Martitime patrol/air-sea rescue flying boat
Crew/Accommodation: Six
Power Plant: Three 1,000 hp BMW Bramo 323R-2 Fafnir air-cooled radials
Dimensions: Span 27m (88.58 ft); length 22m (71.5 ft); wing area 108m^2 (1,162.1 sq ft)
Weights: Empty 9,200kg (20,286 lb); MTOW 17,800kg (39,249 lb)
Performance: Maximum speed 340km/h (211 mph) at 3,000m (9,840 ft); operational ceiling 5,900m (19,352 ft); range 2,900km (1,801 miles)
Load: One 20mm cannon and two 7.9mm machine guns, plus up to 600kg (1,323 lb) of bombs

DORNIER Do 228 (Germany)

Since World War II Dornier has placed its main emphasis on communications and transport aircraft, moving from the single-engined Do 27 and its twin-engined Do 28 version to the larger Do 28D that was later redesignated Do 128. The company then moved to the Do 228 commuter-liner using its patented TNT high-technology wing. The prototype flew in March 1981, and the type has since been produced in basic Do 228-100 form for 15 passengers and stretched Do 228-200 form for 19 passengers.

DORNIER Do 228-200

Role: Light passenger transport
Crew/Accommodation: Two, plus up to 19 passengers
Power Plant: Two 715 shp Garett AiResearch TPE 331-5-252D turboprops
Dimensions: Span 16.97m (55.68 ft); length 16.55m (54.25 ft); wing area 32m^2 (344 sq ft)
Weights: Empty 2,908kg (6,410 lb); MTOW 5,700kg (12,570 lb)
Performance: Maximum speed 432km/h (268 mph) at 3,280m (10,000 ft); operational ceiling 9,020m (29,600 ft); range 1,150km (715 miles) with full bombload
Load: Up to 2,057kg (4,540 lb)

DOUGLAS DC-3 and C-47/DAKOTA
(U.S.A.)

The DC-3 can truly be said to have changed history, for this type opened the era of 'modern' air travel in the mid-1930s, and became the mainstay of the Allies' air transport effort in World War II. Over 10,500 were built in the U.S.A., and at least another 2,000 were produced under licence in the U.S.S.R. as the Lisunov Li-2. The series began with the DC-1 that first flew in July 1933 as a cantilever low-wing monoplane of all-metal construction (except fabric-covered control surfaces) with enclosed accommodation and features such as retractable landing gear and trailing-edge flaps. From this prototype was developed the 14-passenger DC-2 production model, which was built in modest numbers but paved the way for the Douglas Sleeper Transport that first flew in December 1935 as an airliner for transcontinental night flights with 16 passengers in sleeper berths. From this was evolved the 24-passenger DC-3. This latter was produced in five series with either the Wright SGR-1820 or Pratt & Whitney Twin Wasp radial as the standard engine. The type was ordered for the U.S. military as the C-47 Skytrain (U.S. Army) and R4D (U.S. Navy), the British adopting the name Dakota for aircraft supplied under the terms of the Lend-Lease Act. The type was produced in a vast number of variants within the new-build C-47, C-53, C-117 amd R4D series, while impressed aircraft swelled number and designations to a bewildering degree. After the war large numbers of these monumentally reliable aircraft were released cheaply to civil operators, and the series can be credited with the development of air transport in most of the world's remoter regions.

DOUGLAS DC-3A

Role: Passenger transport
Crew/Accommodation: Three, plus two cabin crew and up to 28 passengers
Power Plant: Two 1,200 hp Pratt & Whitney Twin Wasp S1C3-G air cooled radials
Dimensions: Span 28.96m (95 ft); length 19.65m (64.47 ft); wing area 91.7m^2 (987 sq ft)
Weights: Empty 7,650kg (16,865 lb); MTOW 11,431kg (25,200 lb)
Performance: Maximum speed 370km/h (230 mph) at 2,590m (8,500 ft); operational ceiling 7,070m (23,200 ft); range 3,420km (2,125 miles)
Load: Up to 2,350kg (5,180 lb)

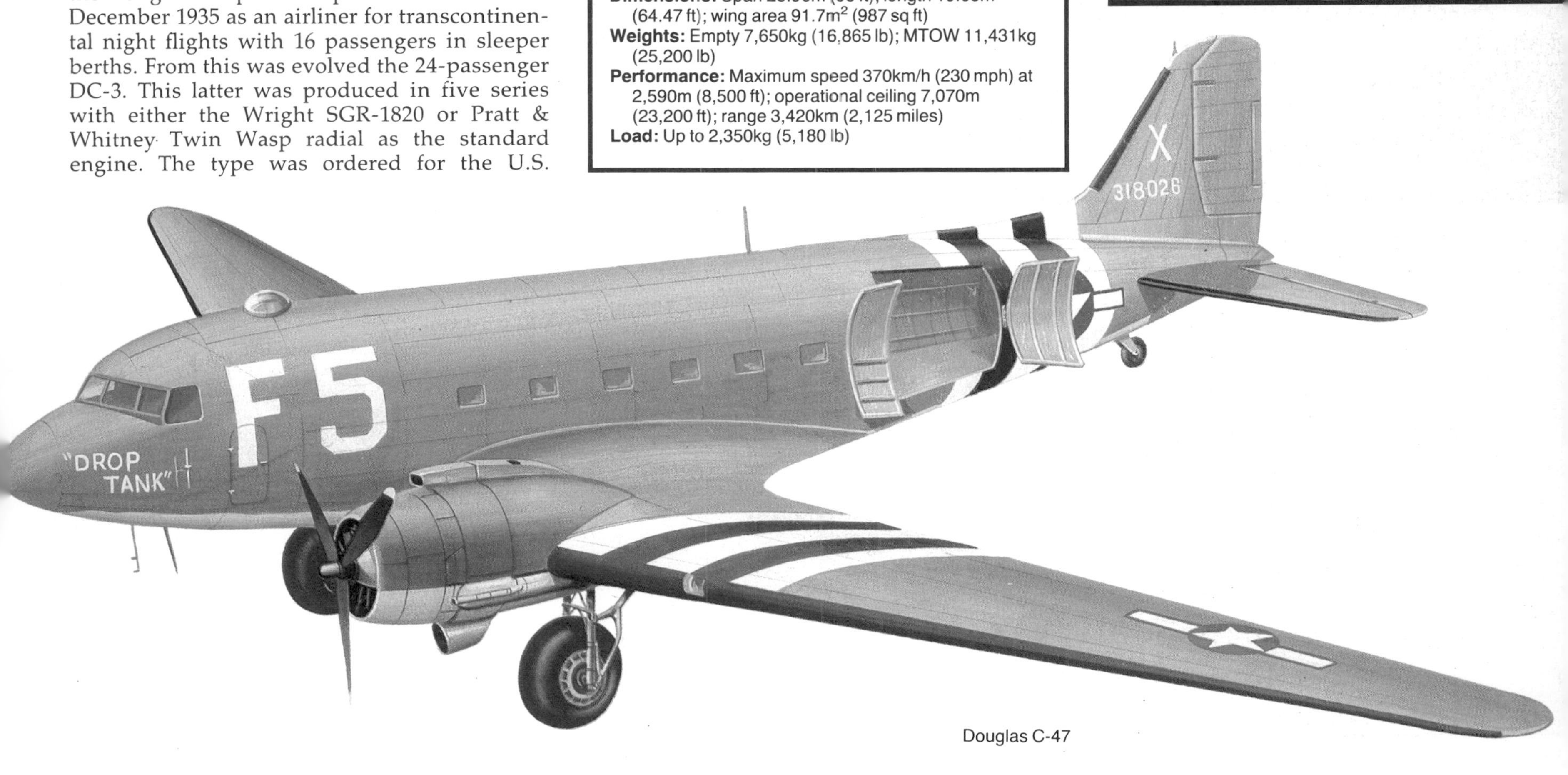

Douglas C-47

DOUGLAS DC-4 and C-54 SKYMASTER (U.S.A.)

Even before the DC-3 had flown, Douglas was planning a longer-range air transport with four engines, tricycle landing gear and greater capacity. The DC-4E pressurized prototype proved too advanced for its time, and the company therefore turned to the unpressurized and otherwise simplified DC-4 that was put into production before the first example had flown. With the U.S.A. caught up into World War II during December 1941, the type became the C-54 (army) and R5D (navy) long-range military transport, and as such first flew in February 1942. The main military versions were the baseline C-54 with R-2000-3 radial engines for 26 passengers, the C-54A (R5D-1) with R-2000-7s for 50 passengers, the C-54B (R5D-2) with integral wing tanks, the C-54D (R5D-3) with R-2000-11s, the C-54E (R5D-4) convertible freight/passenger model with revised fuel tankage, and the C-54G (R5D-5) troop carrier with R-2000-9s. After military service many of these aircraft found their way onto the civil register and performed excellently in the long-range passenger freight roles.

DOUGLAS DC-4 and C-54 SKYMASTER

Role: Long-range passenger transport
Crew/Accommodation: Four, plus three/four cabin crew, plus up to 86 passengers
Power Plant: Four 1,450 hp Pratt & Whitney R.2000 Twin Wasp air-cooled radials
Dimensions: Span 35.81m (117.5 ft); length 28.6m (93.83 ft); wing area 135.35m^2 (1,457 sq ft)
Weights: Empty 16,783kg (37,000 lb); MTOW 33,113kg (73,000 lb)
Performance: Cruise speed 309km/h (192 mph) at 3,050m (10,000 ft); operational ceiling 6,705m (22,000 ft); range 3,220km (2,000 miles) with 9,979kg (22,000 lb) payload
Load: Up to 14,515kg (32,500 lb)

DOUGLAS DC-6 and C-118 LIFTMASTER (U.S.A.)

The success of the C-54 persuaded the U.S. Army to commission a larger transport from the same company, and this XC-112 featured among other improvements a longer, pressurized fuselage married to the wings of the C-54 with four more powerful R-2800 radial engines. The type first flew in February 1946, but with World War II over the U.S. Army no longer required the new type and production was initiated for the civil market as the 86-passenger DC-6. Further development produced the lengthened all-freight DC-6A with 1,790kW (2,400 hp) engines, the equivalent DC-6B for 102 passengers, and the DC-6C convertible freight/passenger transport. The U.S. services eventually opted for the type, and these variants were the U.S. Air Force C-118A and U.S. Navy R6D-1 (later C-118B).

DOUGLAS DC-6A/C-118A

Role: Long-range cargo/passenger transport
Crew/Accommodation: Four, plus three cabin crew and up to 92 passengers
Power Plant: Four 2,500 hp Pratt & Whitney R-2800 52W-CB17 Double Wasp air-cooled radials
Dimensions: Span 38.81m (117.5 ft); length 32.18m (105.58 ft); wing area 135.9m^2 (1,463 sq ft)
Weights: Empty 22,574kg (49,767 lb); MTOW 48,534kg (107,000 lb)
Performance: Cruise speed 507km/h (315 mph) at 6,280m (20,600 ft); operational ceiling 8,840m (29,000 ft); range 4,710km (2,925 miles) with maximum payload
Load: Up to 12,786kg (28,188 lb)

DOUGLAS DC-7
(U.S.A.)

From the DC-6B the company developed the DC-7 with a lengthened fuselage, beefed-up landing gear and Wright R-3350 Turbo-Compound engines as an airliner able to compete with the Lockheed Super Constellation. The type first flew in May 1953 and entered production as a transcontinental transport. To provide transatlantic range Douglas developed the DC-7B with additional fuel capacity, but this proved marginal in its intended role and was superseded by the DC-7C, oftern called the Seven Seas, with more powerful engines, a slightly longer fuselage and, most importantly, an increase in span to provide volume for yet more fuel.

DOUGLAS DC-7C

Role: Long-range passenger transport
Crew/Accommodation: Four and four/five cabin crew, plus up to 105 passengers
Power Plant: Four 3,400 hp Wright R-3350-18EA-1 Turbo-Compound air-cooled radials
Dimensions: Span 38.86m (127.5 ft); length 34.21m (112.25 ft); wing area 152.08m^2 (1,637 sq ft)
Weights: Empty 33,005kg (72,763 lb); MTOW 64,864kg (143,000 lb)
Performance: Cruise speed 571km/h (355 mph) at 5,791m (19,000 ft); operational ceiling 6,615m (21,700 ft); range 7,410km (4,605 miles) with maximum payload
Load: Up to 10,591kg (23,350 lb)

DOUGLAS DC-8
(U.S.A.)

The DC-8 was produced in direct competition with the Boeing Model 707 turbojet-powered airliner, and the late start of the Douglas programme combined with the availability of only a single fuselage length to produce poorer overall sales. The company produced nine test aircraft with three different types of engine, the first which flew in May 1958. Total production of the initial series was 294, and the primary variants were the DC-8-10 domestic model with Pratt & Whitney JT3C-6 turbojets, the similar DC-8-20 with 'hot-and-high' capability, the DC-8-30 intercontinental model with JT4A-9 turbojets, the similar DC-8-40 with Rolls-Royce Conway Mk 509 turbofans, the DC-8-50 with Pratt & Whitney JT3D turbofans and a rear-ranged cabin for 189 passengers, and the DC-8F Jet Trader based on the DC-8-50 but available in all-freight or convertible freight-passenger lay-outs.

From 1967 production was of the JT3D-powered Super Sixty series, of which 262 were produced. This series comprised the DC-8 Super 61 with an 11.18m (36.67 ft) fuselage stretch for 259 passengers, the DC-8 Super 62 with 1.83m (6.0 ft) span and 2.03m (6.67 ft) fuselage length increases for 189 passengers carried over very long ranges, and the DC-8 Super 63 combininig the Super 61 fuselage and Super 62 wing; these models could be delivered in all-passenger, all-freight, or convertible freight/passenger configurations. Finally came the Super Seventy series, which comprised Super 61, 62 and 63 aircraft converted with General Electric/SNECMA CFM56 turbofans with the designations DC-8 Super 71, 72 and 73 respectively.

DOUGLAS DC-8-63

Role: Long-range passenger transport
Crew/Accommodation: Four and four cabin crew, plus up to 251 passengers
Power Plant: Four 8,618kgp (19,000 lb s.t.) Pratt & Whitney JT3D-7 turbofans
Dimensions: Span 45.24m (148.42 ft); length 57.1m (187 ft); wing area 271.93m^2 (2,927 sq ft)
Weights: Empty 71,401kg (157,409 lb); MTOW 158,760kg (350,000 lb)
Performance: Cruise speed 959km/h (517 knots) at 10,973m (36,000 ft); operational ceiling 12,802m (42,000 ft); range 6,301km (3,400 naut. miles) with full payload
Load: Up to 30,126kg (55,415 lb)

McDonnell Douglas MD-80

DOUGLAS DC-9 and MCDONNELL DOUGLAS SUPER 80/MD-80
(U.S.A.)

The DC-9 was planned initially as a medium-range partner to the long-range DC-8, but was then recast as a short-range type to compete with the BAC One-Eleven. Having learned the sales disadvantages of a single-length fuselage with the DC-8, Douglas planned the DC-9 with various fuselage length options, and decided to optimize the efficiency of the wing with engines pod-mounted on the sides of the rear fuselage under a T-tail. The type first flew in February 1965 and has built up an excellent sales record based on efficiency and fuselage length tailored to customer requirements. However, the success of the type also demanded so high a level of production investment that Douglas was forced to merge with McDonnell.

The variants of the initial production series were the DC-9-10 with Pratt & Whitney JT8D turbofans and 90 passengers, the DC-9-15 with uprated engines, the DC-9-20 for 'hot-and-high' operations with more power and a 1.22m (4 ft) increase in span, the DC-9-30 development of the DC-9-20 but with a 4.54m (14.9 ft) fuselage stretch for 119 passengers, and the DC-9-40 with a 1.92m (6.31 ft) further stretch for 132 passengers, and the DC-9-50 development of the DC-9-30 but with more power and a 2.44m (8 ft) further stretch for 139 passengers.

Developments for the military were the C-9A Nightingale aeromedical transport based on the DC-9-30, and the C-9B Skytrain II fleet logistic transport combining features of the DC-9-30 and -40. Production totalled 976, and from 1975 McDonnell Douglas offered the DC-9 Super Eighty series with a longer fuselage and the refanned JT8D (-200 series) turbofan. This first flew in October 1979, and variants are the DC-9 Super 81 (now MD-81) with JT8D-209s and a 4.34m (14.25 ft) fuselage stretch for 172 passengers, the DC-9 Super 82 (now MD-82) with JT8D-217s, the DC-9 Super 83 (now MD-83) with JT8D-219s and extra fuel, the DC-9 Super 87 (now MD-87) with JT8D-217Bs and a fuselage shortened by 5.0m (16 ft 5 inches), and the DC-9 Super 88 (now MD-88) development of the MD-82 with JT8D-217Cs and an electronic flight instrument system combined with a flight-management computer and inertial navigation system. McDonnell Douglas is now considering a further updated MD-90 series with the latest electronic flight systems and advanced turbofan or propfan power plant.

DOUGLAS DC-9-10

Role: Short-range passenger transport
Crew/Accommodation: Two and three cabin crew, plus up to 90 passengers
Power Plant: Two 6,580kgp (14,500 lb s.t.) Pratt & Whitney JT8D-9 turbofans
Dimensions: Span 27.2m (89.42 ft); length 31.8m (104.42 ft); wing area 86.8m^2 (934.3 sq ft)
Weights: Empty 23,060kg (50,848 lb); MTOW 41,142kg (90,700 lb)
Performance: Cruise speed 874km/h (471 knots) at 9,144m (30,000 ft); operational ceiling 12,497m (41,000 ft); range 2,038km (1,100 naut. miles) with maximum payload
Load: Up to 8,707kg (19,200 lb)

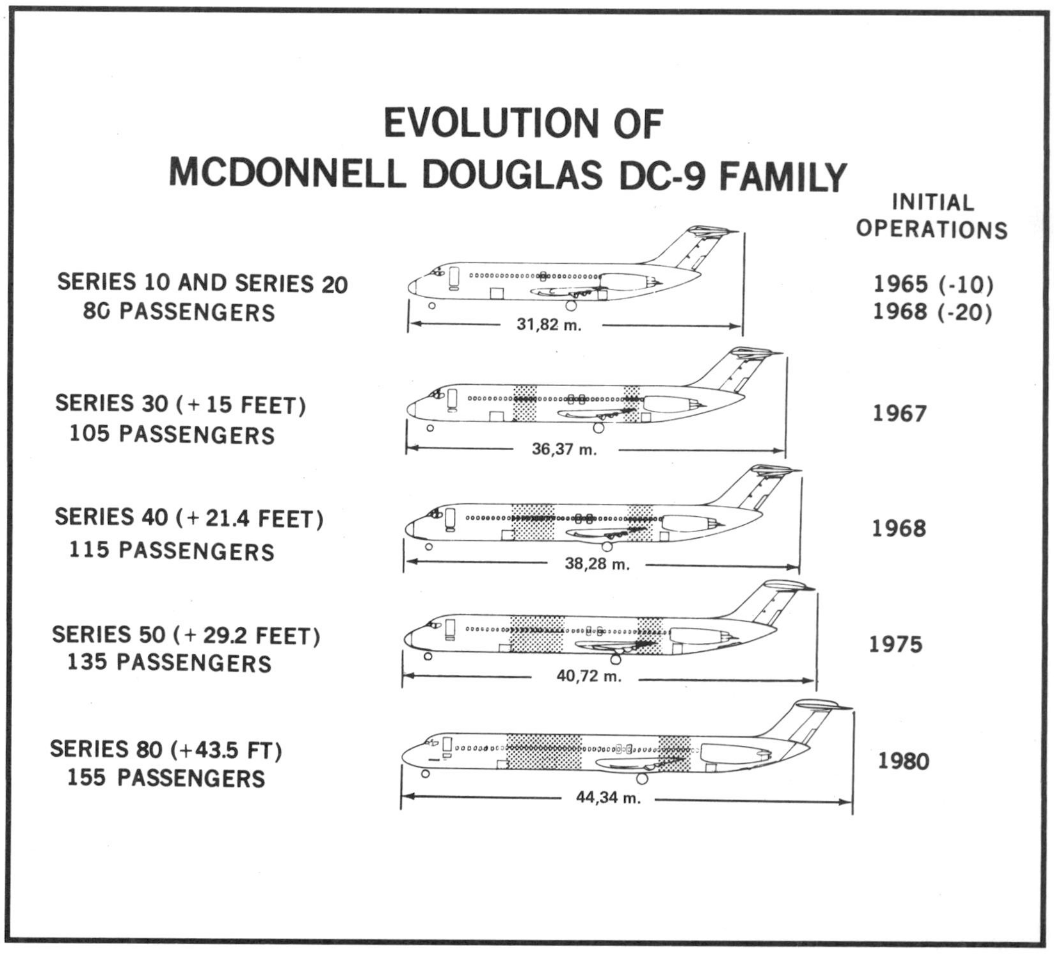

Douglas DC-9-41

DOUGLAS DC-10 and MCDONNELL DOUGLAS MD-11
(U.S.A.)

Design of the DC-10 was started in 1966 to produce a wide-body airliner of middling capacity but long range. The design eventually matured with three engines (one under each wing and the third on a central pylon at the rear of the fuselage with the vertical tail above it), and the first example flew in August 1970. Production totalled 427, the main variants being the DC-10-10 with General Electric CF6-6 turbofans for 380 passengers, the DC-10-10CF convertible freight/passenger model, the DC-10-15 with CF6-50 engines and higher weights, the DC-10-30 intercontinental model with a 3.05m (10 ft) increase in span, extra fuel and a two-wheel additional main landing gear leg between the two standard units, the DC-10-30CF convertible model, the DC-10-30ER extended-range model, the DC-10-40 version of the -30 with Pratt & Whitney JT9D turbofans, and the KC-10A Extender transport/tanker for the U.S. Air Force. The MD-11 is an updated version for the 1990s with drag-reducing winglets on wings extended by 3.05m (10 ft) in span, a fuselage lengthened by 5.66m (18 ft 7 inches) for 405 passengers, a smaller fuel-filled tailplane, advanced avionics and a choice of CF6-80, PW4358 or RB211-524L turbofans.

MCDONNELL DOUGLAS MD-11

Role: Long intermediate-range passenger/cargo transport
Crew/Accommodation: Two, eight cabin crew and up to 323 passengers (2-class, high density seating)
Power Plant: Three 27,896kgp (61,500 lb s.t.) General Electric CF6-80C2-D1F or 27,216kgp (60,000 lb s.t.) Pratt & Whitney PW 4358 or 29,484kgp (65,000 lb s.t.) Rolls-Royce RB211-524L turbofans
Dimensions: Span 51.76m (169.83 ft); length 61.37m (201.33 ft); wing area 342.64^2 (3,688 sq ft)
Weights: Empty 128,462kg (283,131 lb); MTOW 273,290kg (602,500 lb)
Performance: Cruise speed 876km/h (473 knots) at 10,668m (35,000 ft); operational ceiling 12,802m (42,000 ft); range 9,266km (5,000 naut. miles) with 55,792kg (123,000 lb) payload
Load: Up to 70,080kg (154,500 lb) for the Combi version

Both illustrations depict DC-10-30s

DOUGLAS B-18 BOLO
(U.S.A.)

This bomber flew in 1935 as the Douglas Bomber 1 (DB-1), and combined a new fuselage with flying surfaces based on those of the DC-2. The type was ordered into production as the B-18 with 694kW (930 hp) R-1820-45 radial engines. After 133 aircraft, production switched to 217 examples of the B-18A with a lengthened upper nose and 746kW (1,000 hp) R-1820-53 engines. The two variants were replaced in front-line service in 1942, many of the B-18As becoming B-18B (or in Canadian service Digby Mk I) anti-submarine patrollers.

DOUGLAS B-18A BOLO

Role: Medium bomber
Crew/Accommodation: Six
Power Plant: Two 1,000 hp Wright R-1820-53 Cyclone air-cooled radials
Dimensions: Span 27.28m (89.5 ft); length 17.63m (57.83 ft); wing area 89.1m^2 (959 sq ft)
Weights: Empty 7,403kg (16,320 lb); MTOW 12,552kg (27,673 lb)
Performance: Maximum speed 348km/h (216 mph) at 3,050m (10,000 ft); operational ceiling 7,285m (23,900 ft); range 1,450km (990 miles) with 907kg (2,000 lb) bombload
Load: Three .303 inch machine guns, plus up to 1,996kg (4,400 lb) of bombs

DOUGLAS TBD-1 DEVASTATOR

Role: Naval carrierborne torpedo bomber
Crew/Accommodation: Three
Power Plant: One 900 hp Pratt & Whitney R-1830-64 Twin Wasp air-cooled radial
Dimensions: Span 15.24m (50 ft); length 10.67m (35 ft); wing area 39.2m^2 (422 sq ft)
Weights: Empty 2,540kg (5,600 lb); MTOW 4,624kg (10,194 lb)
Performance: Maximum speed 332km/h (206 mph) at 2,440m (8,000 ft); operational ceiling 5,945m (19,500 ft); range 700km (435 miles) with torpedo
Load: Two .303 inch machine guns, plus one 21 inch torpedo, or 544kg (1,200 lb) of bombs

DOUGLAS TBD DEVASTATOR
(U.S.A.)

The TBD was the U.S. Navy's first carrierborne operational monoplane, and resulted from a 1934 requirement for a torpedo bomber. The prototype flew in April 1935, and production of 129 TBD-1 Devastators provided the U.S. Navy with what was for the late 1930s the finest carrierborne torpedo bomber force in the world. However, by 1942 the type was obsolete and relegated to second-line duties.

DOUGLAS DB-7, A-20, P-70 HAVOC and BOSTON SERIES
(U.S.A.)

The Model 7 was a basic two-engined light bomber design that went through a number of important forms during the course of an extensive production programme in World War II. The type originated as a private-venture twin-engined replacement for the U.S. Army's current generation of single-engined attack aircraft, and first flew as the Model 7B in October 1938. Initial orders came from France for a Douglas Bomber 7 (DB-7) variant with more powerful engines and a deeper fuselage, followed by the improved DB-7A. Most of these aircraft were delivered to the U.K. after the fall of France, and were placed in service with the name Boston Mks I and II, though several were converted to Havoc radar-equipped night-fighters. A DB-7B bomber variant with British equipment became the Boston Mk III, and the same basic type was ordered by the U.S. Army as the A-20; these latter were used mainly as reconnaissance aircraft, though a batch was converted to P-70 night-fighter configuration.

DOUGLAS A-20B HAVOC

Role: Light day bomber
Crew/Accommodation: Three
Power Plant: Two 1,600 hp Wright R-2600-11 Double Cyclone air-cooled radials
Dimensions: Span 18.69m (61.33 ft); length 14.48m (47.5 ft); wing area 43.1m^2 (464 sq ft)
Weights: Empty 6,727kg (14,830 lb); MTOW 10,796kg (23,800 lb)
Performance: Maximum speed 563km/h (350 mph) at 3,658m (12,000 ft); operational ceiling 8,717m (28,600 ft); range 1,328km (825 miles) with 454kg (1,000 lb) of bombs
Load: Three .5 inch and one or three .303 inch machine guns, plus up to 1,089kg (2,400 lb) of bombs

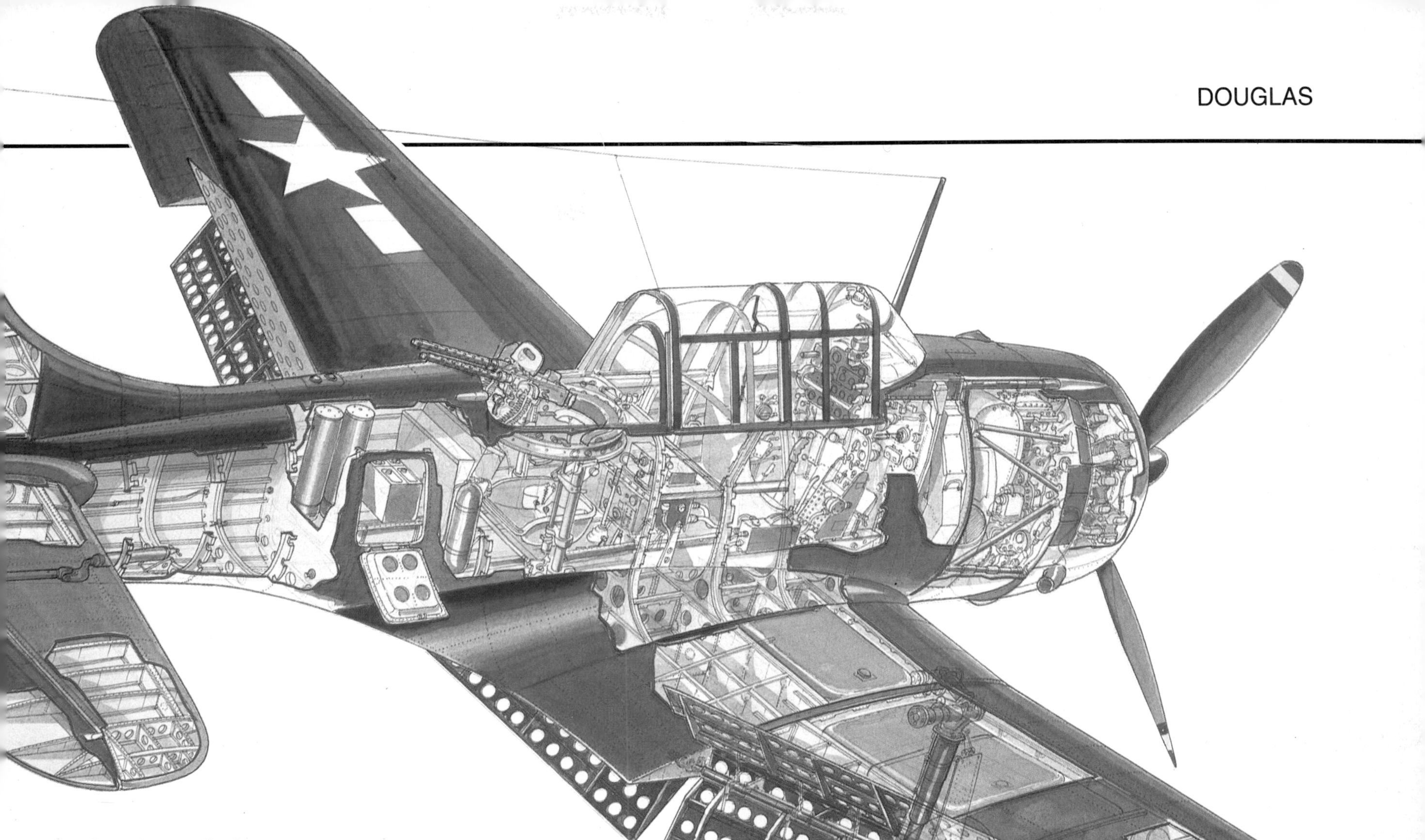

DOUGLAS SBD DAUNTLESS
(U.S.A.)

The Dauntless was the finest dive-bomber produced by the Americans during World War II, and began life as a development of the 1938 Northrop BT-1 after Northrop's acquisition by Douglas. The Douglas development was first flown as the XBT-2 but began to enter U.S. Navy carrierborne and U.S. Marine Corps land-based service as the SBD-1. The SBD-2 had greater fuel capacity and revised armament, the SBD-3 introduced more power and self-sealing fuel tanks, the SBD-4 had a revised electrical system, the SBD-5 had a further increase in power and improved armament, and the SBD-6 had yet more power and increased fuel capacity. The U.S. Army ordered an A-24 version of the SBD-3, further contracts specifying A-24A (SBD-4) and A-24B (SBD-5) aircraft, but these were not successful.

DOUGLAS SBD-5 DAUNTLESS

Role: Naval carrierborne dive bomber
Crew/Accommodation: Two
Power Plant: One 1,200 hp Wright R-1820-60 Cyclone air-cooled radial
Dimensions: Span 12.66m (41.54 ft); length 10.09m (33.1 ft); wing area 30.19m^2 (325 sq ft)
Weights: Empty 2,905kg (6,404 lb); MTOW 4,853kg (10,700 lb)
Performance: Maximum speed 410km/h (255 mph) at 4,265m (14,000 ft); operational ceiling 7,780m (25,530 ft); range 1,795km (1,115 miles) with 726kg (1,600 lb) bombload
Load: Two .5 inch and two .303 inch machine guns, plus up to 1,021kg (2,250 lb) of bombs

DOUGLAS A-26 INVADER
(U.S.A.)

The A-26 resulted from a 1940 U.S. Army requirement for an attack aeroplane, and prototypes were built in three configurations as the XA-26 attack bomber, XA-26A night-fighter and XA-26B attack aeroplane with a 75mm nose cannon; the first of these flew in July 1942 with a bomb-aimer nose. It was a variant of the last that was ordered into production as the A-26B, the nose armament being revised to six 0.5 inch (12.7mm) machine guns and the rest of the gun armament comprising up to 14 similar guns in remotely controlled dorsal and ventral barbettes (two in each) plus underfuselage and underwing packs. The A-26C variant reintroduced a bomber-aimer nose, restricting fixed nose armament to two guns. After World War II these variants became the B-26B and B-26C.

DOUGLAS A-26C INVADER

Role: Fast light bomber
Crew/Accommodation: Three
Power Plant: Two 2,000 hp Pratt & Whitney R-2800-79 Double Wasp air-cooled radials
Dimensions: Span 21.34m (70 ft); length 15.62m (51.25 ft); wing area 50.17m^2 (540 sq ft)
Weights: Empty 10,365kg (22,850 lb); MTOW 15,876kg (35,000 lb)
Performance: Maximum speed 571km/h (355 mph) at 4,570m (15,000 ft); operational ceiling 6,736m (22,100 ft); range 2,255km (1,400 miles) with 1,814kg (4,000 lb) bombload
Load: Six .5 inch machine guns, plus up to 2,722kg (6,000 lb) of bombs

DOUGLAS AD (A-1) SKYRAIDER
(U.S.A.)

One of the classic warplanes of all time, the Skyraider first flew as the massive single-seat XBT2D-1 in March 1945. Despite the imminence of World War II's end, the capabilities of the new aeroplane were such that large-scale production was ordered. This was just as well, for the Skyraider proved an invaluable U.S. tool in the Korean and Vietnam Wars. The type went through a number of major marks; the AD-1 with the 1,864kW (2,500 hp) R-3350-24W radial engine and an armament of two 20mm cannon plus 3,629kg (8,000 lb) of disposable stores, the improved AD-2 with greater fuel capacity, the AD-3 with redesigned canopy and longer-stoke landing gear, the AD-4 with the 2,014kW (2,700 hp) R-3350-26WA and an autopilot, the nuclear-capable AD-4B with four 20mm cannon, the AD-5 anti-submarine search and attack model with a widened fuselage for a side-by-side crew of two, the AD-6 improved version of the AD-4B for highly accurate low-level bombing, and the AD-7 version of the AD-6 with the R-3350-26WB engine and strengthened structure.

DOUGLAS AD-1 SKYRAIDER

Role: Naval carrierborne strike
Crew/Accommodation: One
Power Plant: One 2,500 hp Wright R-3350-24W air-cooled radial
Dimensions: Span 15.24m (50.02 ft); length 12m (39.35 ft); wing area 37.19m^2 (400.3 sq ft)
Weights: Empty 4,749kg (10,470 lb); MTOW 8,178kg (18,030 lb)
Performance: Maximum speed 517km/h (321 mph) at 5,580m (18,300 ft); operational ceiling 7,925m (26,000 ft); range 2,500km (1,554 miles)
Load: Two 20mm cannon, plus up to 2,722kg (6,000 lb) of weapons

DOUGLAS A3D (A-3) SKYWARRIOR and B-66 DESTROYER
(U.S.A.)

The A3D was the heaviest and largest aeroplane to operate from any aircraft-carrier, and was designed to provide the U.S. Navy with a strategic nuclear bombing capability. The XA3D-1 prototype flew in October 1952 with 3,175kg (7,000 lb) thrust Westinghouse XJ40-WE-3 turbojets, but the A3D-1 initial production model had 5,262kg (11,600 lb) thrust Pratt & Whitney J57-P-6 engines, changed in the A3D-2 to 4,763kg (10,500 lb) thrust J57-P-10s. In 1962 these variants became the A-3A and A-3B. The type remains in limited service in the KA-3B tanker and EKA-3B tanker and electronic warfare versions. The U.S. Air Force took a comparable version as the B-66 with Allison J71 turbojets, the main variants being the RB-66A reconnaissance, B-66B bomber, RB-66C electronic reconnaissance and WB-66D weather reconnaissance aircraft.

DOUGLAS A3D-1/A-3A SKYWARRIOR

Role: Naval carrierborne bomber
Crew/Accommodation: Three
Power Plant: Two 5,262kg (11,600 lb s.t.) Pratt & Whitney J57-P-6, plus six 2,041kg (4,500 lb s.t.) JATO rockets for take-off boost
Dimensions: Span 22.1m (72.5 ft); length 23.27m (76.33 ft); wing area 75.44m^2 (812 sq ft)
Weights: Empty 17,876kg (39,409 lb); MTOW 37,195kg (82,000 lb)
Performance: Maximum speed 982km/h (530 knots) at 3,050m (10,000 ft); operational ceiling 12,495m (41,000 ft); range 1,690km (912 naut. miles) with full warload
Load: Two 20mm cannon, plus up to 5,443kg (12,000 lb) of ordnance carried internally

DOUGLAS A4D (A-4) SKYHAWK
(U.S.A.)

Another great aeroplane from the Douglas stable, the Skyhawk first flew in June 1954 as the XA4D-1 successor to the AD Skyraider. The first version was the A4D-1 (from 1962 A-4A) with the 3,493kg (7,700 lb) thrust Wright J65-W-4 turbojet plus an armament of two 20mm cannon and 2,268kg (5,000 lb) of disposable stores. The main successor variants were the A4D-2 (A-4B) with inflight-refuelling capability, the A4D-2N (A-4C) with terrain-following radar, the A4D-5 (A-4E) with the 3,856kg (8,500 lb) thrust Pratt & Whitney J52-P-6 turbojet and two additional hardpoints for a 3,719kg (8,200 lb) disposable load, the A-4F with a dorsal hump for more electronics, the A-4H version of the A-4E for Israel with 30mm cannon and upgraded electronics, the A-4M with an enlarged dorsal hump, and the A-4N development of the A-4M for Israel. There have been a number of TA-4 trainer models. The late 1980s saw a considerable boom in upgraded aircraft, often with a General Electric F404 turbofan.

An 4-AK of the Royal New Zealand Air Force

DOUGLAS A4D-5/A-4E SKYHAWK

Role: Naval carrierborne strike
Crew/Accommodation: One
Power Plant: One 3,856kg (8,500 lb s.t.) Pratt & Whitney J52-P-6A turbojet
Dimensions: Span 8.38m (27.5 ft); length 12.23m (40.125 ft); wing area 24.16m^2 (260 sq ft)
Weights: Empty 4,469kg (9,853 lb); MTOW 11,113kg (24,500 lb)
Performance: Maximum speed 1,083km/h (584 knots) at sea level; operational ceiling 11,460m (37,600 ft); range 1,865km (1,006 naut. miles) with 1,451kg (3,200 lb) bombload
Load: Two 20mm cannon, plus up to 3,719kg (8,200 lb) of weapons

DOUGLAS F4D (F-6) SKYRAY
(U.S.A.)

In common with other American delta-winged aircraft of the period, the Skyray, planned as a carrierborne interceptor, drew extensively in design on captured German research data of World War II. The XF4D-1 prototype first flew in January 1951 with a 2,268kg (5,000 lb) Allison J35-A-17 turbojet, but was subsequently re-engined with the intended Westinghouse XJ40. This engine suffered traumatic development problems, and the F4D-1 finally matured with 6,123kg (13,500 lb) afterburning thrust Pratt & Whitney J57-P-2, later changed to the more powerful J57-P-8. In 1962 the type was redesignated F-6A.

DOUGLAS F4D-1 SKYRAY

Role: Naval carrierborne interceptor
Crew/Accommodation: One
Power Plant: One 7,258kgp (16,000 lb s.t.) Pratt & Whitney J57-P-8B turbojet with reheat
Dimensions: Span 10.21m (33.5 ft); length 13.84m (45.4 ft); wing area 51.74^2 (557 sq ft)
Weights: Empty 7,268kg (16,024 lb); MTOW 12,701kg (28,000 lb) maximum catapult limited
Performance: Maximum speed 1,154km/h (717 mph) at sea level; operational ceiling 11,460m (37,600 ft); range 1,803km (1,120 miles) at altitude
Load: Four 20mm cannon, plus four short-range air-to-air missiles

EMBRAER EMB-110 and EMB-111 BANDEIRANTE
(Brazil)

Designed as a utility light transport, the EMB-110 marked the beginning of Brazil's emergence as a major aircraft-manufacturing power, and first flew in October 1968. The type has been produced in a number of EMB-110 civil variants, and has also been sold to the Brazilian Air Force as the C-95. There is also an EMB-111 coastal patrol version.

EMBRAER EMB-110P1A BANDEIRANTE

Role: Short-range passenger/cargo transport
Crew/Accommodation: Two, plus up to 19 passengers
Power Plant: Two 750 shp Pratt & Whitney PT6A-34 turboprops
Dimensions: Span 15.32m (50.26 ft); length 15.08m (49.47 ft); wing area 29.1m^2 (313 sq ft)
Weights: Empty 3,630kg (8,010 lb); MTOW 5,900kg (13,010 lb)
Performance: Cruise speed 393km/h (244 mph) at 2,438m (8,000 ft); operational ceiling 5,791m (19,000 ft); range 371km (230 miles) with 19 passengers
Load: Up to 1,565 kg (3,450 lb) passenger version, or 1,724 kg (3,800 lb) cargo-carrier

EMBRAER EMB-312 TUCANO
(Brazil)

Now one of the world's best-selling basic trainers, the EMB-312 was planned to a Brazilian Air Force requirement and first flew in August 1980 with a 559kW (750 shp) Pratt & Whitney Canada PT6A-25 turboprop. The type is used by the Brazilian Air Force with the designation T-27. It is also produced under licence by Shorts in Northern Ireland as the RAF's Tucano T.Mk 1 with a number of improvements including an 820kW (1,000 shp) Garrett TPE331 turboprop.

EMBRAER EMB-312 TUCANO

Role: Light strike/advanced trainer
Crew/Accommodation: Two
Power Plant: One 750 shp Pratt & Whitney of Canada PT6A-25C turboprop
Dimensions: Span 11.14m (36.55 ft); length 9.86m (32.35 ft); wing area 19.4m^2 (208.8 sq ft)
Weights: Empty 1,582kg (3,487 lb); MTOW 3,175kg (7,000 lb)
Performance: Maximum speed 469km/h (253 knots) at 2,743m (9,000 ft); operational ceiling 8,686m (28,500 ft); range 1,897km (1,024 naut. miles) on internal fuel only
Load: Up to 1,000 kg (2,204 lb) of external weapons

ENGLISH ELECTRIC CANBERRA
(U.K.)

The Canberra was planned as a nuclear-capable medium bomber with turbojet engines. The type was originally schemed as a high-altitude type, and was therefore designed round a large wing and a crew of two using a radar bombing system. The Canberra in fact matured as a medium/high-altitude type with optical bomb aiming by a third crew member, and first flew in May 1949. For its time the aeroplane was highly advanced and capable, and its development potential ensured that the type enjoyed a long first-line career as well as diversification into other roles. The main bomber stream began with the Canberra B.Mk 2 powered by 2,948kg (6,500 lb) Avon RA.3 Mk 101 turbojets, and then advanced to the B.Mk 6 with greater fuel capacity and 3,357kg (7,400 lb) thrust Avon Mk 109s, the B.Mk 15 conversion of the B.Mk 6 with underwing hardpoints for two 454kg (1,000 lb) bombs for rocket pods, the B.Mk 16 improved version of the B.Mk 15, and the B.Mk 20 Australian-built version of the B.Mk 6; there were also many export versions. The intruder/interdictor series began with the Canberra B(I).Mk 6 version of the B.Mk 6 with underwing bombs and a ventral cannon pack, and then continued with the B(I).Mk 8 multi-role version with air-to-surface missile capability; there were also several export versions.

The reconnaissance models began with the Canberra PR.Mk 3 based on the B.Mk 2, and then moved through variants including the PR.Mk 7 equivalent of the B.Mk 6, and the PR Mk 9 high-altitude model with increased span, extended centre-section chord, powered controls and 4,990kg (11,000 lb) thrust Avon Mk 206s; there were also a few export models. Other streams included trainer, target tug and remotely controlled target drone models, and apart from the first-line PR aircraft the machines that still survive are used mainly for electronic roles. The importance of the Canberra is also attested by the fact that it became the first non-U.S. type to be manufactured under licence in the U.S.A. after World War II.

ENGLISH ELECTRIC CANBERRA B. Mk 2

Role: Bomber reconnaissance
Crew/Accommodation: Two
Power Plant: Two 2,948kgp (6,500 lb s.t.) Rolls-Royce Avon RA.3 Mk 101 turbojets
Dimensions: Span 19.49m (63.96 ft); length 19.96m (65.5 ft); wing area 89.2m^2 (960 sq ft)
Weights: Empty 10,070kg (22,200 lb); MTOW 20,865kg (46,000 lb)
Performance: Maximum speed 917km/h (570 mph) at 12,192m (40,000 ft); operational ceiling 14,630m (48,000 ft); range 4,281km (2,660 miles)
Load: Up to 2,722kg (6,000 lb) of ordnance all carried internally

ENGLISH ELECTRIC LIGHTNING F.Mk 6

Role: Interceptor fighter
Crew/Accommodation: One
Power Plant: Two 7,420 kgp (16,360 lb s.t.) Rolls-Royce Avon 300 turbojets with reheat
Dimensions: Span 10.61m (34.9 ft); length 16.84m (55.25 ft); wing area 44.08m^2 (474.5 sq ft)
Weights: Empty 11,340kg (25,000 lb); MTOW 18,144kg (40,000 lb)
Performance: Maximum speed 2,230km/h (1,203 knots) Mach 2.1 at 10,975m (36,000 ft); operational ceiling 17,375m (57,000 ft); radius 972km (604 miles)
Load: Two Red Top missiles, plus two 30mm Aden cannon

▼

ENGLISH ELECTRIC LIGHTNING
(U.K.)

Derived from the P.1A research aeroplane, the Lightning was the U.K.'s first supersonic fighter, a fast-climbing interceptor that was always limited by poor range and inadequate armament. The P.1B first flew in April 1957, and after a protracted development the type began to enter service in 1959 as the Lightning F.Mk 1 with two 30mm cannon and two Firestreak air-to-air missiles. Subsequent variants were the F.Mk 1 with inflight-refuelling capability, the F.Mk 2 with improved electronics and fully variable afterburners, the F.Mk 3 with 7,420kg (16,360 lb) thrust Avon Mk 300 series engines, provision for overwing drop tanks, a square-topped vertical tail, improved radar, no guns, and Red Top air-to-air missiles. The final variant was the F.Mk 6 with a revised wing and ventral tank that also accommodated a pair of 30mm cannon. There were also Lightning T.Mks 2 and 4 combat-capable trainers, and Mk 50 series fighters and trainers for export.

ETRICH TAUBE (DOVE)
(Austria-Hungary/Germany)

So named because of its dove-like wing planform, the Taube was first flown during 1910 in Austria-Hungary, and entered military service during 1912 in a two-seat form. The type was built in at least 10 Austro-Hungarian and German factories. Austro-Hungarian machines were the A I with the 63.4kW (85 hp) Astro-Daimler engine and the A II with a 89.5kW (120 hp) engine; German machines generally had the 74.6kW (100 hp) Argus or Mercedes engines, and the type remained in service up to 1916.

ETRICH TAUBE

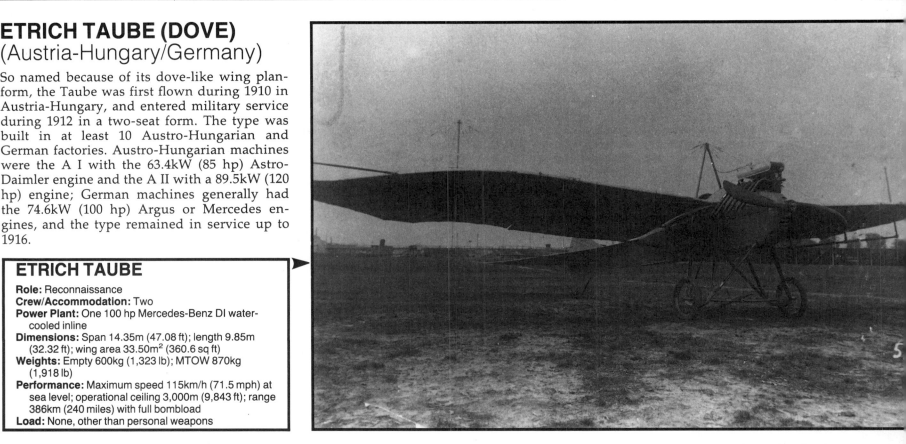

Role: Reconnaissance
Crew/Accommodation: Two
Power Plant: One 100 hp Mercedes-Benz DI water-cooled inline
Dimensions: Span 14.35m (47.08 ft); length 9.85m (32.32 ft); wing area 33.50m^2 (360.6 sq ft)
Weights: Empty 600kg (1,323 lb); MTOW 870kg (1,918 lb)
Performance: Maximum speed 115km/h (71.5 mph) at sea level; operational ceiling 3,000m (9,843 ft); range 386km (240 miles) with full bombload
Load: None, other than personal weapons

EUROFIGHTER EFA
(Italy/Spain/U.K./West Germany)

The EFA is being developed in a collaborative project to produce an advanced combat aeroplane. In overall design the EFA resembles the Dassault-Breguet Rafale, but is a larger and potentially more capable multi-role type with advanced avionics and weapons. The type is due to fly in 1991.

EUROFIGHTER EFA

Role: All-weather multi-mission fighter
Crew/Accommodation: One
Power Plant: Two 9,299kgp (20,500 lb s.t.) Eurojet EJ200 turbofans with reheat
Dimensions: Span 10.50m (34.45 ft); length 14.50m (47.57 ft); wing area 50m^2 (538.2 sq ft)
Weights: Empty 9,750kg (21,495 lb); MTOW 17,000kg (37,480 lb)
Performance: Maximum speed 1,913km/h (1,032 knots) Mach. 1.8+ at 11,000m (36,090 ft); operational ceiling 15,000+m (49,213+ ft); range 556km (300 naut. miles) with full warload
Load: One 27mm cannon, plus up to 4,500kg (9,920 lb) of externally-carried armament

EUROPEAN HELICOPTER INDUSTRIES EH.101
(Italy/U.K.)

The EH.101 first flew in October 1987 and is intended to become the major ship- and land-based anti-submarine and anti-ship helicoper of the British and Italian navies, the former using the Merlin version with Rolls-Royce/Turbomeca RTM.322 turboshafts and the latter a variant modelled more closely on the prototypes with General Electric T700-GE-401A turboshafts. The naval version will have advanced electronics and weapons. The manufacturers also plan a 30-troop battlefield utility model and a civil medium-lift transport.

EUROPEAN HELICOPTER INDUSTRIES EH.101 (ASW VERSION)

Role: Naval shipborne anti-submarine/ship helicopter
Crew/Accommodation: Four
Power Plant: Three 1,714 shp General Electric T700-GE-401A turboshafts
Dimensions: Overall length rotors turning 22.81m (74.83 ft); rotor diameter 18.59m (61 ft)
Weights: Empty 7,121kg (15,700 lb); MTOW 13,000kg (28,660 lb)
Performance: Cruise speed 259km/h (140 knots) at sea level; endurance 5 hours
Load: Up to 960 kg (2,116 lb)

FAIRCHILD 71
(U.S.A.)

The Model 71 remains a good example of Fairchild's forte of the 1920s and early 1930s, namely the rugged high-wing utility monoplane. The Model 71 was in production from 1928 to 1930, and was in essence an updated version of the FC-2W2 with a 313kW (420 hp) Pratt & Whitney Wasp radial engine for the movement of a pilot and up to six passengers. The Model 71 was followed by the product-improved Model 71A, and the two types were also bought in small numbers by the U.S. Army as the C-8 and F-1A (later C-8A). The company's Canadian subsidiary built the Model 71-C and Model 71-CM, the latter with a metal-skinned fuselage. The Super 71 was an enlarged floatplane version with accommodation for eight passengers on a 388kW (520 hp) Wasp.

FAIRCHILD 71-C

Role: Passenger/utility transport
Crew/Accommodation: One, plus up to six passengers
Power Plant: One 420 hp Pratt & Whitney R-1340 Wasp B air-cooled radial
Dimensions: Span 15.24m (50 ft); length 10.01m (32.85 ft); wing area 30.84m^2 (332 sq ft)
Weights: Empty 1,432kg (3,158 lb); MTOW 2,722kg (6,000 lb)
Performance: Cruise speed 171km/h (106 mph) at sea level; operational ceiling 4,724m (15,500 ft); range 322km (200 miles) with full payload
Load: Up to 626kg (1,380 lb).
Note: the length quoted is for the landplane version, as are the weights

FAIRCHILD C-82 PACKET and C-119 FLYING BOXCAR
(U.S.A.)

Design of a specialized freighter for the U.S. Army began in 1941, and the XC-82 prototype was first flown in September 1944. It was a high-wing monoplane with twin booms supporting the empennage so that clamshell rear doors could provide access to the central payload nacelle, which could accommodate 78 passengers (42 paratroops) or freight. The only production version was the C-82A with 1,566kW (2,100 hp) Pratt & Whitney R-2800-34 radial engines, but the basic concept was then developed into the C-119 with more power and the cockpit relocated to the nose of the nacelle. This was ordered into production as the C-119B with a widened fuselage for the carriage of 62 paratroops or freight. Other variants of various improved types followed with revised flying surfaces, rear doors openable in flight, more power and, in some late aircraft, two 1,293kg (2,850 lb) thrust General Electric J85-GE-17 booster turbojets in underwing nacelles.

FAIRCHILD C-119C FLYING BOXCAR

Role: Military bulk freight/paratroop transport
Crew/Accommodation: Four, plus up to 42 paratroops
Power Plant: Two 3,500 hp Pratt & Whitney R-4360-20 Wasp Major air cooled radials
Dimensions: Span 33.32m (109.25 ft); length 26.37m (86.5 ft); wing area 134.4m^2 (1,447 sq ft)
Weights: Empty 18,053kg (39,800 lb); MTOW 33,566kg (74,000 lb)
Performance: Maximum speed 452km/h (281 mph) at 5,486m (18,000 ft); operational ceiling 7,285m (23,900 ft); range 805km (500 miles) with maximum load
Load: Up to 8,346kg (18,400 lb)

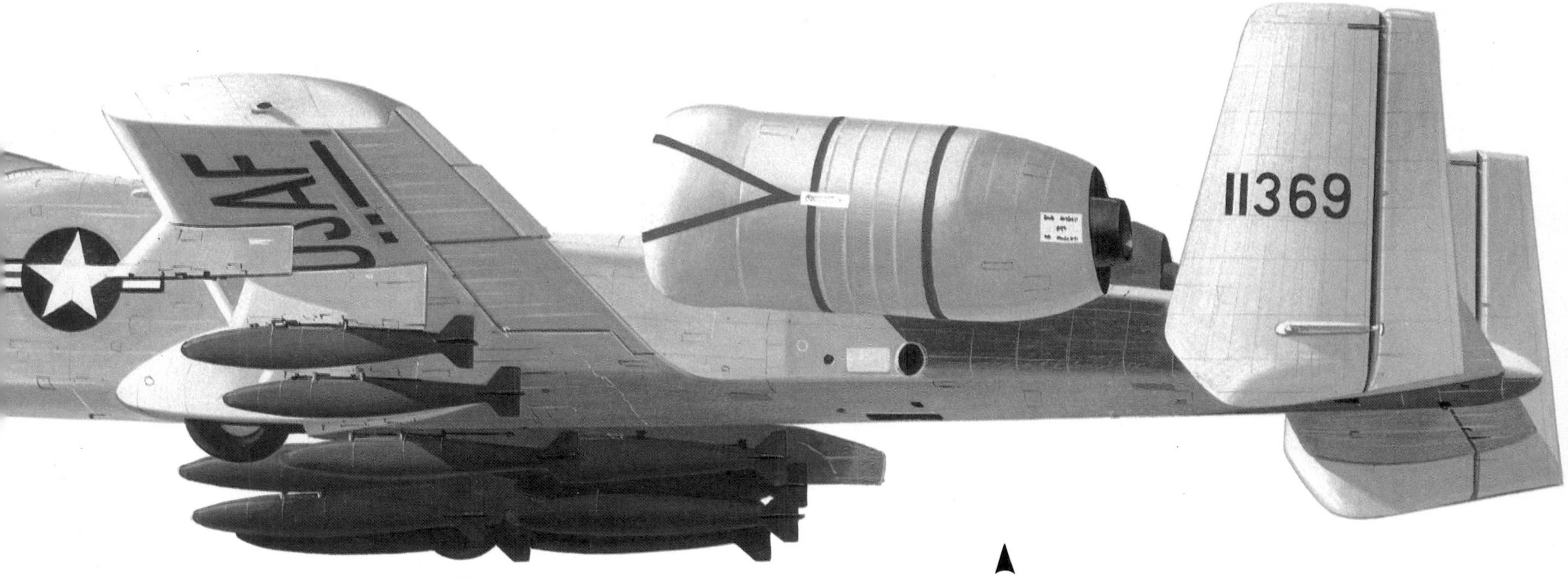

FAIRCHILD REPUBLIC A-10 THUNDERBOLT II
(U.S.A.)

First flown in May 1972 as the YA-10A after Fairchild had become a division of Republic Aviation, the Thunderbolt II is a specialist close-support and anti-tank aeroplane whose peculiar configuration was dictated by its role. All major systems are duplicated, extensive armour is carried, and vulnerable systems such as the engines are shielded as much as possible from ground detection and thus from ground fire. The core of the A-10A production version is the massive GAU-8/A seven-barrel cannon firing depleted uranium anti-tank ammunition, but a wide assortment of other weapons can be carried on no fewer that 11 hardpoints.

FAIRCHILD REPUBLIC A-10 THUNDERBOLT II

Role: Anti-armour close air support
Crew/Accommodation: One
Power Plant: Two 4,112kgp (9,065 lb s.t.) General Electric TF34-GE-100 turbofans
Dimensions: Span 17.53m (57.5 ft); length 16.25m (53.33 ft); wing area 47.01m^2 (506 sq ft)
Weights: Empty 9,006kg (19,856 lb); MTOW 22,221kg (46,786 lb)
Performance: Maximum speed 697km/h (433 mph) at sea level; operational ceiling 10,575m (34,700 ft); radius 974km (605 miles) with 4,327kg (9,540 lb) bombload
Load: One 30mm multi-barrel cannon, plus up to 7,250kg (16,000 lb) of externally-carried weapons

▲ FAIREY FLYCATCHER
(U.K.)

Immensely angular but highly manoeuvrable, the Flycatcher carrierborne fighter was first flown in prototype form during November 1922 and served from 1923 to 1934. The Flycatcher Mk I production version was generally operated with wheeled landing gear, but could be fitted with twin floats.

FAIREY FLYCATCHER

Role: Naval carrierborne fighter
Crew/Accommodation: One
Power Plant: One 400 hp Armstrong Siddeley Jaguar air-cooled radial
Dimensions: Span 8.84m (29 ft); length 7.01m (23 ft); wing area 26.76m² (288 sq ft)
Weights: Empty 926kg (2,038 lb); MTOW 1,375kg (3,028 lb)
Performance: Maximum cruise speed 216km/h (134 mph) at sea level; operational ceiling 6,280m (20,600 ft); endurance 1.82 hours
Load: Two .303 inch machine guns, plus up to 36kg (80 lb) of bombs
Note: the above characteristics apply to the wheel-equipped aircraft, whose performance was generally superior to the floatplane

FAIREY FOX
(U.K.) ▼

The Fox was designed to capitalize on the low frontal area of the Curtiss D-12 inline engine, for whose production Fairey secured a licence in 1923. The type was planned as a high-performance day bomber, and first flew in prototype form during January 1925, achieving a speed some 64km/h (40 mph) greater than that of the current Fairey Fawn with its massive Napier Lion engine. Production followed, the Fox Mk I having the 336kW (450 hp) D-12 engine, before being re-engined, because of RAF doubts about the American engine, with the Rolls-Royce Kestrel as the Fox Mk IA. The main builder of the Fox, however, was Avions Fairey in Belgium, which developed several variants including the definitive Fox Mk VI with the 641kW (860 hp) Hispano-Suiza 12Y inline and an enclosed cockpit.

FAIREY FOX Mk IA

Role: Light day bomber
Crew/Accommodation: Two
Power Plant: One 530 hp Rolls-Royce F XII A/Kestrel II A water-cooled inline
Dimensions: Span 11.48m (37.66 ft); length 8.61m (28.25 ft); wing area 30.1m² (324 sq ft)
Weights: Empty 1,183kg (2,609 lb); MTOW 2,105kg (4,640 lb)
Performance: Maximum speed 257km/h (160 mph) at sea level; operational ceiling 5,182m (17,000 ft); range 805km (500 miles)
Load: Two .303 inch machine guns, plus up to 209kg (460 lb) of bombs

FAIREY SWORDFISH
(U.K.)

Designed as a carrierborne torpedo, strike and reconnaissance aeroplane, the Swordfish was flown in T.S.R.2 prototype form during April 1934. It remained in service right through to the end of World War II, providing one of the finest examples of a type that for a variety of reasons remained valuable despite its technical obsolescence. The Swordfish could be carried on wheel or float landing gear, and the Swordfish Mk I variant entered service with the 515kW (690 hp) Pegasus IIIM radial engine. In 1943 there appeared the Swordfish Mk II with metal-skinned lower wings for the carriage of underwing rockets, and this was followed by the Swordfish Mk III anti-ship and anti-submarine variant with ASV.Mk XV search radar.

FAIREY SWORDFISH Mk I

Role: Naval carrierborne strike reconnaissance
Crew/Accommodation: Three
Power Plant: One 690 hp Bristol Pegasus IIIM3 air-cooled radial
Dimensions: Span 13.87m (45.5 ft); length 10.87m (35.66 ft); wing area 56.4m^2 (607 sq ft)
Weights: Empty 1,905kg (4,195 lb); MTOW 3,505kg (7,720 lb)
Performance: Maximum speed 246km/h (154 mph) at sea level; operational ceiling 5,870m (19,250 ft); endurance 5.7 hours
Load: Two .303 inch machine guns, plus up to a single 730kg (1,610 lb) torpedo
Note: the above characteristics apply to the wheel-equipped aircraft, the performance of the floatplane being somewhat inferior

FAIREY BARRACUDA
(U.K.)

The Barracuda was designed as a carrierborne torpedo- and dive-bomber, and first flew in prototype form during December 1940 with a 969kW (1,300 hp) Rolls-Royce Merlin 30. After early trials the tailplane was moved from its original low-set position to a strut-braced location towards the top of the fin, and in this form a few Barracuda Mk Is were built. These were so underpowered that the 1,223kW (1,640 hp) Merlin 32 was used in the Barracuda Mk II, which was the chief variant. Other models were the Barracuda Mk III with ASV radar in a radome under the rear fuselage, and the ultimate and considerably more potent Barracuda Mk V with the 1,514kW (2,030 hp) Rolls-Royce Griffon 37, a square-cut wing and a larger fin.

FAIREY BARRACUDA Mk II

Role: Naval carrierborne strike
Crew/Accommodation: Three
Power Plant: One 1,640 hp Rolls-Royce Merlin 32 water-cooled inline
Dimensions: Span 14.99m (49.18 ft); length 12.12m (39.75 ft); wing area 37.62m^2 (405 sq ft)
Weights: Empty 4,241kg (9,350 lb); MTOW 6,395kg (14,100 lb)
Performance: Maximum speed 367km/h (228 mph) at 533m (1,750 ft); operational ceiling 5,060m (16,600 ft); range 843km (524 miles) with full warload
Load: Two .303 inch machine guns, plus up to 816kg (1,800 lb) of ordnance

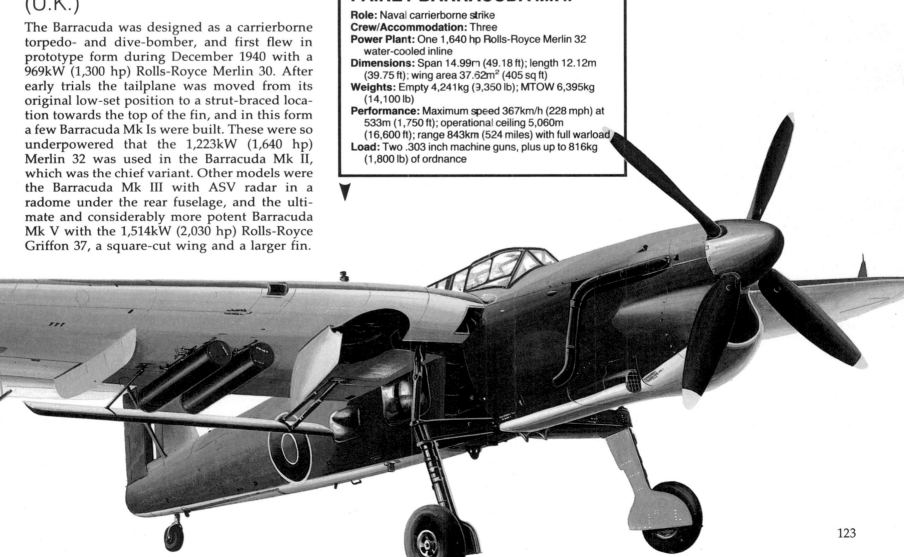

FAIREY FIREFLY
(U.K.)

Designed to a requirement for a carrierborne two-seat reconnaissance fighter and first flown in December 1941, the Firefly was one of the Royal Navy's most successful warplanes of the 1940s. The Firefly Mk I initial production series featured 13.55m (44.5 ft) span wings and the 1,484kW (1,990 hp) Rolls-Royce Griffon XII with a chin radiator, and was produced in F.Mk I fighter, FR.Mk I fighter reconnaissance, NF.Mk I night-fighter and T.Mk I trainer versions. The Firefly Mk IV switched to the 1,566kW (2,100 hp) Griffon 61 with root radiators in a wing spanning 12.55m (41 ft 2 inches), and was produced in F.Mk IV and FR.Mk 4 versions. The Firefly Mk 5 introduced power-folding wings, and was produced in FR.Mk 5, NF.Mk 5, T.Mk 5 and anti-submarine AS Mk 5 versions. The AS. Mk 6 was identical to the AS.Mk 5 other than its use of British rather than American sonobuoys. The final production model was the Firefly AS.Mk 7, which had the original long-span wing and a 1,678kW (2,250 hp) Griffon 59 with a chin radiator. Surplus Fireflies were also converted as remotely controlled target drones for the British surface-to-air missile tests.

FAIREY FIREFLY FR. Mk 5

Role: Fighter reconnaissance
Crew/Accommodation: Two
Power Plant: One 2,250 hp Rolls-Royce Griffon 74 water-cooled inline
Dimensions: Span 12.55m (41.17 ft); length 11.56m (37.91 ft); wing area 30.65m² (330 sq ft)
Weights: Empty 4,389kg (9,674 lb); MTOW 6,114kg (13,479 lb)
Performance: Maximum speed 618km/h (386 mph) at 4,270m (14,000 ft); operational ceiling 8,660m (28,400 ft); range 2,090km (1,300 miles) with long-range tankage
Load: Four 20mm cannon, plus up to 454kg (1,000 lb) of externally underslung bombs

FAIREY GANNET
(U.K.)

Designed as a carrierborne anti-submarine aeroplane and first flown in September 1949, the Gannet was very compact but had a capacious fuselage and the advantages of the Double Mamba power plant. The latter's side-by-side turboprops each drove one of the contra-props, one of these could be closed down for economical cruise. The first model was the Gannet AS.Mk 1 with the 2,200kW (2,950 ehp) Double Mamba Mk 100, and this power plant was also used in the Gannet T.Mk 2 Trainer. The 2,889kW (3,875 ehp) Double Mamba Mk 102 was used in the Gannet AEW.Mk 3 early warner with underfuselage surveillance radar, while the Gannet AS.Mk 4 and T.Mk 5 had the 2,263kW (3,035 ehp) Double Mamba Mk 101.

FAIREY GANNET AS. Mk 4

Role: Naval carrierborne anti-submarine search and strike
Crew/Accommodation: Three
Power Plant: One 3,035 ehp Armstrong Siddeley Double Mamba Mk 101 turboprop
Dimensions: Span 16.56m (54.33 ft); length 13.11m (43 ft); wing area 46.08² (496 sq ft)
Weights: Empty 6,382kg (44,069 lb); MTOW 10,127kg (22,327 lb)
Performance: Maximum speed 482km/h (299 mph) at sea level; operational ceiling 7,620m (25,000 ft); endurance 4.9 hours
Load: No gun armament, but provision to carry up to 907kg (2,000 lb) of ordnance internally

Fairey Gannet AEW Mk.3

FAIREY F.D.2
(U.K.)

The Fairey Delta 2 was designed as a supersonic research aeroplane with a pure delta wing, and first flew in October 1954. The two aircraft undertook a mass of research work, but the type is best remembered for taking the world air speed record past 1,609km/h (1,000 mph) mark on 10 March 1956, when the first aeroplane attained 1,822km/h (1,132 mph). The machine was later revised as the BAC 221 with an ogival wing to test this planform before its use on the Concorde supersonic airliner.

FAIREY F.D.2

Role: Supersonic research
Crew/Accommodation: One
Power Plant: One 4,309 kgp (9,500 lb s.t.) Rolls-Royce Avon RA14R turbojet with reheat — the use of which was limited, but gave 5,386 kgp (11,875 lb s.t.) at 11,580m (38,000 ft)
Dimensions: Span 8.18m (26.83 ft); length 15.74m (51.62 ft); wing area 38.4m² (360 sq ft)
Weights: Empty 5,000kg (11,000 lb); MTOW 6,298kg (13,884 lb)
Performance: Maximum speed 1,822km/h (1,132 mph) or Mach. 1.731 at 11,582m (38,000 ft); operational ceiling 14,021m (46,000 ft); range 1,335km (830 miles) without reheat
Load: Confined to specialized test equipment

FARMAN M.F.7 'LONGHORN' and M.F.11 'SHORTHORN'
(France)

First flown in 1913, the M.F.7 was a simple utility aeroplane with a forward-mounted elevator assembly that resulted in the nickname 'Longhorn'. The type was originally powered by a 52kW (70 hp) Renault engine, but was also built under licence in the U.K. with a 56kW (75 hp) Rolls-Royce Hawk engine. The M.F.11 was a more refined aeroplane that did away with the forward elevator and thus became known as the 'Shorthorn'. The type was first flown with a 52kW (70 hp) Renault, but many other engines of up to 97kW (130 hp) were also used.

FARMAN M.F.11 'SHORTHORN'

Role: Reconnaissance/trainer
Crew/Accommodation: Two
Power Plant: One 80 hp Renault air-cooled inline
Dimensions: Span 15.776m (51.75 ft); length 9.3m (30.50 ft); wing area 52m² (561 sq ft)
Weights: Empty 654kg (1,441 lb); MTOW 928kg (2,046 lb)
Performance: Maximum speed 116km/h (72 mph) at sea level); operational ceiling 2,000+m (6,560+ ft); endurance 3.75 hours)
Load: One .303 inch machine gun (on later aircraft only)

Not illustrated

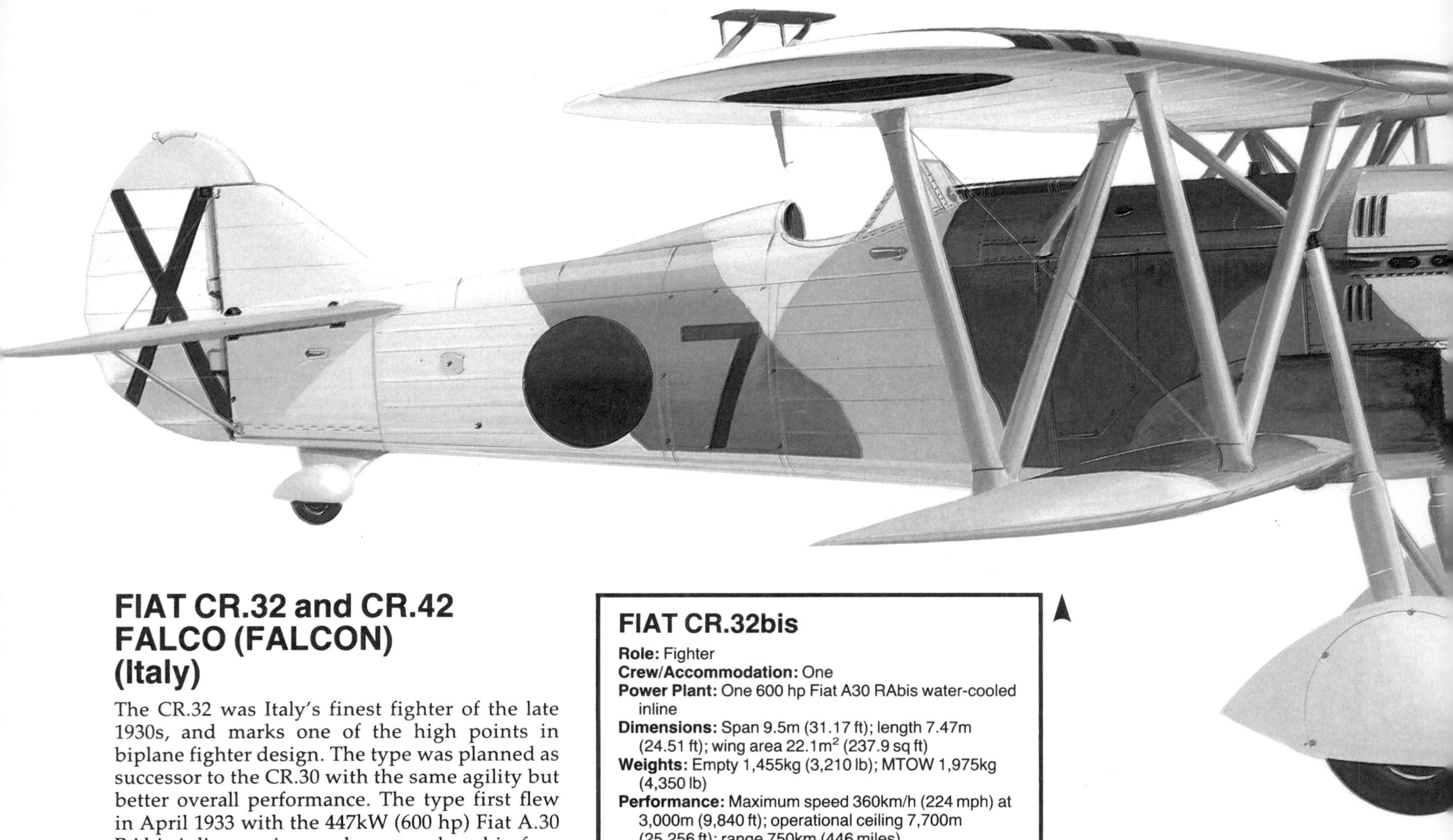

FIAT CR.32 and CR.42 FALCO (FALCON) (Italy)

The CR.32 was Italy's finest fighter of the late 1930s, and marks one of the high points in biplane fighter design. The type was planned as successor to the CR.30 with the same agility but better overall performance. The type first flew in April 1933 with the 447kW (600 hp) Fiat A.30 RAbis inline engine, and was produced in four series. Further development led to the CR.42 with a number of aerodynamic refinements, cantilever main landing gear units and more power in the form of a 626kW (840 hp) Fiat A.74 R1C radial engine.

FIAT CR.32bis

Role: Fighter
Crew/Accommodation: One
Power Plant: One 600 hp Fiat A30 RAbis water-cooled inline
Dimensions: Span 9.5m (31.17 ft); length 7.47m (24.51 ft); wing area 22.1m^2 (237.9 sq ft)
Weights: Empty 1,455kg (3,210 lb); MTOW 1,975kg (4,350 lb)
Performance: Maximum speed 360km/h (224 mph) at 3,000m (9,840 ft); operational ceiling 7,700m (25,256 ft); range 750km (446 miles)
Load: Two 12.7mm and two 7.7mm machine guns, plus provision to carry up to 100kg (220 lb) of bombs

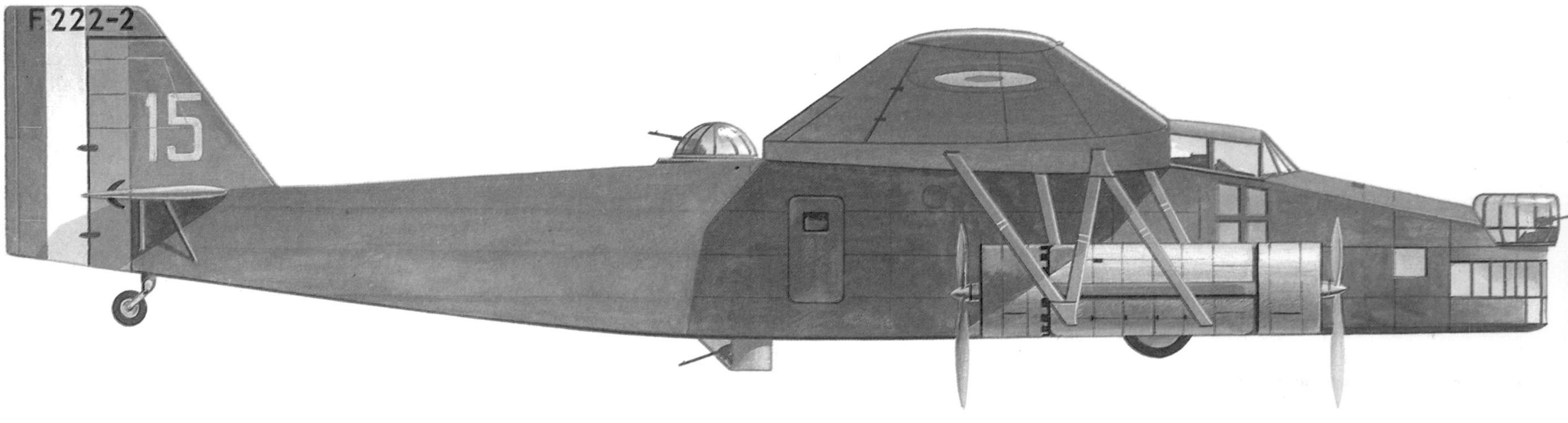

FARMAN F.220 SERIES
(France)

This series was typical of France's apparent predilection for heavy aircraft of singular angularity. The F.220 had a slab-sided fuselage, square-cut flying surfaces, extensive strutting and push/pull engines in nacelles at the tips of the stub wings that also provided the structural base for the wing bracing struts and the massive landing gear units. The prototype was developed from the F.211 and F.212 day/night bombers and first flew in May 1932. Production aircraft were 10 F.221s with four 447kW (600 hp) Hispano-Suiza 12 Lbr engines, 11 F.222.1s with increased fuel capacity and 24 F.222.2s with cleaned-up fuselage contours.

FARMAN F.222.2

Role: Heavy night bomber
Crew/Accommodation: Five
Power Plant: Four 600 hp Hispano-Suiza 12 lbr air-cooled radials
Dimensions: Span 36m (118.11 ft); length 21.45m (70.37 ft); wing area 186m^2 (2,002.1 sq ft)
Weights: Empty 10,500kg (23,122 lb); MTOW 18,700kg (41,226 lb)
Performance: Maximum speed 360km/h (224 mph) at 4,000m (13,120 ft); operational ceiling 8,000m (26,250 ft); range 2,000km (1,240 miles)
Load: Three 7.5mm machine guns, plus up to 4,190kg (9,240 lb) of internally-carried bombs

FARMAN F.60 GOLIATH
(France)

The F.60 was nearing completion as a bomber prototype at the end of World War I, and was then converted as the F.60 airliner. The type first flew in February 1919, and with about 60 built was the single most important airliner used in Europe during the 1920s. The type was also developed as the F.60 Bn night bomber and F.60 Torp torpedo bomber, and also in a number of export and experimental versions.

FARMAN F.60 GOLIATH

Role: Passenger transport
Crew/Accommodation: Two, plus up to 12 passengers
Power Plant: Two 230 hp Salmson 9Z air-cooled radials
Dimensions: Span 26.5m (86.94 ft); length 14.77m (48.46 ft); wing area 160m^2 (1,722 sq ft)
Weights: Empty 2,500kg (5,511 lb); MTOW 4,770kg (10,516 lb)
Performance: Cruise speed 120km/h (74.6 mph) at sea level; operational ceiling 4,000m (13,123 ft); range 400km (249 miles)
Load: Up to 925kg (2,040 lb)

FIAT BR.20 CICOGNA (STORK)
(Italy)

The BR.20 was one of Italy's most important medium bombers of the late 1930s, and first flew in prototype form during February 1936. The type was of commendably clean appearance, but in common with other Italian load-lifting aircraft was handicapped by lack of adequately powerful engines. Production was undertaken in what was for Italian industry comparatively large numbers, and improved versions were the BR.20M and BR.20bis.

FIAT BR.20 CICOGNA

Role: Medium bomber
Crew/Accommodation: Five
Power Plant: Two 1,030 hp Fiat A.80 RC41 air-cooled radials
Dimensions: Span 21.56m (70.73 ft); length 16.17m (53.05 ft); wing area 74m^2 (796.6 sq ft)
Weights: Empty 6,739kg (14,8570 lb); MTOW 10,339kg (22,793 lb)
Performance: Maximum speed 430km/h (267 mph) at 4,000m (13,125 ft); operational ceiling 7,200m (23,620 ft); range 1,240km (770 miles) with 1,000 kg (2,205 lb) bombload
Load: One 12.7mm and two 7.7mm machine guns, plus up to 1,600 kg (3,527 lb) of bombs

FIAT G91
(Italy)

First flown in August 1956, the G91 was the winning design in a NATO-wide cmpetition for a light attack fighter. In the event it was produced only in modest numbers for the Italian and West German air forces, with second-hand aircraft being passed on to Portugal. Initial version was the G91R with a 2,268kg (5,000 lb) thrust Fiat-built Bristol Siddeley Orpheus turbojet. This came in a number of variants with different armament and, in a secondary role, reconnaissance equipment; there was also a G91T operational trainer version. The second version was reworked as the G91Y.

FIAT G91R

Role: Strike fighter/reconnaissance
Crew/Accommodation: One
Power Plant: One 2,268 kgp (5,000 b.s.t.) Bristol Siddeley Orpheus 803 turbojet
Dimensions: Span 8.56m (28.08 ft); length 10.3m (33.79 ft); wing area 16.4m^2 (176.7 sq ft)
Weights: Empty 3,100kg (6,835 lb); MTOW 8,700kg (19,180 lb)
Performance: Maximum speed 1,075km/h (580 knots) Mach 0.877m at sea level; operational ceiling 13,100m (42,979 ft); radius 315m (170 naut. miles) with full warload
Load: Four 12.7mm machine guns, plus up to 907kg (2,000 lb) of externally-carried weapons-fuel

FIESELER Fi 156 STORCH (STORK)
(Germany)

The Fi 156 was Germany's most important army co-operation and battlefield reconnaissance aeroplane of World War II. It was a highly capable STOL type, with slats and slotted ailerons/flaps respectively over the entire leading and trailing edges of the braced high-set wing. The Storch first flew in prototype form during 1936, and was ordered into large-scale production in its initial Fi 156A form. Later variants were the Fi 156C with a raised rear cockpit glazing to allow the installation of a defensive machine gun, and the Fi 156D with the Argus 10P engine.

FIESELER Fi 156 C-1 STORCH

Role: Army co-operation/observation and communications
Crew/Accommodation: Two or one, plus one litter-carried casualty
Power Plant: One 240 hp Argus As 10C air-cooled inline
Dimensions: Span 14.25m (46.75 ft); length 9.9m (32.48 ft); wing area 26m^2 (279.9 sq ft)
Weights: Empty 930kg (2,051 lb); MTOW 1,320kg (2,911 lb)
Performance: Maximum speed 145km/h (90 mph) at sea level; operational ceiling 4,600m (15,092 ft); range 385km (239 miles)
Load: One 7.9mm machine gun, plus provision to evacuate one litterborne casualty

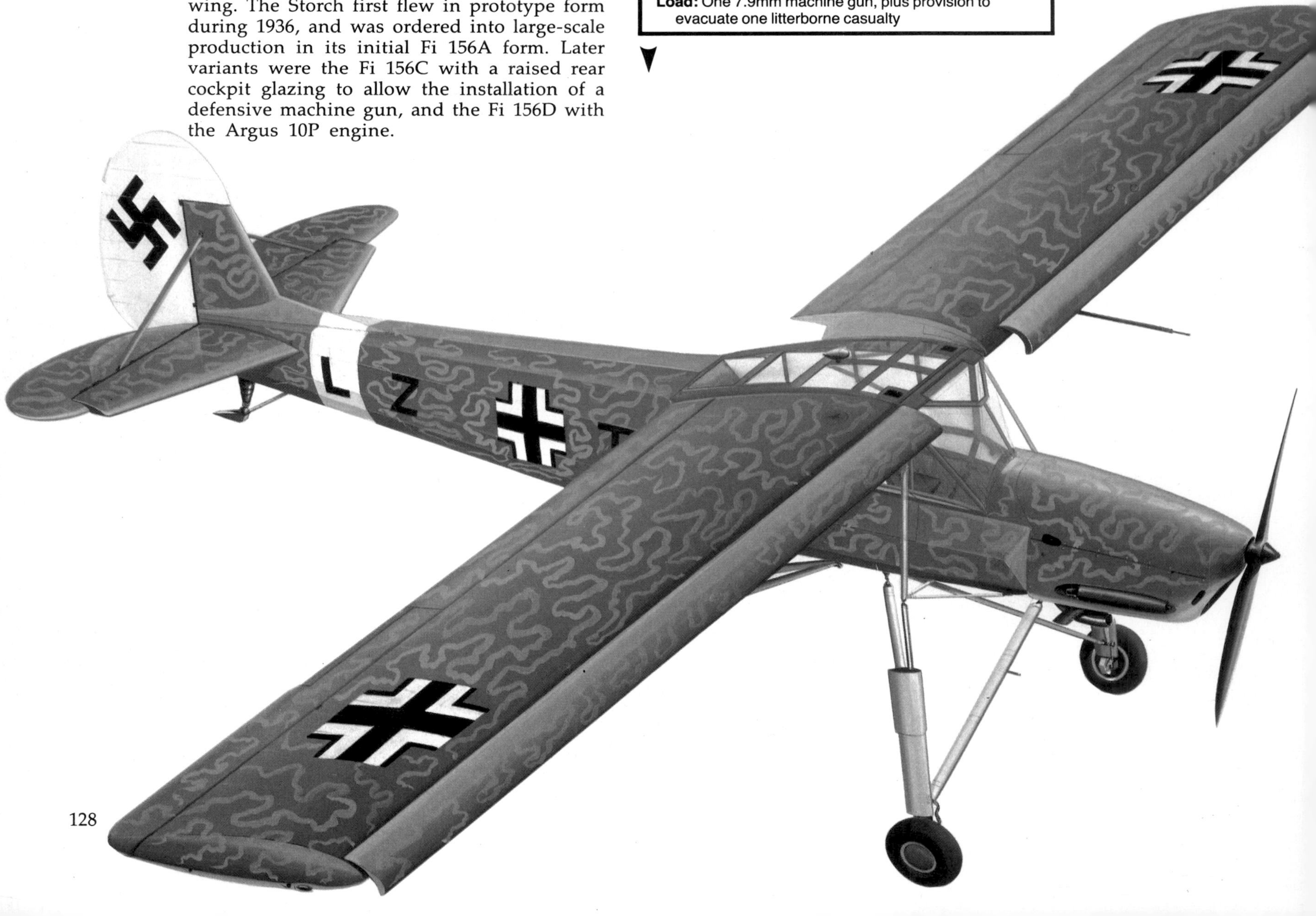

FLETTNER FI 282 KOLIBRI (HUMMING BIRD)
(Germany)

First flown in 1942 and resembling an autogyro rather than a helicopter because of its aeroplane-type tail unit and tricycle landing gear, the Fl 282 was one of history's first operational helicopters. It was a notably compact design using intermeshing rotors turning in opposite directions to do away with the torque problem that would otherwise have demanded a longer tail ending in a rotor. Production of 1,000 such helicopters was planned, but Allied bombing completely disrupted the programme.

FLETTNER FI 282 KOLIBRI

Role: Naval shipborne anti-submarine observation
Crew/Accommodation: Two, including the observer
Power Plant: One 140 hp BMW Bramo Sh 14A air-cooled radial
Dimensions: Overall length rotors turning 11.96m (39.24 ft); rotor diameter 11.96m (39.24 ft)
Weights: Empty 640kg (1,410 lb); MTOW 1,000kg (2,205 lb)
Performance: Cruise speed 109km/h (68 mph) at 1,500m (4,921 ft); operational ceiling 3,292m (10,800 ft); range 170km (106 miles) with observer
Load: None
Note: The operational ceiling quoted above is for two crew mission

FABRICA MILITAR DE AVIONES IA 58 PUCARA
(Argentina)

First flown in glider form during December 1967, and in powered form during August 1969, the Pucara is a simple yet effective counter-insurgency aeroplane. The initial version was the IA 58A, and an improved version is the IA 58C with upgraded electronics and provision for more modern underwing loads including air-to-air and air-to-surface missiles.

FMA IA 58A PUCARA

Role: Strike/close air support
Crew/Accommodation: One
Power Plant: Two 1,022 shp Turbomeca Astazon XVIG turboprops
Dimensions: Span 14.5m (47.57 ft); length 14.25m (46.75 ft); wing area 30.3m^2 (326.1 sq ft)
Weights: Empty 4,037kg (8,900 lb); MTOW 6,800kg (14,990 lb)
Performance: Maximum speed 500km/h (310 mph) at 3,000m (9,843 ft); operational ceiling 10,000m (32,808 ft); radius 351km (218 miles) with full bombload
Load: Two 20mm cannon and four .303 inch machine guns, plus up to 2,000 kg (4,409 lb) of externally-carried weapons/fuel

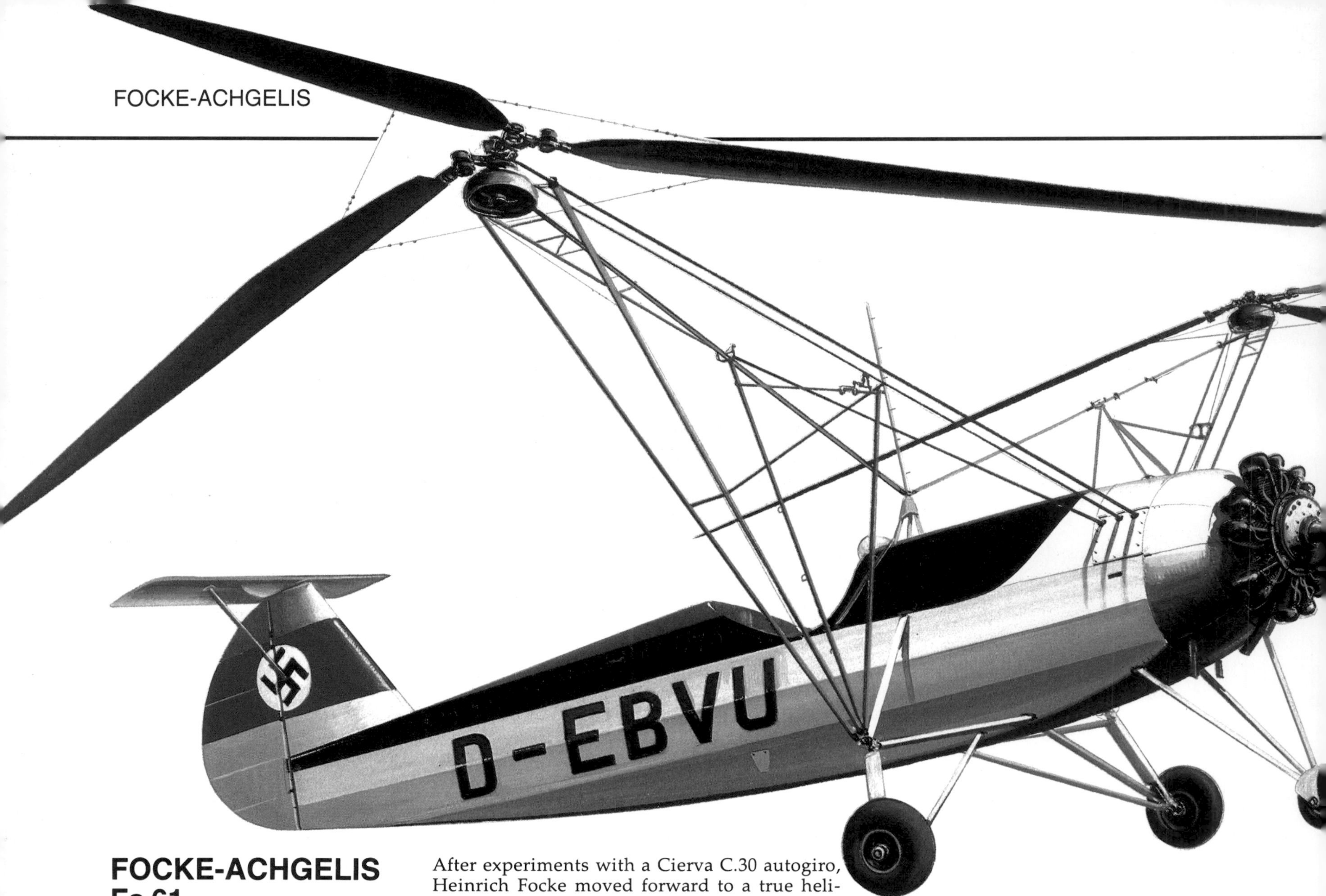

FOCKE-ACHGELIS Fa 61
(Germany)

After experiments with a Cierva C.30 autogiro, Heinrich Focke moved forward to a true helicopter in the form of the Focke-Wulf Fw 61 (sic) that first flew in June 1936. This was based on a conventional fuselage with a 119kW (160 hp) Bramo Sh.14A radial engine driving a small cooling propeller and more importantly, two counter-rotating rotors at the end of long outriggers. The latter also supported the main units of the fixed tricycle landing gear. The type underwent considerable development as the Fa 61 after the creation of Focke-Achgelis, and established a number of important world records as well as forming the basis for subsequent helicopters.

FOCKE-ACHGELIS Fa 61

Role: Developmental helicopter
Crew/Accommodation: One
Power Plant: One 160 hp BMW Bramo Sh. 14A air-cooled radial
Dimensions: Rotor diameter 7 m; (22.97 ft)
Weights: Empty 800kg (1,764 lb); MTOW 953kg (2,100 lb)
Performance: Maximum speed 124km/h (77 mph) at sea level; operational ceiling 2,410m (7,900 ft); range 230km (143 miles)
Load: None

FOCKE-ACHGELIS Fa 223 DRACHE (KITE)
(Germany)

This multi-role helicopter was in essence a scaled-up version of the Fa 61, and this resulted in the overall 24.5m (80.38 ft) width that proved something of a handicap in the tactical operations for which the Fa 223 was designed. The type first flew in the summer of 1940, and the 18 pre-production helicopters completed before Allied bombing terminated the programme, were used in important evaluations of the helicopter in roles as diverse as tactical transport, SAR, reconnaissance, logistic support and anti-submarine patrol.

FOCKE-ACHGELIS Fa 223 DRACHE

Role: Military transport helicopter
Crew/Accommodation: Two, plus up to four troops
Power Plant: One 1,000 hp BMW Bramo Fafnir 323A-3
Dimensions: Rotor diameter 11.9 m; (39.04 ft)
Weights: Empty 3,200kg (7,055 lb); MTOW 5,000kg (11,020 lb)
Performance: Cruise speed 120km/h (75 mph) at sea level; operational ceiling 4,880m (16,010 ft); range 320km (199 miles) with 360kg (794 lb) payload
Load: Up to 908 kg (2,002 lb) externally slung

Not illustrated

FOCKE-WULF Fw 56 STOSSER (FALCON)
(Germany)

The Fw 56 was an attractive parasol-winged monoplane with cantilever main landing gear units, and was first flown in November 1933 as an advanced trainer prototype. Two other prototypes and three Fw 56A-0 pre-production aircraft were used to develop the type. Production of 1,000 or so Fw 56A-1 aircraft provided the Germans and some allies with a useful fighter trainer.

FOCKE-WULF Fw 56 STOSSER

Role: Advanced trainer
Crew/Accommodation: One
Power Plant: One 240 hp Argus As 10C air-cooled inline
Dimensions: Span 10.5m (34.45 ft); length 7.6m (24.93 ft); wing area 14m^2 (150.6 sq ft)
Weights: Empty 695kg (1,532 lb); MTOW 996kg (2,196 lb)
Performance: Maximum speed 278km/h (173 mph) at sea level; operational ceiling 6,200m (20,336 ft); range 400km (250 miles)
Load: Two/one 7.9mm machine guns, plus up to 30kg (66 lb) of bombs

FOCKE-WULF Fw 189 UHU (EAGLE OWL)
(Germany)

The Fw 189 was developed as a battlefield reconnaissance type, and the design with twin booms and an extensively glazed central nacelle was designed to optimize optical reconnaissance capability. The type first flew in prototype form during July 1938, and was soon evaluated with a useful defensive plus a modest offensive armament. The only two production variants were the Fw 189A operational and Fw 189B crew trainer models. Experimental developments included the Fw 189C close-support, Fw 189D twin-float, Fw 189E radial-engined reconnaissance and Fw 189F reconnaissance models, the last with 433kW (580 hp) Argus As 411 inline engines.

FOCKE-WULF Fw 189A-1 UHU

Role: Reconnaissance/communications
Crew/Accommodation: Three
Power Plant: Two 465 hp Argus As 410A-1 air-cooled inlines
Dimensions: Span 18.4m (60.37 ft); length 11.9m (39.04 ft); wing area 38m^2 (409 sq ft)
Weights: Empty 2,805kg (6,185 lb); MTOW 4,175kg (9,193 lb)
Performance: Maximum speed 334km/h (208 mph) at 1,700m (5,578 ft); operational ceiling 7,000m (22,967 ft); range 830km (506 miles)
Load: Three 7.9mm machine guns, plus up to 200kg (440 lb) of bombs

FOCKE-WULF Fw 190 and Ta 152
(Germany)

Though not built in numbers as great as the Bf 109, this was Germany's best fighter of World War II, and resulted from the belief of designer Kurt Tank that careful streamlining could produce a radial-engined fighter with performance equal to that of an inline-engined type without the extra complexity and weight of the latter's water-cooling system. The first prototype flew in June 1939, and an extensive test programme was required to develop the air cooling system and evaluate short- and long-span wings. The latter's additional 1.0m (3 ft 3.7 inches) of span and greater area reduced performance but boosted agility and rate of climb considerably. This wing was selected for the Fw 190A production type in a programme that saw the building of almost 20,000 Fw 190s. The Fw 109A was powered by the BMW 801 radial engine, and was developed in variants up to the Fw 190A-8 with a host of subvariants optimized for the clear- or all-weather interception, ground-attack, torpedo attack and tactical reconnaissance roles with an immensely diverse armament capability.

The Fw 190B series was used to develop high-altitude capability with longer-span wings and a pressurized cockpit, and then with the Daimler-Benz DB 603 inline engine. The Fw 190C was another high-altitude development model with the 1,304kW (1,750 hp) DB 603 engine and a turbocharger. The next operational model was the Fw 190D, which was developed in role-optimized variants between Fw 190D-9 and Fw 190D-13 with the 1,324kW (1,776 hp) Junkers Jumbo 213 inline engine and an annular radiator in a lengthened fuselage. The Fw 190E was a proposed reconnaissance fighter, and the Fw 190F series, which preceded the Fw 190D model, was a specialized ground-attack type based on the radial-engined Fw 190A-4.

Finally, in the main sequence, came the Fw 190G series of radial-engined fighter-bombers evolved from the Fw 190A-5. The Fw 190 was also adapted to create an ultra-high-altitude type: this had long-span wings and was developed as the Ta 152 with the Jumo 213 inline engine, and as the Ta 153 with the DB 603 inline engine. The latter abandoned, the only operational variant of the Ta 152 was the Ta 152H.

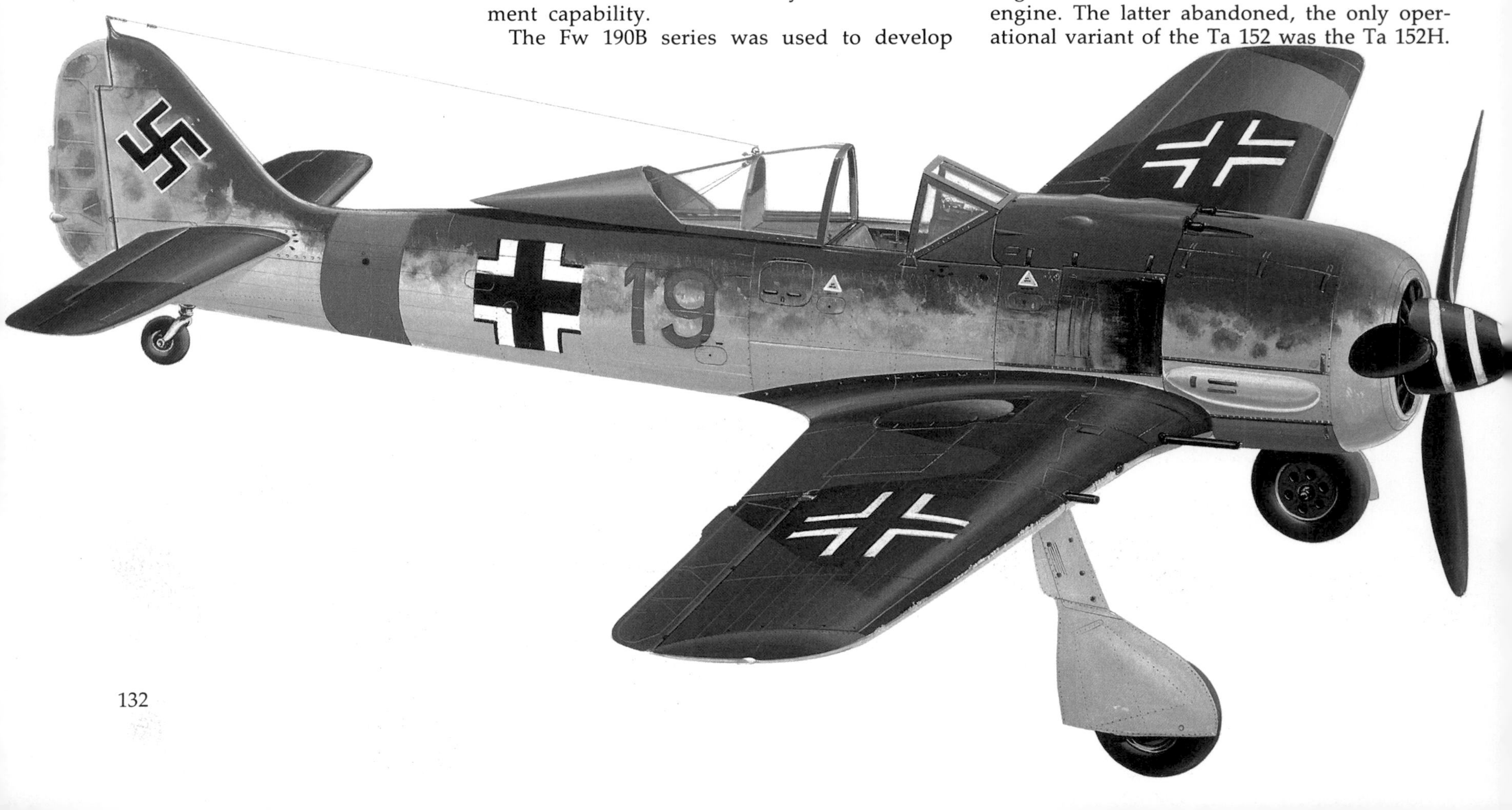

FOCKE-WULF Fw 190 A-8

Role: Fighter
Crew/Accommodation: One
Power Plant: 1,600hp BMW 801C-1 air-cooled radial
Dimensions: Span 10.5m (34.45 ft); length 8.84m (29 ft); wing area 18.3m^2 (196.98 sq ft)
Weights: Empty 3,170kg (7,000 lb); MTOW 4,900kg (10,805 lb)
Performance: Maximum speed 654km/h (408 mph) at 6,000m (19,686 ft); operational ceiling 11,400m (37,403 ft); range 805km (500 miles)
Load: Four 20mm cannon and two 13mm machine guns, plus up to 1,000kg (2,205 lb) of bombs

FOCKE-WULF Fw 200 CONDOR
(Germany)

The Condor was developed as a transatlantic passenger and mail aeroplane, and first flew in this form during July 1937 with 652kW (750 hp) Pratt & Whitney Hornet radial engines. Small numbers were built as airliners, some becoming the personal transports of Nazi VIPs. The type's real claim to fame, however, is as Germany's most important maritime reconnaissance bomber of World War II, the Fw 200C that was built in variants up to the Fw 200C-8 with increasingly sophisticated equipment (including search radar) and weapons (including air-to-surface missiles).

FOCKE-WULF Fw 200C-3 CONDOR

Role: Long-range maritime reconnaissance bomber
Crew/Accommodation: Seven
Power Plant: Four 1,200 hp BMW Bramo 323 R-2 Fafnrir air-cooled radials
Dimensions: Span 32.84m (107.74 ft); length 23.85m (78.25 ft); wing area 118m^2 (1,290 sq ft)
Weights: Empty 17,000kg (37,485 lb); MTOW 22,700kg (50,045 lb)
Performance: Cruise speed 335km/h (208 mph) at 4,000m (13,124 ft); operational ceiling 6,000m (19,685 ft); range 3,560km (2,211 miles)
Load: One 20mm cannon, three 13mm and two 7.9mm machine guns, plus up to 2,100kg (4,630 lb) of bombs

FOCKE-WULF Ta 154
(Germany)

The Ta 154 was schemed as a twin-engined night-fighter of wooden construction to exploit the capabilities of Germany's wood-working industry at a time when the aircraft industry was hard pressed. The first prototype flew in July 1943, and the type proved to have admirable performance and handling together with effective radar and weapons. However, production was halted after the delivery of only a few aircraft after a number of disastrous structural failures. These were found to stem from a chemical reaction between the adhesives used to bond the plywood that formed the primary structural material.

FOCKE-WULF Ta 154 A-1

Role: Night/all-weather fighter
Crew/Accommodation: Two
Power Plant: Two 1,750hp Junkers Jumbo 213A water-cooled inlines
Dimensions: Span 16m (52.5 ft); length 12.1m (39.7 ft); wing area 32.4m^2 (348.6 sq ft)
Weights: Empty 6,405kg (14,123 lb); MTOW 8,250kg (18,191 lb)
Performance: Maximum speed 635km/h (395 mph) at 6,100m (20,008 ft); operational ceiling 10,500m (34,440 ft); endurance 2.75 hours
Load: Two 30mm and two 20mm cannon (interception directed by FuG 202 Lichtenstein BC-1 radar)

FOKKER E SERIES
(Germany)

These aircraft were the world's first true fighters. Though based on the indifferent M.5k monoplane, they proved to be decisive weapons for a time because of their use with a fixed machine gun firing directly ahead through the disc swept by propeller blades protected by an interrupter system. The E I began to enter service in the summer of 1915 with a 60kW (80 hp) Oberursel U.0 rotary engine. This preliminary type was soon superseded by the E II with the 75kW (100 hp) U.I rotary for improved performance; the E III was in essence an improved E II. Finally came the E IV, a 1916 version with two guns and the 119kW (160 hp) U.III engine in an unsuccessful effort to regain the technical lead in the sea-saw fighter race currently being led by Allies.

FOKKER E III

Role: Fighter/reconnaissance
Crew/Accommodation: One
Power Plant: One 100 hp Oberusel U.I. rotary
Dimensions: Span 9.52m (31.23 ft); length 7.3m (23.95 ft); wing area 16m^2 (172.2 sq ft)
Weights: Empty 500kg (1,102 lb); MTOW 635kg (1,400 lb)
Performance: Maximum speed 134km/h (83 mph) at sea level; operational ceiling 3,500m (11,483 ft); endurance 2.75 hours
Load: One-two 7.92mm machine guns

FOKKER D II, D III and D V
(Germany)

Fokker sought to capitalize on the success of his first-generation E-series fighters with a biplane series offering greater agility and performance, and the type was first flown in M.17z prototype form during 1915 as a single-bay biplane. The wing cellule was later revised to a two-bay type, and with an armament of one 7.92mm (0.312 inch) machine gun the type began to enter service in 1916 as the D II with a 75kW (100 hp) Oberursel U.I rotary engine. The D III was an attempt to increase performance by installing a 19kW (160 hp) Oberursel U.III rotary engine, but the additional power was negated by a heavier though stronger airframe with longer-span wings, and twin machine guns. The D V was the D III engined with the U.I rotary, the decline in performance relegating the type to the fighter training role.

FOKKER D III

Role: Fighter
Crew/Accommodation: One
Power Plant: One 160 hp Oberusel U.III air-cooled rotary
Dimensions: Span 9.05m (29.69 ft); length 6.3m (20.67 ft); wing area 20m^2 (215.28 sq ft)
Weights: Empty 452kg (994 lb); MTOW 710kg (1,562 lb)
Performance: Maximum speed 160km/h (99.4 mph) at sea level; operational ceiling 4,000+m (13,120 ft); endurance 1.5 hours
Load: One/two 7.92mm machine gun/s

FOKKER Dr I
(Germany)

The Dr I was produced in response to the Sopwith Triplane, and first flew in June 1917 as a narrow-span cantilever triplane. Interplane struts were added to overcome wing vibration in flight, and the Dr I entered service as a supremely agile dogfighter. The type was built only in modest numbers, its indifferent performance recommending its use only in defensive operations by Germany's finest fighter pilots.

FOKKER Dr I

Role: Fighter
Crew/Accommodation: One
Power Plant: One 110 hp Oberursel U.R. II air-cooled rotary
Dimensions: Span 7.17m (23.52 ft); length 5.77m (18.93 ft); wing area 16m^2 (172.2 sq ft)
Weights: Empty 405kg (893 lb); MTOW 585kg (1,289 lb)
Performance: Maximum speed 185km/h (115 mph) at sea level; operational ceiling 5,975m (19,603 ft); range 210km (130 miles)
Load: Two 7.92mm machine guns

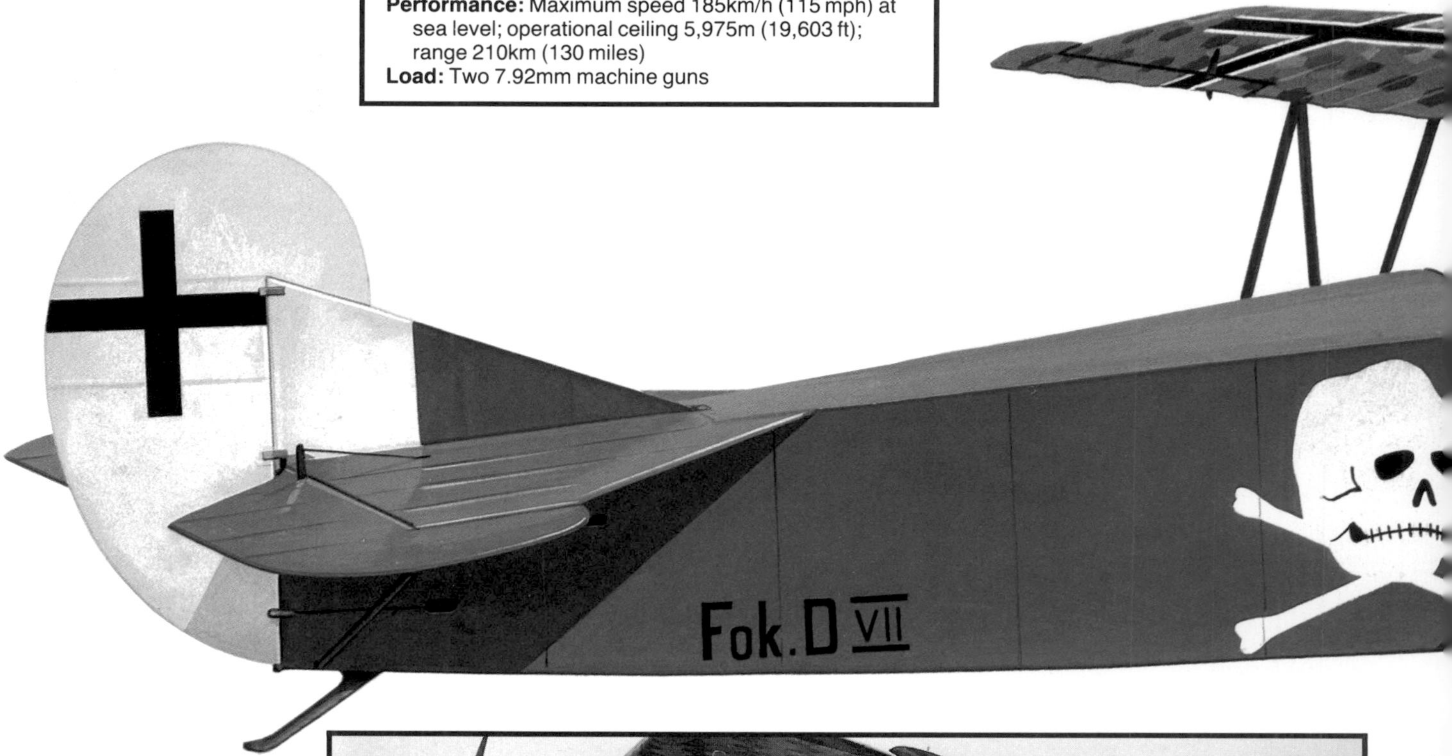

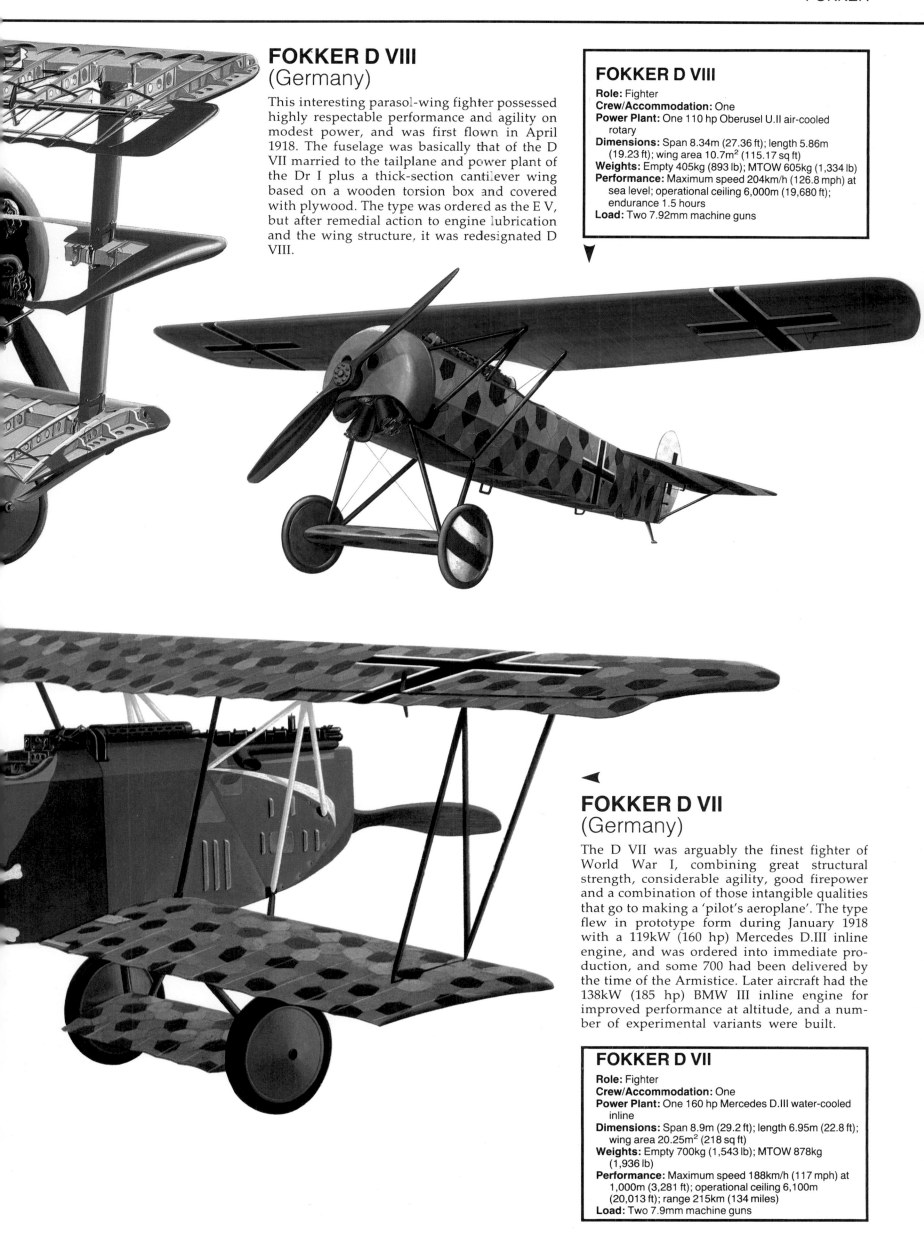

FOKKER D VIII
(Germany)

This interesting parasol-wing fighter possessed highly respectable performance and agility on modest power, and was first flown in April 1918. The fuselage was basically that of the D VII married to the tailplane and power plant of the Dr I plus a thick-section cantilever wing based on a wooden torsion box and covered with plywood. The type was ordered as the E V, but after remedial action to engine lubrication and the wing structure, it was redesignated D VIII.

FOKKER D VIII

Role: Fighter
Crew/Accommodation: One
Power Plant: One 110 hp Oberusel U.II air-cooled rotary
Dimensions: Span 8.34m (27.36 ft); length 5.86m (19.23 ft); wing area 10.7m^2 (115.17 sq ft)
Weights: Empty 405kg (893 lb); MTOW 605kg (1,334 lb)
Performance: Maximum speed 204km/h (126.8 mph) at sea level; operational ceiling 6,000m (19,680 ft); endurance 1.5 hours
Load: Two 7.92mm machine guns

FOKKER D VII
(Germany)

The D VII was arguably the finest fighter of World War I, combining great structural strength, considerable agility, good firepower and a combination of those intangible qualities that go to making a 'pilot's aeroplane'. The type flew in prototype form during January 1918 with a 119kW (160 hp) Mercedes D.III inline engine, and was ordered into immediate production, and some 700 had been delivered by the time of the Armistice. Later aircraft had the 138kW (185 hp) BMW III inline engine for improved performance at altitude, and a number of experimental variants were built.

FOKKER D VII

Role: Fighter
Crew/Accommodation: One
Power Plant: One 160 hp Mercedes D.III water-cooled inline
Dimensions: Span 8.9m (29.2 ft); length 6.95m (22.8 ft); wing area 20.25m^2 (218 sq ft)
Weights: Empty 700kg (1,543 lb); MTOW 878kg (1,936 lb)
Performance: Maximum speed 188km/h (117 mph) at 1,000m (3,281 ft); operational ceiling 6,100m (20,013 ft); range 215km (134 miles)
Load: Two 7.9mm machine guns

FOKKER C V SERIES ▲
(Netherlands)

Fokker was by birth a Dutchman, but before and during World War I he worked in Germany. At the end of the war he decamped to his native Netherlands. The C V was one of the most prolific and important warplanes developed between the two world wars, and was developed in many forms with radial or inline engines and different wings. The first example flew in May 1924 as a two-seat light bomber and reconnaissance type, and was produced initially as three variants with parallel-chord wings: the C V-A had an area of 37.5m^2 (403.66 sq ft), the C V-B of 40.8m^2 (439.18 sq ft) and the C V-C of 46.1m^2 (496.23 sq ft). Then came two variants with tapered sesquiplane wings: the C V-D had an area of 28.8m^2 (310.01 sq ft) with V-struts, and the C V-E of 39.4m^2 (423.04 sq ft) with N-struts. The type was also produced as the twin/float C V-W, and there were several licence-built models.

FOKKER C V-D

Role: Light bomber/reconnaissance
Crew/Accommodation: Two
Power Plant: One 450 hp Hispano Suiza water-cooled inline
Dimensions: Span 12.5m (41.01 ft); length 9.5m (31.17 ft); wing area 28.8m^2 (310 sq ft)
Weights: Empty 1,250kg (2,756 lb); MTOW 1,850kg (4,080 lb)
Performance: Maximum speed 225km/h (140 mph) at sea level; operational ceiling 5,500m (18,050 ft); range 770km (478 miles) with full warload
Load: Two/four 7.9mm machine guns, plus up to 200kg (441 lb) of bombs

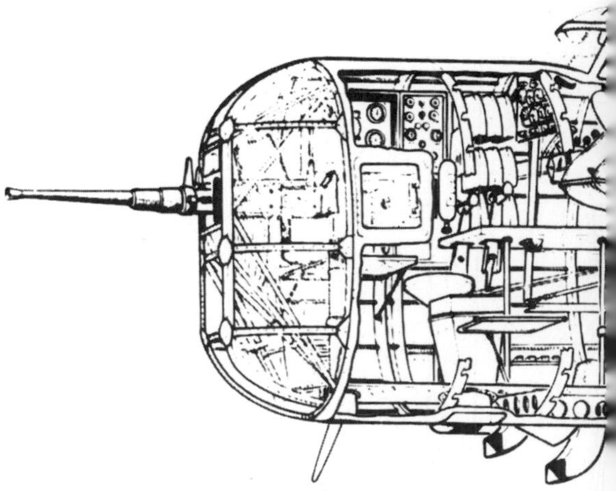

◄

FOKKER F.XXII and F.XXXVI
(Netherlands)

The F.XXXVI was the largest of Fokker's inter-war airliners, and was first flown during June 1934 as a high-wing monoplane with fixed tailwheel landing gear and four 559kW (750 hp) Wright Cyclone radial engines. Accommodation was provided for a crew of four and 32 passengers, the latter carried in four eight-seat cabins that could be rearranged for 16 sleeper passengers. The sole F.XXXVI was followed by the similar but smaller F.XXII that first flew early in 1935 as a 22-passenger transport. Total production of this marque was four.

FOKKER F.XXII

Role: Passenger transport
Crew/Accommodation: Three and one cabin crew, plus up to 22 passengers
Power Plant: Four 500 hp Pratt & Whitney T1D1 Wasp air-cooled radials
Dimensions: Span 30m (98.42 ft); length 21.51m (70.58 ft); wing area 129.98m^2 (1,399 sq ft)
Weights: Empty 8,100kg (17,857 lb); MTOW 13,000kg (28,660 lb)
Performance: Cruise speed 215km/h (134 mph) at 2,400m (7,874 ft); operational ceiling 4,900m (16,076 ft); range 1,500km (932 miles)
Load: Up to 2,370kg (5,225 lb)

FOKKER D XXI
(Netherlands)

By the early 1930s Fokker was becoming increasingly aware of the technical obsolescence of its biplane and high-wing monoplane aircraft, and an important result was the D XXI low-wing monoplane fighter with fixed, but neatly faired, main landing gear units. The type was planned with a Rolls-Royce Kestrel inline, but first flew during March 1936 with a 481kW (645 hp) Mercury VIS radial engine. Production with a more powerful Mercury was undertaken for Finland and the Netherlands, the former licence-building additional aircraft with the 615kW (825 hp) Pratt & Whitney Twin Wasp Jr and with the Bristol Pegasus.

FOKKER D XXI

Role: Interceptor
Crew/Accommodation: One
Power Plant: One 830 hp Bristol Mercury VIII air-cooled radial
Dimensions: Span 11m (36.09 ft); length 8.2m (26.9 ft); wing area 16.2m^2 (174.4 sq ft)
Weights: Empty 1,450kg (3,197 lb); MTOW 2,050kg (4,519 lb)
Performance: Maximum speed 460km/h (286 mph) at 5,100m (16,730 ft); operational ceiling 10,100m (33,135 ft); range 850km (528 miles)
Load: Four 7.9mm machine guns

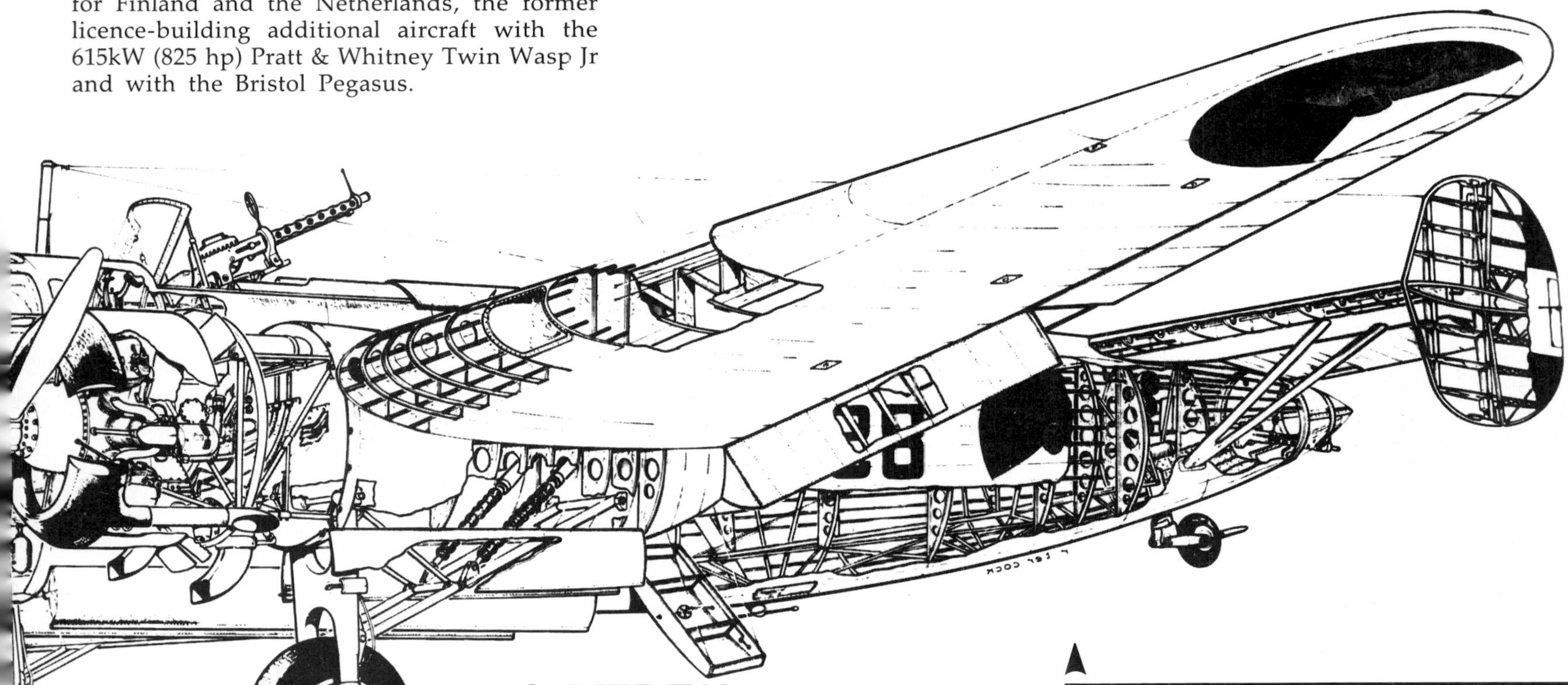

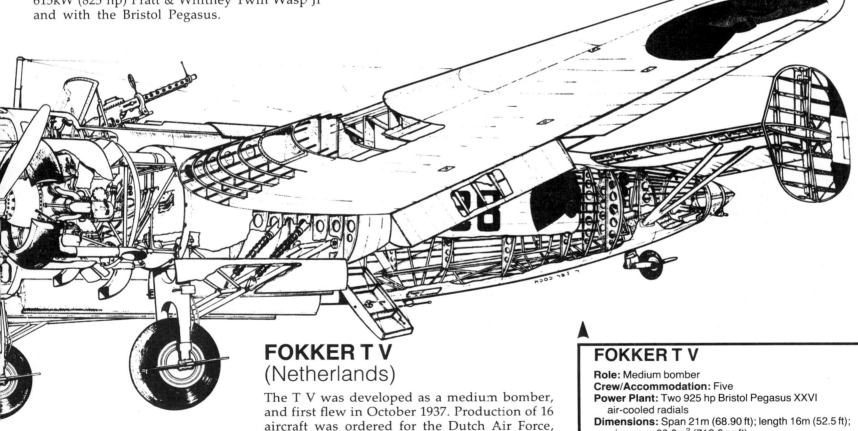

FOKKER T V
(Netherlands)

The T V was developed as a medium bomber, and first flew in October 1937. Production of 16 aircraft was ordered for the Dutch Air Force, and though these were well armed machines with good defensive armament and adequate bombload carried internally, the type was not popular as it was modestly unstable and tricky to fly.

FOKKER T V

Role: Medium bomber
Crew/Accommodation: Five
Power Plant: Two 925 hp Bristol Pegasus XXVI air-cooled radials
Dimensions: Span 21m (68.90 ft); length 16m (52.5 ft); wing area 66.2m^2 (712.6 sq ft)
Weights: Empty 4,640kg (10,230 lb); MTOW 7,235kg (15,950 lb)
Performance: Maximum speed 416km/h (259 mph) at sea level; operational ceiling 7,700m (25,260 ft); range 1,630km (1,012 miles) with 181kg (400 lb) with warload
Load: One 20mm cannon and four 7.9mm machine guns, plus up to 1,000kg (2,205 lb) of bombs

Fokker 50

Fokker F27 Friendship

FOKKER F.27 FRIENDSHIP and FOKKER 50
(Netherlands)

After World War II Fokker sought to recapture a slice of the airliner market with a type comparable with the best of its classic interwar airliners. After considerable deliberation Fokker fixed on a high-wing short/medium-range type powered by a pair of Rolls-Royce Dart turboprops. The first example flew in November 1955, and the type has proved very successful in several variants. Three basic models were licence-built in the U.S.A. as the Fairchild F-27, and the same company also produced stretched models with the designation FH-227.

The durability of the design is attested by the recent development of the Fokker 50, a thoroughly updated version with 1,864kW (2,250 shp) Pratt & Whitney Canada PW125B turboprops driving six-blade propellers.

FOKKER 50

Role: Short-range passenger transport
Crew/Accommodation: Two and two cabin crew, plus up to 58 passengers
Power Plant: Two 2,250 shp Pratt & Whitney PW 125B turboprops
Dimensions: Span 29m (95.15 ft); length 25.25m (82.83 ft); wing area 70m^2 (754 sq ft)
Weights: Empty 12,741kg (28,090 lb); MTOW 18,990kg (41,865 lb)
Performance: Cruise speed 500km/h (27 knots) at 6,096m (20,000 ft); operational ceiling 7,620m (25,000 ft); range 1,125km (607 naut. miles) with 50 passengers
Load: Up to 5,262kg (11,600 lb)

FOKKER F.28 FELLOWSHIP and FOKKER 100
(Netherlands)

The F.28 was designed as a complement to the F.27 with slightly higher capacity and considerably improved performance through the use of a twin-turbofan power plant. Fokker opted for a T-tail configuration and rear-mounted engines to leave the wing uncluttered, and the first example flew in May 1967 with Rolls-Royce Spey turbofans. The first production version was the F.28 Mk 1000 for 65 passengers with two 4,468kg (9,850 lb) thrust Spey Mk 555-15s, and subsequent production models have been the 79-passenger F.28 Mk 2000 with a stretched fuselage, and the F.28 Mks 3000 and 4000 equivalent to the Mks 1000 and 2000 with 4,491kg (9,900 lb) thrust Spey Mk 555-15Ps.

The type has now been updated as the 107-passenger Fokker 100 with 6,282kg (13,850 lb) thrust Rolls-Royce Tay Mk 620-15 turbofans.

FOKKER 100

Role: Short-range jet passenger transport
Crew/Accommodation: Two and four cabin crew, plus up to 119 passengers
Power Plant: Two 6,282kgp (13,850 lb s.t.) Rolls-Royce Tay 620-15 turbofans
Dimensions: Span 28.08m (92.13 ft); length 35.53m (116.57 ft); wing area 93.5m^2 (1,006.5 sq ft)
Weights: Empty 24,355kg (53,695 lb); MTOW 43,090kg (95,000 lb)
Performance: Cruise speed 765km/h (413 knots) at 8,534m (28,000 ft); operational ceiling 10,668m (35,000 ft); range 2,298km (1,240 naut. miles) with 107 passengers
Load: Up to 12,385kg (27,305 lb)

FOLLAND GNAT
(U.K.)

In an effort to break the pattern of increasingly complex, and therefore increasingly heavy and expensive high-performance fighters, Folland sought to prove the virtues of a lightweight and thus highly affordable fighter. The concept was pioneered in the Fo.139 Midge with a 744kg (1,640 lb) thrust Armstrong Siddeley Viper turbojet, and this paved the way for the slightly larger Fo.141 Gnat, This flew in prototype form during July 1955, and in its initial fighter form was bought in modest numbers by Finland and India, the latter subsequently producing the type under licence as the HAL Ajeet. The RAF then ordered an advanced trainer version as the Gnat T.Mk 1 with larger flying surfaces and slightly lengthened two-seater fuselage.

HINDUSTAN AERONAUTICS (FOLLAND) GNAT F.Mk 1

Role: Fighter
Crew/Accommodation: One
Power Plant: One 2,132kgp (4,700 lb s.t.) Bristol Siddeley Orpheus 701 turbojet
Dimensions: Span 6.76m (22.16 ft); length 9.07m (29.75 ft); wing area 12.69m^2 (136.6 sq ft)
Weights: Empty 2,302kg (5,074 lb); MTOW 4,030kg (8,885 lb)
Performance: Maximum speed 1,118km/h (604 knots) Mach 0.98 at 6,096m (20,000 ft); operational ceiling 15,240m (50,000 ft); range 805km (500 miles) with overload fuel
Load: Two 30mm cannon, plus up to 454kg (1,000 lb) of bombs

FORD TRI-MOTOR
(U.S.A.)

This transport played an important part in opening U.S. airways during the late 1920s, and was in general an unremarkable aeroplane for its time with the exception of its corrugated all-metal construction, which has ensured that some examples are still flying. The type first flew in June 1926 as the 4-AT with three 149kW (200 hp) Wright Whirlwind J-4 radial engines and accommodation for two pilots in an open cockpit and eight passengers in an enclosed cabin. The type was produced in many civil and military variants with different and more powerful engines and, on occasion, float landing gear. The civil series included the baseline 4-AT, the enlarged 5-AT for 13/17 passengers, and the re-engined 6-AT.

FORD 4-AT-E TRI-MOTOR

Role: Passenger transport
Crew/Accommodation: Two, plus up to 11 passengers
Power Plant: Three 300 hp Wright J-6 air-cooled radials
Dimensions: Span 22.56m (74 ft); length 15.19m (49.83 ft); wing area 72.93m^2 (785 sq ft)
Weights: Empty 2,948kg (6,500 lb); MTOW 4,595kg (10,130 lb)
Performance: Cruise speed 172km/h (107 mph) at sea level; operational ceiling 5,029m (16,500 ft); range 917km (570 miles)
Load: Up to 782kg (1,725 lb)

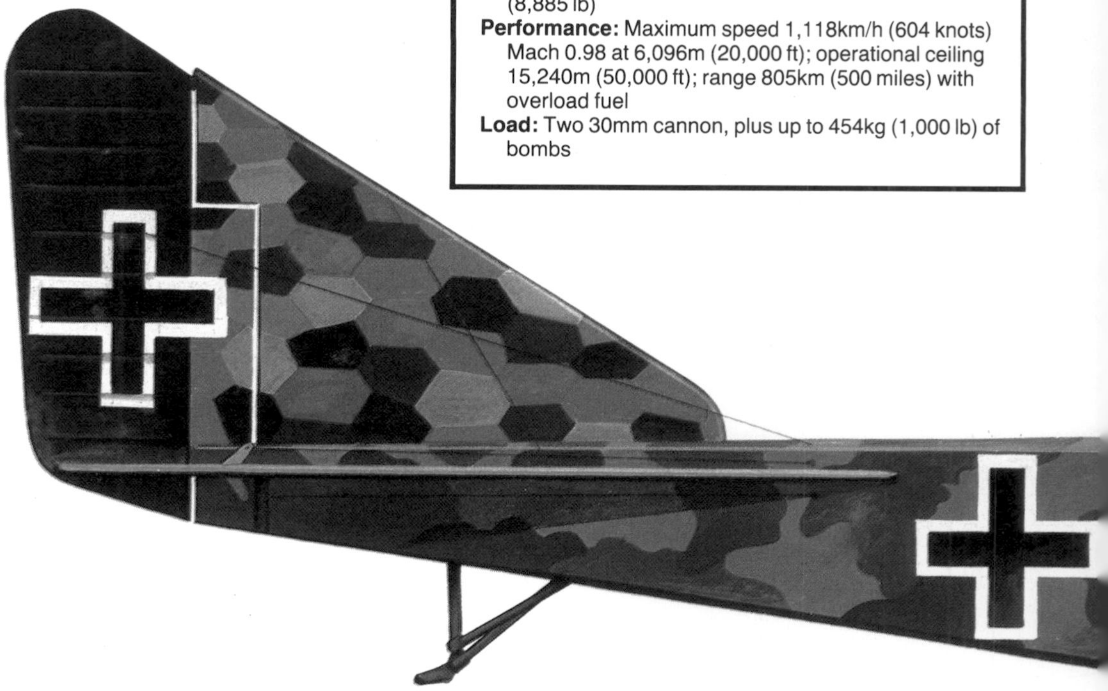

A Ford 5-AT, once the flagship of United Airlines

FOUGA MAGISTER ▲
(France)

This was the first jet trainer to enter large-scale production, and flew in prototype form during July 1952 with the distinctive feature of a V-tail. Substantial production was undertaken of the CM.170-1 land-based variant with two 400kg (880 lb) Turbomeca Marbore IIA turbojets, and lesser production followed of the CM.175 Zephyr navalized model and CM.170-2 Super Magister land-based model with two 480kg (1,058 lb) thrust Marbore VI turbojets. The basic type has a useful light attack capability, and this is improved in the upgraded AMIT Fouga (or Tzukit) developed by Israel Aircraft Industries.

FOUGA CM.170 MAGISTER

Role: Basic/advanced trainer
Crew/Accommodation: Two
Power Plant: Two 400kgp (880 lb s.t.) Turbomeca Mabore IIa turbojets
Dimensions: Span 12.15m (39.83 ft); length 10.06m (33 ft); wing area $17.3m^2$ (186.1 sq ft)
Weights: Empty 2,150kg (4,740 lb); MTOW 3,200kg (7,055 lb)
Performance: Maximum speed 700km/h (435 knots) at sea level; operational ceiling 13,500m (44,291 ft); range 1,250km (775 miles)
Load: Two 7.62mm machine guns

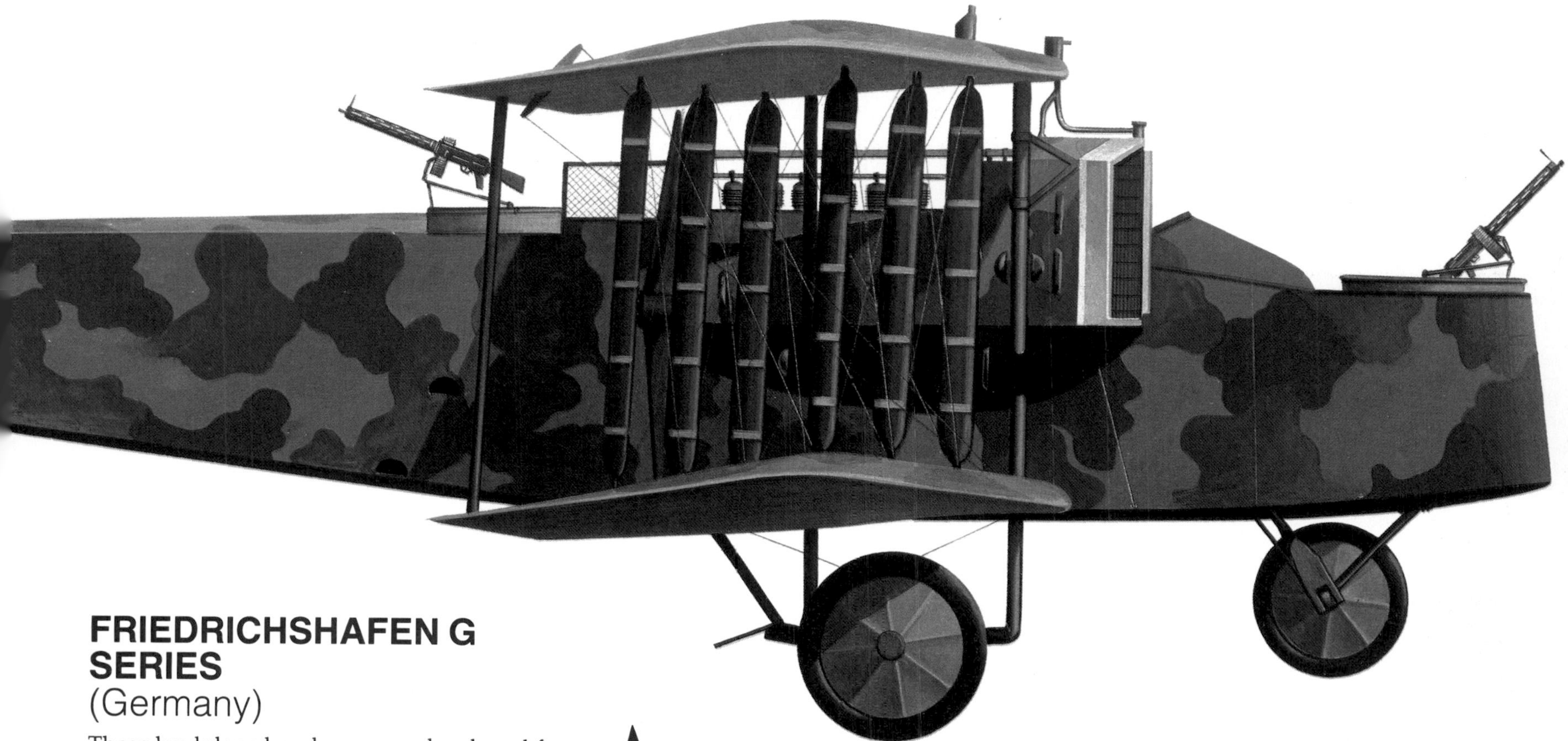

FRIEDRICHSHAFEN G SERIES
(Germany)

These landplane bombers were developed from the G I prototype that first flew in 1915 with two 112kW (150 hp) Benz Bz.III pusher engines. From this was developed the G II service model with reduced span, a monoplane in place of the biplane tail, 149kW (200 hp) Bz. IV engines and improved armament. Only a few G IIs were built before there appeared the much enhanced G III. This embodied the original three-bay large wing, detail modifications and more power in the form of 194kW (260 hp) Mercedes D.IVa engines to make possible a 1,500kg (3,307 lb) bombload. The G IIIa reverted to a biplane tail and the G V had a shorter nose and was equipped with tractor engines.

▲

FRIEDRICHSHAFEN G III

Role: Long-range bomber
Crew/Accommodation: Three
Power Plant: Two 260 hp Mercedes D IVa water-cooled inlines
Dimensions: Span 23.7m (77.76 ft); length 12.8m (41.99 ft); wing area $95m^2$ (1,022.6 sq ft)
Weights: Empty 2,695kg (5,929 lb); MTOW 3,930kg (8,646 lb)
Performance: Maximum speed 135km/h (83.9 mph) at sea level; operational ceiling 5,000m (16,400 ft); endurance 5 hours
Load: Two/three 7.7mm machine guns, plus, typically, 1,500kg (3,307 lb) of bombs

GENERAL AIRCRAFT MONOSPAR ST SERIES
(U.K.)

General Aircraft was formed in 1934 to succeed the Monospar Wing Co, pioneer of wings based on a single tip-to-tip tubular spar. Such a wing was tested on the ST-1, and then proved more fully in the Gloster-built SR-3 before the company embarked on the ST-4 four-seat tourer with two 63kW (85 hp) Pobjoy R radial engines. Other models were the ST-6, with manually retracting main landing gear units, the ST-10 with 67kW (990 hp) Pobjoy Niagara engines, the ST-11 with 97kW (130 hp) de Havilland Gipsy Major engines and manually retracting main landing gear units, and generally similar ST-12 with fixed landing gear. The final development was the ST-25 that appeared in 1935 as an updated version of the ST-10. Production totalled 59 and later aircraft had end-plate vertical tail surfaces.

GENERAL AIRCRAFT MONOSPAR ST-25J

Role: Light, short-range passenger transport
Crew/Accommodation: One, plus up to three passengers
Power Plant: Two 95 hp Pobjoy Niagra III air-cooled radials
Dimensions: Span 12.24m (40.17 ft); length 8.03m (26.33 ft); wing area 20.2m^2 (217 sq ft)
Weights: Empty 762kg (1,680 lb); MTOW 1,304kg (2,875 lb)
Performance: Cruise speed 229km/h (142 mph) at sea level; operational ceiling 6,401m (21,000 ft); range 660km (410 miles)
Load: Up ro 408kg (900 lb)

GENERAL AIRCRAFT HAMILCAR
(U.K.)

The Hamilcar was a glider designed to deliver heavy equipment, including light tanks, needed by airborne forces, and first flew in March 1942. Some 390 Hamilcar Mk I gliders were delivered, and a further 22 aircraft were Hamilcar Mk X conversions with two 720kW (965 hp) Bristol Mercury 31 radial engines.

GENERAL AIRCRAFT HAMILCAR Mk I

Role: Military tank-carrying glider
Crew/Accommodation: Two
Power Plant: None
Dimensions: Span 33.53m (110 ft); length 20.72m (68 ft); wing area 153.98m^2 (1,658 sq ft)
Weights: Empty 8,845kg (19,500 lb); MTOW 16,783kg (36,000 lb)
Performance: Maximum gliding speed 240km/h (150 mph); operational ceiling dependent upon tug aircraft
Load: Up to 7,940kg (17,500 lb)

GENERAL DYNAMICS F-16 FIGHTING FALCON
(U.S.A.)

In the Vietnam War the U.S. Air Force discovered that its fighters were in general handicapped by their size, weight and Mach 2 performance, all of these being factors that seriously eroded reliability and combat agility at increasingly common low and medium altitudes. In 1975 it was announced that the competition for a new lightweight fighter had resulted in victory for a General Dynamics design that entered production as the single-seat F-16A and two-seat F-16B.

The type has gone on to become numerically the most important fighter in the inventory of the Western world. The type is based on blended contours and relaxed stability controlled by a fly-by-wire system whose sidestick joystick is operated by a semi-reclining pilot. The type has matured as an exceptional multi-role fighter, the first two variants having the 10,814kg (23,840 lb) afterburning thrust Pratt & Whitney F100-P-200 turbofan. A host of electronic improvements has been added to create the latest F-16C single-seat and F-16D two-seat variants, which can use the F100 or General Electric F110 turbofan in their most recent and powerful forms. There have been a number of experimental variants, and future developments are centred on the FSX next-generation derivative to be produced in Japan, and a possible close-support model.

GENERAL DYNAMICS F-16A FIGHTING FALCON

Role: Day strike fighter
Crew/Accommodation: One
Power Plant: One 10,814 kgp (23,840 lb s.t.) Pratt & Whitney F100-P-200 turbofan
Dimensions: Span 10.01m (32.83 ft); length 14.52m (47.64 ft); wing area 17.87m^2 (300 sq ft)
Weights: Empty 6,613kg (14,567 lb); MTOW 17,010kg (37,500 lb)
Performance: Maximum speed 2,146km/h (1,158 knots) Mach 2.02 at 12,190m (40,000 ft); operational ceiling 15,850m (52,000 ft); range 580km (313 naut. miles) with 3,000 lb bombload
Load: One 20mm cannon, plus up to 6,894kg (15,200 lb) of bombs

General Dynamics F-16B

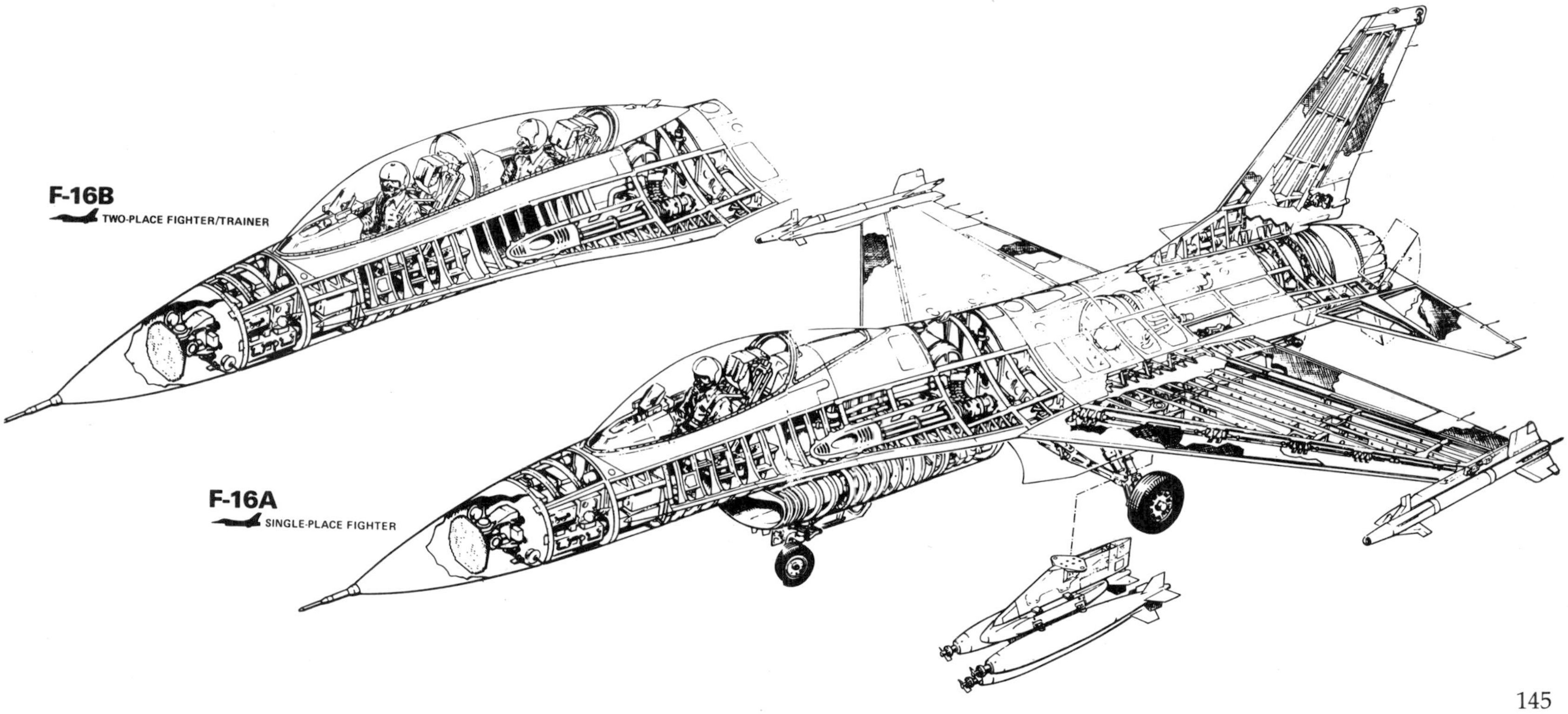

GENERAL DYNAMICS F-111F

Role: Long-range low-level variable-geometry strike
Crew/Accommodation: Two
Power Plant: Two 11,385kgp (25,100 lb s.t.) Pratt & Whitney TF30-P-100 turbofans with reheat
Dimensions: Span spread 19.2m, swept 9.74m, spread 63m, swept 31.95m; length 23.02m (75.52 ft); wing area 48.77m^2 (525 sq ft)
Weights: Empty 23,525kg (47,500 lb); MTOW 45,360kg (100,000 lb)
Performance: Maximum speed 1,471km/h (794 knots) Mach 1.2 at sea level; operational ceiling 17,650m (57,900 ft); range 1,480km (799 naut.miles)
Load: One 20mm multi-barrel cannon, plus up to 11,340kg (25,000 lb) of ordnance/fuel

GENERAL DYNAMICS F-111 ▲
(U.S.A.)

The F-111 was the world's first operational 'swing-wing' aeroplane, and remains in valuable service as the U.S. Air Force's most potent all-weather long-range interdiction platform. The type originated from a 1960 requirement for a variable-geometry tactical aeroplane that could be developed in parallel land- and carrier-based forms. The type first flew in December 1964, but the F-111B naval type was eventually cancelled because of intractable weight and technical problems. The land-based model also suffered teething problems, but despite an indifferent power plant, has matured as an exceptional type whose two most important models are the long-span FB-111A strategic bomber and the F-111F interdictor.

GLOSTER GREBE and GAMECOCK
(U.K.)

The Grebe was the partner of the Armstrong Whitworth Siskin as the RAF's main fighter in the early 1920s. The type first flew in May 1923 as an orthodox biplane with 242kW (325 hp) Armstrong Siddeley Jaguar III radial engine, and the production version was the Grebe Mk II. There was also a Grebe (Dual) trainer version. The Grebe's main liability was its engine, and the generally similar Gamecock was designed as its replacement with a 317kW (425 hp) Bristol Jupiter VI. The type first flew in February 1925 and was ordered for the RAF in its basic Gamecock Mk I form. A few improved Gamecock Mk IIs were also received.

◄

GLOSTER GAMECOCK Mk I

Role: Fighter
Crew/Accommodation: One
Power Plant: One 425 hp Bristol Jupiter VI air-cooled radial
Dimensions: Span 9m (29.9 ft); length 5.99m (19.66 ft); wing area 24.52m^2 (264 sq ft)
Weights: Empty 875.4kg (1,930 lb); MTOW 1,243.7kg (2,742 lb)
Performance: Maximum speed 233km/h (145 mph) at 3,048m (10,000 ft); operational ceiling 6,735m 22,100 ft); endurance 2.5 hours
Load: Two .303 inch machine guns

GLOSTER GLADIATOR
(U.K.)

The Gladiator was the RAF's last biplane fighter, an evolutionary development of the Gauntlet with cantilever main landing gear legs, and enclosed cockpit and trailing-edge flaps on the lower wings. The SS.37 prototype first flew in September 1934 and showed great promise. The type was then trialled with a more powerful engine and ordered as the Gladiator Mk I with the 626kW (840 hp) Mercury IX radial engine. Later variants were the Gladiator Mk II with the 619kW (830 hp) VIIIA plus electric starter, and the Sea Gladiator with equipment for carrier operations.

GLOSTER GLADIATOR Mk I

Role: Fighter
Crew/Accommodation: One
Power Plant: One 840 hp Bristol Mercury IX air-cooled radial
Dimensions: Span 9.83m (32.25 ft); length 8.36m (27.42 ft); wing area 30m^2 (323 sq ft)
Weights: Empty 1,574kg (3,476 lb); MTOW 2,155kg (4,750 lb)
Performance: Maximum speed 402km/h (250 mph) at 4,724m (15,500 ft); operational ceiling 9,997m (32,800 ft); range 660km (410 miles)
Load: Four .303 inch machine guns

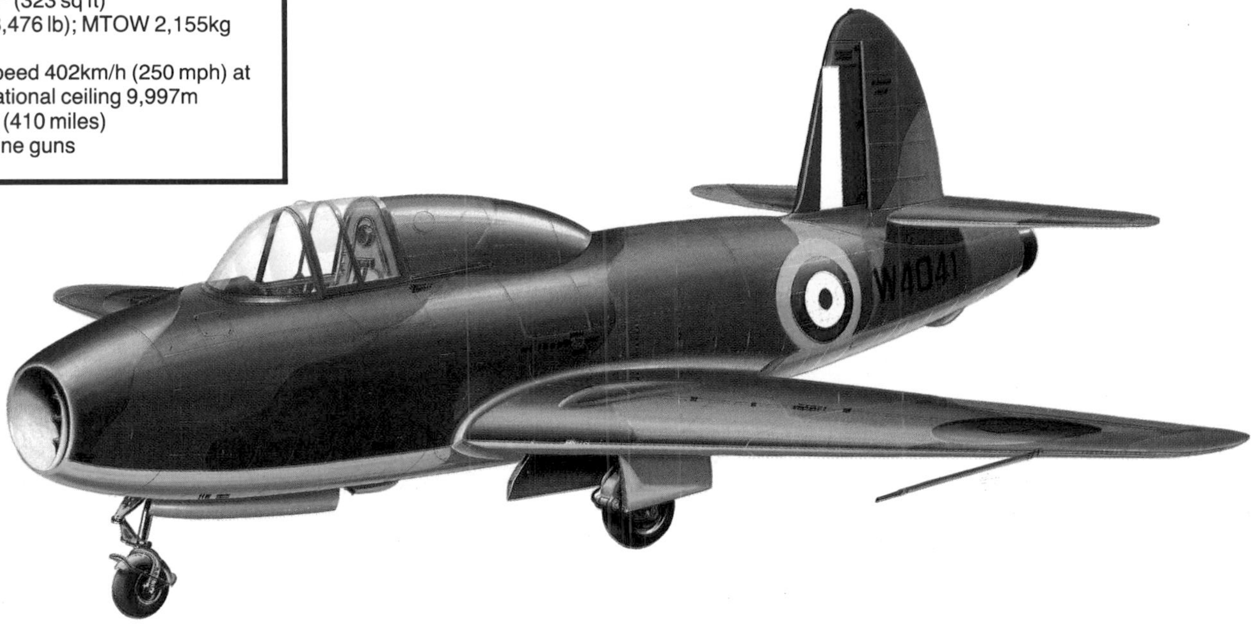

GLOSTER E.28/39
(U.K.)

This was the U.K.'s first jet-powered aeroplane, and resulted from a 1939 requirement for a fighter-type testbed for the new Whittle W.1 centrifugal-flow turbojet. The type first flew in May 1941 with a 390kg (860 lb) thrust Power Jets W.1 engine, and the type proved highly successful in the evaluation of successively more powerful variants of the same basic engine. As it was only experimental it carried no armament.

GLOSTER E.28/39

Role: Experimental jet
Crew/Accommodation: One
Power Plant: One 390kgp (860 lb s.t.) Power Jets W1 turbojet, progressing ultimately to the 907 kgp (2,000 lb s.t.) Power Jets W2/700 turbojet
Dimensions: Span 8.84m (29 ft); length 7.71m (25.3 ft); wing area 13.5m^2 (146.5 sq ft)
Weights: Empty 1,309kg (2,886 lb); MTOW 1,700kg (3,748 lb)
Performance: Maximum speed 750km/h (466 mph) at 3,281m (1,000 ft); operational ceiling 9,756m (32,000 ft); Endurance 45.5 minutes
Load: None

GLOSTER JAVELIN
(U.K.)

This was the RAF's first production delta-winged aeroplane, and also that service's first purpose-designed all-weather fighter. The first prototype flew in November 1951 with two 3,629kg (8,000 lb) thrust Armstrong Siddeley Sapphire ASSA.6 turbojets. The main variants were the Javelin F(AW). Mk 1 with AI. Mk 17 radar, the F(AW). Mk 2 with U.S. APQ-43 radar, the lengthened T.Mk 3 trainer, the F(AW). Mk 4 with a powered slab tailplane, the F(AW). Mk 5 with provision for four Firestreak air-to-air missiles, the F(AW). Mk 6 equivalent of the Mk 5 with APQ-43 radar, the F(AW). Mk 7 with 4,880kg (11,000 lb) thrust ASSA.7 engines, the F(AW). Mk 8 with APQ-43 radar and ASSA.7R engine with primitive afterburning, and the F(AW). Mk 9 conversion of the Mk 7 to Mk 8 standard with AI.Mk 17 radar.

GLOSTER JAVELIN F(AW). Mk 9

Role: All-weather/night fighter
Crew/Accommodation: Two
Power Plant: Two 4,900kgp (11,000 lb s.t.) Armstrong Siddeley Sapphire Sa 7R with limited reheat giving 5,579 (12,300 lb s.t.) at altitude
Dimensions: Span 15.85m (52 ft); length 17.3m (56.75 ft); wing area 86m^2 (927 sq ft)
Weights: Empty 12,610kg (27,800 lb); MTOW 19,578kg (43,165 lb)
Performance: Maximum speed 989km/h (534 knots) Mach 0.926 at 10,699m (35,000 ft); operational ceiling 15,849m (52,000 ft); endurance 1.75 hours with external fuel
Load: Four 30mm cannon, plus four short-range air-to-air missiles

GLOSTER METEOR
(U.K.)

The Meteor was the Allies' only jet fighter to see combat in World War II, and just pipped the Germans' Me 262 to the title of the world's first operational jet aeroplane. Gloster was the logical choice to develop such an aeroplane after its experience with the E.28/39, and the first prototype flew in March 1943. Trials with a number of engine types and variants slowed development of a production variant, but the Meteor F.Mk I finally entered service in July 1944 with 771kg (1,700 lb) thrust Rolls-Royce W.2 Welland turbojets. The Meteor remained in RAF service until the late 1950s with the Derwent turbojet that was introduced on the second production variant, the Meteor F.Mk III. The type underwent significant development, and the main streams were Meteor F.Mks 4 and 8 single-seat fighters, the Meteor FR.Mk 9 reconnaissance fighter, the Meteor NF.Mks 11 to 14 radar-equipped night-fighters, the Meteor PR.Mk 10 photo-reconnaissance aeroplane, and the Meteor T.Mk 7 two-seat trainer. Surplus aircraft were often converted into target tugs or remotely controlled target drones.

GLOSTER METEOR F. Mk 8

Role: Fighter
Crew/Accommodation: One
Power Plant: Two 1,723kgp (3,800 lb s.t.) Rolls-Royce Derwent 9 turbojets
Dimensions: Span 11,33m (37.16 ft); length 13.59m (44.58 ft); wing area 32.5m^2 (350 sq ft)
Weights: Empty 4,846kg (10,684 lb); MTOW 7,121kg (15,700 lb)
Performance: Maximum speed 962km/h (598 mph) at 3,048m (10,000 ft); operational ceiling 13,106m (43,000 ft); endurance 1.2 hours with ventral and wing fuel tanks
Load: Four 20mm cannon

GOTHA G V

Role: Long-range bomber
Crew/Accommodation: Three
Power Plant: Two 260 hp Mercedes D.IVa water-cooled inlines
Dimensions: Span 23.7m (77.76 ft); length 11.86m (38.91 ft); wing area 89.5m² (963.37 sq ft)
Weights: Empty 2,400kg (5,280 lb); MTOW 3,975kg (8,745 lb)
Performance: Maximum speed 140km/h (87 mph) at 3,660m (12,000 ft); operational ceiling 6,500m (21,320 ft); endurance 5.5 hours
Load: Two 7.9mm machine guns, plus up to 500kg (1,102 lb) of bombs

GOTHA G II to G V
(Germany)

This series was Germany's main long-range bomber family of the two-engined type, and began with the G II that first flew in 1915 with 164kW (220 hp) Benz engines and was used mainly on the Western Front. The G III was similar apart from its 194kW (260 hp) Mercedes engines and ventral 'tunnel' for a defensive machine gun. The G IV was of mixed construction and had 194kW Mercedes D.IVa engines. The G Va introduced a compound tail, and the G Vb had a nose wheel in an effort to overcome the earlier models' tendency to nose over on landing.

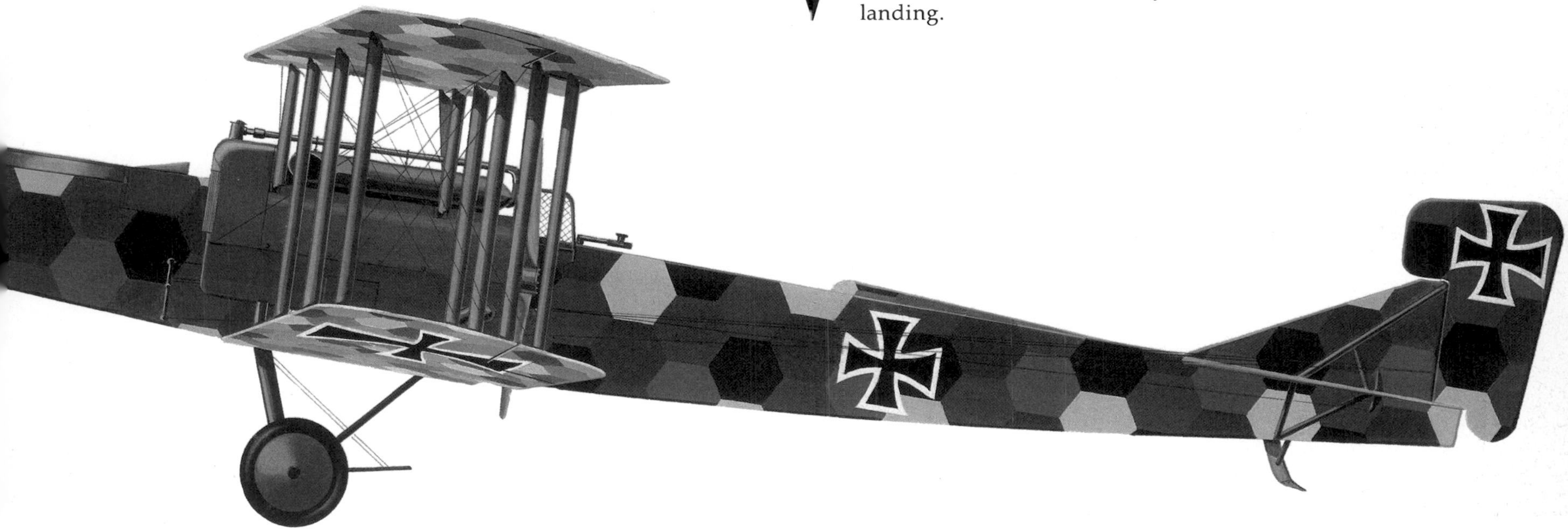

GOTHA Go 242 and Go 244
(Germany)

The Go 242 was developed to replace the DFS 230 as the German air force's main assault glider, and offered accommodation for 21 troops or a light vehicle given access to the hold by the hinged rear door to the central nacelle of this twin-boom type. Two prototypes flown in 1941 were followed by quite extensive production of the Go 242A with skid landing gear, the Go 242B with jettisonable tricycle landing gear and the Go 242C maritime version with a planing hull and outrigger stabilizer floats. Germany also developed a powered version as the Go 244B, a derivative of the Go 242B with two 522kW (700 hp) Gnome-Rhone 14M radial engines. Some 133 Go 242Bs were converted to Go 244Bs.

GOTHA Go 242A-2

Role: Military assault glider
Crew/Accommodation: Two, plus up to 21 troops
Power Plant: None
Dimensions: Span 24,5m (80.38 ft); length 15.8m (51.84 ft); wing area 64.4m² (693.2 sq ft)
Weights: Empty 3,200kg (7,055 lb); MTOW 6,800kg (14,991 lb)
Performance: Maximum gliding speed 290km/h (180 mph) operational ceiling dependent upon tug aircraft
Load: Four 7.9mm machine guns, plus up to 3,450kg (7,606 lb)

GRUMMAN G-5 (FF-1)
(U.S.A.)

This was Grumman's first aeroplane for the U.S. Navy, a two-seat carrierborne fighter of biplane configuration but fitted with retractable landing gear and a long enclosed cockpit. The prototype first flew in December 1931 with a 459kW (616 hp) Wright R-1820E radial engine, and the type was ordered into production as the FF-1. Some 25 were later converted as FF-2 dual-control fighter trainers. The type was also built under licence in Canada as the Goblin (G-23), and a derivative of the U.S. Navy model was the SF-1 (G-6) scout fighter with the 522kW (700 hp) R-1820-78 radial engine and additional fuel but less defensive armament.

GRUMMAN FF-1

Role: Naval carrierborne fighter
Crew/Accommodation: Two
Power Plant: One 700 hp Wright R-1820-78 Cyclone air-cooled radial
Dimensions: Span 10.52m (34.5 ft); length 7.47m (24.5 ft); wing area 28.8m^2 (310 sq ft)
Weights: Empty 1,395kg (3,076 lb); MTOW 2,111kg (4,655 lb)
Performance: Maximum speed 323km/h (201 mph) at sea level; operational ceiling 6,828m (22,400 ft); range 1,178km (732 miles)
Load: Three .303 inch machine guns, plus 91 kg (200 lb) of bombs

GRUMMAN G-4 (JF) and G-15 (J2F DUCK)
(U.S.A.)

This utility amphibian used features of the FF-1 with a single main float neatly fared to the underside of the fuselage and containing tail-wheel landing gear using a neat retraction system. The prototype flew in May 1933 and the type was ordered for the U.S. Navy as the JF-1 with 522kW (700 hp) Pratt & Whitney R-1830 Twin Wasp radial engine; a similar JF-2 version was ordered by the U.S. Coast Guard with the 559kW (750 hp) Wright R-1820 Cyclone, and a similar Cyclone-powered model was taken by the navy as the JF-3. The later J2F differed only in minor details, and was built in successively re-engined versions up to the J2F-6.

GRUMMAN JF-1 DUCK

Role: Naval light utility transport
Crew/Accommodation: Two
Power Plant: One 700 hp Pratt & Whitney R-1830-62 Twin Wasp air-cooled radial
Dimensions: Span 11.89m (39 ft); length 9.93m (32.58 ft); wing area 37.99m^2 (409 sq ft)
Weights: Empty 1,866kg (4,113 lb); MTOW 2,438kg (5,375 lb)
Performance: Maximum speed 270km/h (168 mph) at sea level; operational ceiling 5,486m (18,000 ft); range 1,304km (810 miles)
Load: One .303 inch machine gun, plus 91kg (200 lb) of bombs

GRUMMAN G-8 (F2F) AND G-11 (F3F)
(U.S.A.)

The F2F was a logical evolution of the FF-1's concept into single-seat fighter form, and first flew in prototype form during October 1933. The type proved to possess high performance and adequate handling, and was ordered into production as the F2F-1 with the 485kW (650 hp) Pratt & Whitney R-1535-72 Twin Wasp Jr radial engine. Further development was centred on improvements to handling, and this produced the F3F with a lengthened fuselage and increased-span wings. The prototype flew in March 1935, and resulted in orders for the F3F-1 production model with the 485kW (650 hp) R-1535-84 radial engine. Subsequent development produced the F3F-2 with the 634kW (850 hp) Wright R-1820-22 radial engine driving a controllable-pitch propeller, and finally the F3F-3 with a number of drag-reducing features.

GRUMMAN F3F-1

Role: Naval carrierborne fighter
Crew/Accommodation: One
Power Plant: One 650 hp Pratt & Whitney R-1535-84 air-cooled radial
Dimensions: Span 9.75m (32 ft); length 7.01m (23 ft); wing area 24.25m^2 (261 sq ft)
Weights: Empty 1,301kg (2,868 lb); MTOW 1,857kg (4,094 lb)
Performance: Maximum speed 364km/h (226 mph) at sea level; operational ceiling 8,992m (29,500 ft); range 1,424km (885 miles)
Load: One .5 inch and one .303 inch machine gun, plus 91 kg (200 lb) of bombs

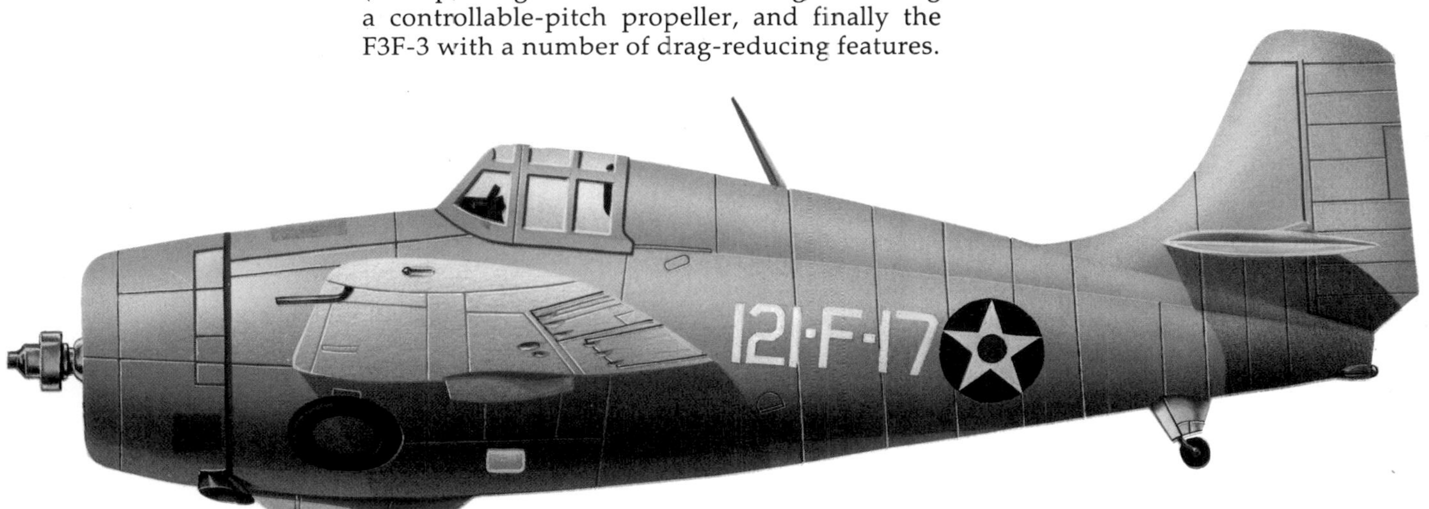

GRUMMAN G-36 (F4F WILDCAT)
(U.S.A.)

The F4F designation was first used for a biplane fighter designed in competition to the Brewster F2A Buffalo monoplane, but re-evaluation of the Grumman proposal led the U.S. Navy to call for an XF4F-2 monoplane prototype, and this first flew in September 1937 with a 783kW (1,050 hp) Pratt & Whitney R-1830-66 Twin Wasp radial engine. This initial model was judged slightly inferior to the Buffalo, but was revised as the XF4F-3 with redesigned tail, a larger wing and the XR-1830-76 radial engine with a two-stage supercharger. Performance and handling were so much improved that the type was ordered as the F4F-3, the British taking a similar version as the Martlet Mk I.

The F4F/Martlet was the Allies' first carrierborne fighter able to meet land-based opponents on anything like equal terms, and proved invaluable during the war years up to 1943. An essentially similar type was produced as the FM-1 and -2 (Martlet Mks V and VI) by the Eastern Aircraft Division of General Motors.

GRUMMAN F4F-4 WILDCAT

Role: Naval carrierborne fighter
Crew/Accommodation: One
Power Plant: One 1,200 hp Pratt & Whitney R-1830-86 Twin Wasp air-cooled radial
Dimensions: Span 11.58m (38 ft); length 8.76m (28.75 ft); wing area 24.16m^2 (260 sq ft)
Weights: Empty 2,624kg (5,785 lb); MTOW 3,607kg (7,952 lb)
Performance: Maximum speed 512km/h (318 mph) at 5,913m (19,400 ft); operational ceiling 10,638m (34,900 ft); range 1,239km (770 miles) on internal fuel only
Load: Six .5 inch machine guns

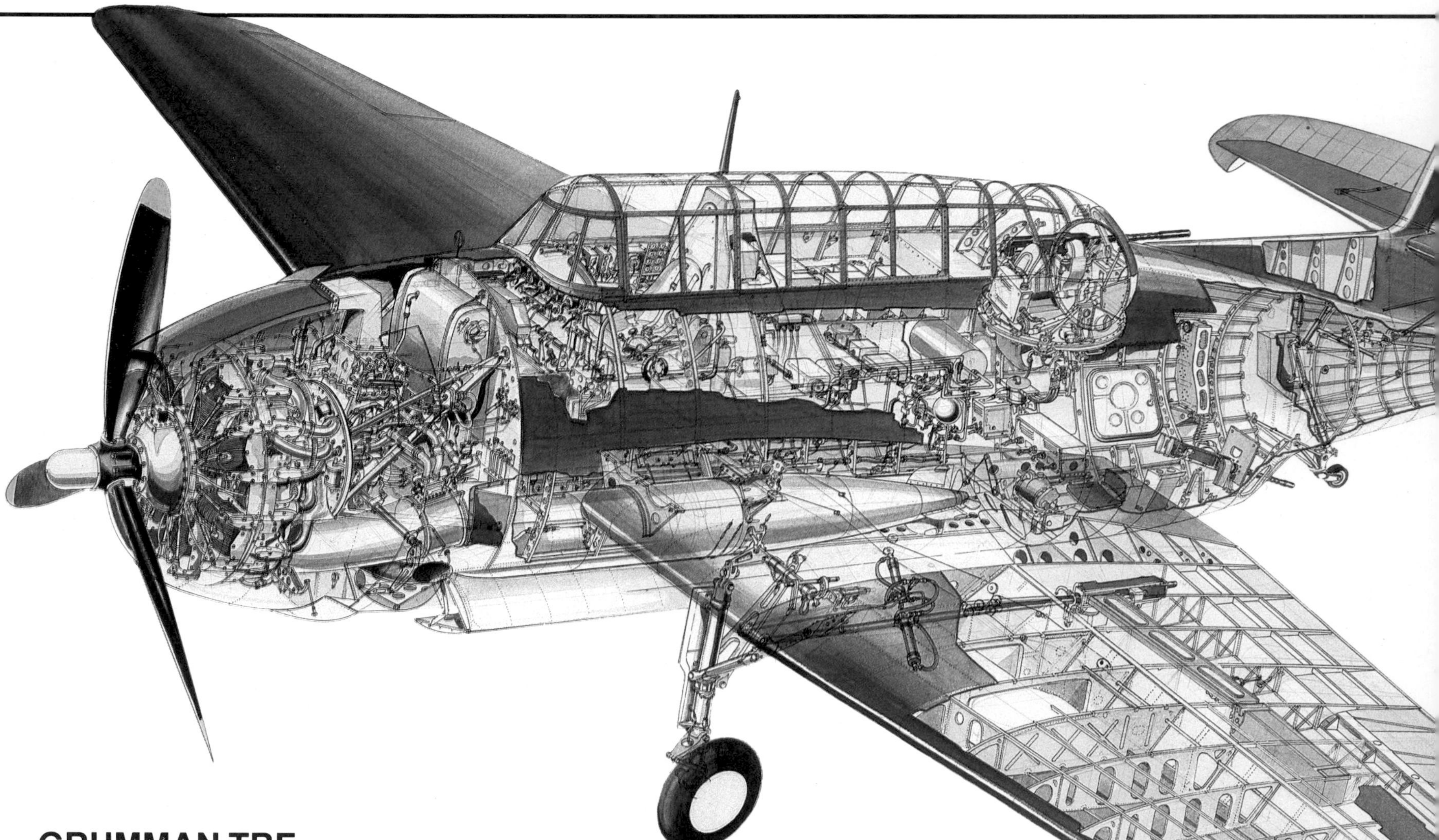

GRUMMAN TBF AVENGER
(U.S.A.)

▲

Despite a disastrous combat debut in which five out of six aircraft were lost, the Avenger was a decisive aeroplane of World War II, and may rightly be regarded as the Allies' premier carrierborne torpedo bomber. The type first flew in August 1941 powered by a 1,268kW (1,700 hp) Wright R-2600-8 Cyclone radial engine, and was ordered into production as the TBF-1 or, with two additional heavy machine gun in the wings plus provision for drop tanks, TBF-1C. The Royal Navy also received the type as the Tarpon Mk I, later changed to Avenger Mk I. The Eastern Aircraft Division of General Motors was also brought into the programme to produce similar TBM-1 and TBM-1C (Avenger Mk II) models, and the only major development was the TBM-3 without the initial model's heavy power-operated dorsal turret. The series was diversified into a host of other roles such as early warning, anti-submarine search/attack, reconnaissance and target towing.

GRUMMAN TBF-1 AVENGER

Role: Naval carrierborne strike
Crew/Accommodation: Three
Power Plant: One 1,700 hp Wright R-2600-8 Double Cyclone air-cooled radial
Dimensions: Span 16.51m (54.16 ft); length 12.23m (40.125 ft); wing area 45.52m^2 (490 sq ft)
Weights: Empty 4,572kg (10,080 lb); MTOW 7,214kg (15,905 lb)
Performance: Maximum speed 436km/h (271 mph) at 3,658m (12,000 ft); operational ceiling 6,828m (22,400 ft); range 1,955km (1,215 miles) with torpedo
Load: One .5 inch and two .303 inch machine guns, plus up to 726 kg (1,600 lb) of internally-stowed torpedo or bombs

GRUMMAN F6F HELLCAT
(U.S.A.)

The Hellcat was the logical successor to the Wildcat with more size and power in a generally similar airframe with a low- rather than mid-set wing. A number of operational improvements suggested by Wildcat experience were incorporated in the type, of which four prototypes were built with different engine installations. The first of these flew in June 1942, and the type selected for production as the F6F-3 was the variant powered by the 1,491kW (2,000 hp) Pratt & Whitney R-2800-20 radial engine. That the basic Hellcat was in all significant respects 'right' is attested by the relatively few variants produced in a large production run. There were a few radar-equipped F6F-3E and F6F-3N night-fighters, and from early 1944 production switched to the F6F-5 with a water-boosted R-2800-10W radial engine for a 10% power boost in combat; there were also a few F6F-5N night-fighter and F6F-5P reconnaissance aircraft.

►

GRUMMAN F6F-5 HELLCAT

Role: Naval carrierborne fighter
Crew/Accommodation: One
Power Plant: One 2,000 hp Pratt & Whitney R-2800-10W Double Wasp air-cooled radial
Dimensions: Span 13.06m (42.83 ft); length 10.31m (33.83 ft); wing area 31.03m^2 (334 sq ft)
Weights: Empty 4,100kg (9,060 lb); MTOW 5,714kg (12,598 lb)
Performance: Maximum speed 612km/h (380 mph) at 7,132m (23,400 ft); operational ceiling 11,369m (37,300 ft); range 1,521km (945 miles)
Load: Two 20mm cannon and four .5 inch machine guns, plus up to 975 kg 2,150 lb of weapons, including one torpedo

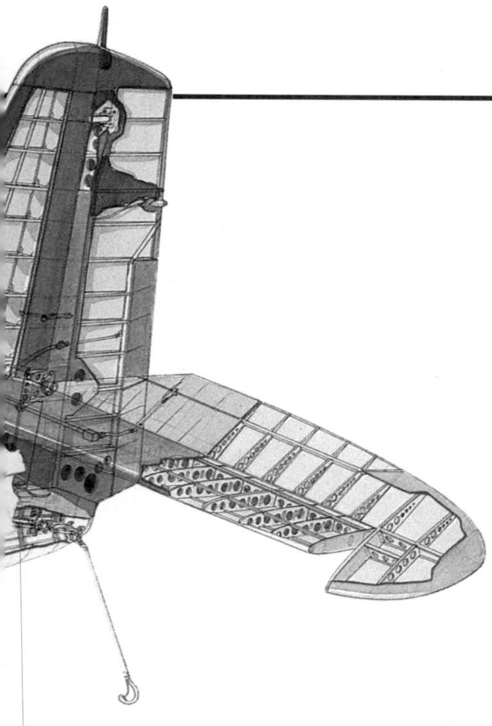

GRUMMAN G-58 (F8F BEARCAT)
(U.S.A.)

Though resembling a cleaned-up Hellcat in configuration, the Bearcat was a smaller but more powerful aeroplane designed to provide U.S. carrier forces with a potent interceptor. The type first flew in August 1944, with deliveries of the F8F-1 version beginning only six months later. With the end of World War II production was curtailed, and the only significant variants were the F8F-1B with cannon rather than machine guns in the wings and the F8F-1N night-fighter with radar. The F8F-2 introduced a taller fin and revised cowling, and there were night-fighter and reconnaissance variants of this model.

GRUMMAN F8F-1 BEARCAT

Role: Naval carrierborne interceptor
Crew/Accommodation: One
Power Plant: One 2,100 Pratt & Whitney R-2800-34W Double Wasp air-cooled radial
Dimensions: Span 10.92m (35.83 ft); length 8.61m (28.25 ft); wing area 22.67m^2 (244 sq ft)
Weights: Empty 3,207kg (7,070 lb); MTOW 5,873kg (12,947 lb)
Performance: Maximum speed 682km/h (421 mph) at 6,005m (19,700 ft); operational ceiling 11,796m (38,700 ft); range 1,778km (1,105 miles) on internal fuel only
Load: Four .5 inch machine guns, plus up to 907 kg (2,000 lb) of bombs or extra fuel

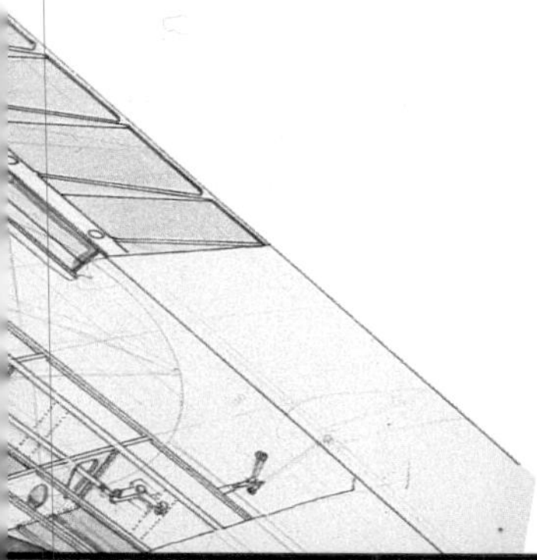

Grumman TF-9J

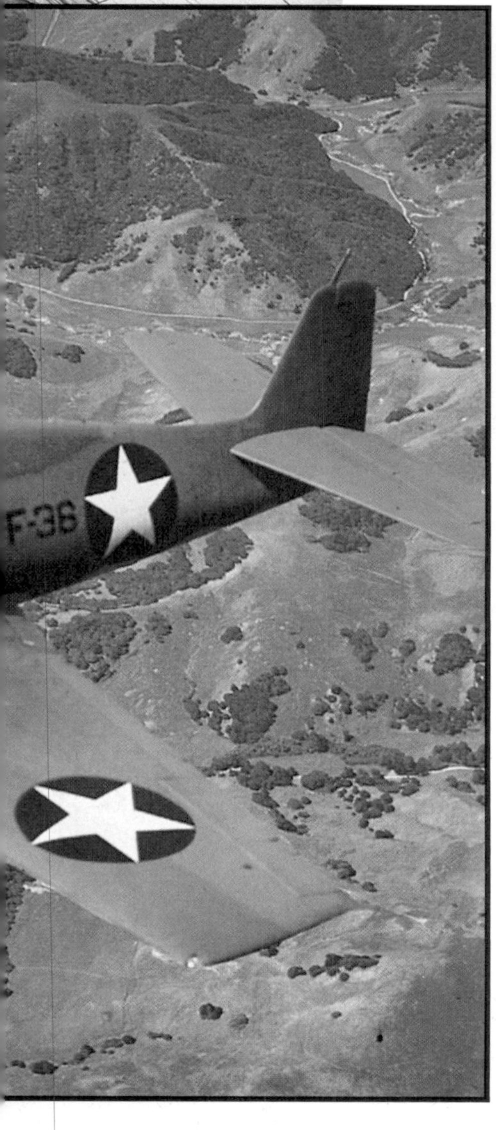

GRUMMAN G-79 (F9F PANTHER and COUGAR)
(U.S.A.)

The straight-wing Panther was the U.S. Navy's most important fighter of the Korean War, and first flew in November 1948. Prototypes were trialled with the Allison J33 and the Pratt & Whitney J42 (licence-built Rolls-Royce Nene) turbojets. Both engines were selected for production models, the F9F-2 having the J42-P-6 and the F9F-3 and J33-A-8, though the unreliability of the latter led to re-engining with the J42. A few F9F-4s were built with the J33-A-16, but the major variant was the F9F-5 with the Pratt & Whitney J48-P-2, a licence-built version of the Rolls-Royce Tay. There were also F9F-2B fighter-bomber and F9F-5P reconnaissance models. Grumman then introduced a radically improved version as the swept-wing Cougar with a more powerful J48-P-8 turbojet in the initial F6F-6 variant that was redesignated F-9F in 1962. Major variants were the improved F9F-7 (F-9H), the F9F-8 (F-9J) with a longer fuselage and broader-chord wing and the F9F-8T (TF-9J) tandem-seat trainer.

GRUMMAN F9F-5 PANTHER

Role: Naval carrierborne fighter
Crew/Accommodation: One
Power Plant: One 3,175 kgp (7,000 lb s.t.) Pratt & Whitney J48-P-2 turbojet
Dimensions: Span 11,58m (37.99 ft); length 11.84m (38.85 ft); wing area 23.22m^2 (250 sq ft)
Weights: Empty 4,603kg (10,147 lb); MTOW 09,344g (20,600 lb) for carrier launch
Performance: Maximum speed 972km/h (604 mph) at sea level; operational ceiling 13,045m (42,800 ft); range 2,093km (1,300 miles) at attitude
Load: Four 20mm cannon, plus up to 1,361 kg (3,000 lb) of bombs

GRUMMAN G-89 (S2F and S-2 TRACKER)
(U.S.A.)

The Tracker was designed as a compact yet capable carrierborne anti-submarine hunter/killer, and first flew in December 1952. The engine and airframe combination remained essentially unaltered through the type's production life, variants up to the S2F-3 (from 1962 S-2D) having steadily upgraded mission electronics and the S-2G conversion featuring state-of-the-art equipment of the early 1960s. Some surviving aircraft were being re-engined with turboprops in the late 1980s. Other variants were the TF-1 (C-1) Trader nine-passenger carrier onboard delivery and the WF-2 (E-1) Tracer airborne early warning aircraft, the latter with a huge 'guppy' radome above the fuselage.

GRUMMAN S2F-1/S-2A TRACKER

Role: Naval carrierborne anti-submarine
Crew/Accommodation: Four
Power Plant: Two, 1,525 hp Wright R-1820-82 Cyclone air-cooled radials
Dimensions: Span 21.23m (69.66 ft); length 12.88m (42.25 ft); wing area 45.1m^2 (485 sq ft)
Weights: Empty 8,505kg (18,750 lb); MTOW 13,222kg (29,150 lb)
Performance: Maximum speed 426km/h (230 knots) at 450m (1,500 ft); operational ceiling 6,400m (21,000 ft); range 2,095km (1,129 naut. miles)
Load: Up to 2,182 kg (4,810 lb)

GRUMMAN G-134 (OV-1 MOHAWK)
(U.S.A.)

The Mohawk resulted from a joint U.S. Army and Marine Corps requirement for a battlefield reconnaissance aeroplane with STOL capability, and first flew in April 1959. The Marine Corps pulled out of the programme, and production aircraft were delivered to the Army with the designation OV-1. The current model is the OV-1D with infra-red surveillance and side-looking radar equipment, while specialized derivatives are the RV-1 electronic intelligence and EV-1 electronic surveillance models.

GRUMMAN OV-1D MOHAWK

Role: Short rough field capable observation
Crew/Accommodation: Two
Power Plant: Two 1,400 shp Lycoming T53-L-701 turboprops
Dimensions: Span 14.63m (48 ft); length 12.5m (41 ft); wing area 33.45m^2 (360 sq ft)
Weights: Empty 5,467kg (12,054 lb); MTOW 8,214kg (18,109 lb)
Performance: Maximum speed 491km/h (305 mph) at 3,050m (10,000 ft); operational ceiling 7,620m (25,000 ft); range 1,627km (1,011 miles)
Load: None, other than special-to-mission onboard equipment

GRUMMAN G-128 (A-6 INTRUDER)
(U.S.A.)

The Intruder was designed as a long-range attack aeroplane able to operate in all weathers and deliver large warloads with pinpoint accuracy. Despite its firmly subsonic performance and a first flight as early as April 1960, the type is still the U.S. Navy's most important attacker as a result of steady improvement of the electronic suite. The latest variant is the A-6E/TRAM with a unique combination of navigation and attack systems.

GRUMMAN A-6E INTRUDER

Role: Naval carrierborne all-weather heavy strike (bomber)
Crew/Accommodation: Two
Power Plant: Two, 4,218kgp (9,300 lb s.t.) Pratt & Whitney J52-P-8B turbojets
Dimensions: Span 16.15m (53 ft); length 16.7m (54.75 ft); wing area 49.15m² (529 sq ft)
Weights: Empty 12,000kg (26,456 lb); MTOW 27,395kg (60,395 lb)
Performance: Maximum speed 1,038km/h (560 knots) Mach 0.85 at sea level; operational ceiling 12,954m (42,500 ft); radius 595km (316 naut. miles) 6,350kg (14,00 lb) bombload
Load: Up to 8,165 kg (18,000 lb) of weapons — all externally carried

GRUMMAN EA-6 PROWLER
(U.S.A.)

After experience with the A-6 converted to the electronic warfare escort role with the designation EA-6A, the U.S. Navy called for a specialized electronic warfare variant. This appeared as the EA-6B with its fuselage lengthened by 1.37m (4 ft 6 inches) to allow the insertion of a stretched cockpit to accommodate, in addition to the standard two crew, two specialist operators for the ALQ-99 system of radar analysers and jammers. The EA-6B has undergone enormous electronic development, and this process continues in an effort to ensure the Prowler remains capable of handling all electronic threats until well into the next century.

GRUMMAN EA-6B PROWLER

Role: Naval carrierborne, all-weather electronic warfare
Crew/Accommodation: Four
Power Plant: Two 5,080kgp (11,200 lb s.t.) Pratt & Whitney J52-P-40B turbojets
Dimensions: Span 16.15m (53 ft); length 18.1m (59.25 ft); wing area 49.15m² (529 sq ft)
Weights: Empty 14,589kg (32,162 lb); MTOW 27,493kg (60,610 lb)
Performance: Maximum speed 1,002km/h (541 knots) Mach 0.82 at sea level; operational ceiling 11,582m (38,000 ft); radius 639km (345 naut. miles) with full 5-jammer warload
Load: Up to in excess of 11,340 kg (25,000 lb) of electronic broad-band jammers and other specialized electronic warfare systems

GRUMMAN/GULFSTREAM AEROSPACE GULFSTREAM I
(U.S.A.)

This twin-turboprop executive transport was planned in the mid-1950s, and first flew in August 1958. Production amounted to some 200 24-passenger aircraft including several TC-4 crew trainers for the U.S. Navy, and several have since been converted to Gulfstream I-C commuterliners by the lengthening of the fuselage by 3.25m (10 ft 8 inches) to provide accommodation for 37 passengers. ▼

GRUMMAN/GULFSTREAM AEROSPACE GULFSTREAM I

Role: Executive/passenger transport
Crew/Accommodation: Two, plus up to 12 (executive) or 24 (commercial) passengers
Power Plant: Two, 2,190 shp Rolls-Royce Dart 529 turboprops
Dimensions: Span 23.88m (78.33 ft); length 20.93m (68.67 ft); wing area 56.7m^2 (610.3 sq ft)
Weights: Empty 8,493kg (18,723 lb); MTOW 16,329kg (36,000 lb)
Performance: Maximum speed 563km/h (304 knots) at 7,620m (25,000 ft); operational ceiling 9,266m (30,400 ft); range 4,088km (2,205 naut. miles) with 1,243 kg (2,740 lb) payload
Load: Up to 2,286 kg (5,040 lb)

GRUMMAN E-2 HAWKEYE ▲
(U.S.A.)

To succeed the E-1 Tracer, Grumman developed the more capable E-2. This has turboprop power, significantly more advanced radar in a rotodome above the wing and rear fuselage, and an internal processing system allowing a small tactical crew to watch and control all air activity within a large radius. This first flew as the W2F-1 during April 1960, but entered service as the E-2A. Electronic and radar improvements have characterized development up to current E-2C. There is also a C-2 Greyhound carrier onboard delivery variant with provision for 39 passengers.

GRUMMAN E-2C HAWKEYE

Role: Naval carrierborne airborne early warning and control
Crew/Accommodation: Five
Power Plant: Two 4,910 hp General Electric T56A-425 turboprops
Dimensions: Span 24.6m (80.6 ft); length 17.5m (57.6 ft); wing area 65.03m^2 (700 sq ft)
Weights: Empty 17,212kg (37,945 lb); MTOW 23,503kg (51,817 lb)
Performance: Maximum speed 602km/h (325 knots) at 4,877m (16,000 ft); operational ceiling 9,388m (30,800 ft); endurance 6.1 hours
Load: None, other than special-to-task onboard equipment

GRUMMAN G-303 (F-14 TOMCAT)
(U.S.A.)

The Tomcat was designed as a fleet defence fighter to supersede the cancelled F-111B, and was planned round the same weapon system, namely the AWG-9 radar fire-control system and AIM-54 Phoenix long-range air-to-air missile. Grumman was also able to use its experience as prime contractor for the F-111B in the development of the Tomcat's variable-geometry wing. The type first flew in December 1970 and proved immensely capable in its F-14A initial form. The one limitation has been the Pratt & Whitney TF30 afterburning turbofan, also inherited from the F-111B, and this has been replaced in the current F-14A Plus variant by the General Electric F100. This will also be used in the F-14D together with a mass of electronic improvements in the offensive and defensive suites.

GRUMMAN F-14A TOMCAT

Role: Naval carrierborne fighter
Crew/Accommodation: Two
Power Plant: Two, 9,480 kgp (20,900 lb s.t.) Pratt & Whitney TF30-P-312A turbofans with reheat
Dimensions: Span unswept 19.6m swept 11.7m (64.1 ft, 38.2 ft); length 19.1m (62.7 ft); wing area 52.5m^2 (565 sq ft)
Weights: Empty 18,036kg (39,762 lb); MTOW 31,945kg (70,426 lb)
Performance: Maximum speed 2,498km/h (1,348 knots, Mach 2.35 at 11,276m (37,000 ft); operational ceiling 18,228+m (60,000 ft); radius of action unrefuelled 1,232km (665 miles)
Load: Up to 2,913 kg (6,423 lb) of externally mounted missilles, typically 6 AIM-554 Phoenix and 2 AIM-9 Sidewinders, plus an internal 6-barrel 20mm General Electric M61 Vulcan cannon

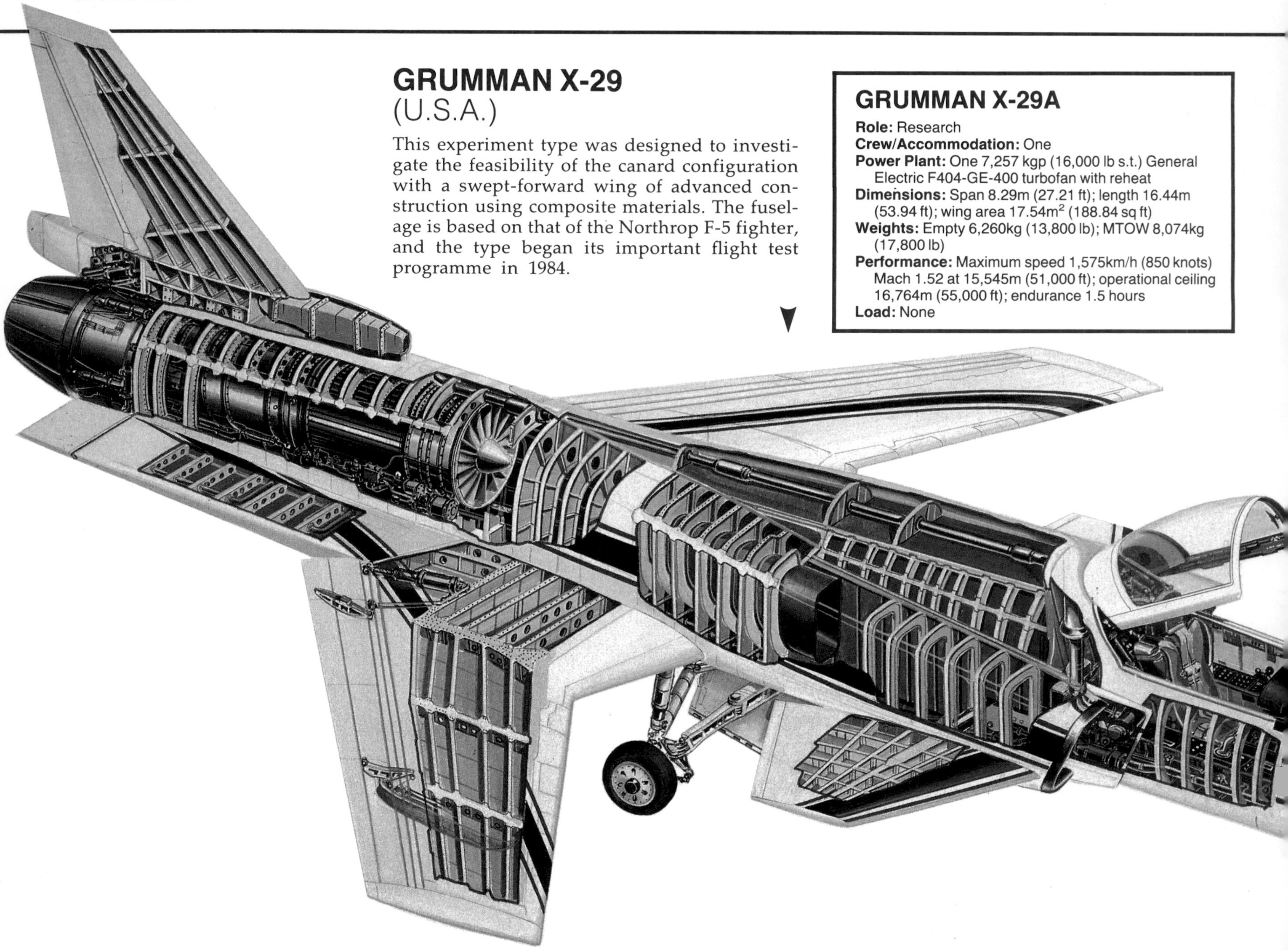

GRUMMAN X-29
(U.S.A.)

This experiment type was designed to investigate the feasibility of the canard configuration with a swept-forward wing of advanced construction using composite materials. The fuselage is based on that of the Northrop F-5 fighter, and the type began its important flight test programme in 1984.

GRUMMAN X-29A

Role: Research
Crew/Accommodation: One
Power Plant: One 7,257 kgp (16,000 lb s.t.) General Electric F404-GE-400 turbofan with reheat
Dimensions: Span 8.29m (27.21 ft); length 16.44m (53.94 ft); wing area 17.54m^2 (188.84 sq ft)
Weights: Empty 6,260kg (13,800 lb); MTOW 8,074kg (17,800 lb)
Performance: Maximum speed 1,575km/h (850 knots) Mach 1.52 at 15,545m (51,000 ft); operational ceiling 16,764m (55,000 ft); endurance 1.5 hours
Load: None

GRUMMAN/ GULFSTREAM AEROSPACE GULFSTREAM II, III and IV
(U.S.A.)

The success of the Gulfstream I persuaded the manufacturer that there was market space for a more advanced long-range executive transport with swept flying surfaces and twin-turbofan power plant. The first such Gulfstream II flew in October 1966 with two 5,171kg (11,400 lb) Rolls-Royce Spey Mk 511-8 engines. More than 250 of these aircraft were built, many of them being retrofitted as the Gulfstream II-B with the advanced wing of the Gulfstream III. This first flew in December 1979 and features a lengthened 19-passenger fuselage as well as the revised wing with drag-reducing winglets at its tips. Further development has produced the Gulfstream IV with 6,282kg (13,850 lb) thrust Rolls-Royce Tay Mk 611-8 turbofans.

GRUMMAN/GULFSTREAM AEROSPACE GULFSTREAM IV

Role: Executive jet transport
Crew/Accommodation: Two/three, plus up to 15 passengers
Power Plant: Two, 6,282kgp (13,850 lb s.t.) Rolls-Royce Tay Mk 611-8 turbofans
Dimensions: Span 23.72m (77.83 ft); length 26.9m (88.83 ft); wing area 88.3m^2 (950.4 sq ft)
Weights: Empty 15,967kg (35,200 lb); MTOW 35,523kg (71,700 lb)
Performance: Cruise speed 936km/h (505 knots) Mach 0.88 at 10,975m (36,000 ft); operational ceiling 15,545m (51,000 ft); range 7,968km (4,300 naut. miles) with 11 people
Load: Up to 1,814kg (4,000 lb)

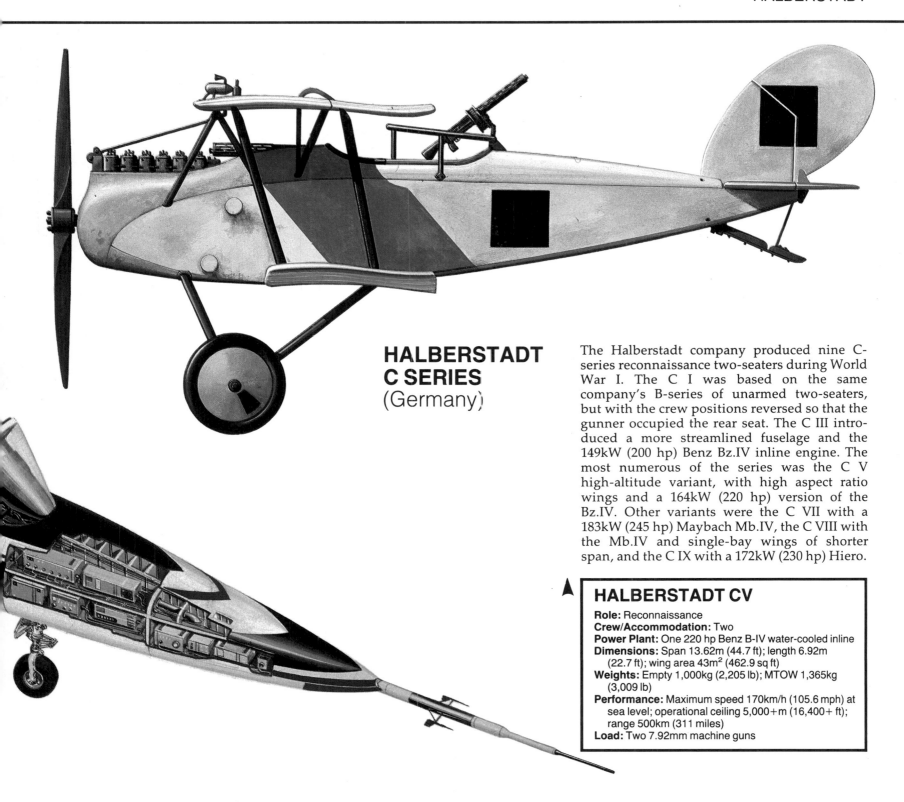

HALBERSTADT C SERIES
(Germany)

The Halberstadt company produced nine C-series reconnaissance two-seaters during World War I. The C I was based on the same company's B-series of unarmed two-seaters, but with the crew positions reversed so that the gunner occupied the rear seat. The C III introduced a more streamlined fuselage and the 149kW (200 hp) Benz Bz.IV inline engine. The most numerous of the series was the C V high-altitude variant, with high aspect ratio wings and a 164kW (220 hp) version of the Bz.IV. Other variants were the C VII with a 183kW (245 hp) Maybach Mb.IV, the C VIII with the Mb.IV and single-bay wings of shorter span, and the C IX with a 172kW (230 hp) Hiero.

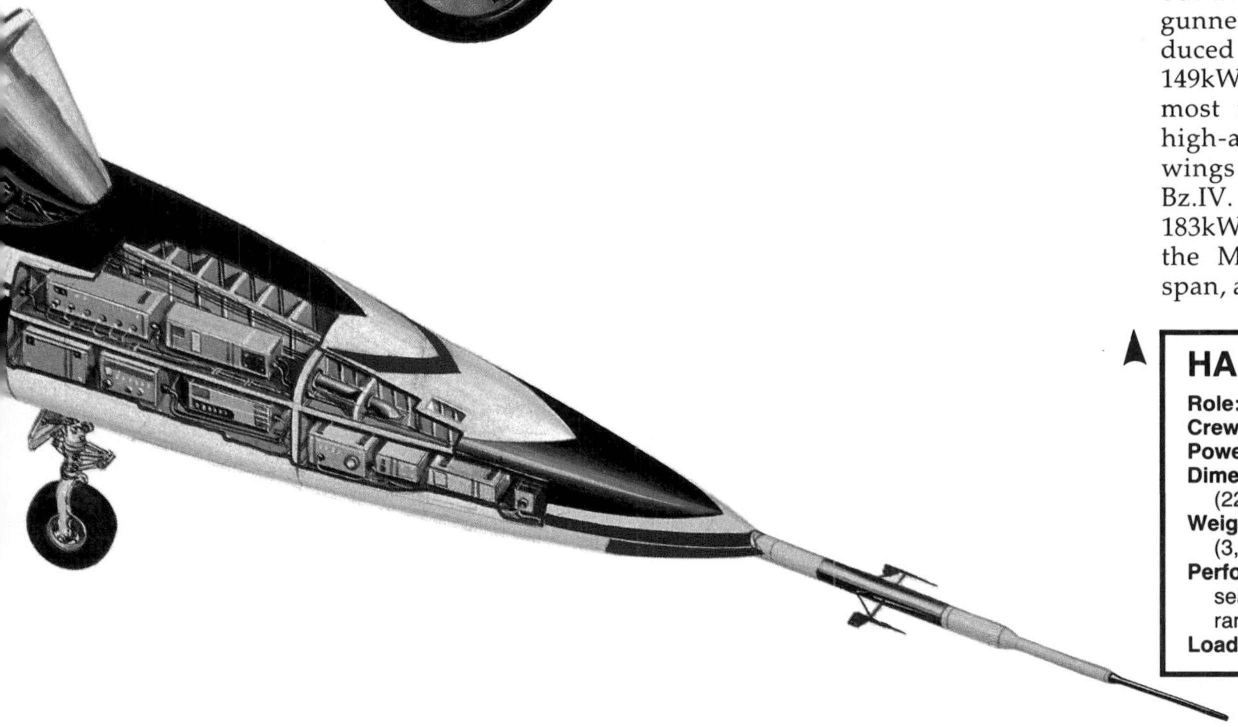

HALBERSTADT CV

Role: Reconnaissance
Crew/Accommodation: Two
Power Plant: One 220 hp Benz B-IV water-cooled inline
Dimensions: Span 13.62m (44.7 ft); length 6.92m (22.7 ft); wing area 43m^2 (462.9 sq ft)
Weights: Empty 1,000kg (2,205 lb); MTOW 1,365kg (3,009 lb)
Performance: Maximum speed 170km/h (105.6 mph) at sea level; operational ceiling 5,000+m (16,400+ ft); range 500km (311 miles)
Load: Two 7.92mm machine guns

HALBERSTADT CL SERIES
(Germany)

The CL designation indicated a two-seat reconnaissance fighter able to escort the larger and heavier C-series types, and in this category the two main Halberstadt offerings were the CL II and CL IV. The CL II appeared in 1917 with a 119kW (160 hp) Mercedes D.III inline engine, replaced by a 138kW (185 hp) BMW IIIa in the CL IIa, and had a single cockpit for the pilot and observer/gunner. Further refinement of the design resulted in the CL IV with a redesigned empennage and shorter fuselage.

HALBERSTADT CL II

Role: Escort fighter
Crew/Accommodation: Two
Power Plant: One 160 hp Mercedes D.IIIa water-cooled inline
Dimensions: Span 10.77m (35.33 ft); length 7.3m (23.96 ft); wing area 30.7m^2 (351.98 sq ft)
Weights: Empty 953kg (2,100 lb); MTOW 1,133kg (2,498 lb)
Performance: Maximum speed 165km/h (102.5 mph) at sea level; operational ceiling 5,000+m (16,400+ ft); range 450km (279 miles)
Load: Two/three 7.92mm machine guns, plus rear cockpit-stowed 10kg (22 lb) hand-dropped bombs

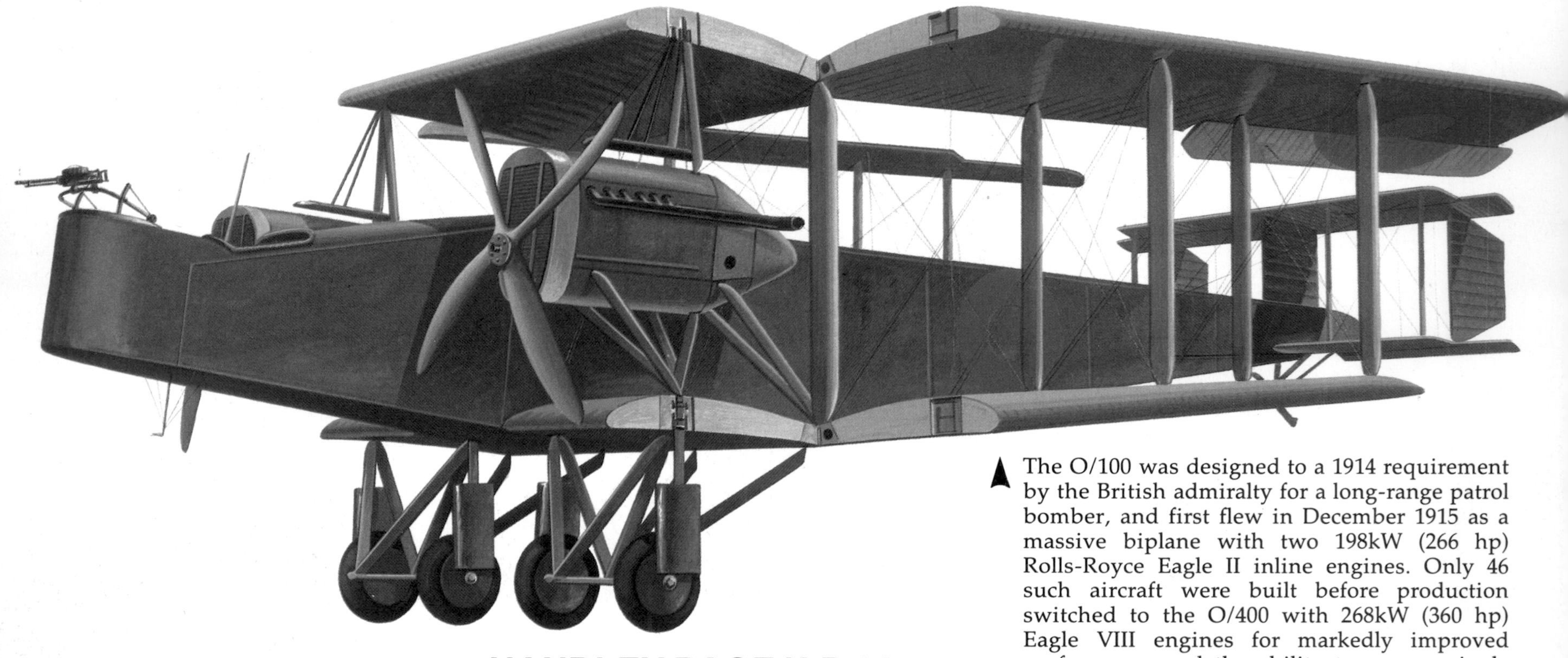

HANDLEY PAGE H.P.11 (O/100) and H.P.12 (O/400)
(U.K.)

▲ The O/100 was designed to a 1914 requirement by the British admiralty for a long-range patrol bomber, and first flew in December 1915 as a massive biplane with two 198kW (266 hp) Rolls-Royce Eagle II inline engines. Only 46 such aircraft were built before production switched to the O/400 with 268kW (360 hp) Eagle VIII engines for markedly improved performance and the ability to carry a single 748kg (1,650 lb) bomb as an alternative to a 907kg (2,000 lb) load of smaller bombs. A fair number of the aircraft were modified as civil transports after World War I.

HANDLEY PAGE O/400

Role: Heavy bomber
Crew/Accommodation: Three/four
Power Plant: Two 360 hp Rolls-Royce Eagle VIII water-cooled inlines
Dimensions: Span 30.48m (100 ft); length 19.16m (62.85 ft); wing area 153.11m^2 (1,648 sq ft)
Weights: Empty 3,856kg (8,502 lb); MTOW 6,060kg (13,360 lb)
Performance: Maximum speed 136km/h (84.5 mph) at 1,981m (6,500 ft); operational ceiling 2,591m (8,500 ft); endurance 8.5 hours
Load: Five .303 inch machine guns, plus to 1,497kg (3,300 lb) of bombs

HANDLEY PAGE H.P. 15 (V/1500)
(U.K.)

► The V/1500 was designed to provide the British air forces with a bomber able to attack targets in Germany from bases in the U.K., and first flew in May 1918 with tandem push/pull pairs of 280kW (375 hp) Rolls-Royce Eagle VIII inline engines. Only a few had been completed by the end of World War I, when production contracts were cancelled, but these saw useful if limited service after the war as bombers and record-breaking aircraft.

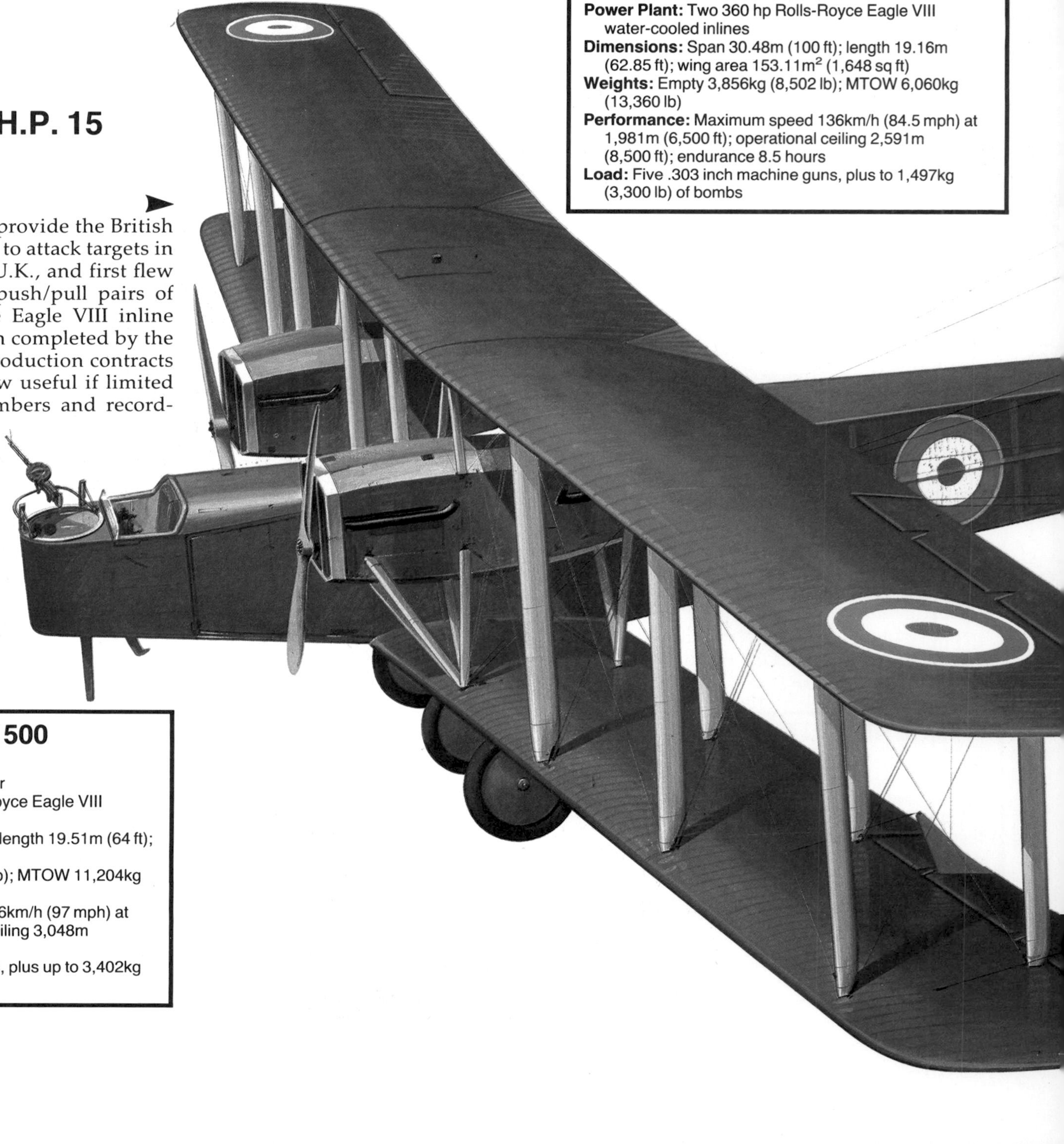

HANDLEY PAGE V/1500

Role: Heavy bomber
Crew/Accommodation: Three/four
Power Plant: Four 375 hp Rolls-Royce Eagle VIII water-cooled inlines
Dimensions: Span 38.4m (126 ft); length 19.51m (64 ft); wing area 278.7m^2 (3,000 sq ft)
Weights: Empty 7,353kg (16,210 lb); MTOW 11,204kg (24,700 lb)
Performance: Maximum speed 156km/h (97 mph) at 2,667m (8,750 ft); operational ceiling 3,048m (10,000 ft); endurance 14 hours
Load: Five .303 inch machine guns, plus up to 3,402kg (7,500 lb) of bombs

HANDLEY PAGE H.P.42 and H.P.45
(U.K.)

The H.P.42 was designed to an Imperial Airways requirement for a ruggedly reliable airliner for use on British empire routes. The airline ordered four each of the H.P.42E and H.P.42W (later H.P.45) versions optimized for eastern and western operations respectively, and the first of these flew in November 1930. The H.P.42E had 365kW (490 hp) Bristol Jupiter XIF radials mounted two each on the upper and lower wings, while the H.P.42W had 414kW (555 hp) Jupiter XFBM engines; the H.P.42E seated 24 in two 12-seat cabins, while the H.P.42W had less baggage accommodation and could thus carry 38 passengers in 18- and 20-seat cabins. ►

HANDLEY PAGE H.P. 42E

Role: Intermediate-range passenger transport
Crew/Accommodation: Three and one cabin crew, plus up to 24 pasengers
Power Plant: Four 490 hp Bristol Jupiter XIF air-cooled radials
Dimensions: Span 39.62m (130 ft); length 27.33m (89.66 ft); wing area 227.8m^2 (2,990 sq ft)
Weights: Empty 9,111kg (20,086 lb); MTOW 12,701kg (28,000 lb)
Performance: Maximum speed 193km/h (120 mph) at 610m (2,000 ft); operational ceiling 2,743m (9,000 ft); range 805km (500 miles)
Load: Up to 3,175kg (7,000 lb)

HANDLEY PAGE H.P.50 HEYFORD ▲
(U.K.)

Designed as a heavy night bomber and first flown in June 1930, the Heyford was of unusual and ungainly appearance with the fuselage attached to the upper wing of the biplane wing cellule. The main units of the massive main landing gear were attached to the lower wing, whose thickened centre section provided stowage for the bomb load. Some 124 aircraft were built in Mks I, Mk IA, Mk II and Mk III variants which differed in the installed variant of the Rolls-Royce Kestrel inline engine.

HANDLEY PAGE H.P.50 HEYFORD Mk IA

Role: Heavy night bomber
Crew/Accommodation: Four
Power Plant: Two 525 hp Rolls-Royce Kestrel IIIS water-cooled inlines
Dimensions: Span 22.86m (75 ft); length 17.68m (58 ft); wing area 136.57m^2 (1,470 sq ft)
Weights: Empty 4,173kg (9,200 lb); MTOW 7,666kg (16,900 lb)
Performance: Maximum speed 229km/h (142 mph) at sea level; operational ceiling 6,401m (21,000 ft); range 1,481km (920 miles) with 635kg (1,400 lb) warload
Load: Three .303 inch machine guns, plus up to 1,588kg (3,500 lb) of bombs

HANDLEY PAGE H.P.53 HAMPDEN and H.P.53 HEREFORD (U.K.)

The Hampden was designed to the same specification as the Wellington. The former emerged for its first flight in June 1936 with a pod-and-boom fuselage of extreme slenderness, and the latter with a very portly fuselage; the slender fuselage was always an operational liability. Both types were ordered into production, the Hampden because of its very high performance coupled with a low landing speed. In combat the 1,430 Hampden Mk Is proved to have inadequate defensive armament, and 157 aircraft were modified as Hampden TB.Mk I interim torpedo bombers. The Hereford was a version with 746kW (1,000 hp) Napier Dagger VIII inline engines, but only 152 were built.

HANDLEY PAGE HAMPDEN Mk I

Role: Bomber
Crew/Accommodation: Four
Power Plant: Two 965 hp Bristol Pegasus XVIII air-cooled radials
Dimensions: Span 21.09m (69.16 ft); length 16.32m (53.58 ft); wing area 62.1m^2 (668 sq ft)
Weights: Empty 5,789kg (12,764 lb); MTOW 10,206kg (22,500 lb)
Performance: Maximum speed 397km/h (247 mph) at 4,206m (13,800 ft); operational ceiling 5,791m (19,000 ft); range 1,770km (1,100 miles) with 1,814kg (4,000 lb) bombload
Load: Up to 2,268kg (5,000 lb) of which up to 80 per cent was carried internally

Handley Page Halifax B. Mk.III

HANDLEY PAGE H.P.57 HALIFAX (U.K.)

The Halifax was one of the RAF's trio of four-engined night bombers in World War II, and while not as important in this role as the Lancaster, was more important in secondary roles such as maritime reconnaissance, transport and airborne forces' support. The type was planned with two Rolls-Royce Vulture inline engines before being recast with four Rolls-Royce Merlins, and first flew in October 1939. The type entered service as the Halifax B.Mk I with 954kW (1,280 hp) Merlin Xs, and later bombers were the B.Mk II with Merlin XXs or 22s and a two-gun dorsal turret, the B.Mk III with 1,204kW (1,615 hp) Bristol Hercules VI or XVI radial engines, the B.Mk V based on the Mk II with revised landing gear, the B.Mk VI based on the Mk III but with 1,249kW (1,675 hp) Hercules 100s, and the B.Mk VII that reverted to Hercules XVIs.

There were also bomber subvariants with important modifications. The other variants retained the same mark number as the equivalent bomber, and in the transport type these were the C.Mks II, VI and VII, in the maritime role GR.Mks II, V and VI, and in the airborne support role the A.Mks II, V and VII. Post-war development produced the C.Mk 8 and A.Mk 9.

HANDLEY PAGE HALIFAX B.Mk III

Role: Heavy night bomber
Crew/Accommodation: Seven
Power Plant: Four 1,615 hp Bristol Hercules XVI air-cooled radials
Dimensions: Span 30.12m (98.83 ft); length 21.82m (71.58 ft); wing area 116.3m^2 (1,250 sq ft)
Weights: Empty 17,346kg (38,240 lb); MTOW 29,484kg (65,000 lb)
Performance: Maximum speed 454km/h (282 mph) at 4,115m (13,500 ft); operational ceiling 7,315m (24,000 ft); range 2,030km (1,260 miles) with full warload
Load: Nine .303 inch machine guns, plus up to 5,897kg (13,000 lb) of internally-stowed bombload

HANDLEY PAGE H.P.81 HERMES (U.K.)

The Hermes was virtually a civil version of the RAF Hastings transport, and first flew in December 1945 as a medium-range type, but was delayed by problems to the extent that the Hermes IV initial-production variant with tricycle landing gear began to enter service only in 1950. BOAC operated the 24 aircraft for only two years, but most were sold to smaller airlines which kept the type in operation for some time more, often upgrading the machines to Hermes IVA standard with Hercules 773 radial engines in place of the original Hercules 763 units.

HANDLEY PAGE HERMES IV

Role: Intermediate-range passenger transport
Crew/Accommodation: Five and two cabin crew, plus up to 74 passengers
Power Plant: Four 2,000 hp Bristol Hercules 763 air-cooled radials
Dimensions: Span 34.4m (113 ft); length 29.53m (96.86 ft); wing area 130.8m² (1,408 sq ft)
Weights: Empty 25,110kg (55,350 lb); MTOW 37,194kg (82,000 lb)
Performance: Maximum speed 571km/h (357 mph) at 6,100m (20,000 ft); operational ceiling 7,315m (24,000 ft); range 3,219km (2,000 miles) with full payload
Load: Up to 3,629kg (8,000 lb)

HANDLEY PAGE H.P. 80 VICTOR (U.K.)

This was the last of the U.K.'s trio of nuclear 'V-bombers' to enter service, but is now the only one still in service, albeit as a tanker. The type was planned with a pod-and-boom fuselage supporting crescent-shaped flying surfaces, and first flew in December 1952. The type entered service as the Victor B. Mk 1 with Armstrong Siddeley Sapphire Mk 200 series turbojets. The radically improved Victor B.Mk 2 had a larger wing and 9,344kg (20,600 lb) thrust Rolls-Royce Conway Mk 201 turbofans, and was revised as a launch platform for the Blue Steel stand-off nuclear missile. Later conversions were the S.Mk 2 maritime reconnaissance and K.Mk 2 tanker models.

HANDLEY PAGE VICTOR K.Mk 2

Role: Air-to-air refueller
Crew/Accommodation: Five
Power Plant: Four 9,344 kgp (20,600 lb s.t.) Rolls-Royce Conway Mk.201 turbo fans
Dimensions: Span 35.69m (117 ft); length 35.02m (114.92 ft); wing area 204.38m² (2,200 sq ft)
Weights: Empty 33,550kg (110,000 lb); MTOW 101,150kg (223,000 lb)
Performance: Maximum speed 1,020km/h (550 knots) Mach 0.96 at 11,000m (36,090 ft); operational ceiling 15,850m (52,000 ft); range 7,403km (3,995 naut. miles) unrefuelled
Load: Up to 15,876kg (35,000 lb)

HANNOVER CL SERIES
(Germany)

Hannover responded vigorously to the German Air Force's 1917 CL-class specification for an escort/reconnaissance two-seater. The CL II was a sturdy plywood-covered biplane with an Argus As.III inline engine and a compact biplane tail whose small dimensions increased the gunner's fields of fire. In the improved CL III the As.III was replaced by a 119kW (160 hp) Mercedes D.III, though shortages meant that many had the original As. III under the designation CL IIIa. The larger CL IV, intended for the high-altitude role, appeared only as a prototype. The CL V was similar to the CL III, but with a 138kW (185 hp) BMW IIIa inline engine.

HANNOVER CL IIIa

Role: Escort fighter/ground attack
Crew/Accommodation: Two
Power Plant: One 180 hp Argus As III water-cooled inline
Dimensions: Span 11.7m (30.39 ft); length 7.58m (24.87 ft); wing area 32.7m^2 (351.98 sq ft)
Weights: Empty 800kg (0,000 lb); MTOW 1,080kg (00,000 lb)
Performance: Maximum speed 165km/h (102.5 mph) at 5,000m (16,400 ft); operational ceiling 7,500m (24,606 ft); range 435km (270 miles)
Load: Two 7.92mm machine guns, plus light anti-personnel weapons hand-dropped by the observer-gunner

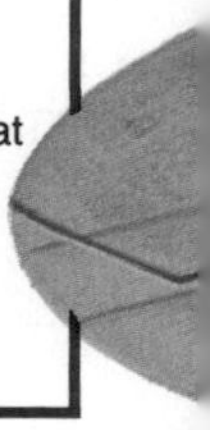

HANRIOT HD.1
(France)

Though overshadowed by the Spad S.7, the HD.1 was a trim and extremely agile fighter notable for the beautiful harmonization of its controls. First flown in June 1916, the HD.1 was ordered mainly for the Belgian Air Force, and was also extensively built under licence in Italy, its overall combat-worthiness being degraded only by very modest power and thus low performance.

HANRIOT HD.1

Role: Fighter
Crew/Accommodation: One
Power Plant: One 110 hp Le Rhone air-cooled rotary
Dimensions: Span 8.7m (28.54 ft); length 5.85m (19.19 ft); wing area 17.5m^2 (188.4 sq ft)
Weights: Empty 395kg (871 lb); MTOW 575kg (1,268 lb)
Performance: Maximum speed 182km/h (113 mph) at sea level; operational ceiling 6,401m (21,000 ft); endurance 1.5 hours
Load: One .303 inch machine gun

HANSA-BRANDENBURG W 29
(Germany)

Designed by Ernst Heinkel and first flown in 1917, the W 29 floatplane fighter was clearly based on the earlier W 12 with its underslung vertical tail to provide the gunner with excellent fields of fire, but embodying a monoplane wing for higher performance. The aircraft were powered by the 112 or 138kW (160 or 185 hp) Benz Bz. III or IIIa inline engines, and a larger version was the W 33 with the 183kW (245 hp) Maybach Mb.IV inline engine.

HANSA-BRANDENBURG W 29

Role: Naval reconnaissance floatplane
Crew/Accommodation: Two
Power Plant: One 160 hp Benz III water-cooled, inline
Dimensions: Span 13.5m (44.29 ft); length 9.4m (30.75 ft); wing area 32.2m^2 (346.6 sq ft)
Weights: Empty 1,100kg (2,425 lb); MTOW 1,470kg (3,241 lb)
Performance: Maximum speed 175km/h (109 mph) at sea level; operational ceiling 5,000m (16,404 ft); endurance 4.5 hours
Load: Two/three 7.92mm machine guns

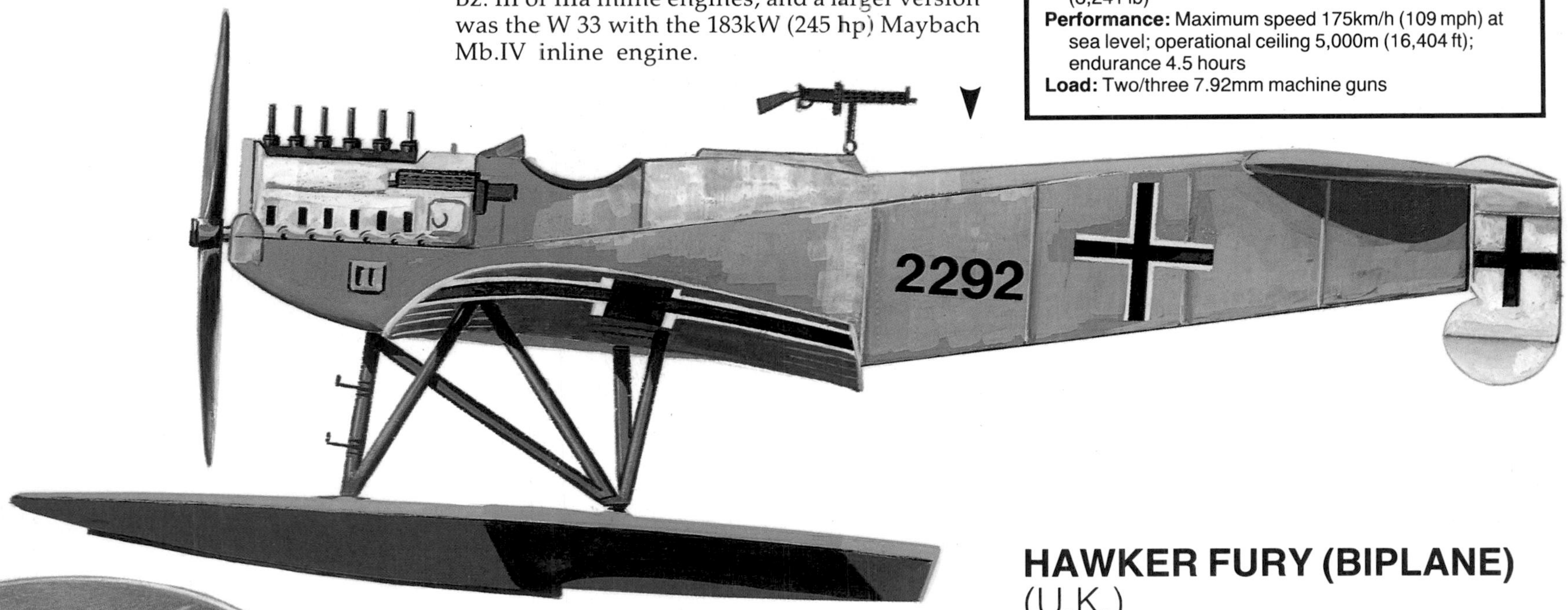

HAWKER FURY (BIPLANE)
(U.K.)

This single-seat fighter resulted from a 1927 requirement and first flew with the specified 336kW (450 hp) Bristol Jupiter radial engine. Hawker then recast the design with a Rolls-Royce F.XIS, and in this form the design was ordered as the Fury Mk I with the 391kW (525 hp) Kestrel IIs. Export aircraft were delivered with other inline or radial engines. The Fury Mk II introduced wheel spats, and with the 477kW (640 hp) Kestrel VI had higher speed at the expense of a slight reduction in range. The Nimrod was a naval equivalent.

Hawker Fury Mk.II

HAWKER FURY Mk II

Role: Interceptor
Crew/Accommodation: One
Power Plant: One 525 hp Rolls-Royce Kestrel IIS water-cooled inline
Dimensions: Span 9.15m (30 ft); length 8.13m (26.67 ft); wing area 23.4m^2 (251.8 sq ft)
Weights: Empty 1,190kg (2,623 lb); MTOW 1,583kg (3,490 lb)
Performance: Maximum speed 309km/h (192 mph) at 1,525m (5,000 ft); operational ceiling 8,534m (28,000 ft); range 491km (305 miles)
Load: Two .303 inch machine guns

HAWKER HART
(U.K.)

Designed to a 1926 requirement for a light day bomber, the Hart pioneered an elegant basic design that characterized a host of important Hawker aircraft of the late 1920s and early 1930s. The prototype was first flown in June 1928 with a Rolls-Royce F.XIB inline engine in a well streamlined nose installation, and returned outstanding performance figures.The Hart Mk I bomber remained essentially unaltered through its production life. The Hart Mk II was a dual-control trainer and target tug; other variants were the Demon two-seat figher, the dual-control Hart Trainer, and the Hart Communications liaison type.

HAWKER HART Mk I

Role: Light day bomber
Crew/Accommodation: Two
Power Plant: One 525 hp Rolls-Royce Kestrel IB water-cooled inline
Dimensions: Span 11.35m (37.25 ft); length 8.94m (29.33 ft); wing area 32.33m^2 (348 sq ft)
Weights: Empty 1,148kg (2,530 lb); MTOW 2,066kg (4,554 lb)
Performance: Maximum speed 296km/h (184 mph) at 1,524m (5,000 ft); operational ceiling 6,507m (21,350 ft); range 756km (470 miles) with full warload
Load: Two .303mm machine guns, plus up to 236kg (520 lb) of bombs

HAWKER HURRICANE
(U.K.)

This was the U.K.'s first monoplane fighter of the semi-modern type with retractable landing gear, flaps and an enclosed cockpit, but retaining features such as a fabric-covered alloy tube structure. The type first flew in November 1935 and, in its Hurricane Mk I form with the 768kW (1,030 hp) Rolls-Royce Merlin II inline engine, was the RAF's most important and successful fighter of the Battle of Britain. Adoption of the 954kW (1,280 hp) Merlin XX resulted in the Hurricane Mk II, which was produced in variants such as the Mk IIA with eight 7.7mm (0.303 inch) machine guns, Mk IIB with 12 such guns and provision for underwing bombs, Mk IIC based on the Mk IIB but with four 20mm cannon, and Mk IID with two 40mm cannon in the anti-tank role. The final version was the Hurricane Mk IV with the 1,208kW (1,620 hp) Merlin 24 or 27 and a universal wing to allow the use of any of the standard armament combinations. The type was also produced in Sea Hurricane forms based on land-based variants, and was produced in Canada as the Hurricane Mks X to XIIA.

HAWKER HURRICANE Mk IIB

Role: Fighter bomber
Crew/Accommodation: One
Power Plant: One 1,280 hp Rolls-Royce Merlin XX water-cooled inline
Dimensions: Span 12.19m (40 ft); length 9.75m (32 ft); wing area 23.9m^2 (257.5 sq ft)
Weights: Empty 2,495kg (5,500 lb); MTOW 3,311kg (7,300 lb)
Performance: Maximum speed 772km/h (342 mph) at 6,706m (22,000 ft); operational ceiling 10,973m (36,000 ft); range 772.5km (480 miles) on internal fuel only
Load: Twelve .303 inch machine guns, plus up to 454kg (1,000 lb) bombload

HAWKER TYPHOON
(U.K.)

The Typhoon was planned as a Hurricane successor, and was trialled in various forms with the Napier Sabre and Rolls-Royce Vulture inline engines. The first prototype flew in February 1940, and the variant ordered into production was that with the Sabre IIA. The type was disappointing as an interceptor because of its slow rate of climb, but after early problems with tail failure matured as a quite exceptional fighter-bomber with potent underwing armament of bombs or rockets. The Typhoon Mk IA had 12 7.7mm (0.303 inch) machine guns while the considerably more numerous Typhoon Mk IB had four 20mm cannon.

HAWKER TYPHOON Mk IB

Role: Strike fighter
Crew/Accommodation: One
Power Plant: One 2,260 hp Napier Sabre IIC water-cooled inline
Dimensions: Span 12.67m (41.57 ft); length 9.73m (31.92 ft); wing area 23.13m^2 (249 sq ft)
Weights: Empty 4,445kg (9,800 lb); MTOW 6,341kg (13,980 lb)
Performance: Maximum speed 652km/h (405 mph) at 5,485m (18,000 ft); operational ceiling 10,363m (34,000 ft); range 982km (610 miles)
Load: Four 20mm cannon, plus up to 907kg (2,000 lb) of bombs or rockets

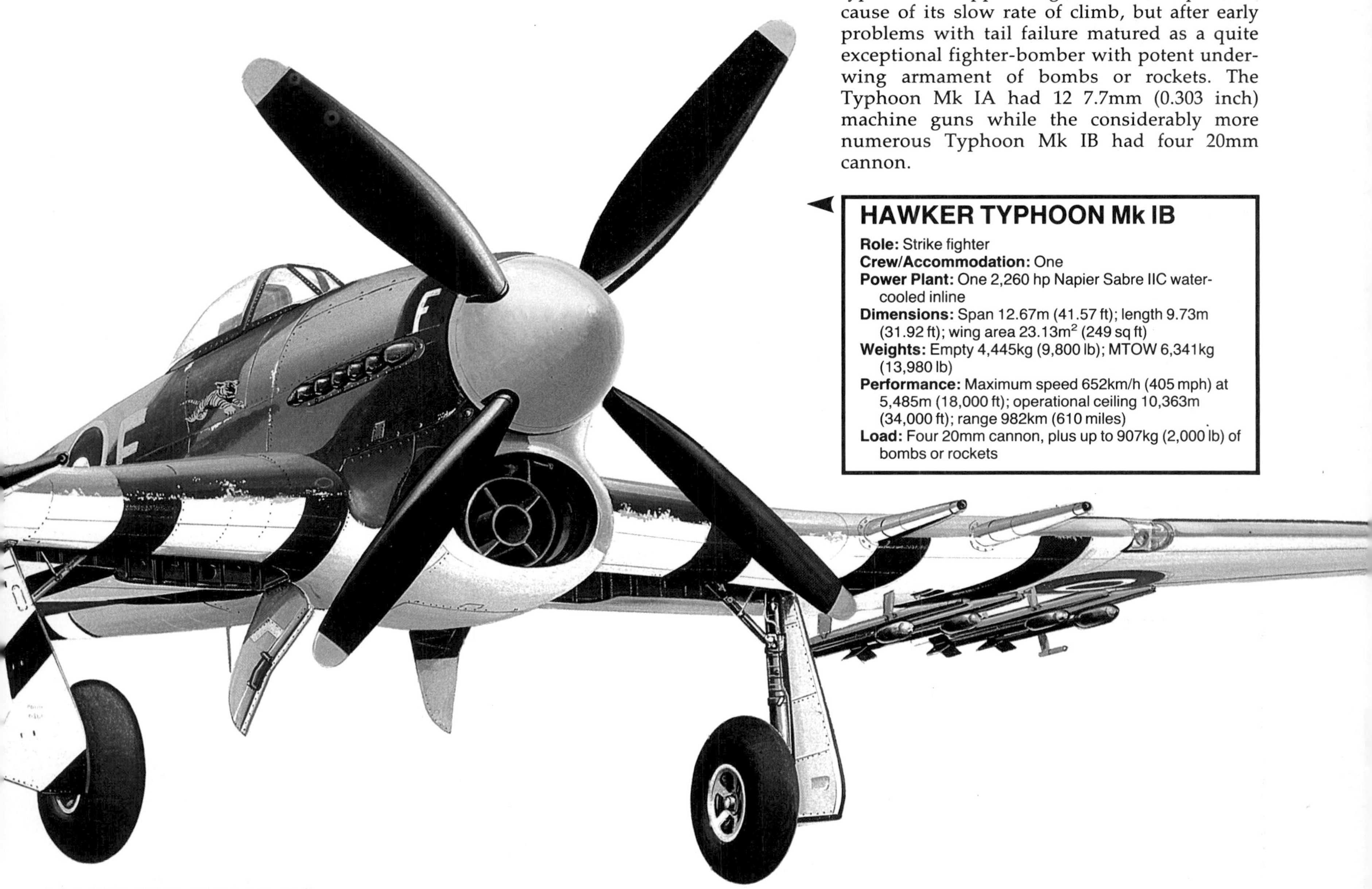

HAWKER TEMPEST
(U.K.)

The Tempest was in essence a version of the Typhoon with a thinner wing of elliptical planform to improve climb and altitude performance. The type was again developed to prototype form with different engines (the Bristol Centaurus radial, and the Napier Sabre and Rolls-Royce Griffon inline engines), and first flew in February 1943. Versions with the 1,879kW (2,520 hp) Centaurus V and 1,626kW (2,180 hp) Sabre IIA were ordered as the Tempest Mk II and V, entering service respectively just before and after the end of World War II. There was also an improved Tempest Mk VI with the 1,745kW (2,340 hp) Sabre V.

HAWKER TEMPEST Mk V

Role: Strike fighter
Crew/Accommodation: One
Power Plant: One 2,180 hp Napier Sabre IIA water-cooled inline
Dimensions: Span 12.49m (41 ft); length 10.26m (33.67 ft); wing area 28.05m^2 (302 sq ft)
Weights: Empty 4,196kg (9,250 lb); MTOW 6,187kg (13,640 lb)
Performance: Maximum speed 700km/h (435 mph) at 5,180m (17,000 ft); operational ceiling 11,125m (36,500 ft); range 1,191km (740 miles) on internal fuel only
Load: Four 20mm cannon, plus up to 907kg (2,000 lb) of bombs or rockets

HAWKER HUNTER
(U.K.)

The Hunter was the most successful of British post-World War II fighters, and still serves in modest numbers as a first-line type with smaller air forces. The Hunter first flew in prototype form during July 1951, and in its early production forms was engined with Rolls-Royce Avon or Armstrong Siddeley Sapphire turbojets as the Hunter F.Mks 1 and 4 or Hunter F.Mks 2 and 5 respectively. Later variants had more fuel and underwing armament capability. The Hunter F.Mk 6 introduced the Avon Mk 200 series turbojet, and was later developed as the Hunter FGA.Mk 9 definitive ground-attack fighter. There were also tactical reconnaissance and side-by-side trainer variants produced in Hunter FR.Mk 10 and PR.Mk 11, and Hunter T.Mks 7 and 8 forms for the RAF and Royal Navy respectively. Export derivatives were numbered in the Hunter Mks 50, 60, 70 and 80 series.

HAWKER HUNTER F.Mk 6

Role: Day fighter
Crew/Accommodation: One
Power Plant: One 4,605 kgp (10,150 lb s.t.) Rolls-Royce Avon Mk 207 turbojet
Dimensions: Span 10.25m (33.33 ft); length 13.97m (45.83 ft); wing area 32.42m^2 (349 sq ft)
Weights: Empty 6,405kg (14,122 lb); MTOW 8,051kg (17,750 lb)
Performance: Maximum speed 1,002km/h (623 mph) at 10,975m (36,000 ft); operational ceiling 14,630m (48,000 ft); range 789km (490 miles) on internal fuel only
Load: Four 30mm cannon

Top: Hawker Hunter FGA 9

Hawker Hunter T.Mk.7 ▲

Hawker Hunter F.Mk.5 ▼

▲ HEINKEL He 51
(Germany)

Flown in prototype form during November 1932 as the He 49 civil advanced trainer, the He 51 was Germany's last biplane fighter and a worthy if unexceptional type. The initial production model was the He 51A with the 559kW (750 hp) BMW VI 7,3Z inline engine, and this was followed by the structurally strengthened He 51B (including a floatplane variant) and the He 51C with ground-attack capability.

HEINKEL He 51A-1

Role: Fighter
Crew/Accommodation: One
Power Plant: One 750 hp BMW V17,3Z water-cooled inline
Dimensions: Span 11m (36.09 ft); length 8.4m (27.56 ft); wing area 27.2m^2 (292.78 sq ft)
Weights: Empty 1,615kg (3,560 lb); MTOW 1,900kg (4,189 lb)
Performance: Maximum speed 330km/h (205 mph) at 500m (1,640 ft); operational ceiling 7,700m (25,262 ft); range 390km (242 miles) with internal fuel only
Load: Two 7.9mm machine guns

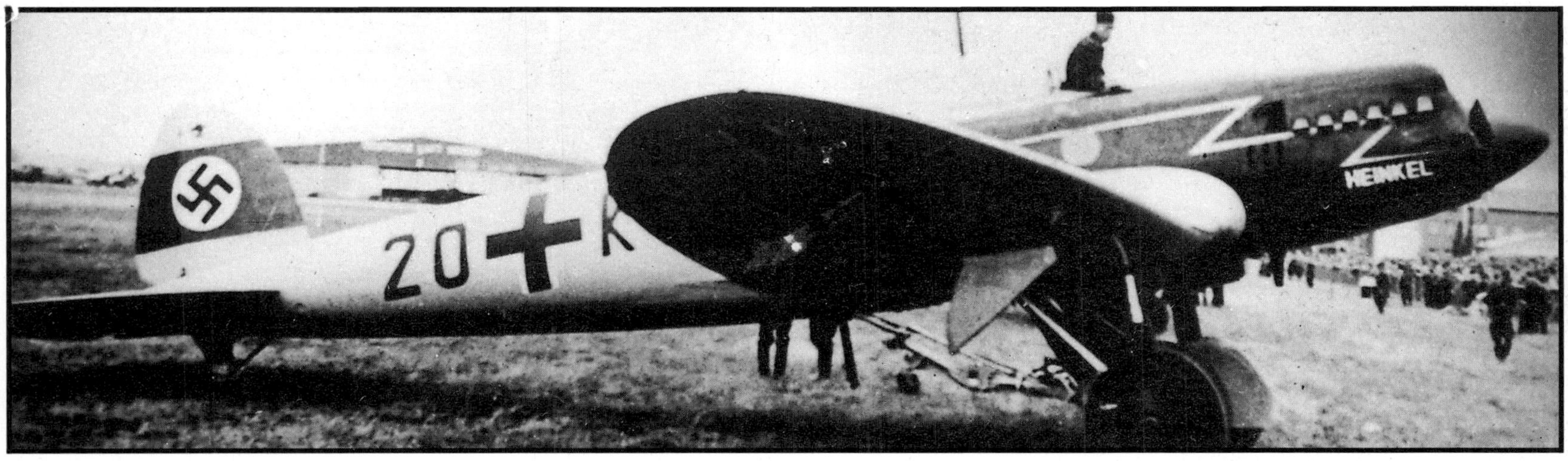

HEINKEL He 70 BLITZ (LIGHTNING)
(Germany)

▲ The He 70, a high-speed transport for mail and four passengers, first flew in December 1932. The He 70A production variant for Lufthansa had the 470kW (630 hp) BMW VI 6,0Z inline engine. With more power in the form of the 559kW (750 hp) BMW VI 7,3 engine, a similar He 70D variant was ordered as an air force communications type. Further development resulted in the He 70E high-speed bomber with a 300kg (661 lb) warload, the He 70F long-range reconnaissance aeroplane, and the He 170 export bomber.

HEINKEL He 70A BLITZ

Role: High speed passenger transport
Crew/Accommodation: Two, plus up to four passengers
Power Plant: One 630 hp BMW VI 6,0Z water-cooled inline
Dimensions: Span 14.8m (48.56 ft); length 11.7m (38.39 ft); wing area 36.5m^2 (392.9 sq ft)
Weights: Empty 2,300kg (5,072 lb); MTOW 3,420kg (7,541 lb)
Performance: Cruise speed 335km/h (208 mph) at sea level; operational ceiling 5,250m (17,220 ft); range 1,000km (621 miles)
Load: Up to 320kg (705 lb)

HEINKEL He 100
(Germany)

This was designed as an alternative to the Bf 109 with greater performance and a structure optimized for mass production. The first prototype flew in January 1938 and revealed superlative performance plus tricky handling. Later prototypes were used for record-breaking and evolution of a true fighter. It matured as the He 100D, but only small numbers were built.

HEINKEL He 100D

Role: Fighter
Crew/Accommodation: One
Power Plant: One 1,000 hp Daimler-Benz DB601 water-cooled inline
Dimensions: Span 9.4m (30.4 ft); length 8.18m (26.84 ft); wing area 14.5m^2 (156 sq ft)
Weights: Empty 2,070kg (4,564 lb); MTOW 2,500kg (5,512 lb)
Performance: Maximum speed 670km/h (416 mph) at 4,000m (13,120 ft); operational ceiling 10,500m (34,440 ft); range 900km (559 miles)
Load: One 20mm cannon and two 7.9mm machine guns

HEINKEL He 111
(Germany)

This was Germany's most important bomber of World War II, but was designed supposedly as an airliner. The first prototype flew in February 1935, and considerable development was necessary before the definitive He 111B began to enter military service with Daimler-Benz DB 600 inline engines, also used for the six He 111C 10-passenger airliners. The He 111E used the 746kW (1,000 hp) Junkers Jumo 211A and was developed in five subvariants. The He 111F combined the wing of the He 111G with Jumo 211A-3 engines, while the He 111G introduced a wing of straight rather than curved taper. The He 111H became the most extensively built model in subvariants up to the He 111H-23 with increasingly powerful engines and heavier armament. The He 111J was a torpedo bomber. The He 111P was introduced in 1939, and pioneered an asymmetric and extensively glazed forward cockpit in place of the original design. The oddest variant was the He 111Z heavy glider tug, which was two He 111H-6 bombers joined by a revised wing section incorporating a fifth Jumo 211F engine.

HEINKEL He 111H-16

Role: Bomber
Crew/Accommodation: Five
Power Plant Two 1,350 hp Junkers Jumo 211F-2 water-cooled, inlines
Dimensions: Span 22.6m (74.15 ft); length 16.4m (53.81 ft); wing area 86.5m^2 (931 sq ft)
Weights: Empty 8,680kg (19,136 lb); MTOW 14,000kg (30,865 lb)
Performance: Maximum speed 435km/h (270 mph) at 6,000m (19,685 ft); operational ceiling 6,700m (21,982 ft); range 1,950km (1,212 miles) with maximum bombload
Load: Two 20mm cannon and five 13mm machine guns, plus up to 3,600kg (7,937 lb) of bombs

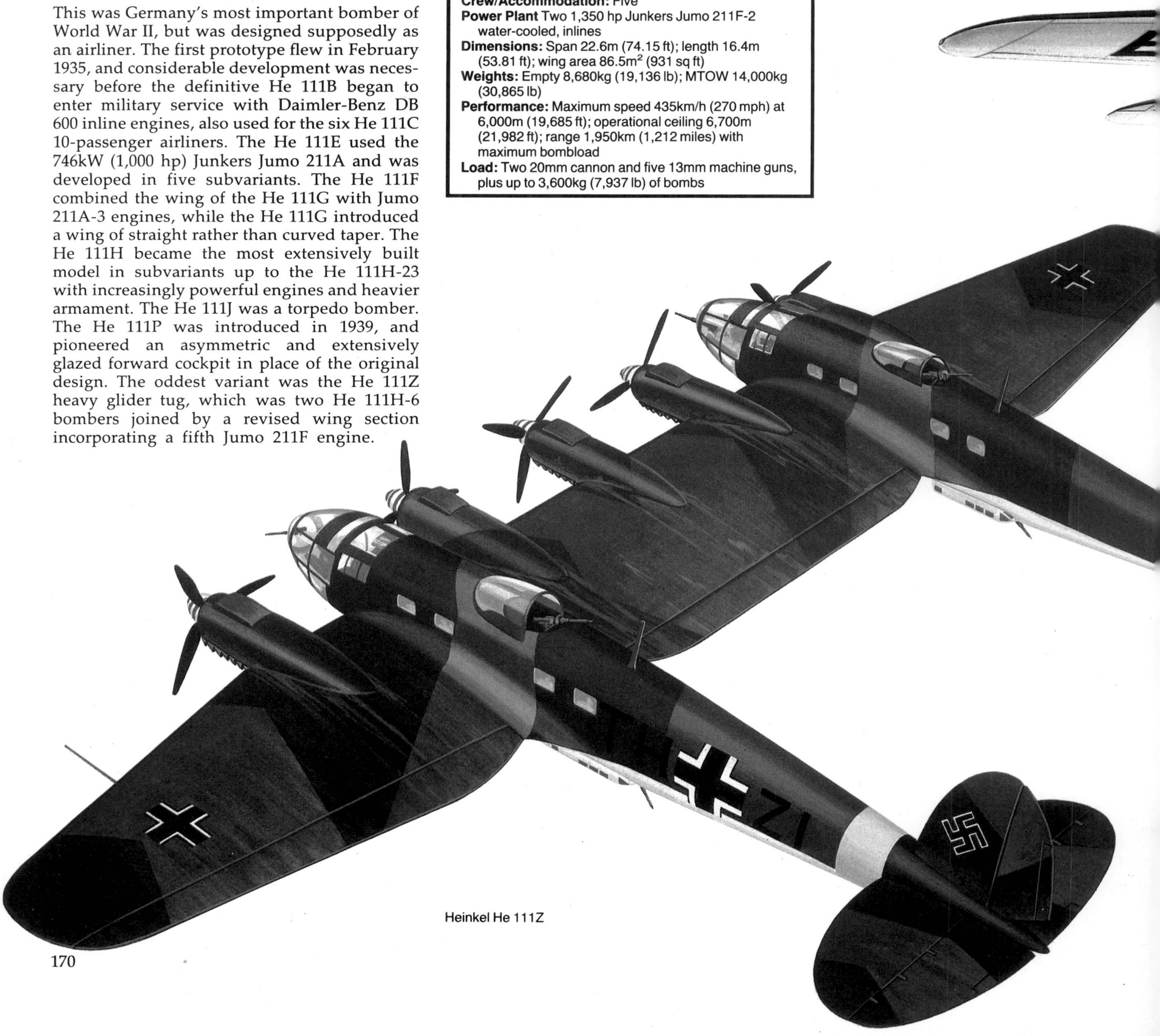

Heinkel He 111Z

HEINKEL He 115
(Germany)

This coastal reconnaissance floatplane was developed as successor to the He 59, and first flew in 1936. The original He 115A was followed by the He 115B with additional fuel and reinforced floats for operations from ice, the He 115C with heavier armament, and the He 115E with further revision of the armament.

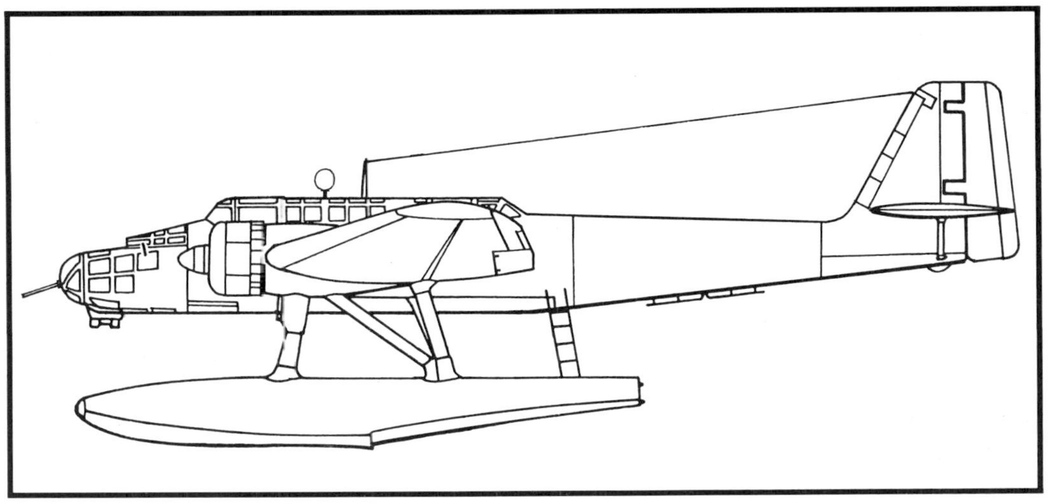

HEINKEL He 115B-1

Role: Bomber/torpedo floatplane
Crew/Accommodation: Three
Power Plant: Two 960 hp BMW 132K air-cooled radials
Dimensions: Span 22.2m (72.18 ft); length 17.3m (56.76 ft); wing area 87.5m^2 (934.5 sq ft)
Weights: Empty 5,300kg (11,687 lb); MTOW 10,400kg (22,932 lb)
Performance: Maximum speed 355km/h (220 mph) at 3,400m (11,152 ft); operational ceiling 5,500m (18,040 ft); range 2,000km (1,242 miles) with 1,000kg (2,205 lb) bombload
Load: Three 13mm machine guns, plus up to 1,500kg (3,307 lb) of bombs

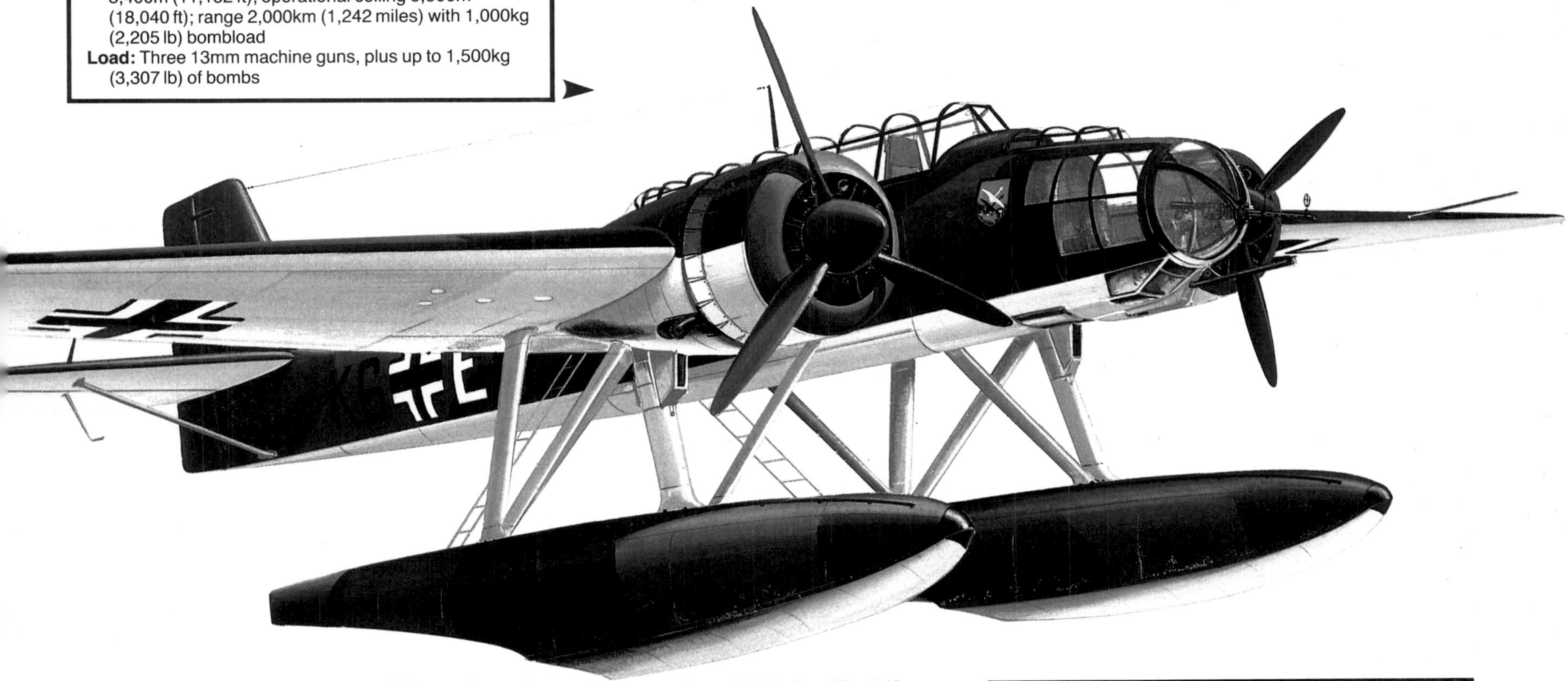

HEINKEL He 162 SALAMANDER
(Germany)

First flown in December 1944, the He 162 was an extremely ambitious attempt to create an easily flown interceptor of basically wooden construction that could be built in vast numbers by the German wood-working industry in the dire closing days of World War II. The location of the single turbojet in a piggyback position was selected for ease of installation and maintenance, and some 800 He 162As were almost complete at the wars end.

HEINKEL He 162A-2

Role: Interceptor
Crew/Accommodation: One
Power Plant: One 800kg (1,760 lb s.t.) BMW 109-003 E turbojet
Dimensions: Span 7.2m (23.62 ft); length 9m (29.53 ft); wing area 11.15m^2 (120 sq ft)
Weights: Empty 1,750kg (3,859 lb); MTOW 2,700kg (5,953 lb)
Performance: Maximum speed 835km/h (522 mph) at 6,000m (19,680 ft); operational ceiling 11,000m (36,080 ft); range 1,000km (628 miles) at MTOW
Load: Two 20mm cannon

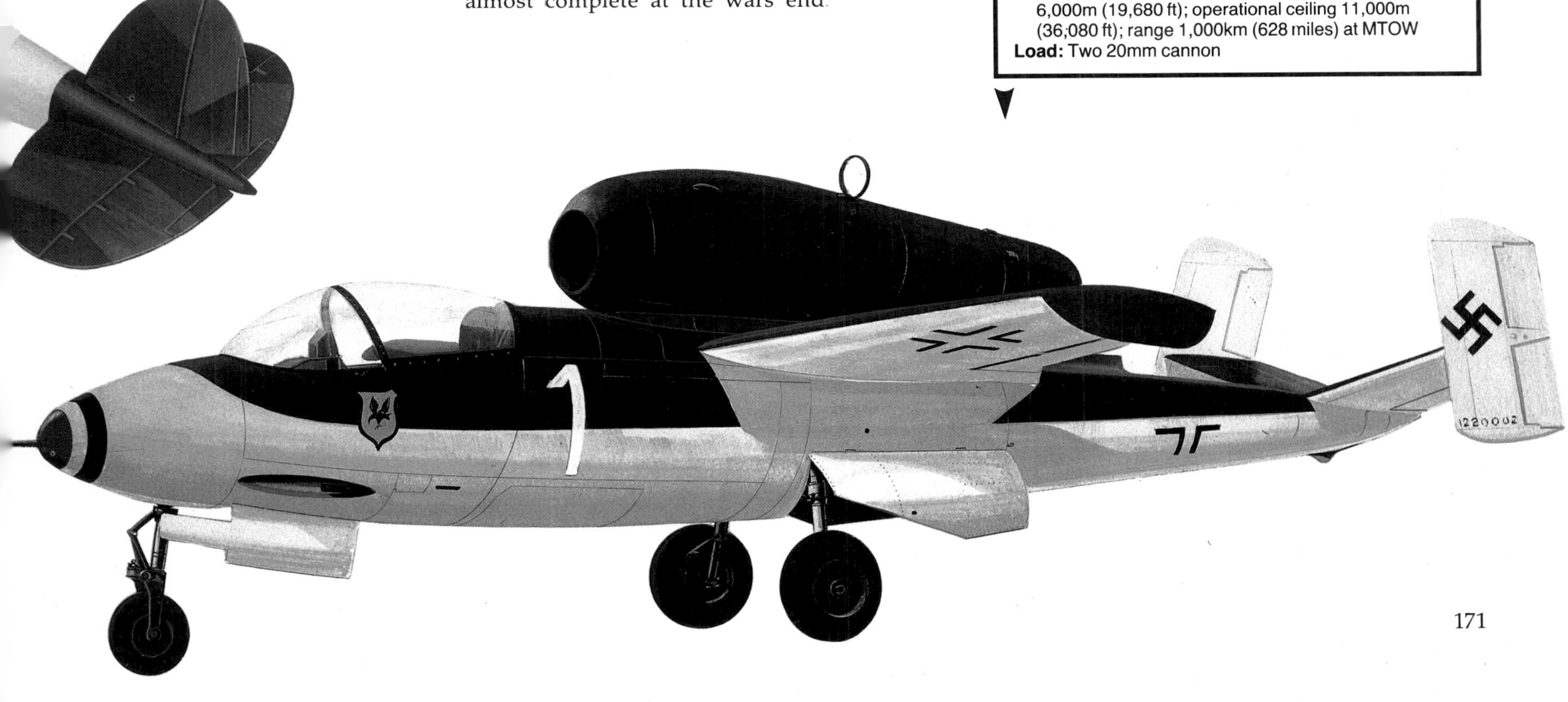

HEINKEL He 177 GREIF (GRIFFON)
(Germany)

The He 177 was an ingenious and ambitious attempt to create a long-range strategic bomber, but was ultimately defeated by the technical problems of the power plant. The first prototype flew in November 1939, but the development programme was seriously affected by problems with the coupled DB601 inline engines, which showed an alarming tendency to overheat and catch fire. The type also suffered from structural deficiencies that were never wholly cured.

HEINKEL He 177A GREIF

Role: Long range bomber
Crew/Accommodation: Six
Power Plant: Two 2,950 hp Daimler-Benz, DB610A/B water-cooled inlines
Dimensions: Span 31.44m (103.15 ft); length 19.4m (63.65 ft); wing area 102m^2 (1,098 sq ft)
Weights: Empty 16,900kg (37,038 lb); MTOW 31,000kg (68,343 lb)
Performance: Maximum speed 488km/h (303 mph) at 6,000m (19,686 ft); operational ceiling 8,000m (26,248 ft); range 5,000km (3,107 miles) with 3,140kg (6,922 lb) bombload
Load: Two 20mm cannon, three 13mm machine guns, three 7.9 machine guns, plus up to 3,140kg (6,923 lb) of bombs or missiles

HEINKEL He 178
(Germany)

The He 178 was the world's first turbojet-powered aeroplane, and planned specifically as the low-speed testbed for its engine. The type first flew on 27 August 1939.

HEINKEL He 178

Role: Experimental jet
Crew/Accommodation: One
Power Plant: One 360kgp (795 lb s.t.) Heinkel HeS 3b turbojet
Dimensions: Span 7.2m (23.62 ft); length 7.48m (24.54 ft); wing area 9.1m^2 (97.9 sq ft)
Weights: Empty 1,620kg (3,565 lb); MTOW 1,998kg (4,396 lb)
Performance: Maximum speed 700km/h (434 mph) at sea level
Load: None

HEINKEL He 219 UHU (OWL)
(Germany)

This superb and potentially decisive night-fighter was late entering service because of political reasons rather than any technical deficiencies. The first prototype flew in November 1942. The programme was cancelled in May 1944, but small-scale deliveries continued to be made in variants up to the He 219A-7 with advanced radar and very heavy armament.

HEINKEL He 219A-7 UHU

Role: Night/all-weather fighter
Crew/Accommodation: Two
Power Plant: Two 1,900 hp Daimler-Benz DB603G water-cooled inlines
Dimensions: Span 18.5m (60.7 ft); length 15.33m (50.30 ft); wing area 44.5m² (479 sq ft)
Weights: Empty 11,200kg (24,692 lb); MTOW 15,300kg (33,730 lb)
Performance: Maximum speed 670km/h (416 mph) at 7,000m (22,967 ft); operational ceiling 12,700m (41,668 ft); range 1,545km (960 miles) internal fuel only
Load: Six 30mm and two 20mm cannon (directed by FuG 220/Lichtenstein SN-2 radar

HEINKEL He 280
(Germany)

The He 280 was designed to succeed the lightweight He 178 as a turbojet testbed, but was also planned with the notion that it could become the basis of a fighter in which the use of twin underwing nacelles would facilitate the problems of engine installation and maintenance. The type was first tested as a glider, and completed its first powered flight in April 1941. Thereafter the He 280 was test flown with several engine types, but lacked the potential of the Me 262 and was never seriously considered for fighter development.

HEINKEL He 280

Role: Fighter
Crew/Accommodation: One
Power Plant: One 750kg (1,650 lb s.t.) Heinkel HeS 8A turbojet
Dimensions: Span 12.2m (40.03 ft); length 10.4m (34.12 ft); wing area 21.5m² (231.4 sq ft)
Weights: Empty 3,215kg (7,073 lb); MTOW 4,310kg (9,482 lb)
Performance: Maximum speed 900km/h (559 mph) at 6,000m (19,680 ft); operational ceiling 11,500m (37,720 ft); range 650km (404 miles) at 6,000m (19,685 ft) altitude
Load: Three 20mm cannon

HENSCHEL Hs 123
(Germany)

The Hs 123 was designed to meet a 1933 dive-bomber specification, and first flew in August 1935. The type entered limited production as the Hs 123A-1 and was generally used in the close-support role. The proposed Hs 123B would have had the 716kW (960 hp) BMW 132K engine for the carriage of heavier armament.

HENSCHEL Hs 123A-1

Role: Dive bomber
Crew/Accommodation: One
Power Plant: One 880 hp BMW 132Dc air-cooled radial
Dimensions: Span 10.5m (34.45 ft); length 8.33m (27.33 ft); wing area 24.8m^2 (267.4 sq ft)
Weights: Empty 1,500kg (3,307 lb); MTOW 2,215kg (4,884 lb)
Performance: Maximum speed 341km/h (212 mph) at 1,200m (3,937 ft); operational ceiling 9,000m (29,529 ft); range 860km (534 miles)
Load: Two 7.9mm machine guns, plus up to 200kg (441 lb) of weapons

HENSCHEL Hs 129
(Germany)

The Hs 129 was designed as a specialized successor to the Hs 123 with heavier armament, armour protection and a less vulnerable twin-engined power plant. The type first flew in the spring of 1939 with 347kW (465 hp) Argus As 410 inline engines, but was clearly underpowered and restricted tactically by its cramped cockpit. The production version, therefore, became the improved Hs 129B with more powerful radial engines and a steadily more devastating armament whose weight and drag seriously eroded performance.

HENSCHEL Hs 129B-2

Role: Close air support/anti-armour
Crew/Accommodation: One
Power Plant: Two 700 hp Gnome-Rhone 14M 4/5 air-cooled radials
Dimensions: Span 14.2m (46.59 ft); length 9.75m (31.99 ft); wing area 29m^2 (312.2 sq ft)
Weights: Empty 3,984kg (8,400 lb); MTOW 5,109kg (11,266 lb)
Performance: Maximum speed 355km/h (220 mph) at sea level; operational ceiling 9,000m (29,528 ft); range 560km (348 miles)
Load: One 30mm and two 20mm cannon, two 7.9mm machine guns, plus up to 300kg (661 lb) of bombs

HILLER MODEL 360 and UH-12 (H-23 RAVEN)

(U.S.A.)

This light helicopter was comparable in many ways with the Bell Model 47, and though developed as the Model 360 was produced as the UH-12 after Hiller's purchase of United Helicopters. The type was produced in a large number of forms for the civil market, and was also built as the H-23 Raven series for the U.S. military.

HILLER MODEL 12E/H-23G RAVEN

Role: Light utility/communications helicopter
Crew/Accommodation: One, plus up to two passengers/troops
Power Plant: One 305 hp Lycoming VO-540-B1D air-cooled flat-opposed
Dimensions: Overall length rotors turning 12.46m (40.74 ft); rotor diameter 10.80m (35.58 ft)
Weights: Empty 798kg (1,759 lb); MTOW 1,497kg (3,300 lb)
Performance: Cruise speed 145km/h (90 mph) at sea level; operational ceiling 4,630m (15,200 ft); range 725km (450 miles)
Load: Up to 57kg (125 lb) excluding passengers

HILLER MODEL 1100

(U.S.A.)

This helicopter was built in competition to the Bell Model 205 and Hughes Model 369 for the U.S. Army's 1961 Light Observation Helicopter programme. Though unsuccessful in that competition, the type entered production as the FH-1110 for the civil market. It was successfully sold to a number of overseas foreign purchasers.

HILLER MODEL 1100

Role: Light utility helicopter
Crew/Accommodation: One, plus up to four passengers
Power Plant: One 317 shp Allison 250-C18 turboshaft
Dimensions: Overall length rotors turning 12.13m (39.79 ft); rotor diameter 10.80m (35.42 ft)
Weights: Empty 633kg (1,396 lb); MTOW 1,247kg (2,750 lb)
Performance: Cruise speed 204km/h (127 mph) at 1,525m (5,000 ft); operational ceiling 4,325m (14,200 ft); range 560km (348 miles) with full payload
Load: Up to 68kg (150 lb) excluding passengers

HOWARD DGA SERIES
(U.S.A.)

▲ Under the DGA (Damned Good Airplane) designation there appeared during the 1930s a series of classic high-wing lightplanes. The DGA-1 first flew in 1933, and was followed by the DGA-3, -4 and -5 racers, the last of these a four-seater that formed the basis for the commercial DGA-6. This was developed further with different engines as the DGA-9 and DGA-11 to -15. The latter was built in useful numbers for the civil and military markets, in the latter having designations GH (U.S. Navy) and UC-70 (U.S. Army).

HOWARD DGA-15-P (UC-70)

Role: Executive/light transport
Crew/Accommodation: One, plus up to four passengers
Power Plant: One 450 hp Pratt & Whitney R-985-33 Wasp Junior air-cooled radial
Dimensions: Span 11.58m (38 ft); length 7.92m (26 ft); wing area 19.51m^2 (210 sq ft)
Weights: Empty 1,383kg (3,050 lb); MTOW 2,041kg (4,500 lb)
Performance: Maximum speed 282km/h (175 mph) at 1,859m (6,100 ft); operational ceiling 6,553m (21,500 ft); range 1,408km (875 miles)
Load: Up to 372kg (819 lb)

ILYUSHIN DB-3
(U.S.S.R.)

Developed as a long-range medium bomber, this type gave clear indication that Soviet aeronautical achievements of the early and mid-1930s were great. The TsKB-26 prototype flew in 1935 with 597kW (800 hp) Gnome-Rhone 14K radial engines, and was developed via the TsKB-30 into the DB-3 with M-85 radials. Later developments were the DB-3M with M-87 engines, the DB-3T torpedo bomber and the DB-3PT floatplane torpedo bomber.

ILYUSHIN DB-3M

Role: Bomber
Crew/Accommodation: Three
Power Plant: Two 1,100 hp M-87 air-cooled radials
Dimensions: Span 21.44m (70.33 ft); length 14.80m (48.56 ft); wing area 66.7m^2 (717.9 sq ft)
Weights: Empty 5,800kg (12,790 lb); MTOW 8,380kg (18,975 lb)
Performance: Maximum speed 429km/h (267 mph) at 6,700m (21,982 ft); operational ceiling 9,700m (31,824 ft); range 1,577km (980 miles) with full warload
Load: Three 12.7mm machine guns, plus up to 2,500 kg (5,512 lb) of bombs, or bombs and a single torpedo

ILYUSHIN Il-2
(U.S.S.R)

This was probably the finest ground-attack aeroplane of World War II and was built to the extent of more than 36,000 aircraft. The type began life as the TsKB-55 prototype that was considered too heavy because of its massive armour 'bath' structural core. It was thus developed into the TsKB-57 that flew in October 1940 with a 1,268kW (1,700 hp) AM-38 inline engine. This entered service as the single-seat Il-2, which was found to be vulnerable to rear attack and therefore refined as the Il-2M with the cockpit extended aft for a rear gunner. Later in the war the type was also produced in aerodynamically improved versions with greater firepower; there was also an Il-2T torpedo bomber variant.

ILYUSHIN Il-2M

Role: Strike/close air support
Crew/Accommodation: Two
Power Plant: One 1,700 hp AM-38F water-cooled inline
Dimensions: Span 14.6m (47.9 ft); length 11.6m (38.06 ft); wing area 38.5m^2 (414.41 sq ft)
Weights: Empty 4,525kg (9,976 lb); MTOW 6,360kg (14,021 lb)
Performance: Maximum speed 404km/h (251 mph) at 1,500m (4,921 ft); operational ceiling 6,000m (19,685 ft); range 765km (475 miles) with full warload
Load: Two 23mm cannon and two 7.62mm machine guns, plus up to 600kg (1,321 lb) of bombs or anti-armour rockets

ILYUSHIN Il-18 and Il-38
(U.S.S.R.)

The Il-18 was the U.S.S.R.'s equivalent to the Lockheed Electra turboprop airliner, and first flew in July 1957. The type was developed in a number of civil forms with more passengers and greater range, and also formed the basis for the Il-38 anti-submarine aeroplane with its wing moved forward to keep the centre of gravity in the right position despite the addition of heavy mission electronics in the forward fuselage.

ILYUSHIN Il-18D

Role: Intermediate/short range passenger transport
Crew/Accommodation: Five and four cabin crew, plus up to 122 passengers
Power Plant: Four 4,250 shp Ivchenkov AI-20M turboprops
Dimensions: Span 37.4m (122.71 ft); length 35.9m (117.75 ft); wing area 140m^2 (1,507 sq ft)
Weights: Empty 35,000kg (77,160 lb); MTOW 64,000kg (141,000 lb)
Performance: Cruise speed 675km/h (419 mph) at 6,000m (18,288 ft); operational ceiling 10,000m (32,808 ft); range 3,700km (2,300 miles) with maximum payload
Load: Up to 13,500kg (29,762 lb)

ILYUSHIN Il-28
(U.S.S.R.)

First flown in August 1948, the Il-28 was the U.S.S.R.'s most important tactical day bomber of the 1950s, and still serves in modest numbers with several Soviet clients. Variants are the Il-28R reconnaissance platform, the Il-28T torpedo bomber and the Il-28U conversion trainer with a second cockpit behind and above the standard cockpit.

ILYUSHIN Il-28 'BEAGLE'

Role: Bomber reconnaissance
Crew/Accommodation: Three
Power Plant: Two 2,700kgp (5,952 lb s.t.) Klimov VK-1 turbojets
Dimensions: Span 21.45m (70.37 ft); length 17.65m (57.91 ft); wing area 60.8m^2 (654.4 sq ft)
Weights: Empty 12,890kg (28,417 lb); MTOW 21,000kg (46,297 lb)
Performance: Maximum speed 800km/h (497 mph) at sea level); operational ceiling 12,300m (40,355 ft); range 960+km (597+ miles) with full warload
Load: Four 23mm cannon, plus up to 1,000kg (2,205 lb) of bombs.
Note: the above figures for span excludes the wingtip-mounted fuel tanks that can be carried

▼

ILYUSHIN Il-62
(U.S.S.R.)

▲

The Il-62 was developed as a long-range airliner and first flew in January 1963. With its large wing, rear-mounted engines and T-tail the design was clearly influenced by that of the Vickers VC10. The initial version was supplemented by the Il-62M with uprated Soloviev D-30KU turbofans, and the Il-62MK with strengthening for operation at higher weights.

ILYUSHIN Il-62M 'CLASSIC'

Role: Long-range passenger transport
Crew/Accommodation: Five, four cabin crew, plus up to 186 passengers
Power Plant: Four 11,500kgp (25,350 lb s.t.) Soloviev D-30KU turbofans
Dimensions: Span 43.2m (141.75 ft); length 53.12m (174.28 ft); wing area 279.6m^2 (3,010 sq ft)
Weights: Empty 69,400kg (153,000 lb); MTOW 165,000kg (363,760 lb)
Performance: Cruise speed 900km/h (485 knots) at 12,000m (39,370 ft); operational ceiling 13,000+m (42,650 +)ft; range 8,000km (4,317 naut. miles) with full payload
Load: Up to 23,000kg (50,700 lb)

ILYUSHIN Il-76 and Il-78
(U.S.S.R)

The Il-76 first flew in March 1971 as a heavy freighter designed for the military and for the civil resources exploitation of Siberia's under-developed regions. The type is of typical high-wing transport configuration with multi-wheel landing gear and a rear ramp/door arrangement. It has been developed in several versions including the 'Mainstay' airborne early warning aeroplane with an overfuselage rotodome for the surveillance radar, and the Il-78 inflight-refuelling tanker.

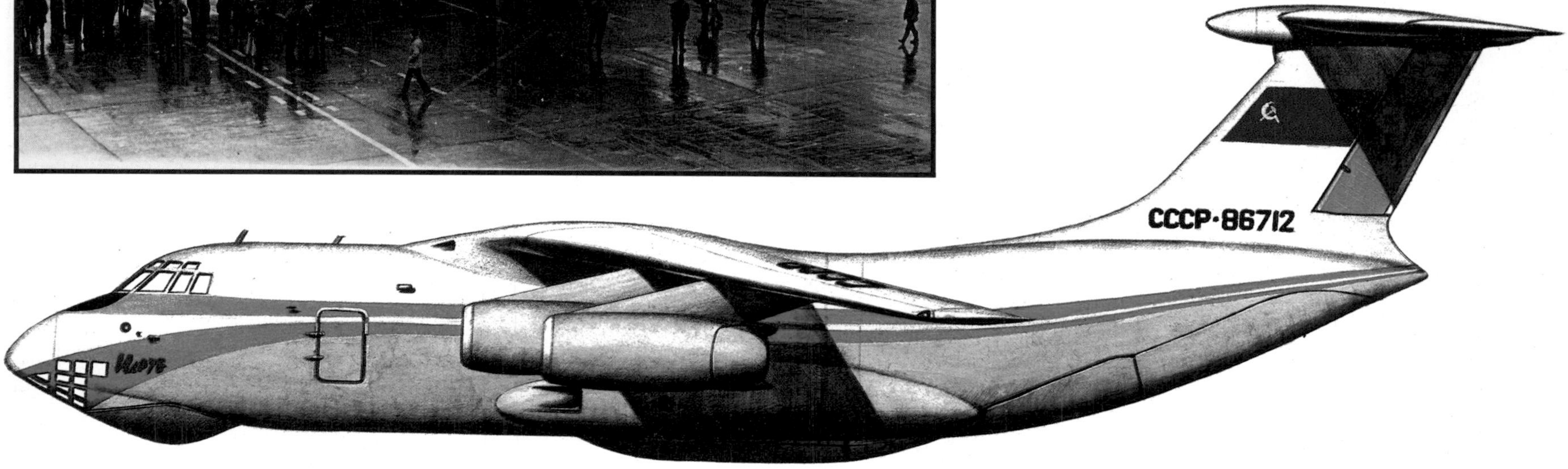

ILYUSHIN Il-76T 'CANDID'

Role: Military and civil bulk cargo transport
Crew/Accommodation: Four and two loadmasters, plus up to 140 troops
Power Plant: Four 12,000kgp (26,455 lb s.t.) Soloviev D-30KP turbofans
Dimensions: Span 50.5m (165.67 ft); length 46.59m (152.85 ft); wing area 300m² (3,229.2 sq ft)
Weights: MTOW 170,000kg (374,785 lb)
Performance: Cruise speed 800km/h (432 knots) at 11,000m (36,090 ft); operational ceiling 13,000m (425,651 ft); range 5,000km (3,170 naut. miles) with full payload
Load: Up to 40,000kg (88,185 lb)

ILYUSHIN Il-86
(U.S.S.R.)

This is the U.S.S.R.'s first wide-body airliner, and flew in prototype form during December 1976. The type's most unusual feature is the use of three airstairs providing access to the lower fuselage, where coats and baggage are left before passengers walk upstairs to the main cabin. There is also a much improved Il-86-300, a long-range variant that has since become the Il-96.

ILYUSHIN Il-86 'CAMBER'

Role: Intermediate-range passenger transport
Crew/Accommodation: Three/four, plus up to 350 passengers
Power Plant: Four 13,000kgp (28,600 lb s.t.) Kuznetsov NK-86 turbofans
Dimensions: Span 48.06m (157.68 ft); length 59.54m (195.33 ft); wing area 320m² (3,444 sq ft)
Weights: MTOW 206,000kg (454,150 lb)
Performance: Maximum speed 950km/h (513 knots) at 9,000m (29,530 ft); operational ceiling 13,000+m (42,650+ ft); range 3,600km (1,942 naut. miles) with full payload
Load: Up to 40,000kg (88,185 lb) including belly cargo

ISRAEL AIRCRAFT INDUSTRIES ARAVA
(Israel)

This STOL type was designed with the civil (Series 100) and military (Series 200) air needs of third-world countries as well as of Israel in mind, and is therefore a rugged type with fixed tricycle landing gear and simple maintenance requirements. The type first flew in November 1969. Civil variants are the IAI 101 freight and IAI 102 20-passenger utility aircraft, and the military models are the IAI 24-troop transport and the lengthened-fuselage IAI 202 30-troop transport.

ISRAEL AIRCRAFT INDUSTRIES 201 ARAVA

Role: Short take-off and landing utility transport
Crew/Accommodation: Two, plus up to 24 troops
Power Plant: Two 750 shp Pratt & Whitney Canada PT6A-34 turboprops
Dimensions: Span 20.96m (68.75 ft); length 13.03m (42.75 ft); wing area 43.68m^2 (470.2 sq ft)
Weights: Empty 3,999kg (8,816 lb); MTOW 6,803kg (15,000 lb)
Performance: Cruise speed 326km/h (203 mph) at 3,048m (10,000 ft); operational ceiling 7,620m (25,000 ft); range 280km (174 miles) with full payload
Load: Up to 2,351kg (5,184 lb)

ISRAEL AIRCRAFT INDUSTRIES KFIR (LION CUB)
(Israel)

Though the Dassault Mirage 5 was developed to an Israeli requirement, France refused for political reasons to deliver the ordered aircraft. Israel had foreseen such an eventuality and therefore developed an equivalent type as the Kfir with Israeli electronics and superior performance offered by an American turbojet in a revised rear fuselage. The first Kfir probably flew in 1974, and the initial Kfir-C1 was followed by the Kfir-C2 with canard foreplanes for improved agility and field performance; an equivalent Kfir-TC2 was produced as a conversion trainer and electronic warfare aeroplane. Further development with an uprated engine, modernized cockpit and advanced electronics led to the single-seat Kfir-C7 and two-seat Kfir-TC7 variants. The type has been exported in several differing forms with an assortment of names.

ISRAEL AIRCRAFT INDUSTRIES KFIR-C2

Role: Strike fighter
Crew/Accommodation: One
Power Plant: One 8,120kgp (17,900 lb s.t.) Bet Shemesh/General Electric J79-GE-17 turbojet with reheat
Dimensions: Span 8.22m (26.97 ft); length 15.65m (51.35 ft); wing area 34.80m^2 (374.6 sq ft)
Weights: Empty 7,290kg (16,072 lb); MTOW 14,700kg (32,408 lb)
Performance: Maximum speed 2,440km/h (1,317 knots) at 10,975m (36,000 ft); operational ceiling 15,000+m (49,213+ ft); radius 520km (281 naut. miles)
Load: Two 30mm cannon, plus up to 4,295kg (9,468 lb) of externally carried weapons

Two views of the popular DR.1050 Ambassadeur

JODEL LIGHTPLANES
(France)

Jodel is an acronym for Edouard Joly and Jean Delamontez, who joined forces in 1946 to produce simple lightplanes of wooden construction suitable for the homebuilder. The marque's most popular aircraft have been the single-seat D.9 Bebe with a Poinsard or Volkswagen engine, and two side-by-side two-seaters with the Continental A65 engine: the open-cockpit D.112 and cabin D.11.

JODEL D.9 BEBE

Role: Light sportsplane
Crew/Accommodation: One
Power Plant: One 25-45 hp air-cooled flat-opposed of various makes including Poinsard and Volkswagen
Dimensions: Span 7m (22.97 ft); length 5.45m (17.88 ft); wing area 9m^2 (96.8 sq ft)
Weights: Empty 190kg (420 lb); MTOW 320kg (705 lb)
Performance: Cruise speed 137km/h (85 mph) at sea level; operational ceiling 3,500m (11,483 ft); range 400km (250 miles)
Load: None

JUNKERS J 1
(Germany)

Hugo Junkers was an early advocate of all-metal construction for aircraft, and in December 1915 there flew his first such aeroplane, the J 1. This was a cantilever monoplane with fixed tail-wheel landing gear and a skinning of thin sheet iron, and was powered by an 80kW (120 hp) Mercedes D.II inline engine. Extra development produced the J 2 armed variant, but the further improved J 3 was not completed. The types had been sufficiently impressive, however, for the German Air Force to request an armoured ground-attack biplane of the same structural concept. This was designed as the J 4 and entered service as the J 1, a sesquiplane covered with corrugated light alloy skinning and notable for its considerable strength.

JUNKERS J 1

Role: Research aircraft
Crew/Accommodation: Two
Power Plant: One 120 hp Daimler Mercedes D.II water-cooled inline
Dimensions: Span 12.95m (42.5 ft); length 8.64m (28.33 ft); wing area 23.88m^2 (257 sq ft)
Weights: Empty 898kg (1,980 lb); MTOW 1,020kg (2,247 lb)
Performance: Maximum speed 170km/h (105.5 mph) at sea level
Load: None

JUNKERS J 10
(Germany)

The J 7 was planned as a low-wing cantilever monoplane and first flew in 1917. The type was developed into the J 9 with a longer fuselage, and this entered production as the D I single-seat fighter with a 138kW (185 hp) BMW inline engine. The basic concept was then scaled up to create the two-seat J 8, precursor of the J 10 that entered production with a 134kW (180 hp) Mercedes D.IIIa inline engine, as the CL I escort/reconnaissance fighter. There were also a few examples of the CLS I floatplane version.

JUNKERS J 10/CL I

Role: Ground attack
Crew/Accommodation: Two
Power Plant: One 180 hp Mercedes D.IIIa water-cooled inline
Dimensions: Span 12.05m (36.1 ft); length 7.9m (25.88 ft); wing area 21.65m^2 (233 sq ft)
Weights: Empty 735kg (1,620 lb); MTOW 1,064kg (2,345 lb)
Performance: Maximum speed 190km/h (118 mph) at sea level; operational ceiling 5,180m (17,000 ft); endurance 2.0 hours
Load: Three 7.9mm machine guns, plus provision to stow stick grenades and other light anti-personnel weapons

JUNKERS F 13
(Germany)

From the J 10 Junkers developed Europe's single most important transport of the 1920s, the classic F 13. The type first flew in June 1919 with an open cockpit for two pilots and an enclosed cabin for four passengers. Later the cockpit was also enclosed. Production amounted to some 320 aircraft in no fewer than 60 variants with different engines and other modifications. The fact that the type was immensely strong and could operate from wheel, ski or float landing gear, made it popular with operators in remoter areas.

JUNKERS F 13

Role: Light passenger transport
Crew/Accommodation: One, plus up to four passengers
Power Plant: One 185 hp BMW III A water-cooled in-line
Dimensions: Span 14.47m (47.47 ft); length 9.6m (31.5 ft); wing area 39m^2 (419.8 sq ft)
Weights: Empty 1,150kg (2,535 lb); MTOW 1,650kg (3,638 lb)
Performance: Cruise speed 140km/h (75.5 mph) at sea level; operational ceiling 3,000m (9,843 ft); range 725km (450 miles)
Load: Up to 320kg (705 lb) payload

JUNKERS W 33 and W 34
(Germany)

The W 33 was an evolution of the F 13 and first flew in 1926 as a rugged mailplane whose payload bay could be converted for the carriage of six passengers. The W 33 variant was powered by an inline engine (generally the Junkers L 5) while the designation W 34 was applied to the same basic type powered by a radial engine. As with the F 13, the type could be operated on wheel, ski or float landing gear.

JUNKERS W 34 hi

Role: Light military/civil passenger transport/trainer
Crew/Accommodation: One, plus up to 7 passengers
Power Plant: One 660 hp BMW 132 air-cooled radial
Dimensions: Span 17.75m (58.23 ft); length 10.27m (33.69 ft); wing area 43m^2 (462.8 sq ft)
Weights: Empty 1,700kg (3,748 lb); MTOW 3,200kg (7,056 lb)
Performance: Cruise speed 233km/h (145 mph) at sea level; operational ceiling 6,300m (20,670 ft); range 900km (559 miles)
Load: Up to 630kg (1,390 lb)

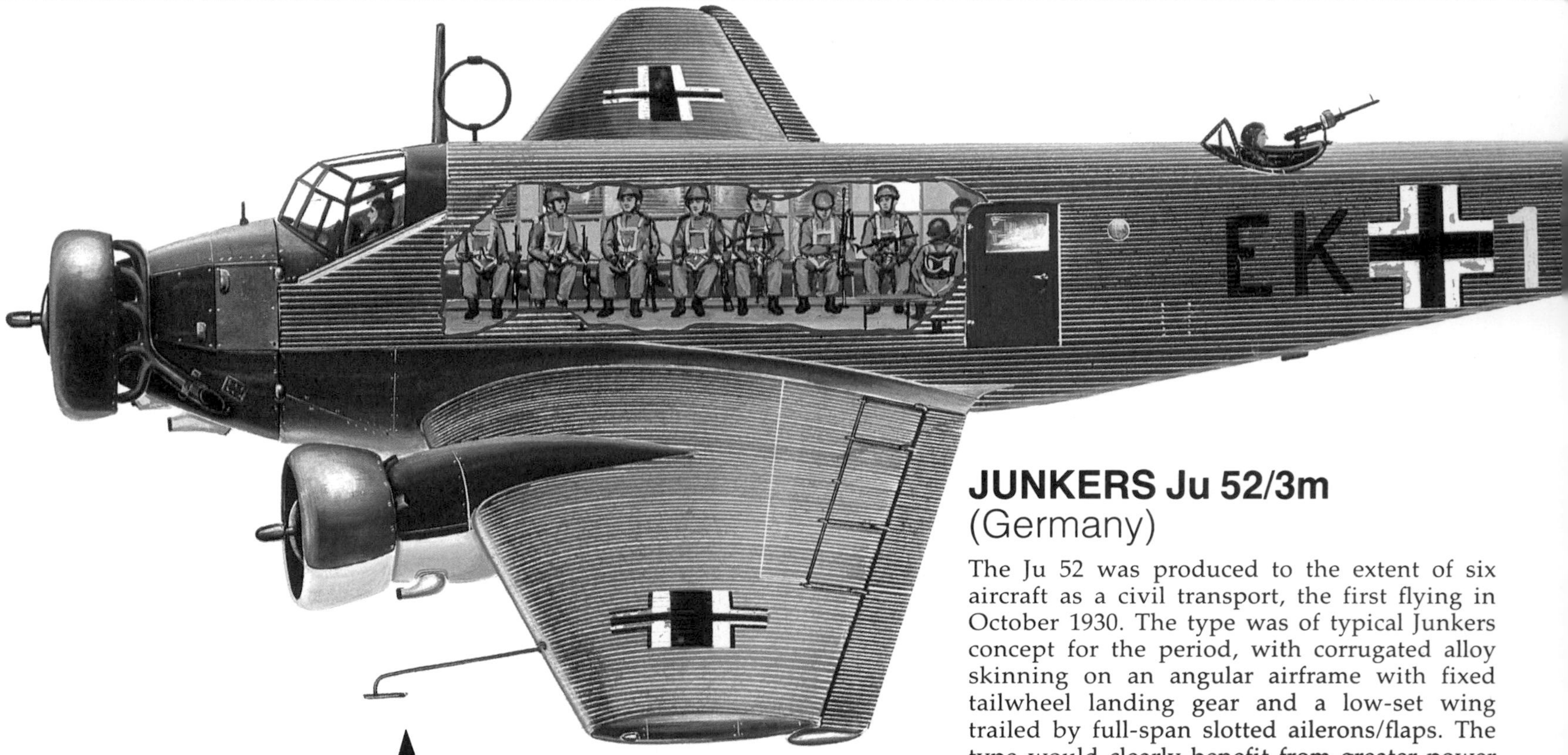

JUNKERS Ju 52/3m
(Germany)

The Ju 52 was produced to the extent of six aircraft as a civil transport, the first flying in October 1930. The type was of typical Junkers concept for the period, with corrugated alloy skinning on an angular airframe with fixed tailwheel landing gear and a low-set wing trailed by full-span slotted ailerons/flaps. The type would clearly benefit from greater power and thus there emerged the Ju 52/3m tri-motor version, initially flown in April 1931 with 410kW (550 hp) Pratt & Whitney Hornet radial engines. The type was produced in several civil variants, but then became the Ju 52/3mge and Ju 52/3mg3e interim bombers for the German Air Force pending the arrival of purpose-designed aircraft. Thereafter the type was built in large numbers as Germany's main transport and airborne forces aeroplane of World War II. There were also later Ju 252 and Ju 352 developments which had more power and were equipped with retractable landing gear.

JUNKERS Ju 52/3mg4e

Role: Military transport (land or water-based)
Crew/Accommodation: Three, plus up to 18 troops
Power Plant: Three BMW 132T-2 air-cooled radials
Dimensions: Span 29.25m (95.97 ft); length 18.9m (62 ft); wing area 110.5m^2 (1,189.4 sq ft)
Weights: Empty 6,510kg (14,354 lb); MTOW 10,500kg (23,157 lb)
Performance: Cruise speed 200km/h (124 mph) at sea level; operational ceiling 5,000m (18,046 ft); range 915km (568 miles) with full payload
Load: Three 7.9mm machine guns and up to 2,000kg (4,409 lb) payload

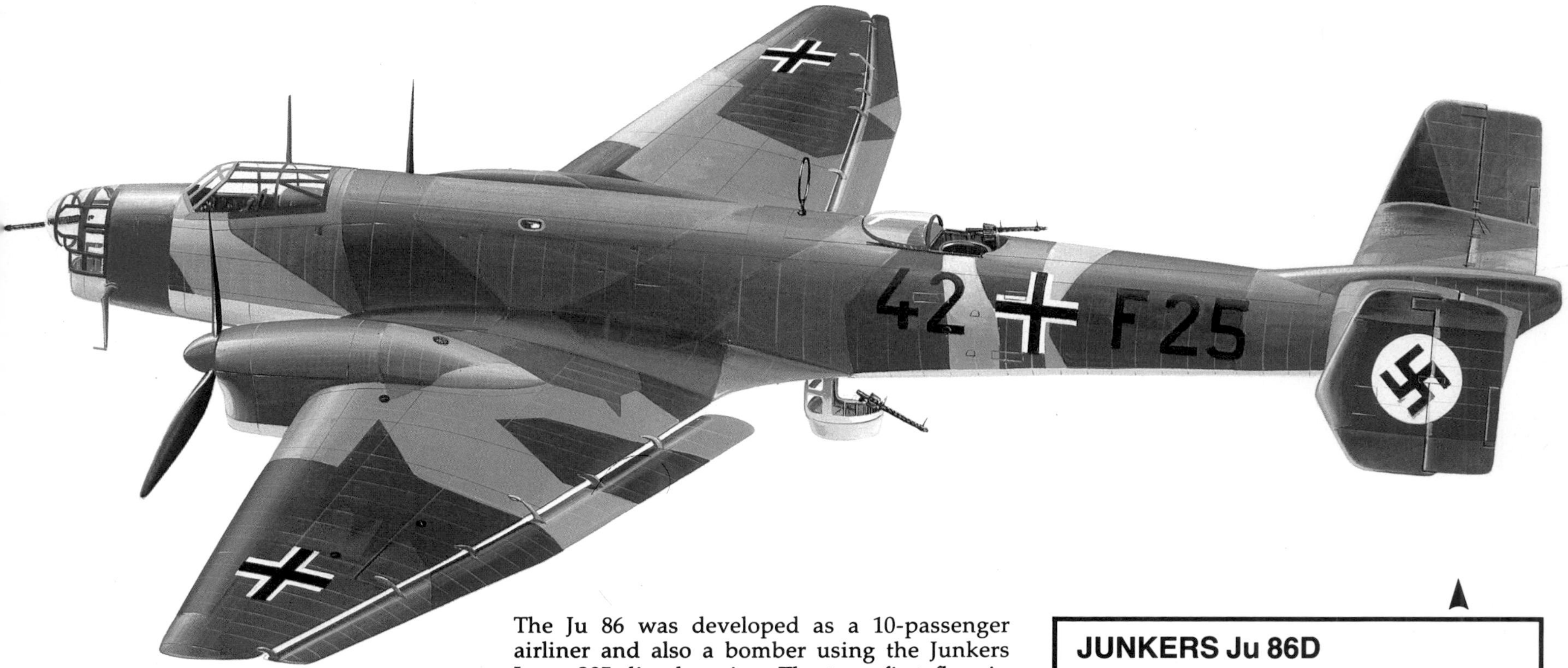

JUNKERS Ju 86
(Germany)

The Ju 86 was developed as a 10-passenger airliner and also a bomber using the Junkers Jumo 205 diesel engine. The type first flew in 1934, and initial deliveries were of the Ju 86A pre-production bomber and Ju 86B airliner. The Ju 86D full-production bomber was not a great success because of its power plant, and production was thus switched to the Ju 86E with BMW 132 radial engines, and the Ju 86G with a glazed nose. Further development centred on high-altitude versions: the Ju 86P bomber and reconnaissance aeroplane with a two-seat pressure cabin, and the Ju 86R with a much larger wing. There were also several variants for a number of foreign buyers.

JUNKERS Ju 86D

Role: Medium bomber
Crew/Accommodation: Four
Power Plant: Two 600 hp Junkers Jumo 205 water-cooled diesels
Dimensions: Span 22.5m (73.9 ft); length 17.6m (57.7 ft); wing area 82m^2 (882.3 sq ft)
Weights: Empty 5,355kg (11,795 lb); MTOW 8,200kg (18,080 lb)
Performance: Maximum speed 300km/h (186 mph) at sea level; operational ceiling 5,900m (19,360 ft); range 1,000km (621 miles) with full bombload
Load: Three 7.9mm machine guns, plus up to 800kg (1,764 lb) of bombs

JUNKERS Ju 87
(Germany)

The Ju 87 was planned as a dedicated dive-bomber, and proved a decisive weapon in the opening campaigns of World War II. The type first flew late in 1935 and began to enter service as the Ju 87A with the 477kW (640 hp) Jumo 210C. Later variants included the Ju 87B with the 895kW (1,200 hp) Jumo 211D and revised wheel spats, the Ju 87D with 1,051kW (1,410 hp) Jumo 211J in the dive-bomber and ground-attack roles, the Ju 87G anti-tank model, the Ju 87H conversion of the Ju 87D as a dual-control trainer, and the Ju 87R version of the Ju 87B in the anti-ship role.

JUNKERS Ju 87B

Role: Dive bomber
Crew/Accommodation: Two
Power Plant: One 1,200 hp Junkers Jumo 211Da water-cooled inline
Dimensions: Span 13.8m (45.3 ft); length 11m (36.83 ft); wing area 31.9m² (343.3 sq ft)
Weights: Empty 2,750kg (6,063 lb); MTOW 4,250kg (9,321 lb)
Performance: Maximum speed 380km/h (237 mph) at 4,000m (13,124 ft); operational ceiling 8,100m (26,575 ft); range 600km (372 miles) with full warload
Load: Three 7.9mm machine guns, plus 1,000 kg (2,205 lb) bombload

JUNKERS Ju 88
(Germany)

The Ju 88 can be considered Germany's equivalent to the British Mosquito, and was with the latter the most versatile warplane of World War II. The type was schemed as a high-speed bomber and first flew in December 1936 with 746kW (1,000 hp) Daimler-Benz DB 600A inline engines. Subsequently these were changed to Junkers Jumo 211s of the same rating. With this latter engine type the Ju 88A entered widespread service, being built in variants up to the

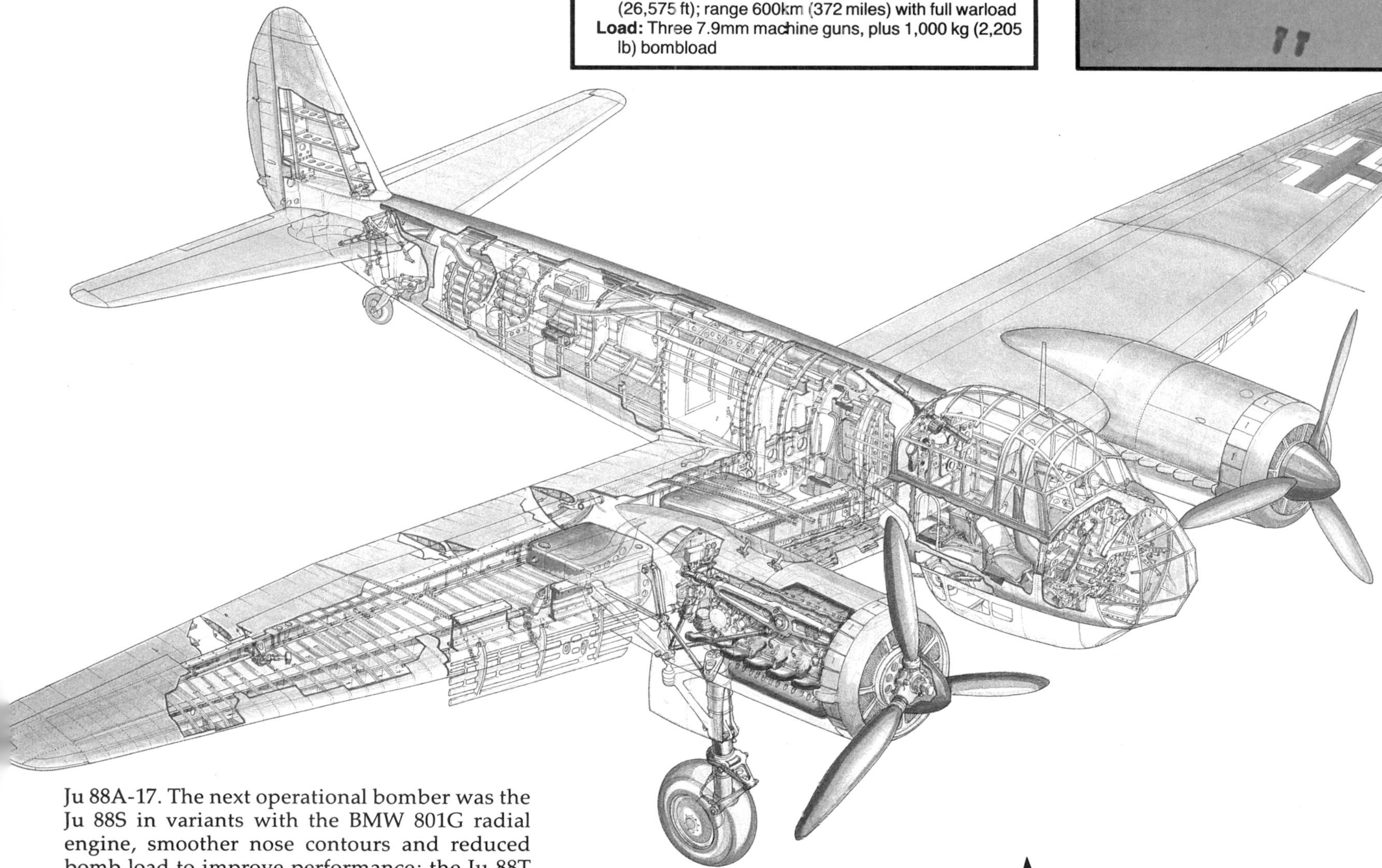

Ju 88A-17. The next operational bomber was the Ju 88S in variants with the BMW 801G radial engine, smoother nose contours and reduced bomb load to improve performance; the Ju 88T was a companion reconnaissance model. From the Ju 88A was developed the Ju 88C heavy fighter: this had BMW 801A radial engines and a solid nose for the heavy gun armament, together with radar in a few night-fighter variants. The definitive night-fighter series was the Ju 88G, together with the improved Ju 88R version of the Ju 88C. Other series were the Ju 88D long-range reconnaissance, Ju 88H longer-range reconnaissance and Ju 88P anti-tank aircraft. Total production of the Ju 88 series came to about 15,000. Further development of the same basic concept also yielded the high-performance Ju 188 type and the high-altitude Ju 388 series.

JUNKERS Ju 88A-4

Role: Light fast bomber
Crew/Accommodation: Four
Power Plant: Two 1,340 hp Junkers Jumo 211J-1 water-cooled inlines
Dimensions: Span 20m (65.63 ft); length 14.4m (47.23 ft); wing area 54.5m² (586.6 sq ft)
Weights: Empty 9,860kg (21,737 lb); MTOW 14,000kg (30,870 lb)
Performance: Maximum speed 470km/h (292 mph) at 5,300m (17,390 ft); operational ceiling 8,200m (26,900 ft); range 1,790km (1,112 miles) with full bombload
Load: Two 13mm and three 7.9mm machine guns, plus up to 2,000 kg (4,409 lb) bombload

JUNKERS Ju 90
(Germany)

From the Ju 89 four-engined bomber project, cancelled in 1937, Junkers developed the Ju 90 airliner. Four prototypes were followed by 10 Ju 90B airliners with accommodation for 40 passengers. Eight of these were accepted by Lufthansa and though the other two were ordered by South African Airways they were never delivered, being used instead to develop the Ju 290.

The prototype Junkers Ju 90 ▼

JUNKERS Ju 90B

Role: Long-range passenger transport
Crew/Accommodation: Four and two cabin crew, plus up to 38 passengers
Power Plant: Four 830 hp BMW 132H-1 air-cooled radials
Dimensions: Span 35.02m (114.9 ft); length 26.3m (86.3 ft); wing area 184m^2 (1,981 sq ft)
Weights: Empty 16,000kg (35,274 lb); MTOW 23,000kg (50,706 lb)
Performance: Cruise speed 320km/h (199 mph) at 3,000m (9,842 ft); operational ceiling 5,500m (18,044 ft); range 1,247km (775 miles) with maximum payload
Load: Up to 4,000 kg (8,818 lb)

JUNKERS Ju 290
(Germany)

Ju 290 was the designation given to the Ju 90S engined with BMW 801 radial engines when the planned BMW 139 radials failed to materialize. The type first flew in 1941, and the Ju 90A series was produced as a maritime aeroplane (reconnaissance, bombing and missile launching) as well as a 50-passenger transport. In 1944 there appeared the Ju 290B prototype of a long-range high-altitude heavy bomber, and a scaled-up derivative that was extensively flight tested but not placed in production was the Ju 390 with a span of 55.35m (181 ft 7.25 inches) and a power plant of six 1,268kW (1,700 hp) BMW 801D radial engines.

JUNKERS Ju 290A-5

Role: Maritime patrol
Crew/Accommodation: Nine
Power Plant: Four 1,700 hp BMW 801D air-cooled radials
Dimensions: Span 42m (137.8 ft); length 28.64m (93.96 ft); wing area 205.3m^2 (2,191.5 sq ft)
Weights: Empty 33,000kg (72,752 lb); MTOW 45,000kg (99,225 lb)
Performance: Maximum speed 440km/h (273 mph) at 6,000m (19,686 ft); operational ceiling 6,000m (19,686 ft); range 6,150km (3,820 miles) maximum
Load: Six 20mm cannon, one 13mm machine gun

KAMAN H-2 SEASPRITE
(U.S.A.)

This trim helicopter was produced in response to a 1956 U.S. Navy utility helicopter requirement, and has since been built in substantial numbers. Powered originally by a single T58 turboshaft, early variants were used in the utility and the armed or unarmed rescue roles. The SH-2D became the primary anti-submarine helicopter of smaller warships and was then developed into the SH-2F with twin T58 turboshafts, and then into the definitive SH-2G with two General Electric T700 turboshafts.

KAMAN SH-2F SEASPRITE

Role: Naval shipborne anti-submarine helicopter
Crew/Accommodation: Three
Power Plant: Two 1,350 shp General Electric T58-GE-8F turboshafts
Dimensions: Overall length rotors turning 16m (52.6 ft); rotor diameter 13.4m (44 ft)
Weights: Empty 3,193kg (7,040 lb); MTOW 6,124kg (13,500 lb)
Performance: Maximum speed 266km/h (143 knots) at sea level; operational ceiling 6,858m (22,500 ft); range 679km (366 naut. miles)
Load: Two lightweight anti-submarine torpedoes carried out to 65km (35 naut. miles) from ship, with 1.2 hour loiter on station

KAMOV Ka-25
(U.S.S.R.)

The Ka-25 is of typical Kamov helicopter configuration with a compact yet capacious fuselage surmounted by twin contra-rotating rotors. The type has a good power/weight ratio and thus a high level of performance. After a first flight in 1960, it was developed in three shipboard variants denoted by NATO as the 'Hormone-A' anti-submarine, 'Hormone-B' missile-targeting and 'Hormone-C' utility variants.

KAMOV Ka-25 'HORMONE-A'

Role: Naval shipborne anti-submarine helicopter
Crew/Accommodation: Four
Power Plant: Two 1,800 shp Glushenkov GTD-3 turboshafts
Dimensions: Overall length rotors turning 15.85m (52 ft); rotor diameter 15.85m (52 ft)
Weights: Empty 4,765kg (10,500 lb); MTOW 7,494kg (16,500 lb)
Performance: Maximum speed 241km/h (130 knots) at sea level; operational ceiling 3,350m (11,000 ft); radius 200km (108 naut. miles) with full warload
Load: Up to 1,800 kg (3,969 lb)

KAMOV Ka-27 and Ka-32
(U.S.S.R.)

This was developed as a larger, more powerful and more capacious derivative of the Ka-25, and first flew in about 1979. The type has since been built in Ka-27 military and Ka-32 civil variants, the former including the 'Helix-A' anti-submarine, 'Helix-B' naval infantry assault and 'Helix-D' rescue/utility variants.

KAMOV Ka-27 'HELIX A'

Role: Naval shipborne anti-submarine helicopter
Crew/Accommodation: Four/five
Power Plant: Two 2,205 shp Isotov TV3-117V turboshafts
Dimensions: Overall length rotors turning 15.9m (52.17 ft); rotor diameter 15.9m (52.17 ft)
Weights: Empty 5,000kg (11,023 lb); MTOW 9,525kg (21,000 lb) estimated
Performance: Maximum speed 241km/h (130 knots) at sea level; operational ceiling 5,000m (16,404 ft); range 580km (313 naut. miles) with full warload
Load: in excess of 1,800 kg (3,968 lb)

KAWANISHI H6K
(Japan)

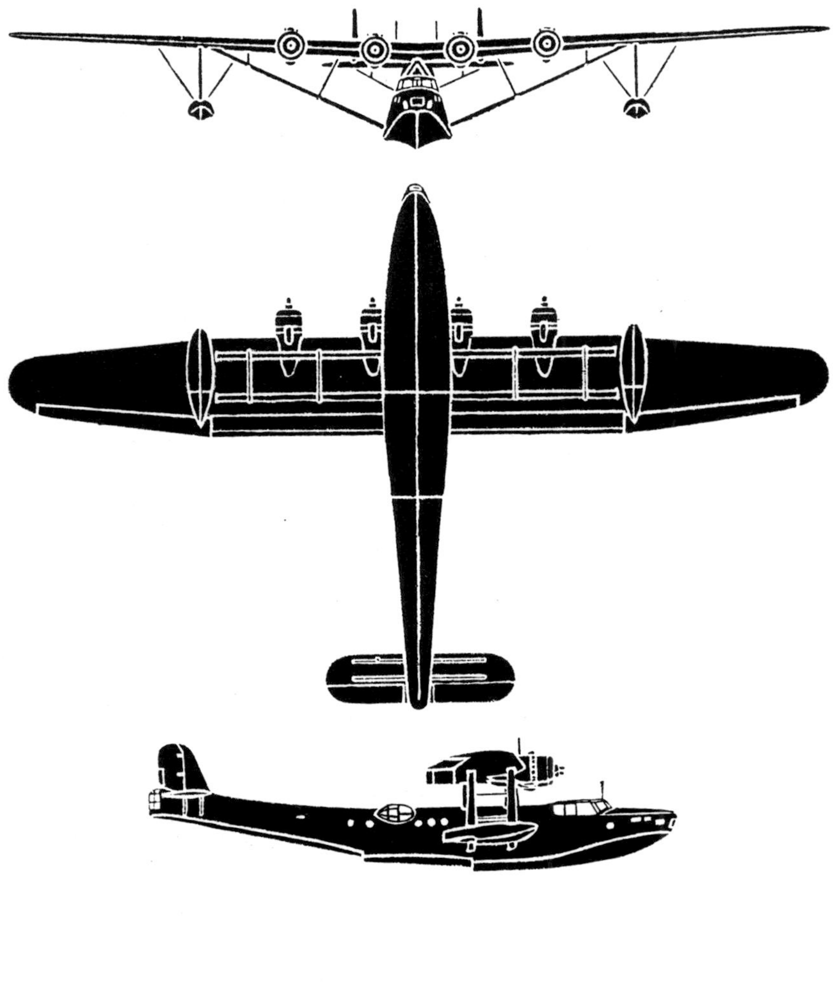

The H6K was designed as a high-performance reconnaissance flying boat, and emerged as an elegant machine with a slender two-step hull surmounted by a large strut-braced parasol wing. The Type S prototype first flew in July 1936 and was considered underpowered, so the 626 kW (840 hp) Nakajima Hikari radial engines were replaced by 746 kW (1,000 hp) Mitsubishi Kinsei 43 radial engines to create the H6K1 prototype. The H6K2 production version was similar, and variants that followed were the H6K2-L unarmed transport for Japan Air Lines, the H6K3 VIP transport, the H6K4 major production version with greater fuel capacity, heavier armament and 798 kW (1,070 hp) Kinsei 46s, the H6K4-L unarmed transport, and the H6K5 with 969 kW (1,300 hp) Kinsei 51 or 53s and revised armament.

KAWANISHI H6K4 'MAVIS'

Role: Long-range patrol bomber flying boat
Crew/Accommodation: Nine
Power Plant: Four 930 hp Mitsubishi Kinsei 43 air-cooled radial
Dimensions: Span 40m (131.23 ft); length 25.63m (84.09 ft); wing area 170m^2 (1,830 sq ft)
Weights: Empty 11,619kg (25,615 lb); MTOW 23,000kg (50,705 lb) special overload
Performance: Cruise speed 222km/h (138 mph) at 2,000 (6,560 ft); operational ceiling 9,610m (31,530 ft); range 6,084km (3,780 miles) maximum
Load: One 20mm cannon and four 7.7mm machine guns plus up to 1,600 kg (3,527 lb) in the shape of two torpedoes

KAWANISHI H8K
(Japan)

Probably the finest flying boat to see extensive service in World War II, the H8K resembled the British Sunderland in overall configuration, and first flew in H8K1 prototype form with 1,141kW (1,530 hp) Mitsubishi MK4A radial engines during December 1940. Trials showed that the type was unstable on the water and required extensive remedial action. Immediately apparent, however, were the high airborne performance and, by comparison with other Japanese warplanes of the period, operationally desirable features such as heavy armament, good protection and partially self-sealing fuel tankage. The type began to enter service in 1942, the major production version being the H8K2 of which 112 were built with fully protected fuel tankage, heavier armament, surface-search radar and 1,380kW (1,850 hp) MK4Q radial engines. There was also an H8K2-L transport with reduced armament, and the H8K3 designation covered two prototypes with retractable stabilizing floats and a retractable dorsal turret, these being redesignated H8K4 when fitted with 1,361kW (1,825 hp) MK4T-B Kasai 25b radial engines.

KAWANISHI H8K2 'EMILY'
Role: Long-range maritime patrol bomber flying boat
Crew/Accommodation: Ten
Power Plant: Four 1,850 hp Mitsubishi MK4Q air-cooled radials
Dimensions: Span 38m (124.67 ft); length 28.13m (92.33 ft); wing area 160m^2 (1,722 sq ft)
Weights: Empty 18,380kg (40,512 lb); MTOW 32,500kg (71,650 lb)
Performance: Maximum speed 454km/h (282 mph) at 5,000m (16,404 ft); operational ceiling 8,770m (28,773 ft); range 4,000km (2,486 miles) with full warload
Load: Five 20mm cannon and three 7.7 machine guns, plus up to 2,000 kg (4,408 lb) or ordnance

KAWANISHI N1K KYOFU (MIGHTY WIND)
(Japan)

Designed from 1940 as a fighter able to protect and support amphibious landings, the N1K was schemed as a substantial seaplane with single main/two stabilizing floats and a powerful radial engine driving contra-rotating propellers that would mitigate torque problems during take-off and landing. The type began to enter service in 1943 as the N1K1, but the type's *raison d'etre* had disappeared by this stage of the war and production was terminated with the 97th machine. The N1K2 with a more powerful engine remained only a project.

KAWANISHI N1K1 'REX'
Role: Fighter floatplane
Crew/Accommodation: One
Power Plant: One 1,460 hp Mitsibishi Kasei 14 air-cooled radial
Dimensions: Span 12.m (39.37 ft); length 10.59m (34.74 ft); wing area 23.5m^2 (252.9 sq ft)
Weights: Empty 2,700kg (5,952 lb); MTOW 3,712kg (8,184 lb)
Performance: Maximum speed 482km/h (300 mph) at 5,700m (18,701 ft); operational ceiling 10,560m (34,646 ft); range 1,690km (1,050 miles) with full bombload
Load: Two 20mm cannon, two 7.7mm machine guns, plus up to 60 kg (132 lb) of bombs

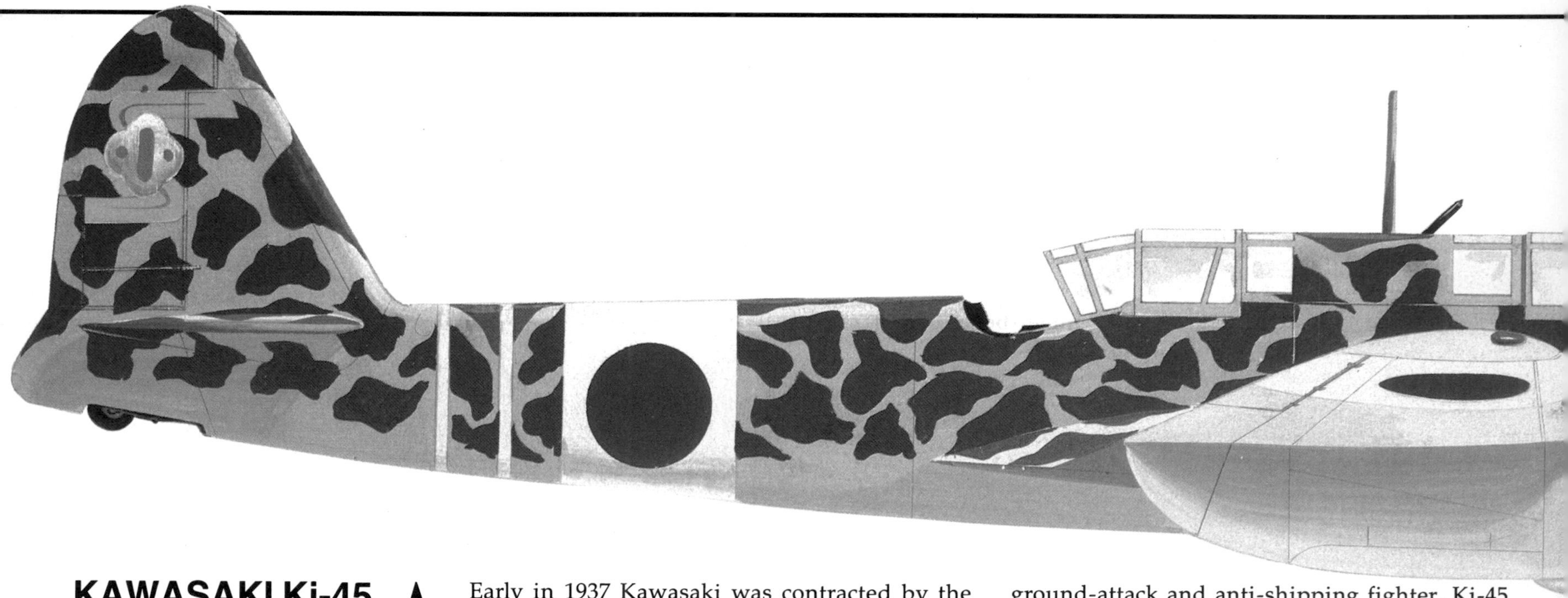

KAWASAKI Ki-45 ▲ TORYU (DRAGON SLAYER)
(Japan)

Early in 1937 Kawasaki was contracted by the Japanese army to initiate development of a twin-engined fighter suitable for long-range operations over the Pacific. The design started life as the Ki-38, but was redesignated Ki-45 after the incorporation of major design changes. The first prototype flew in January 1939 with 611kW (820 hp) Nakajima Ha-20B radial engines, but was clearly underpowered. The power plant was revised to a pair of 746kW (1,000 hp) Nakajima Ha-25s, but development problems with this engine meant that production of the Ki-45 KAI started only in September 1941 for the Toryu's combat debut in August 1942. The type's forte was bomber destruction, and as the U.S. bomber effort began to develop a nocturnal aspect, the Toryu was developed as a night-fighter. The type was used in most Japanese army theatres of war, and total production amounted to 1,698 aircraft in variants such as the Ki-45 KAIa fighter, Ki-45 KAIb ground-attack and anti-shipping fighter, Ki-45 KAIc night-fighter with obliquely upward-firing 20mm cannon and Ha-102 radial engines, and Ki-45 KAId anti-shipping fighter with heavy nose armament and Ha-102s.

KAWASAKI Ki-45-I 'NICK'

Role: Long-range/night fighter
Crew/Accommodation: Two
Power Plant: Two 1,000 hp Nakajima Ha-25 air-cooled radials
Dimensions: Span 15.02m (49.28 ft); length 11m (36.09 ft); wing area 32m^2 (344.4 sq ft)
Weights: Empty 4,000kg (8,818 lb); MTOW 5,500kg (12,125 lb)
Performance: Maximum speed 540km/h (336 mph) at 6,000m (19,685 ft); operational ceiling 10,000m (32,808 ft); range 2,000km (1,243 miles) fighter mission
Load: One 20mm cannon, two 12.7mm and one 7.7mm machine gun, plus up to 500kg (1,102 lb) of bombs

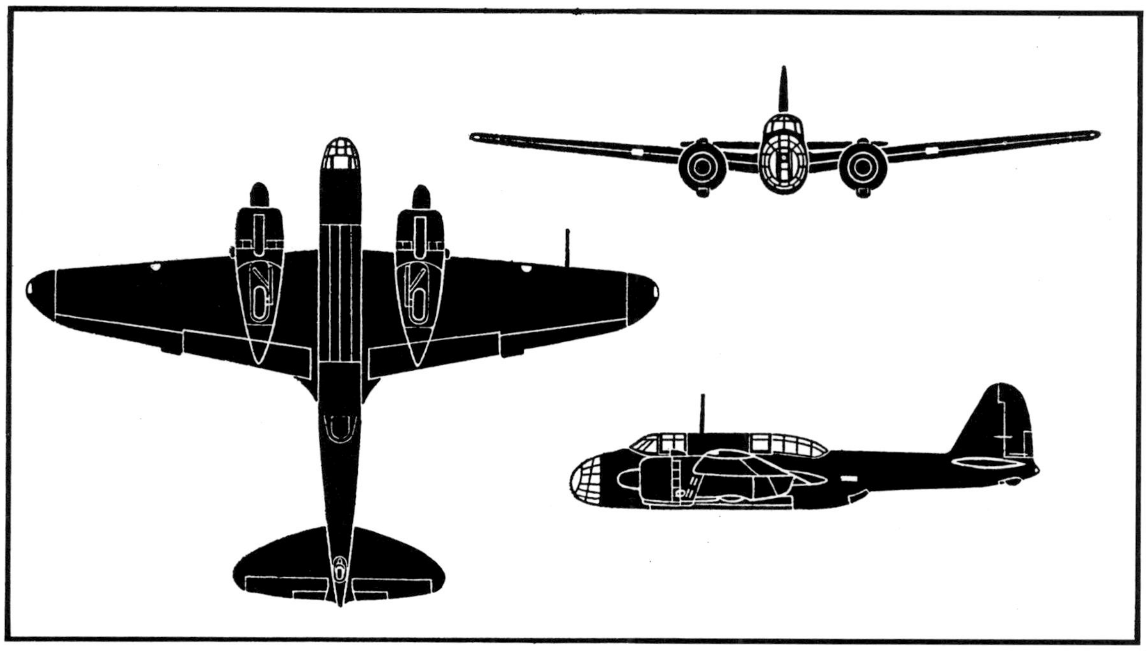

▲ KAWASAKI Ki-48
(Japan)

This light bomber for the Japanese army first flew in July 1939 and entered service as the Ki-48-Ia with two 708kW (950 hp) Nakajima Ha-25 radial engines, no armour and only light defensive armament. The type proved woefully deficient in combat and the 557 Ki-48-Ia and -Ib aircraft were followed by 1,408 examples of the Ki-48-II (in -IIa, -IIb and -IIc subvariants) with heavier armament and improved protection, though even this basic variant was hopelessly outmatched by Allied fighters when it appeared in 1942. From 1944 the type was relegated to second-line and *kamikaze* use.

KAWASAKI Ki-48-II 'LILY'

Role: Light bomber
Crew/Accommodation: Four
Power Plant: Two 1,130 hp Nakajima Ha-115 air-cooled radials
Dimensions: Span 17.47m (57.32 ft); length 12.75m (41.83 ft); wing area 40m^2 (430.5 sq ft)
Weights: Empty 4,550kg (10,031 lb); MTOW 6,750kg (14,881 lb)
Performance: Maximum speed 505km/h (314 mph) at 5,600m (18,373 ft); operational ceiling 10,000m (32,808 ft); range 1,675km (1,041 miles) with full bombload
Load: Four 7.7mm machine guns, plus up to 300kg (661 lb) of bombs

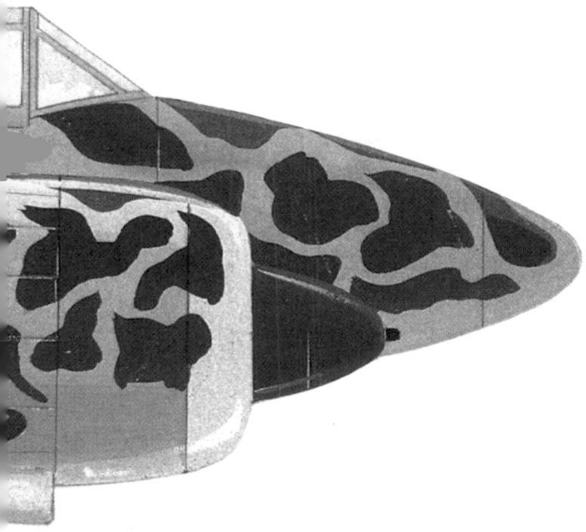

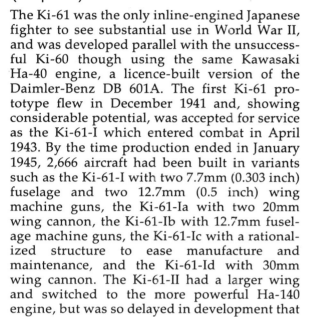

KAWASAKI Ki-61 HIEN (SWALLOW) and Ki-100
(Japan)

The Ki-61 was the only inline-engined Japanese fighter to see substantial use in World War II, and was developed parallel with the unsuccessful Ki-60 though using the same Kawasaki Ha-40 engine, a licence-built version of the Daimler-Benz DB 601A. The first Ki-61 prototype flew in December 1941 and, showing considerable potential, was accepted for service as the Ki-61-I which entered combat in April 1943. By the time production ended in January 1945, 2,666 aircraft had been built in variants such as the Ki-61-I with two 7.7mm (0.303 inch) fuselage and two 12.7mm (0.5 inch) wing machine guns, the Ki-61-Ia with two 20mm wing cannon, the Ki-61-Ib with 12.7mm fuselage machine guns, the Ki-61-Ic with a rationalized structure to ease manufacture and maintenance, and the Ki-61-Id with 30mm wing cannon. The Ki-61-II had a larger wing and switched to the more powerful Ha-140 engine, but was so delayed in development that only 99 had been produced before U.S.A.A.F. bombing destroyed engine production capacity.

Variants were the Ki-61-II KAI with the Ki-61-I wing, the Ki-61-IIa with the armament of the Ki-61-Ic, and the Ki-61-IIb with four 20mm wing cannon. With the Ha-140 engine unavailable for a comparatively large number of completed Ki-61-II airframes, the Japanese army ordered adaptation of the type to take the Mitsubishi Ha-112-II radial engine, whose 1,119kW (1,500 hp) rating was identical to that of the Ha-140. The resulting Ki-100 first flew in February 1945 and proved to be an outstanding interceptor, perhaps Japan's best fighter of World War II.

The army ordered conversion of the 272 Ki-61-II airframes as Ki-100-Ia 'production' aircraft. New production amounted to 99 Ki-100-Ib aircraft with the cut-down rear fuselage and bubble canopy developed for the proposed Ki-61-III fighter. The designation Ki-100-II was used for three prototypes with the Mitsubishi Ha-112-IIru turbocharged radial engine for improved high-altitude performance.

KAWASAKI Ki-100-II 'TONY'

Role: Fighter
Crew/Accommodation: One
Power Plant: One 1,500 hp Mitsubishi Ha-112-II air-cooled radial
Dimensions: Span 12m (39.37 ft); length 8.82m (28.94 ft); wing area 20m^2 (215.3 sq ft)
Weights: Empty 2,525kg (5,567 lb); MTOW 3,495kg (7,705 lb)
Performance: Maximum speed 590km/h (367 mph) at 10,000m (32,808 ft); operational ceiling 11,500m (37,730 ft); range 1,800km (1,118 miles)
Load: Two 20mm cannon and two 12.7mm machine guns

KAWASAKI C-1
(Japan)

This twin-turbofan STOL tactical transport was designed from 1966 to meet a Japanese requirement for a Curtiss C-46 Commando replacement, and first flew as the XC-1 in November 1970. Production totalled only 28 aircraft, and of these one has been converted as an ECM trainer.

KAWASAKI C-1A

Role: Military short field-capable cargo/troop transport
Crew/Accommodation: Five, plus up to 60 troops
Power Plant: Two 6,575kgp (14,500b s.t.) Pratt & Whitney JT8D9 turbofans
Dimensions: Span 29m (95.14 ft); length 30.60m (100.39 ft); wing area 120.5m^2 (1,297 sq ft)
Weights: Empty 24,300kg (53,572 lb); MTOW 45,000kg (99,210 lb)
Performance: Maximum speed 815km/h (440 knots) at 7,620m (25,000 ft); operational ceiling 11,580m (38,000 ft); range 1,300km (701 naut. miles) with 7,900kg (17,416 lb) payload
Load: Up to 11,900 kg (26,235 lb)

KEYSTONE B-4
(U.S.A.)

In its original Huff-Daland and later Keystone guises, this company was responsible for the U.S. Army's most important heavy bombers of the 1920s. The series began with the XLB-1 of 1923, and there then appeared increasingly large and powerful bombers in variants up to the LB-12. Further orders were placed for the LB-13 and LB-14: the LB-13s were delivered as five Y1B-4 and two Y1B-6 aircraft with Pratt & Whitney and Wright radial engines respectively, and the LB-14s were delivered as three Y1B-5s with Wright radial engines. These pre-production types were followed by 25 B-4A (Y1B-4), 27 B-5A (Y1B-5) and 39 B-6A (Y1B-6) aircraft with the R-1860-7, R-1750-3 and R-1820-1 engines respectively.

KEYSTONE B-4A PANTHER

Role: Bomber
Crew/Accommodation: Five
Power Plant: Two 575 hp Pratt & Whitney R-1860-7 Hornet B air-cooled radials
Dimensions: Span 22.76m (74.66 ft); length 14.88m (48.83 ft); wing area 106.38m^2 (1,145.0 sq ft)
Weights: Empty 3,607kg (7,951 lb); MTOW 5,992kg (13,209 lb)
Performance: Maximum speed 195km/h (121 mph) at sea level; operational ceiling 4,267m (14,000 ft); range 1,376km (855 miles)
Load: Three .303 inch machine guns, plus up to 1,134 kg (2,500 lb) of bombs

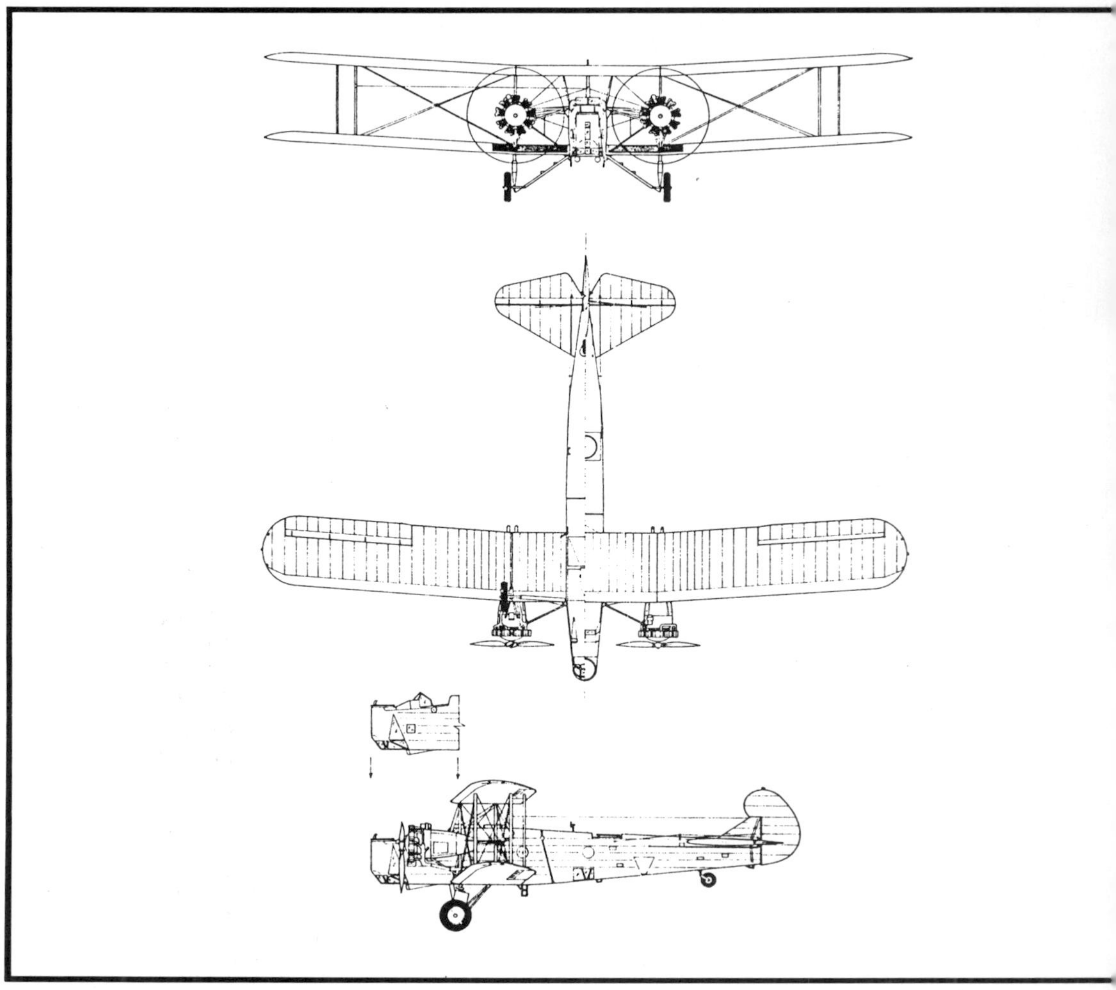

LATECOERE 28
(France)

The Latecoere 28 was developed from the Latecoere 26 mailplane as a braced high-wing monoplane transport, and first flew in the late 1920s as the Latecoere 28.0 with the 373kW (500 hp) Renault 12Jbr inline engine. Some 17 Latecoere 28.0 aircraft were followed by 29 examples of the Latecoere 28.1 with the 373kW (500 hp) Hispano-Suiza 12 Hbxr inline, one Latecoere 28.2 mailplane, five Latecoere 28.3 long-range mailplanes, and a number of limited-production versions up to the Latecoere 28.9.

LATECOERE 28-3

Role: Passenger/mail transport
Crew/Accommodation: Two, plus up to eight passengers
Power Plant: One 660 hp Hispano-Suiza 12 Lbr water-cooled inline.
Dimensions: Span 19.25m (63.16 ft); length 13.645m (44.77 ft); wing area 58.2m^2 (626.5 sq ft)
Weights: Empty 2,637kg (5,813 lb); MTOW 5,017kg (11,060 lb)
Performance: Cruise speed 200km/h (124 mph) at 3,000m (9,845 ft); operational ceiling 5,400m (17,717 ft); range 1,400km (870 miles) with full payload
Load: Up to 620 kg (1,367 lb)

LAVOCHKIN LaGG-3
(U.S.S.R.)

Designed by Lavochkin with the aid of Gorbunov and Gudkov, the LaGG-1 flew in March 1940 as a mainly wooden fighter with an M-105 inline engine. The prototype was revised with a supercharged engine (driving a three- rather than two-blade propellor) and leading-edge slats, the dramatic improvement in performance leading to orders for the LaGG-3 production model. Some 6,500 LaGG-3 fighters were built with a number of armament installations, and these performed adequately if not ably in the early campaigns of the German invasion of the U.S.S.R. in July 1941.

LAVOCHKIN LaGG-3

Role: Fighter
Crew/Accommodation: One
Power Plant: One 1,210 hp Klimov M-105PF water-cooled inline
Dimensions: Span 9.8m (32.15 ft); length 8.82m (28.94 ft); wing area 17.5m² (188.4 sq ft)
Weights: Empty 2,620kg (5,776 lb); MTOW 3,190kg (7,032 lb)
Performance: Maximum speed 560km/h (348 mph) at 5,000m (16,404 ft); operational ceiling 9,600m (31,496 ft); range 650km (404 miles)
Load: One 20mm cannon and two 12.7mm machine guns, plus up to 200kg (440 lb) of bombs or rockets

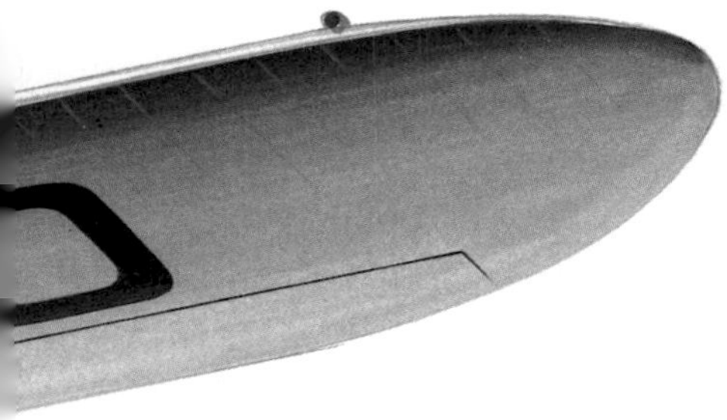

Lavochkin La-7

LAVOCHKIN La-5 and La-7
(U.S.S.R.)

The LaGG-3 rightly regarded as an interim fighter: its wooden construction meant that it made minimum demands on the already overworked portions of the U.S.S.R. aviation industry producing metal aircraft. But performance was no more than adequate, and in 1941 the designer revised the aeroplane to take the ASh-82 radial engine, which boosted performance in overall terms and also made the type considerably more sprightly at altitude. This LaGG-5 superseded the LaGG-3 in May 1942, and was then revised as the La-5 with a cut-down rear fuselage and 360° vision canopy. The La-5 was another step in the right direction, but was still inferior to the Germans' Messerschmitt Bf 106G. Lavochkin therefore cleaned up the airframe, reduced structural weight and fuel capacity, and introduced metal spars as well as the boosted ASh-82RFN radial engine. The resultant La-5FN appeared in combat late in 1942 and proved excellent: about 10,000 were produced for service into the period immediately after World War II. Some 5,500 La-7 aircraft were also produced: these were essentially similar to the La-5 apart from additional reduction in weight and drag to improve high-altitude performance. Two-seat trainer versions were the La-5UTI and La-7UTI.

LAVOCHKIN La-5FN

Role: Fighter bomber
Crew/Accommodation: One
Power Plant: One 1,850 hp Shvetsov ASh-82RFN air-cooled radial
Dimensions: Span 9.8m (32.16 ft); length 8.67m (28.44 ft); wing area 17.59m² (189.3 sq ft)
Weights: Empty 2,605kg (5,743 lb); MTOW 3,360kg (7,408 lb)
Performance: Maximum speed 648km/h (403 mph) at 6,400m (20,997 ft); operational ceiling 9,500m (31,116 ft); range 765km (475 miles) in fighter role
Load: Two 23mm cannon, plus up to 300kg (662 lb) of bombs or rockets

LAVOCHKIN La-9 and La-11 (U.S.S.R.)

Further development of the La-5's basic concept resulted in the La-9 that entered service late in 1944. Retaining the La-5's engine and configuration, the La-9 had an improved cockpit, squarer-cut flying surfaces and a higher proportion of alloy in its structure. Armament was four 23mm cannon, though in the La-11 escort fighter version, this was reduced to three cannon to allow greater fuel capacity.

LAVOCHKIN La-11

Role: Fighter
Crew/Accommodation: One
Power Plant: One 1,870 hp Shvetsov ASh-82FNV air-cooled radial
Dimensions: Span 9.8m (32.16 ft); length 8.66m (28.41 ft); wing area 17.59m^2 (189.3 sq ft)
Weights: Empty 2,770kg (6,107 lb); MTOW 3,996kg (8,810 lb)
Performance: Maximum speed 674km/h (419 mph) at 6,200m (20,341 ft); operational ceiling 10,250m (33,629 ft); range 2,550km (1,585 miles)
Load: Three 23mm cannon

Not illustrated

▲ GATES LEARJET 20 SERIES (U.S.A.)

The Lear Jet (later Learjet) series is derived somewhat tenuously from the Swiss P-16 fighter project cancelled in the 1950s, and has matured as one of today's classic 'bizjets'. The first of the series was the Lear Jet 23 designed for a maximum of seven passengers carried over long range at high speed by two 1,293kg (2,850 lb) thrust General Electric CJ610 turbojets. This first flew in October 1963, and for certification with a single pilot was limited to a maximum weight of 5,670kg (12,500 lb).

Many potential customers preferred a two-pilot crew, and this resulted in the development of the Learjet 24, which first flew in February 1964 with features such as a greater pressurization differential to allow higher-altitude cruise. There were several versions with higher-power engines and increased fuel capacity, and in 1976 there appeared the Learjet 24 with more fuel and a modified wing to improve stall and landing speeds. The Learjet 25 of 1966 introduced a longer fuselage for eight passengers, and the Learjet 28 and 29 Longhorns of 1979 pioneered a winglet-tipped wing of greater span, the Learjet 29 trading two passengers for additional fuel.

GATES LEARJET 25

Role: Executive transport
Crew/Accommodation: Two, plus up to eight passengers
Power Plant: Two 1,338kgp (2,950 lb s.t.) General Electric CJ610-6 turbojets
Dimensions: Span 10.86m (35.65 ft); length 14.51m (47.60 ft); wing area 21.53m^2 (231.8 sq ft)
Weights: Empty 3,295kg (7,265 lb); MTOW 6,804kg (15,000 lb)
Performance: Maximum speed 818km/h (441 knots) at 12,192m (40,000 ft); operational ceiling 13,716m (45,000 ft); range 3,315km (2,060 miles) with full payload
Load: Up to 1,497kg (3,260 lb)

GATES LEARJET 30 SERIES
(U.S.A.)

This series was a close relative of the Learjet 20 series but introduced turbofan power in the form of two 1,588kg (3,500 lb) thrust Garrett TFE731s, as well as a longer fuselage and greater span. The type first flew in 1973, and the production versions were the Learjet 35 able to carry eight passengers over transcontinental range and the Learjet 36 designed to carry six passengers over intercontinental range. Introduction of a cambered wing resulted in the Learjets 35A and 36A.

GATES LEARJET 35A

Role: Executive transport
Crew/Accommodation: Two, plus up to ten passengers
Power Plant: Two 1,588 kgp (3,500 lb s.t.) Garrett AiResearch TFE 731-3-2B turbofans
Dimensions: Span 12.04m (39.5 ft); length 14.83m (48.67 ft); wing area 23.53m² (253.3 sq ft)
Weights: Empty 4,468kg (9,850 lb); MTOW 7,711kg (17,000 lb)
Performance: Maximum speed 860km/h (464 knots) at 12,497m (41,000 ft); operational ceiling 13,716m (45,000 ft); range 4,262km (2,300 naut. miles) with 7 passengers
Load: Up to 1,656kg (3,650 lb)

▼

GATES LEARJET 50 SERIES
(U.S.A.)

Though planned as a sequence of three aircraft carrying up to 11 passengers over very long ranges, this series eventually comprised just the Learjet Longhorn 55 that first flew in April 1979 with winglets on the wing and a larger fuselage than those of its predecessors.

GATES LEARJET 55C

Role: Executive transport
Crew/Accommodation: Two plus up to ten passengers
Power Plant: Two, 1,678kgp (3,700 lb s.t.) Garrett AiResearch TFE 731-3A-2B turbofans
Dimensions: Span 13.34m (43.79 ft); length 16.79m (55.125 ft); wing area 24.57m² (264.5 sq ft)
Weights: Empty 5,725kg (12,622 lb); MTOW 9,526kg (21,000 lb)
Performance: Cruise speed 853km/h (460 knots) Mach 0.8 at 12,497m (41,000 ft); operational ceiling 15,545m (51,000 ft); range 4,170km (2,250 naut. miles) with 4 passengers
Load: Up to 1,079kg (2,378 lb)

▲

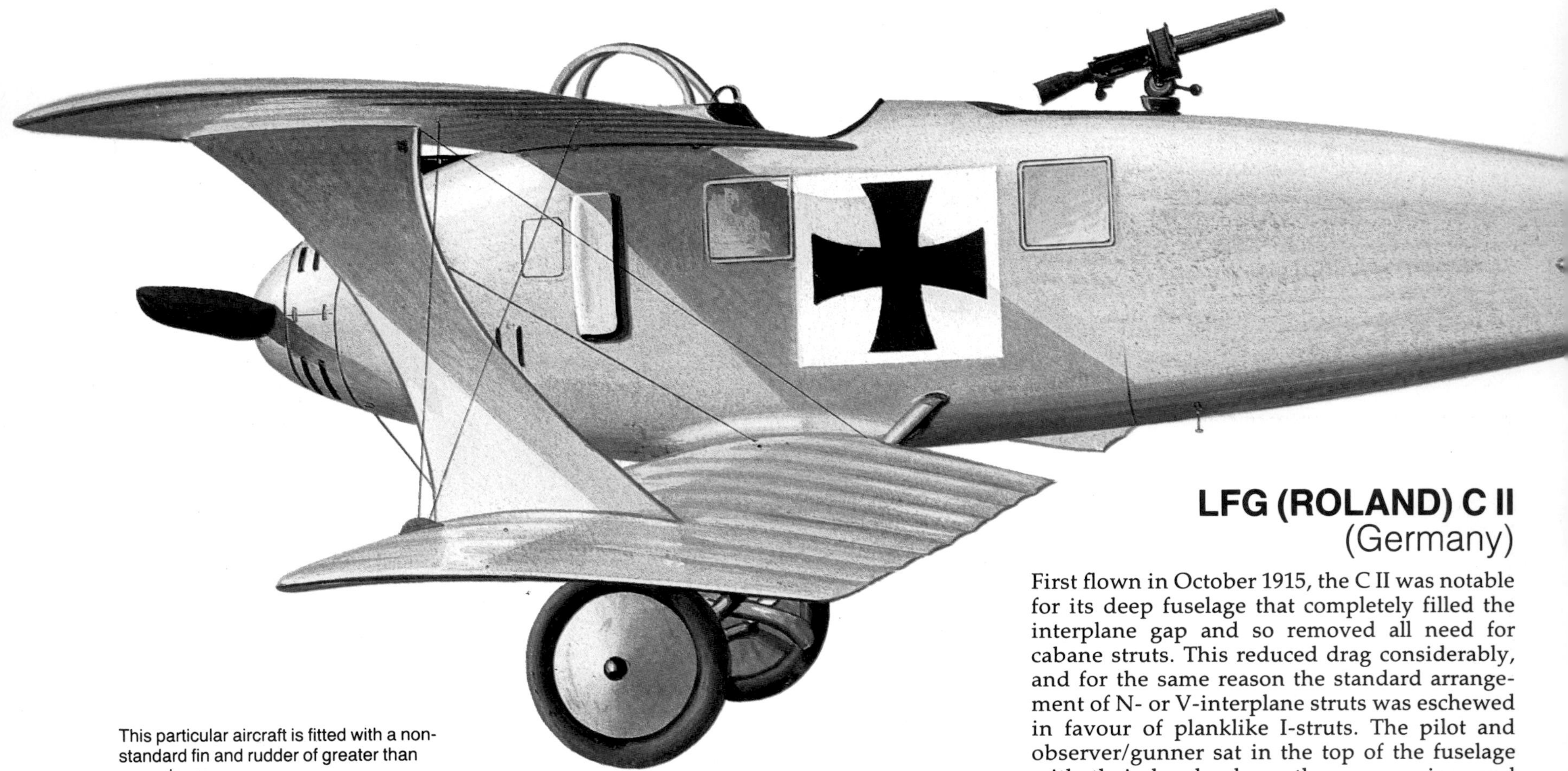

This particular aircraft is fitted with a non-standard fin and rudder of greater than normal area.

LFG (ROLAND) C II
(Germany)

First flown in October 1915, the C II was notable for its deep fuselage that completely filled the interplane gap and so removed all need for cabane struts. This reduced drag considerably, and for the same reason the standard arrangement of N- or V-interplane struts was eschewed in favour of planklike I-struts. The pilot and observer/gunner sat in the top of the fuselage with their heads above the upper wing, and while this provided an excellent field of vision in the upper hemisphere it restricted their downward fields of vision. The C II entered service early in 1916, and the only variant in the 300 or so aircraft built was the C IIa with revised and strengthened wingtips.

LFG (ROLAND) C II

Role: Escort fighter/reconnaissance
Crew/Accommodation: Two
Power Plant: One 160 hp Mercedes D III water-cooled inline
Dimensions: Span 10.3m (33.79 ft); length 7.7m (25.26 ft); wing area 26m^2 (279.9 sq ft)
Weights: Empty 764kg (1,684 lb); MTOW 1,284kg (2,831 lb)
Performance: Maximum speed 165km/h (102.5 mph) at sea level; operational ceiling 4,000+m (13,123+ ft); endurance 5 hours
Load: One/two 7.92mm machine guns

LFG (ROLAND) D I, D II and D III
(Germany)

In the D I single-seat fighter LFG opted for the same type of deep fuselage layout as pioneered in the C II, and the first example flew in July 1916 with 119kW (160 hp) Mercedes D.III inline engines, which was also used in the improved D II, with a revised tail unit and reduced drag. The D IIa was similar to the D II apart from its use of the 134kW (180 hp) Argus AS.III engine. The aircraft began to enter service in early 1917, but were not popular for their poor fields of vision and heavy controls. The D III therefore adopted a more conventional fuselage with the upper-wing centre section supported by cabane struts, but this variant was inferior to contemporary Albatroses and produced only in small numbers.

LFG (ROLAND) D IIa

Role: Fighter
Crew/Accommodation: One
Power Plant: One 180 hp Argus AS III water-cooled inline
Dimensions: Span 8.9m (29.2 ft); length 6.95m (22.8 ft); wing area 22m^2 (236.8 sq ft)
Weights: Empty 635kg (1,397 lb); MTOW 795kg (1,749 lb)
Performance: Maximum speed 180km/h (111.8 mph) at sea level; operational ceiling 5,000+m (16,400+ ft); endurance 1.33 hours
Load: Two 7.92mm machine guns

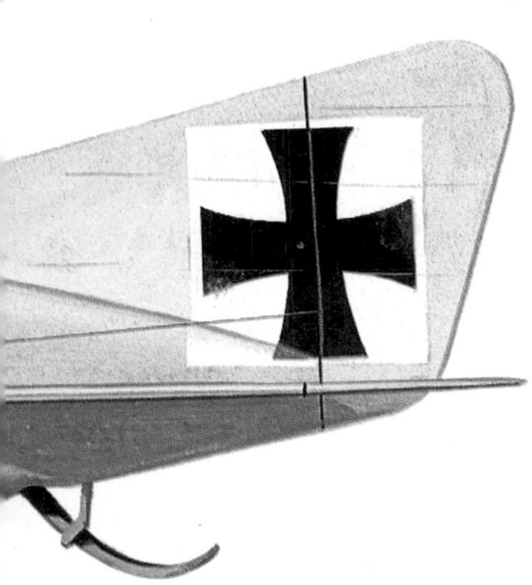

LIORE ET OLIVIER LeO 45
(France)

Aerodynamically one of the most refined aircraft to see service in World War II, the LeO 45 series of warplanes resulted from a 1934 specification for a fast day bomber, and first flew in January 1937 with Gnome-Rhone 14Aa radial engines. Production was authorized for 1,783 LeO 451 bombers, but only 452 had been flown by the time France capitulated in June 1940, with another 225 being completed for the Vichy regime. Several variants were planned or delivered in prototype form, and the baseline version remained in service into the late 1940s.

LIORE-ET-OLIVIER LeO 451

Role: Medium bomber
Crew/Accommodation: Four
Power Plant: Two, 1,140 hp Gnome-Rhone 14N 48/49 air-cooled radials
Dimensions: Span 22.52m (73.88 ft); length 17.17m (56.33 ft); wing area 68m^2 (732 sq ft)
Weights: Empty 7,813kg (17,225 lb); MTOW 11,400kg (25,133 lb)
Performance: Maximum speed 495km/h (307 mph) at 4,800m (15,750 ft); operational ceiling 9,000m (29,530 ft); range 2,300km (1,430 miles) with 500kg (1,102 lb) of bombs
Load: One 20mm cannon and two 7.5mm machine guns, plus up to 1,000kg (2,205 lb) of bombs

LVG C II
(Germany)

In 1915 the C I appeared as one of Germany's first armed reconnaissance aircraft. The type was based on the unarmed B II with a 112kW (150 hp) Benz Bz.III inline engine, but only a few were produced before the C I was overtaken by the much improved C II with a more powerful engine and structural refinements. Production of the C I and C II totalled about 300 aircraft.

LVG C II

Role: Reconnaissance
Crew/Accommodation: Two
Power Plant: One 160 hp Mercedes DIII water-cooled inline
Dimensions: Span 12.85m (42.16 ft); length 8.1m (26.57 ft); wing area 37.6m^2 (404.7 sq ft)
Weights: Empty 845kg (1,863 lb); MTOW 1,405kg (3,097 lb)
Performance: Maximum speed 130km/h (81 mph) at sea level; operational ceiling 3,000+m (9,843+ ft); endurance 4 hours
Load: One/two 7.92mm machine guns

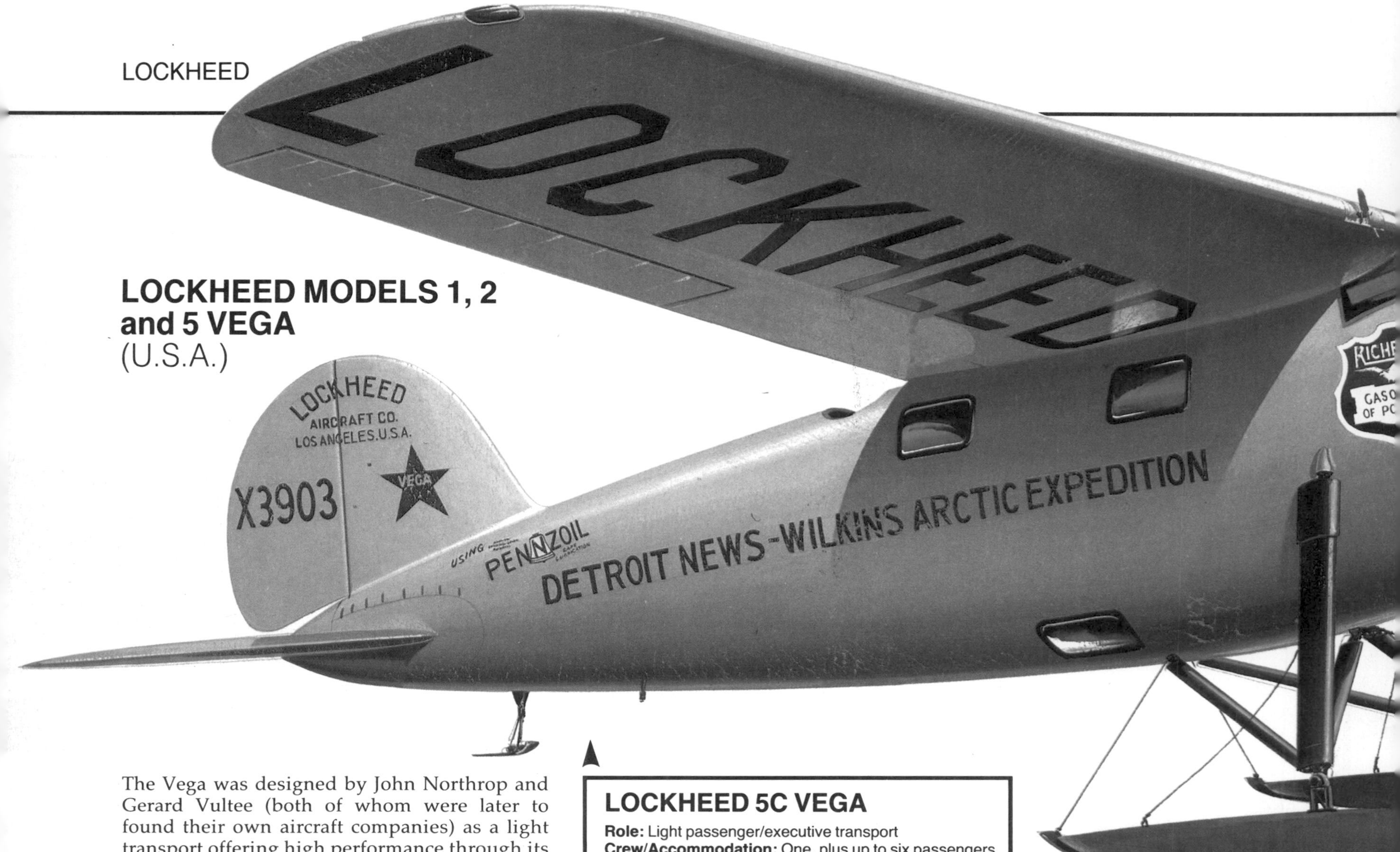

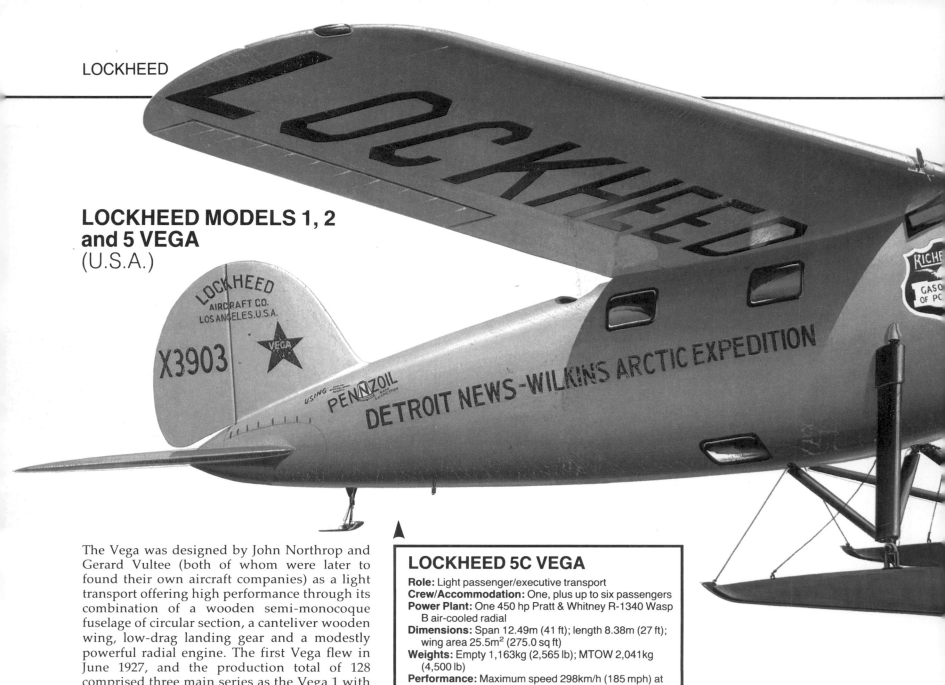

LOCKHEED MODELS 1, 2 and 5 VEGA
(U.S.A.)

The Vega was designed by John Northrop and Gerard Vultee (both of whom were later to found their own aircraft companies) as a light transport offering high performance through its combination of a wooden semi-monocoque fuselage of circular section, a canteliver wooden wing, low-drag landing gear and a modestly powerful radial engine. The first Vega flew in June 1927, and the production total of 128 comprised three main series as the Vega 1 with the Wright Whirlwind J-5, the Vega 2 with the Whirlwind J-6 and the Vega 5 with the Pratt & Whitney Wasp. Vegas were used for a number of notable long-distance flights.

LOCKHEED 5C VEGA

Role: Light passenger/executive transport
Crew/Accommodation: One, plus up to six passengers
Power Plant: One 450 hp Pratt & Whitney R-1340 Wasp B air-cooled radial
Dimensions: Span 12.49m (41 ft); length 8.38m (27 ft); wing area 25.5m^2 (275.0 sq ft)
Weights: Empty 1,163kg (2,565 lb); MTOW 2,041kg (4,500 lb)
Performance: Maximum speed 298km/h (185 mph) at sea level; operational ceiling 4,570m (15,000 ft); range 1,450km (900 miles)
Load: Up to 490kg (1,080 lb).
Note: dimensions and weights are for the landplane version

LOCKHEED MODEL 8 SIRIUS
(U.S.A.)

This was developed to meet a requirement of Charles Lindbergh for a long-range aeroplane in which to undertake air-route survey flights for Pan American Airways. The Sirius that first flew in November 1939 was essentialy the fuselage of the Vega married to a new low-set cantilever wooden wing and nicely spatted fixed landing gear. The success of Lindbergh's aeroplane, which was flown with a variety of engines and also on twin-float alighting gear, led to orders for another 13 aircraft: these were four similar Sirius 8s, eight Sirius 8As with an enlarged tailplane, and finally a single Sirius 8B four-passenger transport.

LOCKHEED 8 SIRIUS

Role: High performance sportsplane
Crew/Accommodation: Two
Power Plant: One 450 hp Pratt & Whitney R-1340 Wasp air-cooled radial
Dimensions: Span 13.04m (42.78 ft); length 8.25m (27.07 ft); wing area 27.32m^2 (294.1 sq ft)
Weights: Empty 1,945kg (4,289 lb); MTOW 3,220kg (7,099 lb)
Performance: Cruise speed 241km/h (150 mph) at sea level; operational ceiling 7,955m (26,100 ft); range 1,570km (975 miles)
Load: Up to 91 kg (200 lb).
Note: figures are for the landplane version

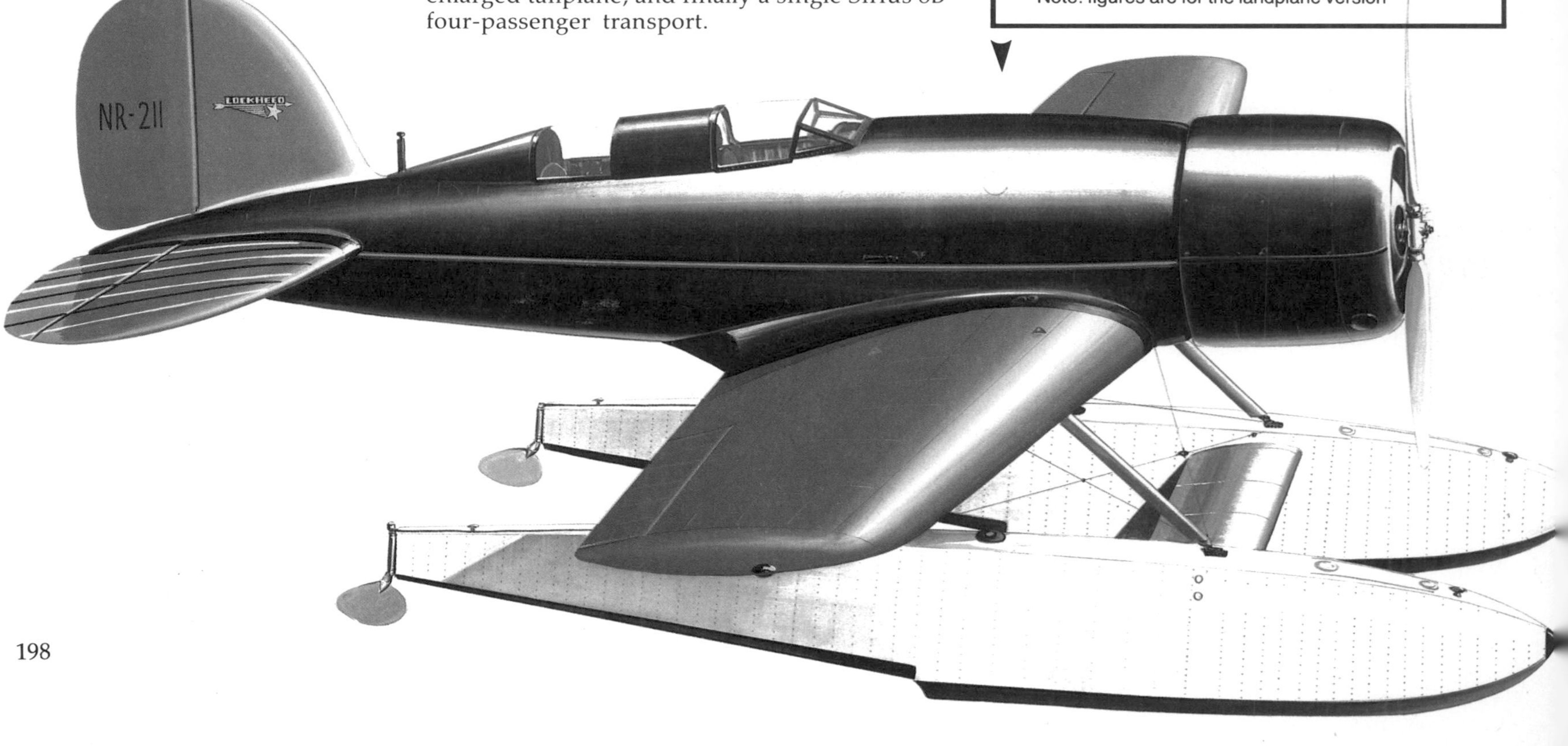

LOCKHEED MODEL 10 ELECTRA
(U.S.A.)

This was Lockheed's first important aeroplane for the transport market, and was an advanced type with all-metal construction, retractable landing gear and other advanced features. The first machine flew in February 1934 with a pair of Pratt & Whitney Wasp Junior radial engines, and production totalled 148 aircraft in major variants such as the Electra 10-A with Wasp Juniors, the Electra 10-B with Wright Whirlwinds, the Electra 10-C with Wasp SCs and the Electra 10-D with Wasp S3H1s. Nought came of the projected Electra 10-D military transport, but many civil Electras were later impressed with the designation C-36. The L-12 Electra Junior was a scaled-down version for feederline operators, and 130 were built, many also being impressed with the designation C-40.

LOCKHEED L10-A ELECTRA

Role: Passenger transport
Crew/Accommodation: Two, plus up to ten passengers
Power Plant: Two 450 hp Pratt & Whitney R-1340 Wasp Junior SB air-cooled radials
Dimensions: Span 16.76m (55 ft); length 11.76m (38.58 ft); wing area 42.59m^2 (458.5 sq ft)
Weights: Empty 2,927kg (6,454 lb); MTOW 4,672kg (10,300 lb)
Performance: Maximum speed 306km/h (190 mph) at 1,525m (5,000 ft); operational ceiling 5,915m (19,400 ft); range 1,305km (810 miles)
Load: Up to 816kg (1,800 lb)

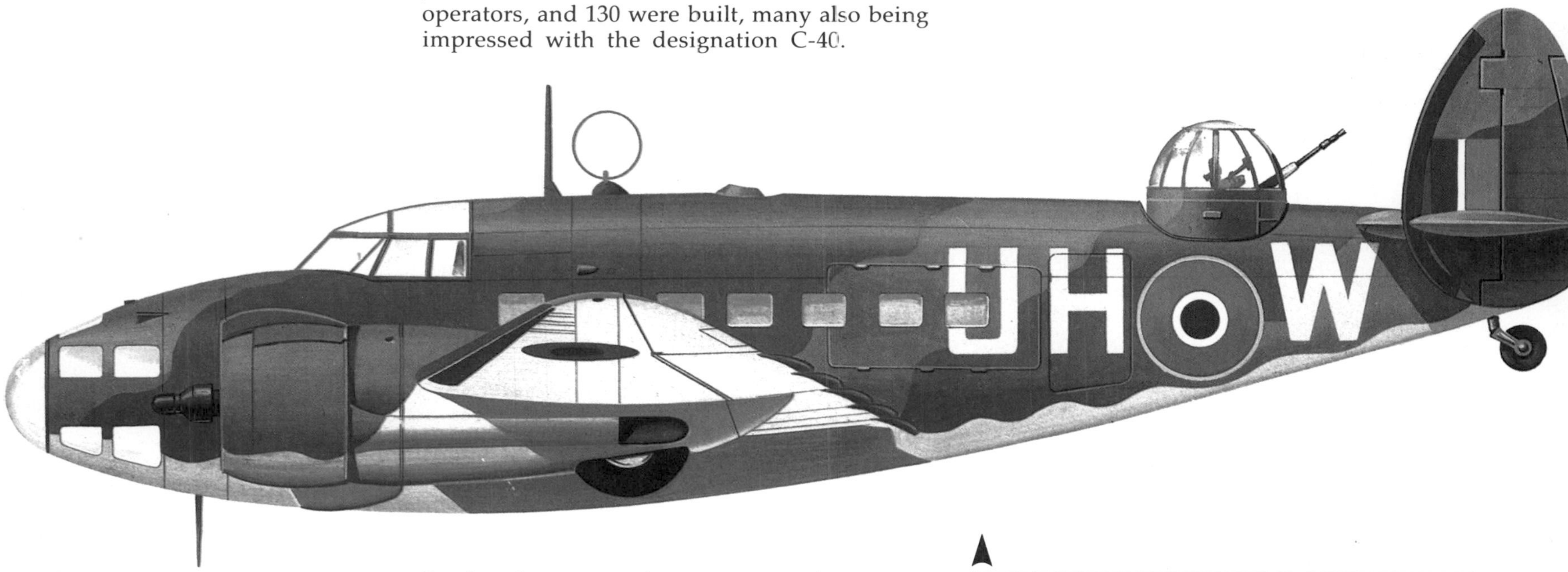

LOCKHEED MODEL B14L HUDSON
(U.S.A.)

Produced to a British requirement for a maritime patroller and navigation trainer, the Hudson was in essence a militarized Super Electra. The first such aeroplane was flown in December 1938, and by the time production ended in May 1943 some 2,941 aircraft had been produced. RAF versions went up to the Hudson Mk VI with steadily more power and improved military capability, while the U.S. Army and U.S. Navy took aircraft similiar to the Hudson Mk IIIA with the designations A-29 and PBO-1. Later army versions were the A-29A and A-29B. The U.S. Army also procured AT-18 gunnery training aircraft and AT-18A navigation trainers.

LOCKHEED HUDSON Mk I

Role: General reconnaissance/light bomber
Crew/Accommodation: Four
Power Plant: Two 1,100 hp Wright R-1820-G102A Cyclone air-cooled radials
Dimensions: Span 19.96m (65.5 ft); length 13.5m (44.29 ft); wing area 51.19m^2 (551 sq ft)
Weights: Empty 5,484kg (12,091 lb); MTOW 8,845kg (19,500 lb)
Performance: Maximum speed 357km/h (222 mph) at 2,408m (7,900 ft); operational ceiling 6,400m (21,000 ft); endurance 8.75 hours with reserves
Load: Seven .303mm machine guns, plus up to 726kg (1,600 lb) of bombs

LOCKHEED MODEL 22 (P-38 ▲ LIGHTNING) SERIES
(U.S.A.)

The Lightning was one of the more important fighters of World War II and, though not as nimble a machine as single-engined types, found its metier in the long-range role with heavy armament and high performance. The machine resulted from a 1937 high-performance fighter specification, and the design team opted for an unconventional central nacelle/twin boom layout with two inline engines trailed by the booms that accommodated the turbochargers and supported the wide-span tailplane and oval vertical surfaces. The XP-38 prototype flew in January 1939 with 716kW (960 hp) Allison V-1710-11/15 engines driving opposite-rotating propellers. Development was protracted, and the P-38 did not enter service until late 1941. Production totalled 10,037 in variants up to the P-38M with steadily improved performance and armament, and variants included the F-4 and F-5 photo-reconnaissance series. The baseline P-38 series was used in the interceptor, long-range escort fighter, fighter-bomber, night-fighter and radar-equipped bomber leader roles.

LOCKHEED P-38L LIGHTNING

Role: Long-range fighter bomber
Crew/Accommodation: One
Power Plant: Two 1,475 hp Allison V-1710-111 water-cooled inlines
Dimensions: Span 15.85m (52 ft); length 11.53m (37.83 ft); wing area 30.47m^2 (327 sq ft)
Weights: Empty 5,806kg (12,800 lb); MTOW 9,798kg (21,600 lb)
Performance: Maximum speed 666km/h (414 mph) at 7,620m (25,000 ft); operational ceiling 13,410m (44,000 ft); range 725km (450 miles)with 1,451kg (3,200 lb) of bombs
Load: One 20mm cannon and four .5 inch machine guns, plus up to 1,451kg (3,200 lb) of bombs

LOCKHEED MODEL 18 LODESTAR
(U.S.A.)

The Lodestar was basically a lengthened version of the L-14 Super Electra, whose sales of only 112 were a great disappointment to the company. The fuselage of the Lodestar was 1.68m (5 ft 6 inches) longer than that of the Super Electra to provide accommodation of between 18 and 26 rather than the 14 of the Super Electra. The Lodestar first flew in September 1939, but the type again failed the company's expectations, only 96 aircraft being ordered by civil operators. Then the U.S. forces' need for light transports increased before and during World War II, however, and 529 aircraft were built to military orders with the designations R50 (U.S. Navy) and C-60 (U.S. Army); large numbers of civil Lodestars were also impressed.

LOCKHEED L18 LODESTAR

Role: Intermediate-range passenger transport
Crew/Accommodation: Two and one cabin crew, plus up to 17 passengers
Power Plant: Two 1,200 hp of various Pratt & Whitney or Wright air-cooled radials
Dimensions: Span 19.96m (65.5 ft); length 15.19m (49.84 ft); wing area 51.19m^2 (551 sq ft)
Weights: Empty 5,103kg (11,250 lb); MTOW 8,709kg (19,200 lb)
Performance: Cruise speed 317km/h (197 mph) at 2,440m (8,000 ft); operational ceiling 6,220m (20,400 ft); range 2,895km (1,800 miles) with normal fuel
Load: Up to 1,905kg (4,200 lb)

LOCKHEED MODEL 37 (VENTURA, PV HARPOON and B-34) SERIES
(U.S.A.)

Just as the Hudson was derived from the Super Electra, the Ventura was a military counterpart to the Lodestar offering greater range and payload. The type was ordered during 1940 by the British, and by the time production ended in September 1945 some 3,028 aircraft had been built in RAF versions up to the Ventura GR.Mk V, in U.S. Navy versions up to the PV-3 Harpoon, and in U.S. Army versions up to the B-34B; there was also a limited-production B-37 version.

LOCKHEED PV-2 HARPOON

Role: Maritime patrol bomber
Crew/Accommodation: Four/five
Power Plant: Two 2,000 hp Pratt & Whitney R-2800-31 Double Wasp air-cooled radials
Dimensions: Span 22.86m (75 ft); length 15.87m (52.07 ft); wing area 63.73m^2 (686 sq ft)
Weights: Empty 9,538kg (21,028 lb); MTOW 16,329kg (36,000 lb)
Performance: Maximum speed 454km/h (282 mph) at 4,175m (13,700 ft); operational ceiling 7,285m (23,900 ft); range 2,880km (1,790 miles)
Load: Nine .5 inch machine guns, plus up to 2,722kg (6,000 lb) of bombs

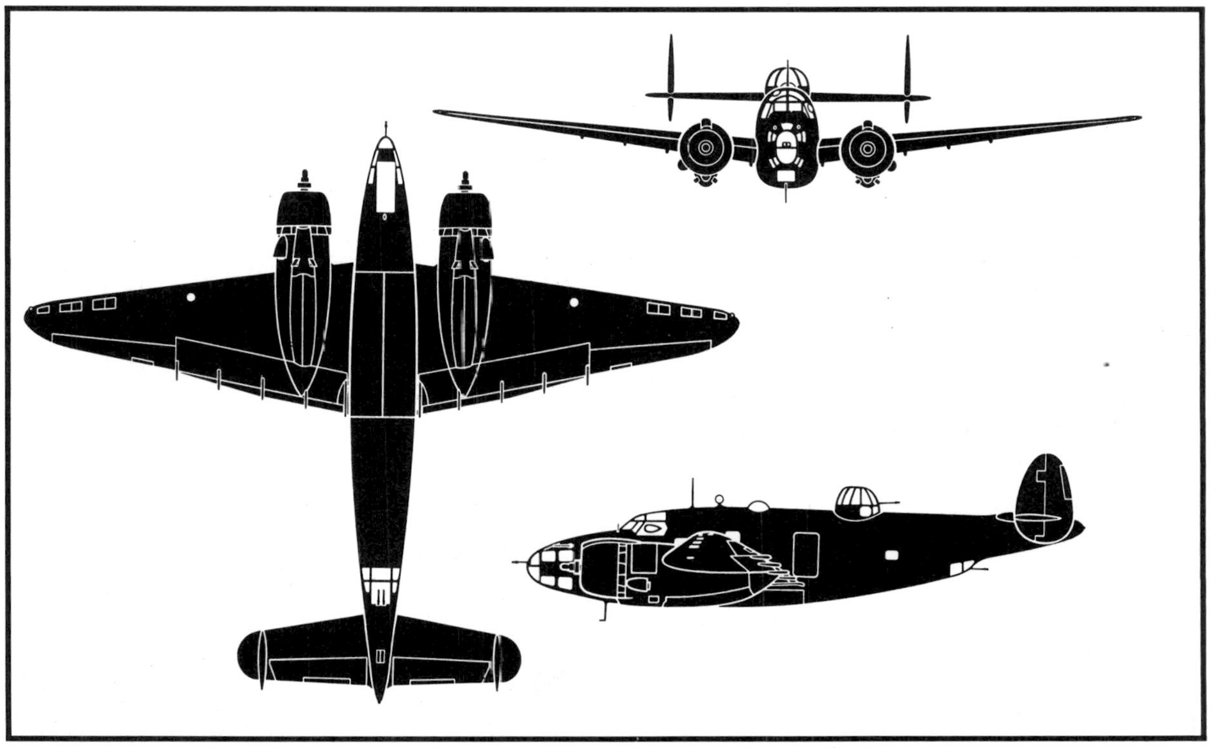

LOCKHEED MODEL 49 CONSTELLATION, SUPER CONSTALLATION and STARLINER SERIES
(U.S.A.)

This was surely one of the classic aircraft of all time, developed as an elegant yet efficient airliner, but also of great military importance as the basis of the world's first long-range airborne early warning and electronic warfare aeroplane. The design was originated in 1939 to provide Pan American Airways, and Transcontinental and Western Air with an advanced airliner for use on long-range domestic routes. The Lockheed design was centred on refined aerodynamics, pressurized accommodation and high power for sustained high-altitude cruise at high speed, and a tricycle landing gear was incorporated for optimum field performance and passenger comfort on the ground. The type first flew in January 1943, and civil production was overtaken by the needs of the military during World War II: the L-49 thus became the U.S.Army's C-69, of which 22 were completed before Japan's capitulation and the cancellation of military orders. Some aircraft then on the production line were completed as 60-seat L-049 airliners, but the first true civil version was the 81-seat L-649 with 1,864kW (2,500 hp) Wright 749C-18BD-1 radial engines. Further airliners were the L-749 with addition fuel, the L-1049 Super Constellation with the fuselage lengthened by 5.59m (18 ft 4 inches) for the accommodation of 109 passengers, and the L-1649 Starliner with a new, longer-span wing and 2,535kW (3,500 hp) Wright 988TC-18EA-2 radial engines fed from increased fuel tankage for true intercontinental range.

Production of the series totalled 856 including military variants that included the C-121 transport version of the L-749, the R70 transport version of the L-1049, and the PO-1 and WV-2 Warning Star airborne early warning aircraft. These R7O, PO-1 and WV-2 aircraft were later redesignated in the C-121 series that expanded to include a large number of EC-121 electronic warfare aircraft.

LOCKHEED L749 CONSTELLATION

Role: Long-range passenger transport
Crew/Accommodation: Four and two cabin crew, plus up to 81 passengers
Power Plant: Four 2,500 hp Wright 749C-18BD-1 Double Cyclone air-cooled radials
Dimensions: Span 37.49m (123 ft); length 29.03m (95.25 ft); wing area 153.3m^2 (1,650 sq ft)
Weights: Empty 27,648kg (60,954 lb); MTOW 47,627kg (105,000 lb)
Performance: Cruise speed 557km/h (346 mph) at 9,072m (20,000 ft); operational ceiling 10,886m (24,000 ft); range 3,219km (2,000 miles) with 6,124kg (13,500 lb) payload plus reserves
Load: Up to 6,690 kg (14,750 lb) with 6,124kg (13,500 lb)

LOCKHEED MODEL 26 (P-2 NEPTUNE)
(U.S.A.)

The Neptune was planned as a Harpoon replacement as early as 1941, but the success of the Harpoon and other maritime patrollers meant that it was 1944 before detail design was inaugurated, and the XP2V-1 prototype flew in May 1945 with two 1,715kW (2,300 hp) Wright R-3350-8 Duplex Cyclone radial engines. The P2V-1 began to enter service in March 1947, and this classic aeroplane was built to the extent of 1,181 aircraft in several important and many more lesser variants all characterized by high speed, great range, heavy weapons load and ever more sophisticated mission electronics. The series was redesignated P-2 in 1962, and the main variants after the P2V-1 were the P2V-2 with sonobuoy provision and armament revisions, the P2V-3 with more powerful engines, the P2V-3 (P-2D) airborne early warning aeroplane, the P2V-5 (P-2E) with more powerful engines, two underwing booster jets and magnetic anomaly detection equipment, the P2V-6 (P-2F) multi-role aeroplane, and the P2V-7 (P-2H) with higher-rated turbojets.

LOCKHEED P2V-7/P-2H NEPTUNE

Role: Long-range maritime patrol and anti/submarine
Crew/Accommodation: Nine (US Navy), or ten (European)
Power Plant: Two 3,500 hp Wright R-3350-35W Turbc Cyclone air-cooled radials, plus two, 1,542kgp (3,400 lb s.t.) Westinghouse 134-WE turbojets
Dimensions: Span 31.65m (103.83 ft); length 27.94m (91.66 ft); wing area 92.9m^2 (1,000 sq ft)
Weights: Empty 19,278kg (42,500 lb); MTOW 34,019kg (75,000 lb)
Performance: Maximum speed 648km/h (403 mph) at 6,095m (20,000 ft); operational ceiling 9,145m (30,000 ft); range 4,450km (2,765 miles) at 230 mph
Load: Two .5 inch machine guns (subsequently deleted), plus up to 3,629kg (8,000 lb) of lightweight torpedoes, depth charges or mines carried internally

LOCKHEED MODEL 80 (F-80 SHOOTING STAR and T-33) SERIES
(U.S.A.)

This was the best Allied jet fighter to emerge from World War II, though the type was in fact just too late for combat use in that conflict. The design was launched in June 1943 on the basis of the British Halford H.1B turbojet, and the first XP-80 prototype with this engine flew in January 1944 as a sleek, low-wing monoplane with tricycle landing gear and a 360 degree vision canopy. The P-80A version began to enter service in January 1945, and just 45 had been delivered before the end of World War II. Production plans for 5,000 aircraft were then savagely curtailed, but the development of later versions with markedly improved capabilities eventually restored the position to the extent that 5,691 of the series were finally built. The baseline fighter was redesignated in the F-series after World War II, and variants up to the F-80C were produced. The versatility of the design also resulted in F-14 (later RF-80) photo-reconnaissance, TF-80 (later T-33A) air force and TO-1/2 (later TV-2) navy flying trainers, AT-33A weapons trainers, T2V SeaStar advanced flying trainer and many other variants.

LOCKHEED F-80B SHOOTING STAR

Role: Day fighter
Crew/Accommodation: One
Power Plant: One 2,041kgp (4,500 lb s.t.) Allison J33-A-21 turbojet
Dimensions: Span 11.81m (38.75 ft); length 10.49m (34.42 ft); wing area 22.07m^2 (237.6 sq ft)
Weights: Empty 3,709kg (8,176 lb); MTOW 7,257kg (16,000 lb)
Performance: Maximum speed 929km/h (577 mph) at 1,830m (6,000 ft); operational ceiling 13,870m (45,500 ft); range 1,270km (790 miles) without drop tanks
Load: Six .5 inch machine guns

LOCKHEED MODEL 80 (F-94 STARFIRE) SERIES
(U.S.A.)

The Starfire was a development of the TF-80C (T-33A) two-seat trainer's airframe as an interim all-weather fighter with radar and an afterburning turbojet. Development began in 1947, and the first YF-94 flew in April 1949. After initial teething problems the type proved remarkably successful, and 854 aircraft were built in variants up to the F-94C with an armament of unguided rockets.

LOCKHEED F-94C STARFIRE

Role: All-weather/night fighter
Crew/Accommodation: Two
Power Plant: One 3,969kgp (8,750 lb s.t.) Pratt & Whitney J48-P 5 turbojet with reheat
Dimensions: Span 11.38m (37.33 ft); length 13.56m (44.5 ft); wing area 21.63m^2 (232.8 sq ft)
Weights: Empty 5,764kg (12,708 lb); MTOW 10,970kg (24,184 lb)
Performance: Maximum speed 1,030km/h (640 mph) at sea level; operational ceiling 15,665m (51,400 ft); range 1,295km (805 miles) with full warload
Load: Forty-eight 2.75 inch Mighty Mouse rockets

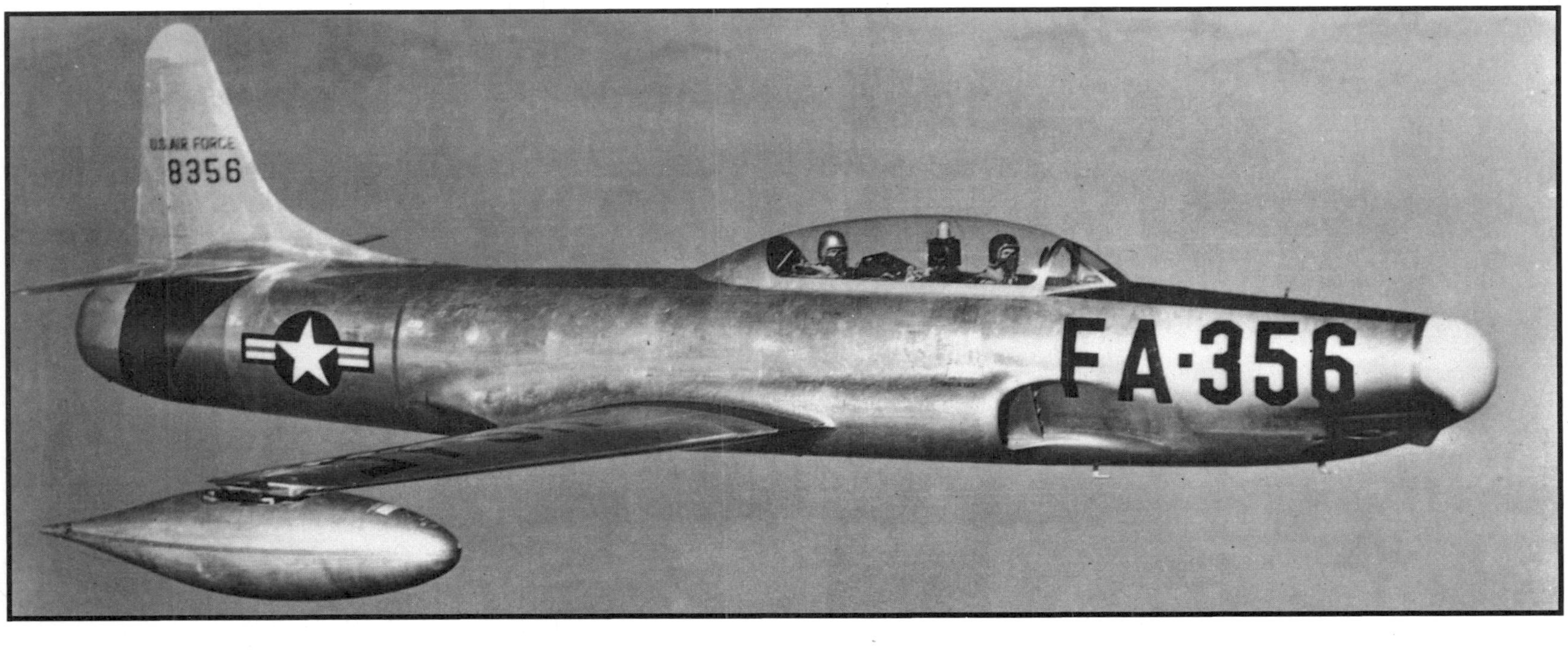

LOCKHEED MODEL 82 (C-130 HERCULES) SERIES
(U.S.A.)

The Hercules is still the airlifter against which all other turboprop tactical transports are measured. The type pioneered the modern airlifter layout with a high wing, capacious fuselage with a rectangular-section hold, and an integral rear ramp/door that allows the straight-in loading/unloading of bulky items under the upswept tail. For operations into and out of semi-prepared airstrips a multi-wheel landing gear arrangement was adopted with its main units accommodated in external fuselage fairings so that no demands are made on hold area and volume, yet the hold floor and opened ramp are held level at truckbed height to facilitate loading and unloading. The type was designed in response to a 1951 requirement, and first flew in YC-130 prototype form during August 1954 with 2,796kW (3,750 shp) Allison T56-A-1A turboprops.

Over 1,700 aircraft have been delivered and the type remains in production having been developed in major variants for its initial C-130A form to the C-130B with more fuel and a high maximum weight, the C-130E with 3,020kW (4,050 shp) T56-A-7 turboprops, greater internal fuel capacity and provision for external fuel tanks, the C-130H with airframe and system improvements as well as 3,362kW (4,508 shp) T56-A-15 turboprops, and the C-130H-30 with a lengthened fuselage. There are a host of variants for tasks as diverse as arctic operations, drone and spacecraft recovery, in-flight-refuelling, special forces insertion and extraction, airborne command post, and communication with submerged submarines. The type is also produced in L-100 civil form.

LOCKHEED C-130H HERCULES

Role: Land-based, rough field-capable transport/tanker
Crew/Accommodation: Four crew with up to 92 troops
Power Plant: Four 4,508 shp Allison T56-A-15 turboprops
Dimensions: Span 40.4m (132.6 ft); length 29.8m (97.75 ft); wing area 162.1m^2 (1,745 sq ft)
Weights: Empty 34,397kg (75,832 lb); MTOW 79,380kg (175,000 lb)
Performance: Maximum speed 621km/h (335 knots) at 3,658m (12,000 ft); operational ceiling 10,060m (33,000 ft); range 7,410km (3,995 naut. miles) with 9,070 kg (20,000 lb) payload
Load: 19,850kg (43,761 lb)

LOCKHEED MODEL 83 (F-104 STARFIGHTER) SERIES
(U.S.A.)

The Starfighter resulted from the U.S. Air Force's experiences in the Korean War, where the need for a fast-climbing interceptor became clear. The type was planned round the most powerful available axial-flow turbojet in a long and basically cylindrical fuselage with unswept and diminutive wings plus a large T-tail assembly. The XF-104 first flew in March 1954 with an interim engine, the 4,627kg (10,200 lb) Wright XJ65-W-6, and four years of troubled development followed before the F-104A entered service with a longer fuselage accommodating the J79-GE-3 engine. The U.S.A.F. eventually ordered 296 examples of the Starfighter in variants up to the F-104D in the inteceptor and tactical strike roles, but the commercial success of the type was then ensured by the adoption of the much-improved F-104G multi-role type by a NATO consortium. This resulted in the largely licensed production of another 1,986 aircraft up to 1983. The F-104G itself spawned the TF-104 trainer and RF-104 reconnaisance variants, and Italy developed the special F-104S variant as a dedicated interceptor with more capable radar and medium-range Sparrow and Aspide air-to-air missiles.

LOCKHEED F-104A STARFIGHTER

Role: Interceptor
Crew/Accommodation: One
Power Plant: One 6,713kgp (14,800 lb s.t.) General Electric J79-GE-3B turbojet with reheat
Dimensions: Span 6.63m (21.75 ft); length 16.66m (54.66 ft); wing area 18.2m^2 (196.1 sq ft)
Weights: Empty 6,071kg (13,384 lb); MTOW 11,271kg (25,840 lb)
Performance: Maximum speed 1,669km/h (1,037 mph) at 15,240m (50,000 ft); operational ceiling 19,750m (64,795 ft); range 1,175km (730 miles) with full warload
Load: One 20mm multi-barrel cannon and two short-range air-to-air missiles

LOCKHEED U-2 and TR-1
(U.S.A.)

The U-2 is in essence a turbojet-powered sailplane designed for subsonic cruise over very long ranges at extremely high altitude. Produced in the greatest secrecy at Lockheed's so-called 'Skunk Works', the U-2 first flew in August 1955 with the Pratt & Whitney J57 turbojet, which was also used in the U-2A 'production' version. The airframe and engine were poorly matched, however, and in the U-2B the J75 engine was used. The U-2C introduced additional fuel capacity and a lengthened nose for electronic intelligence equipment, and the U-2R featured provision for more reconnaissance equipment in wing leading-edge pods. The latest version is the TR-1 intended for the location of enemy rear-area installations that can then be destroyed by interdiction aircraft.

LOCKHEED U-2R

Role: Long range, high altitude reconnaissance
Crew/Accommodation: One
Power Plant: One 7,711kgp (17,000 lb s.t.) Pratt & Whitney J75-P-13B turbojet
Dimensions: Span 31.39m (103 ft); length 19.17m (62.9 ft); wing area 92.9m^2 (1,000 sq ft)
Weights: Empty 6,850kg (15,101 lb); MTOW 18,598kg (41,000 lb)
Performance: Maximum speed 756km/h (408 knots) Mach 0.72 at 23,927m (78,500 ft); operational ceiling 23,927m (78,500 ft); endurance more than 15 hours at Mach 0.72
Load: Over 1,700kg (3,750 lb) of photographic and electronic intelligence gathering sensors

LOCKHEED SR-71 'BLACKBIRD'
(U.S.A.)

Currently the world's fastest and highest-flying 'conventional' aeroplane, the SR-71 is an extraordinary machine designed for the strategic reconnaissance role with a mass of classified sensors. The airframe is slender delta-winged layout with large fuselage chines that blend in the leading edges of the wing, and is built largely of titanium and stainless steel to deal with the high temperatures created by air friction. Power is provided by two special bleed turbojets which at high speed provide only a small part of the motive power in the form of direct jet thrust from the nozzles (18 per cent), the bulk of the power being provided by inlet suction (54 per cent) and thrust from the special outlets at the rear of the multiple-flow nacelles (28 per cent). The type was developed from the A-11 drone-launching reconnaissance platform first revealed in 1964, and the following YF-12 interceptor that reached only the experimental stage. The SR-71A entered service in 1966 and is due to be retired late in 1989. There are also SR-71B and SR-71C trainers.

LOCKHEED SR-71A

Role: Long-range high supersonic reconnaissance
Crew/Accommodation: Two
Power Plant: Two 14,742kgp (32,500 lb s.t.) Pratt & Whitney J58 turbo-ramjets
Dimensions: Span 16.94m (55.58 ft); length 32.74m (107.41 ft); wing area 149.1m^2 (1,605 sq ft)
Weights: Empty 30,618kg (67,500 lb); MTOW 78,020kg (172,000 lb)
Performance: Cruise speed 3,661km/h (1,976 knots) Mach 3.35 at 24,385m (80,000 ft); operational ceiling 25,908m (85,000 ft); range 5,230km (2,822 naut. miles) unrefuelled
Load: Up to around 9,072kg (20,000 lb) of specialized sensors

LOCKHEED MODEL 85 (P-3 ORION)
(U.S.A.)

The Orion was designed in 1957 as a replacement for the Neptune, the airframe/power plant combination being that of the relatively unsuccessful Model 188 Electra turboprop-powered airliner shortened by 2.24m (7 ft 4 inches) and modified to include a weapons bay in the lower fuselage. The YP3V-1 prototype first flew in November 1959, and the initial P3V-1 production variant was redesignated P-3A in 1962. More than 700 Orions have been delivered in steadily improved P-3B and P-3C variants with radically upgraded mission electronics. So important is the type that the concept has been extended into the new P-7 variant with new-generation turboprops and a host of improvements to the airframe, weapons and electronics.

LOCKHEED P-3C ORION

Role: Long-range maritime patrol and anti submarine
Crew/Accommodation: Ten
Power Plant: Four 4,910 shp Allison T56A-114 turboprops
Dimensions: Span 30.4m (99.7 ft); length 35.6m (116.8 ft); wing area 120.8m^2 (1,300 sq ft)
Weights: Empty 27,892kg (61,491 lb); MTOW 64.410kg (142,000 lb)
Performance: Maximum speed 761km/h (411 knots) at 4,572m (15,000 ft); operational ceiling 8,625m (28,300 ft); radius 2,494km (1,346 miles) including 3 hours on patrol
Load: Up to 8,733kg (19,252 lb) of weapons and sonobuoys

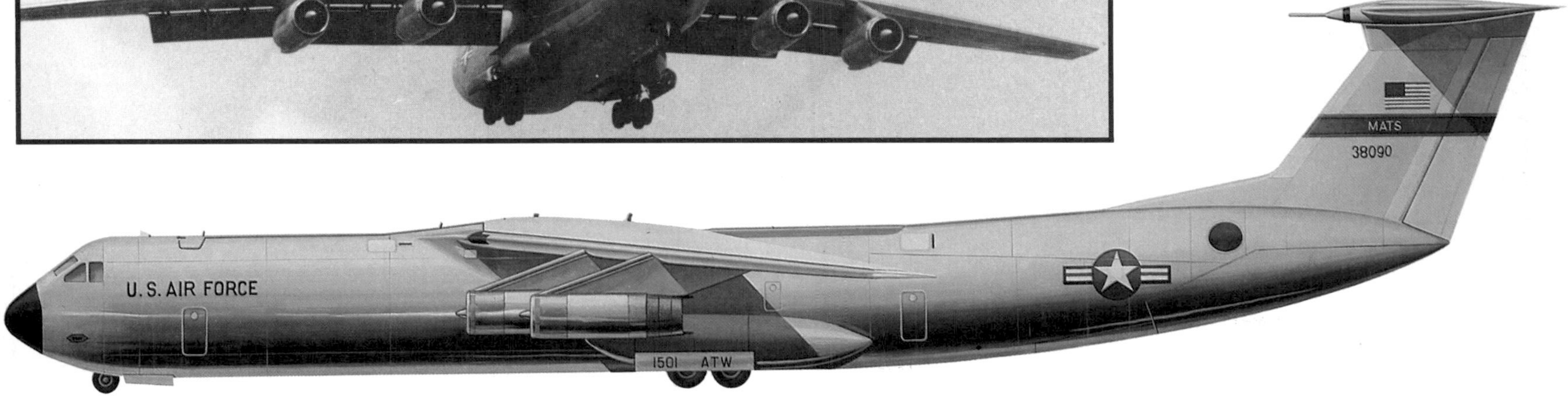

LOCKHEED C-141 STARLIFTER
(U.S.A.)

The C-141A was designed to provide America with a global transport capability, and 284 were built for the U.S. Air Force. The type entered service in April 1965, and immediately proved successful. The C-141A's single operational failing was its comparatively small hold, which meant that in general all available volume had been filled before the aeroplane had reached its maximum payload. Surviving aircraft have therefore been rebuilt to C-141B standard with their fuselages lengthened by 7.11m (23ft 4 inches) to increase hold volume.

LOCKHEED C-141B STARLIFTER

Role: Military long-range cargo transport
Crew/Accommodation: Five, plus up to 138 troops
Power Plant: Four 9, 525kgp (21,000 lb s.t.) Pratt & Whitney TF33-P-7 turbofans
Dimensions: Span 48.77m (160 ft); length 51.29m (168.27 ft); wing area 299.9m^2 (3,228 sq ft)
Weights: Empty 69,558kg (153,350 lb); MTOW 146,556kg (323,100 lb)
Performance: Cruise speed 769km/h (415 knots) at 7,440m (24,400 ft); operational ceiling 15,760m (51,700 ft); range 5,150km (2,779 naut. miles) with maximum payload
Load: Up to 40, 439kg (89,152 lb).
Note: the C-141B carries provision for inflight refuelling

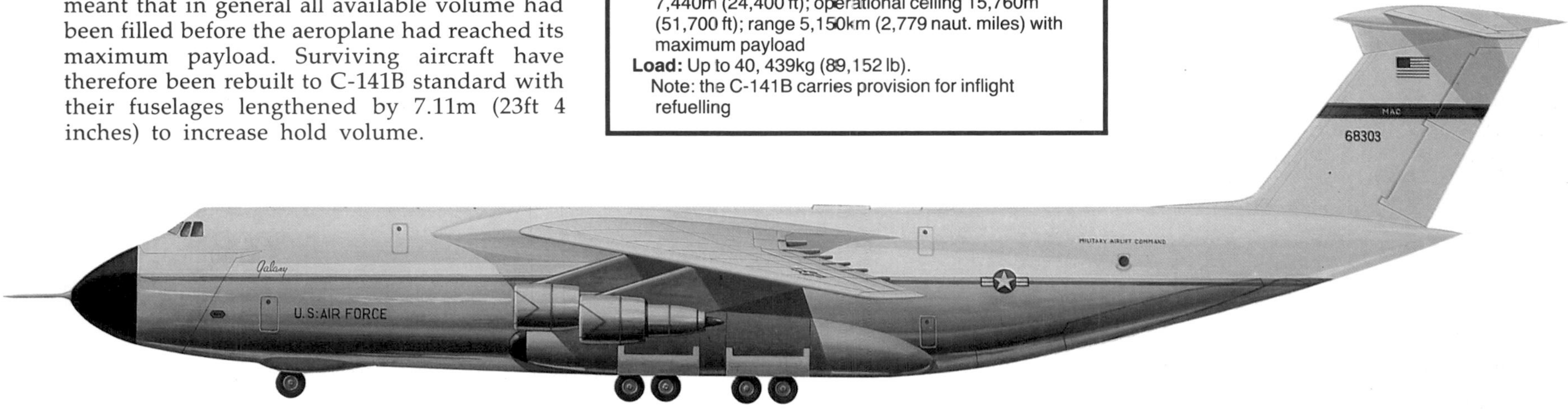

LOCKHEED C-5 GALAXY
(U.S.A.)

The Galaxy was produced to meet a requirement of the early 1960s for a long-range strategic airlifter to complement the C-141, and the C-5A first flew in June 1968. Production comprised 81 aircraft to equip four squadrons. The type has since been upgraded with more power, better systems and a strengthened wing, and another 50 aircraft have been built to this standard as C-5Bs.

LOCKHEED C-5 GALAXY

Role: Military long-range, heavy cargo transport
Crew/Accommodation: Five, with provision for relief crew and up to 75 troops on upper deck as well as 290 troops on main deck in place of cargo
Power Plant: Four 18,643 kgp (41,100 lb s.t.) General Electric TF39-GE-1 turbofans
Dimensions: Span 67.88m (222.7 ft); length 75.53m (247.8 ft); wing area 576m^2 (6,200 sq ft)
Weights: Empty 145,603kg (321,000 lb); MTOW 348,812kg (769,000 lb)
Performance: Cruise speed 814km/h (439 knots) at 7,620m (25,000 ft); operational ceiling 14,540m (47,700 ft); range 3,015km (1,627 naut. miles) with maximum payload
Load: Up to 120,200kg (265,000 lb)

LOCKHEED MODEL 93 (L-1011 TRISTAR)
(U.S.A.)

The TriStar was schemed to meet an American Airlines' requirement for a wide-body airliner optimized for the short/medium-range opereational bracket, and was planned in parallel with its engine, the Rolls-Royce RB211 turbofan. Development problems with the engine broke Rolls-Royce financially and nearly broke Lockheed, both companies having to be rescued by their respective governments. The first TriStar flew in November 1970 and the L-1011-1 variant entered service in April 1972. Production totalled 247 aircraft in L-1011-1, longer-range L-1011-100, 'hot-and-high' L-1011-200 and long-range advanced L-1011-500 variants.

LOCKHEED L-1011 TRISTAR

Role: Intermediate-range passenger transport
Crew/Accommodation: Three, six cabin crew, plus up to 400 passengers (charter)
Power Plant: Three 19,051kgp (42,000 lb s.t.) Rolls-Royce RB211-22 turbofans
Dimensions: Span 47.35m (155.33 ft); length 54.46m (178.66 ft); wing area 321.1m^2 (3,456 sq ft)
Weights: Empty 106,265kg (234,275 lb); MTOW 195,045kg (430,000 lb)
Performance: Cruise speed 796km/h (495 mph) at 9,1450m (30,000 ft); operational ceiling 12,800m (42,000 ft); range 4,635km (2,880 miles) with maximum payload
Load: Up to 41,152kg (90,725 lb)

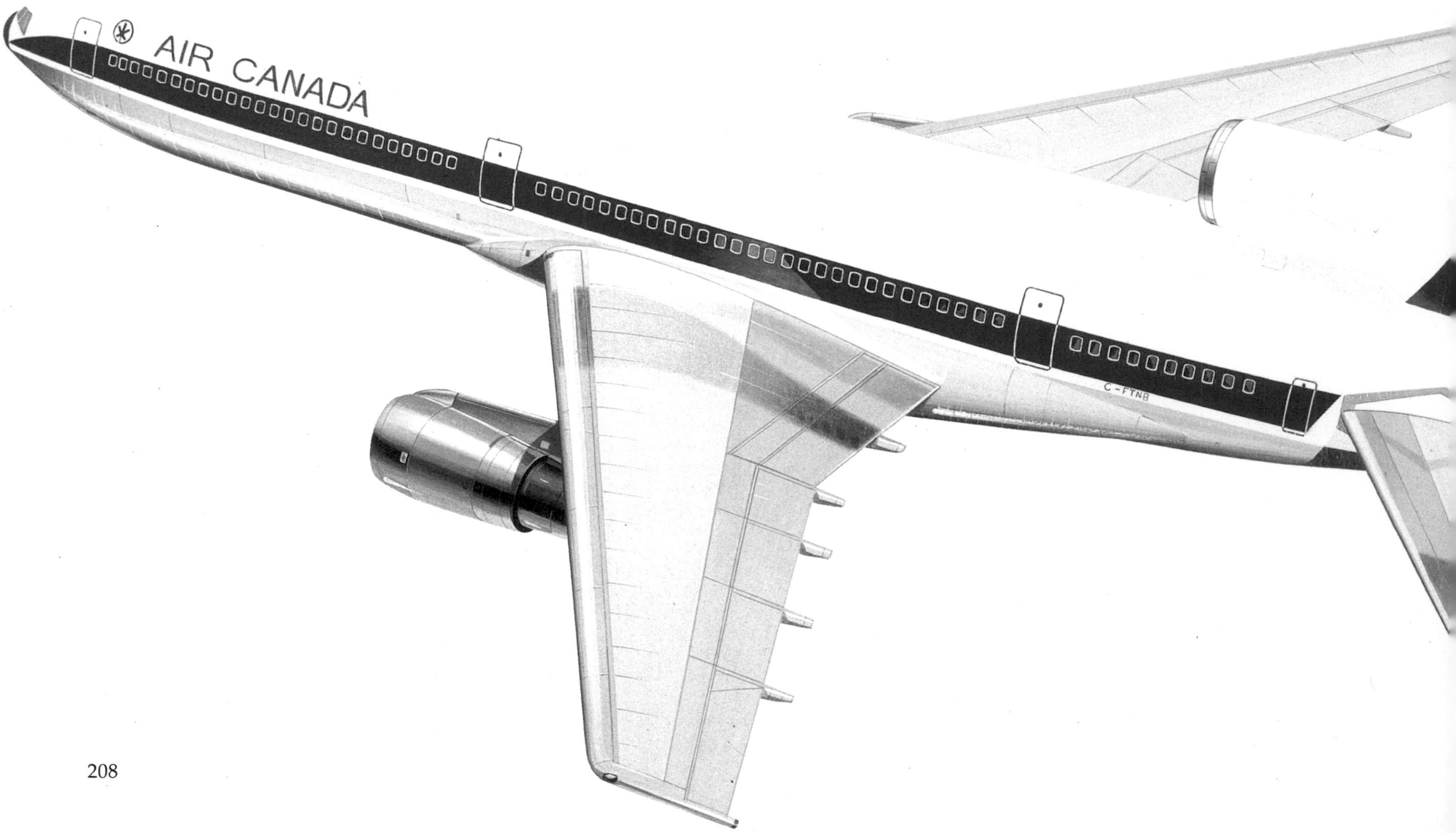

The Viking is a remarkable example of packing maximum operational capability into minimum airframe, and the result is a machine with the capability of the P-3 Orion in an aeroplane able to operate from aircraft-carriers. The type was planned as replacement for the piston-engined Lockheed S-2 Tracker, and first flew in January 1972. Production of 187 S-3A aircraft has been completed, and most of these are to be upgraded to S-3B standard with systems improvements and the ability to operate the Harpoon anti-ship missile. Limited-number variants produced by conversion are the ES-3A electronic reconnaissance, KS-3A inflight-refuelling tanker and US-3A carrier onboard delivery aircraft.

LOCKHEED S-3A VIKING

Role: Naval carrierborne anti-submarine/ship
Crew/Accommodation: Four
Power Plant: Two 4,207kgp (9,275 lb s.t.) General Electric TF34-GE-400A turbofans
Dimensions: Span 20.93m (68.66 ft); length 16.26m (53.33 ft); wing area 55.56m^2 (598 sq ft)
Weights: Empty 12,088kg (26,650 lb); MTOW 23,832kg (52,540 lb)
Performance: Maximum speed 834km/h (450 knots) Mach 0.76 at 8,534m (28,000 ft); operational ceiling 10,670+m (35,000+ ft); radius 2,131km (1,150 naut. miles) without refuelling
Load: Up to 2,269kg (5,000 lb) including up to four lightweight torpedoes, or two anti-ship cruise missiles (on S-3B only)

▲ LOCKHEED MODEL 94 (S-3 VIKING)
(U.S.A.)

▲

LOCKHEED F-117
(U.S.A.)

Initially thought to be designated F-19, this 'stealth' aeroplane was revealed in 1989 as the F-117, an angular and firmly subsonic type apparently designed for the subsonic strike role against heavily defended targets. The layout and structure of the aeroplane are designed to absorb, refract or attenuate the enemy's radar emissions, while the carefully shielded engines produce minimum heat for detection by infra-red systems, allowing the aeroplane to penetrate enemy airspace undetected for the delivery of a single precision-guided weapon.

LOCKHEED F-117A

Role: Low detectability operations
Crew/Accommodation: One
Power Plant: Two 4,900kgp (10,800 lb s.t.) General Electric F404-GE-400 turbofans
Dimensions: Span 12.2m* (40 ft*); length 67m* (22 ft); wing area 00.0m^2 (000.00 sq ft)
Weights: Empty 10,000kg* (22,050 lb*); MTOW 15,000kg* (33,070 lb*)
Performance: Maximum speed 1,038km/h* (645 mph*) at sea level; range 556-741*km (345-460 miles*)

*Estimated

MACCHI M.5 and M.7
(Italy)

The M.5 was developed as a sesquiplane flying boat fighter in late 1916, and the Tipo M prototype first flew in 1917, being followed by an improved Tipo Ma prototype. These 'boats were steadily perfected and the result was the M.5 production type of which 244 were built for service to the end of World War I. The M.7 retained the same basic layout, but with slightly smaller dimensions and the 194kW (260 hp) Isotta-Fraschini V.6B inline engine in place of the M.5's 119kW (160 hp) Isotta-Fraschini V.4B offered higher performance and agility. Only limited production was possible in World War I, but after the war there appeared the M.7bis racer and the radically revised M.7ter fighter that was built in three subvariants but only in comparatively small numbers.

MACCHI M.5

Role: Fighter flying boat
Crew/Accommodation: One
Power Plant: One 160 hp Isotta-Fraschini V.4B water-cooled inline
Dimensions: Span 11.9m (39.04 ft); length 8.08m (26.51 ft); wing area 28m^2 (301.4 sq ft)
Weights: Empty 720kg (1,587 lb); MTOW 990kg (2,183 lb)
Performance: Maximum speed 189km/h (117 mph) at sea level; operational ceiling 3,000+m (9,843+ ft); endurance 3.66 hours
Load: One 7.7mm machine gun

MACCHI MC.200 SAETTA (LIGHTNING), MC.202 FOLGORE (THUNDERBOLT) and MC.205 VELTRO (GREYHOUND)
(Italy)

In 1936 the Italian Air Force belatedly realized that the day of its beloved biplane fighter was over, and requested the development of a 'modern' monoplane fighter with stressed-skin metal construction, a low-seat cantilever monoplane wing, an enclosed cockpit and retractable landing gear. Macchi's response was the MC.200 that first flew in December 1937 with the 649kW (870 hp) Fiat A.74 RC 38 radial engine. The type was declared superior to its competitors during 1938 and ordered into production to a total of 1,153 aircraft in variants that 'progressed' from an enclosed to an open and finally a sem-enclosed cockpit.

The MC.200 was a beautiful aeroplane to fly, but clearly lacked the performance to deal with the higher-performance British fighters. No Italian inline engine could offer the required performance, so the MC.202 that flew in August 1940 with an enclosed cockpit used an imported Daimler-Benz DB 601A engine. About 1,500 production aircraft followed, initially with imported engines but later with licence-built Alfa-Romeo RA.100 RC 41-I Monsone engines rated at 876kW (1,175 hp).

The MC.205 was a development of the MC.202 with the 1,100kW (1,475 hp) DB 605A engine and considerably heavier armament. The MC.205 was first flown in April 1942, but production had then to await the availability of the licensed DB 605A, the RA.1050 RC 58 Tifone, so deliveries started only in mid-1943. Production amounted to 252, and most of these aircraft served with the fascist republic established in northern Italy after the effective division of the country by the September 1943 armistice with the Allies.

MACCHI MC.205V VELTRO SERIES II

Role: Fighter
Crew/Accommodation: One
Power Plant: One 1,475 hp Fiat-built Daimler-Benz DB605A water-cooled inline
Dimensions: Span 10.58m (34.71 ft); length 8.85m (29.04 ft); wing area 16.8m^2 (180.8 sq ft)
Weights: Empty 2,581kg (5,690 lb); MTOW 3,224kg (7,108 lb)
Performance: Maximum speed 642km/h (399 mph) at 7,200m (2,620 ft); operational ceiling 11,000m (36,090 ft); range 950km (590 miles)
Load: Two 20mm cannon, plus up to 320kg (706 lb) of bombs

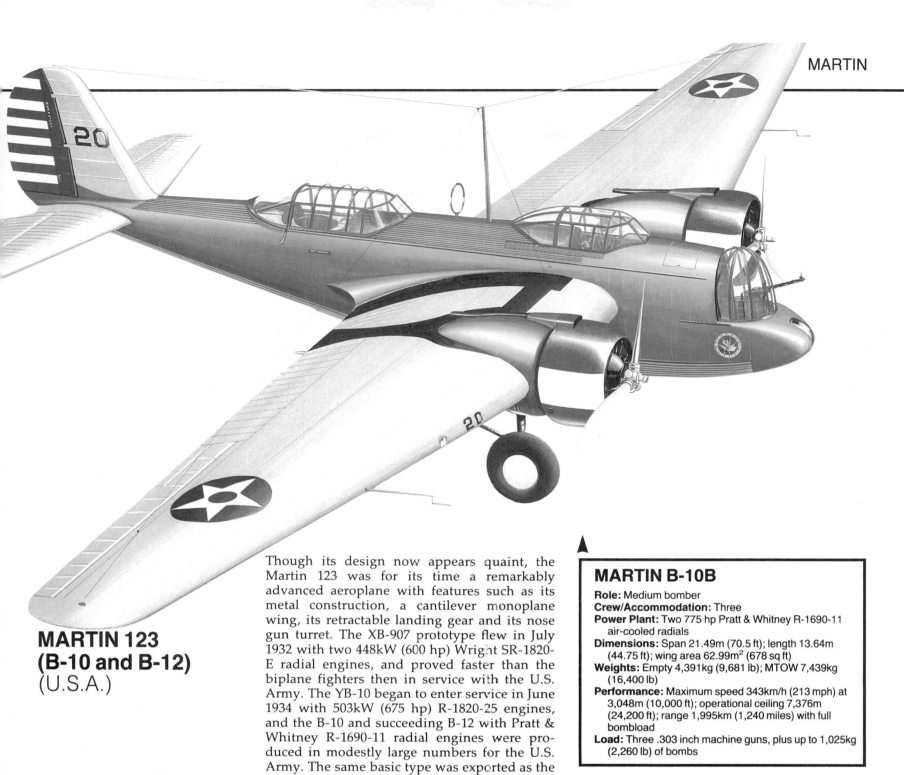

MARTIN 123 (B-10 and B-12)
(U.S.A.)

Though its design now appears quaint, the Martin 123 was for its time a remarkably advanced aeroplane with features such as its metal construction, a cantilever monoplane wing, its retractable landing gear and its nose gun turret. The XB-907 prototype flew in July 1932 with two 448kW (600 hp) Wright SR-1820-E radial engines, and proved faster than the biplane fighters then in service with the U.S. Army. The YB-10 began to enter service in June 1934 with 503kW (675 hp) R-1820-25 engines, and the B-10 and succeeding B-12 with Pratt & Whitney R-1690-11 radial engines were produced in modestly large numbers for the U.S. Army. The same basic type was exported as the Martin 139 principally to Argentina, the Netherlands, Siam and Turkey.

MARTIN B-10B

Role: Medium bomber
Crew/Accommodation: Three
Power Plant: Two 775 hp Pratt & Whitney R-1690-11 air-cooled radials
Dimensions: Span 21.49m (70.5 ft); length 13.64m (44.75 ft); wing area 62.99m^2 (678 sq ft)
Weights: Empty 4,391kg (9,681 lb); MTOW 7,439kg (16,400 lb)
Performance: Maximum speed 343km/h (213 mph) at 3,048m (10,000 ft); operational ceiling 7,376m (24,200 ft); range 1,995km (1,240 miles) with full bombload
Load: Three .303 inch machine guns, plus up to 1,025kg (2,260 lb) of bombs

MARTIN 130 and 156
(U.S.A.)

The Martin 130 was designed as a flying boat airliner capable of transoceanic flights. Three were built for Pan American Airways in 1935, the accommodation being interchangeable between day and night layouts for 48 and 18 passengers respectively. The 'boats entered service in 1936 on the route across the Pacific from San Francisco in California to Manila in the Philippines. Experience with these 'boats led to the Martin 156 with basically the same hull and wings but a twin-finned empennage and 746kW (1,000 hp) Wright GR-1820-G2 Cyclone radial engines. This offered improved performance with 46 passengers, but was competitively bested by the Boeing 314.

MARTIN 130

Role: Long-range passenger transport flying boat
Crew/Accommodation: Five and two cabin crew, plus up to 41 passengers
Power Plant: Four 830 hp Pratt & Whitney R-1830 Twin Wasp air-cooled radials
Dimensions: Span 39.62m (130 ft); length 27.69m (90.88 ft); wing area 215.08m^2 (2,315 sq ft)
Weights: Empty 10,478kg (23,100 lb); MTOW 23,700kg (52,252 lb)
Performance: Cruise speed 209km/h (130 mph) at 2,134m (7,000 ft); operational ceiling 5,547m (18,200 ft); range 5,150km (3,200 miles) with 14 passengers
Load: Up to 3,105kg (6,845 lb).
Note: in actual Pan Am service, the aircraft was equipped to carry 14 passengers and mail over the long Pacific route

MARTIN PBM-3C MARINER

Role: Long-range patrol bomber flying boat
Crew/Accommodation: Seven-eight
Power Plant: Two 1,700 hp Wright R-2600-12 Cyclone air-cooled radials
Dimensions: Span 35.97m (118 ft); length 24.38m (80 ft); wing area 130.81m^2 (1,408 sq ft)
Weights: Empty 14,687kg (32,378 lb); MTOW 26,309kg (58,000 lb)
Performance: Maximum speed 319km/h (198 mph) at 3,962m (13,000 ft); operational ceiling 5,151m (16,900 ft); range 3,439km (2,137 miles) with full warload
Load: Eight .5 inch machine guns, plus up to 1,179kg (2,600 lb) of depth charges

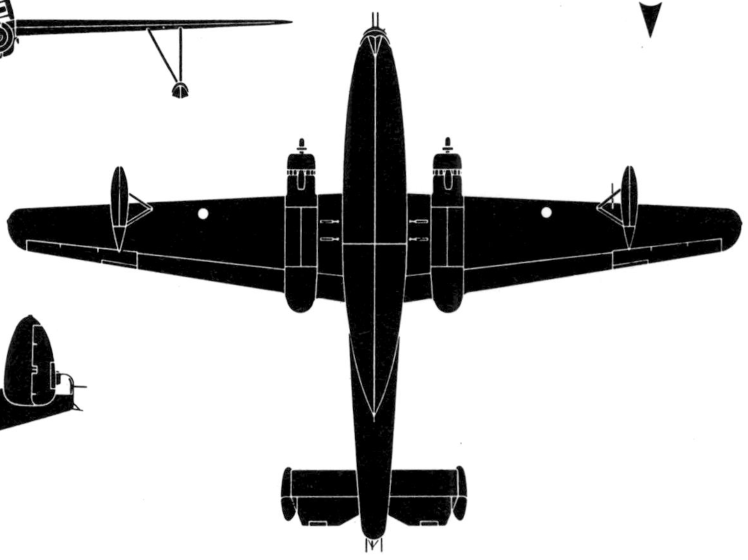

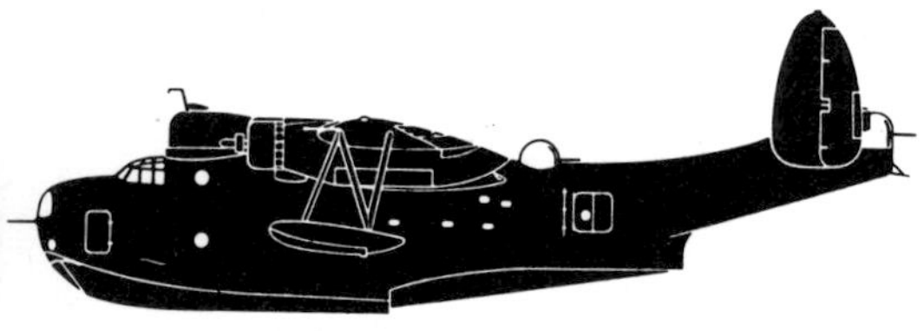

MARTIN 162 (PBM MARINER)
(U.S.A.)

The Mariner was designed in 1937 to a U.S. Navy requirement for a patrol flying boat, and first flew in February 1939 as the XPBM-1 prototype with two 1193kW (1,600 hp) Wright R-2600-6 Cyclone radial engines in wing-mounted nacelles large enough to incorporate weapon bays for a 907kg (2,000 lb) bombload. Production totalled 1,366 'boats, the first operational type being the PBM-1 that introduced the dihedralled tailplane and inward-canted fins that became the type's most distinctive feature.

Later production variants were the PBM-3B for the U.K. (Mariner GR.Mk I) with non-retractable stabilizing floats and 1,268kW (1,700 hp) R-2600-12 engines, the PBM-3C with armour protection and a 1,814kg (4,000 lb) bombload, the PBM-3D with 1,417kW (1,900 hp) R-2600-22 engines, surface search radar, self-sealing fuel tanks and a 3,628kg (8,000 lb) bombload, the PBM-3R unarmed transport with provision for 20 passengers and freight, the PBM-3S anti-submarine version with provision for four depth charges, and the PBM-5 major production version in a number of subvariants.

MARTIN 167 MARYLAND
(U.S.A.)

Maryland was the name given to the reconnaissance bomber ordered by France before the outbreak of World War II. The type had been developed for the U.S. Army as the XA-22 attack bomber that first flew in March 1939, but was rejected by its proposed initial operator. Only 140 had been delivered to France before its surrender in June 1940, the other 75 of the French orders supplementing 75 British aircraft as Maryland Mk Is with 783kW (1,050 hp) Pratt & Whitney R-1830-SC3G radial engines. The 150 Maryland Mk IIs had 895kW (1,200 hp) R-1830-S3C4Gs and other improvements, being used mainly for reconnaissance in the Mediterranean theatre and for target towing.

MARTIN 167 MARYLAND Mk I

Role: Light bomber
Crew/Accommodation: Three
Power Plant: Two 1,050 hp Pratt & Whitney R-1830-SC3G Wasp air-cooled radials
Dimensions: Span 18.69m (60.33 ft); length 14.22m (46.66 ft); wing area 54.21m^2 (583.5 sq ft)
Weights: Empty 5,067kg (11,170 lb); MTOW 7,711kg (17,000 lb)
Performance: Maximum speed 451km/h (280 mph) at 1,524m (5,000 ft); operational ceiling 6,096m (20,000 ft); range 1,207km (750 miles) with full bombload
Load: Six .303 inch machine guns, plus up to 816kg (1,800 lb) of bombs

MARTIN 187 BALTIMORE
(U.S.A.)

The Baltimore was a development of the Maryland to meet a British requirement and first flew in 1940 with the more powerful Wright R-2600 radial engine and a wider fuselage to allow crew members to change places in flight. Production totalled 1,575 aircraft in variants up to the Baltimore Mk V with steadily greater power and armament.

MARTIN 187 A-30A BALTIMORE Mk V

Role: Light bomber
Crew/Accommodation: Four
Power Plant: Two 1,700 hp Wright R-2600-29 Cyclone air-cooled radials
Dimensions: Span 18.69m (61.33 ft); length 14.78m (48.5 ft); wing area 50.03m^2 (538.5 sq ft)
Weights: Empty 5,171kg (11,400 lb); MTOW 7,725kg (17,031 lb)
Performance: Maximum speed 505km/h (314 mph) at 4,572m (15,000 ft); operational ceiling 8,708m (28,570 ft); range 1,014km (630 miles) with full bombload
Load: Two .5 inch and six .303 inch machine guns, plus up to 907kg (2,000 lb) of bombs

MARTIN 179 (B-26 MARAUDER)
(U.S.A.)

The Marauder was designed to meet a particularly difficult 1939 specification for a high-performance medium bomber, and was ordered 'off the drawing board' straight into production without any prototype or even pre-production aircraft. The first B-26 flew in November 1940 with two 1,380kW (1,850 hp) Pratt & Whitney R-2800-5 radial engines as a highly streamlined mid-wing monoplane with tricycle landing gear. The type was able to meet the performance requirements of its specification, but low-speed handling was decidedly poor. Total production was 4,708 and, in addition to 201 B-26s, the main variants were the B-26A (139 aircraft) which introduced features to overcome the handling problem, though these were negated by inevitably increased weight; the B-26C (1,883) which introduced 1,491kW (2,000 hp) R-2800-41 engines and, from the 642nd aircraft, a wing increased in span by 1.83m (6 ft); the B-26C (1,210) generally similar to the b-26B, but from a different production line; the B-26F (300) which introduced an increased wing incidence angle to improve field performance; and the B-26G (893) generally similar to the B-26F. There were two target tug/gunnery trainer variants produced by converting bombers, and also the new-build TB-26G crew trainer. A few aircraft were used by the U.S. Navy under designation JM.

MARTIN B-26B MARAUDER

Role: Medium bomber
Crew/Accommodation: Seven
Power Plant: Two 2,000 hp Pratt & Whitney R-2800-41 Double Wasp air-cooled radials
Dimensions: Span 21.64m (71 ft); length 17.75m (58.25 ft); wing area 61.13m^2 (658 sq ft)
Weights: Empty 10,660kg (23,500 lb); MTOW 17,328kg (38,200 lb)
Performance: Maximum speed 454km/h (282 mph) at 4,572m (15,000 ft); operational ceiling 4,572+m (15,000+ ft); range 1,086km (675 miles) with maximum bombload
Load: Twelve .5 inch machine guns, plus up to 1,815kg (4,000 lb) of internally carried bombs, or one externally carried torpedo

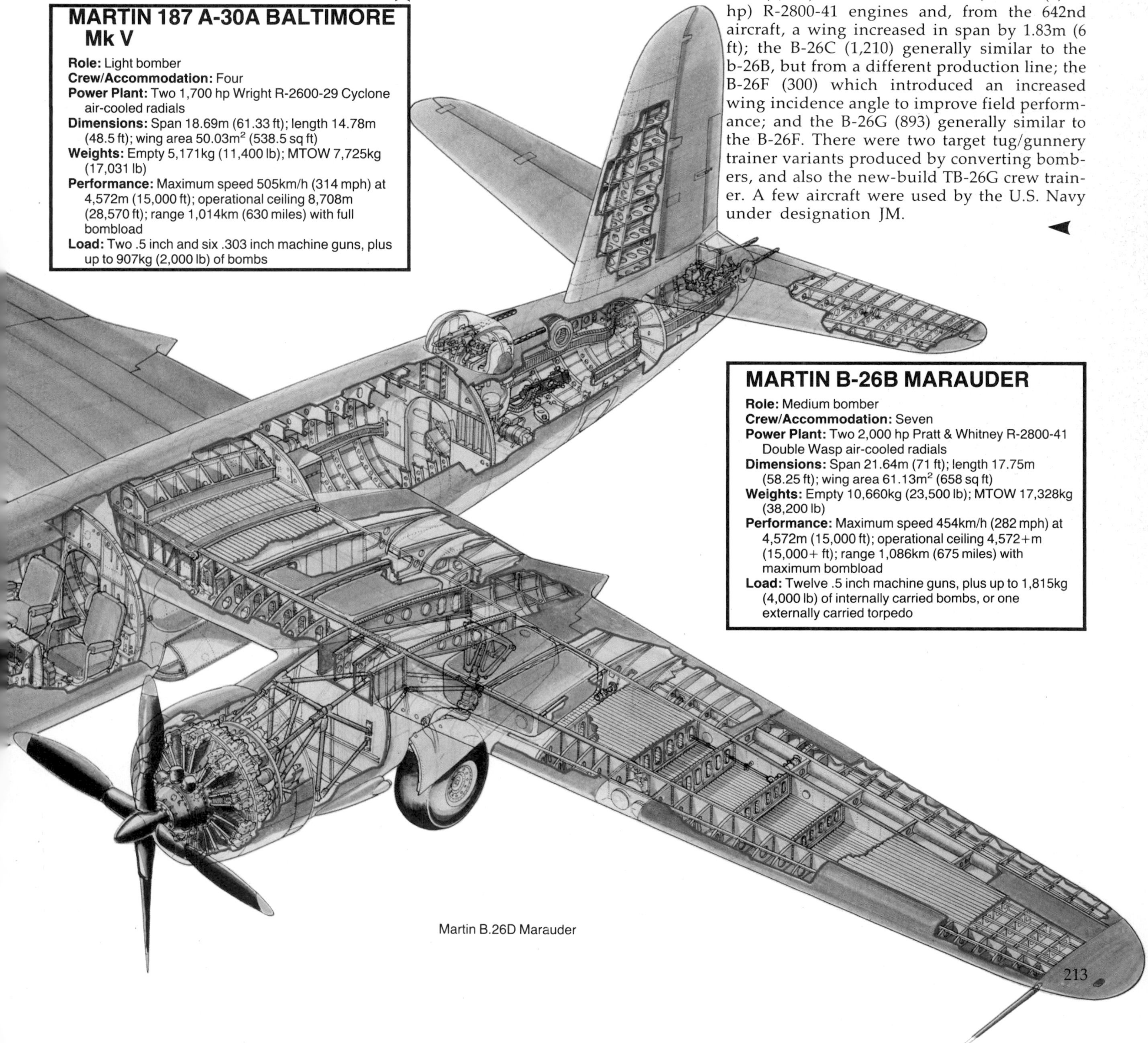

Martin B.26D Marauder

MARTIN 237 (P5M MARLIN) (U.S.A.)

The Marlin combined the wings and upper hull of the Mariner with a new lower hull and power plant to produce a markedly superior flying boat. A PBM-5 was modified as the XP5M-1 prototype that first flew in May 1948 with 2,424kW (3,250 hp) Wright R-3350 radial engines, radar-directed nose and tail turrets, and a powered dorsal turret. Production of 160 P5M-1 (later P-5A) 'boats was undertaken, this variant having 2,535kW (3,400 hp) R-3350-30WA engines and later adapted into several subvariants. Finally came 115 P5M-2 (P-5B) 'boats with 2,573kW (3,450 hp) R-3350-32WA engines, with a T-tail, a modified hull and a permanent rather than retrofitted magnetic anomaly detection installation.

MARTIN P5M-2 MARLIN

Role: Long-range patrol bomber flying boat
Crew/Accommodation: Eleven
Power Plant: Two 3,450 hp Wright R-3350-32WA Turbo-Compound air-cooled radials
Dimensions: Span 36.02m (118.16 ft); length 30.81m (101.08 ft); wing area 130.62m^2 (1,406 sq ft)
Weights: Empty 22,444kg (49,480 lb); MTOW 38,556kg (85,000 lb)
Performance: Maximum speed 404km/h (251 mph) at sea level; operational ceiling 7,315m (24,000 ft); range 3,299km (2,050 miles) with full warload
Load: No guns, but up to 3,629kg (8,000 lb) of torpedoes, mines, depth charges or bombs

MARTIN B-57 (U.S.A.)

This was the American version of the British English Electric Canberra light bomber. The main variants in the 403-aircraft production run were the B-57A used mainly for training, the extensively revised and 'Americanized, B-57B night intruder with tandem seating under a fighter-type canopy, the B-57C dual-control trainer and the B-57E target tug. Comparatively simple reconnaissance conversions were the RB-57A and RB-57B; the RB-57D was an altogether more advanced high-altitude type with two 4,990kg (11,000 lb) thrust Pratt & Whitney J57-P-37A turbojets and a wing increased in span to 32.31m (106 ft). The most extreme development, however, was the RB-57F produced by General Dynamics with two 8,165kg (18,000 lb) thrust Pratt & Whitney TF33-P-11 turbofans supplemented by two 1,497kg (3,300 lb) thrust Pratt & Whitney J60-P-9 turbojets in pods under the much enlarged wing, which now spanned 37.19m (122 ft) for flight at extreme altitude.

MARTIN B-57B

Role: Night strke
Crew/Accommodation: Two
Power Plant: Two 3,275kgp (7,200 lb s.t.) Wright J65-W-1 turbojets
Dimensions: Span 19.5m (63.96 ft); length 19.96m (65.5 ft); wing area 89.19m^2 (960 sq ft)
Weights: Empty 12,247kg (27,000 lb); MTOW 25,832kg (56,950 lb)
Performance: Maximum speed 853km/h (530 mph) at sea level; operational ceiling 14,630m (48,000 ft); range 3,701km (2,300 miles) with full warload
Load: Four 20mm cannon, plus up to 2,268kg (5,000 lb) of bombs

MARTIN-BAKER MB.5
(U.K.)

This was arguably the finest piston-engined fighter ever designed, a compact yet agile and excellently armed high-performance machine optimized for easy production and maintenance. The type first flew in May 1944, and it was only the imminent arrival of turbojet-engined fighters that prevented large-scale production towards the end of the war.

MARTIN-BAKER MB.5

Role: Day fighter
Crew/Accommodation: One
Power Plant: One 1,900 hp Rolls-Royce Griffon 83 water-cooled inline
Dimensions: Span 10.67m (35 ft); length 11.3m (37.07 ft); wing area 24.4m^2 (262.64 sq ft)
Weights: Empty 4,192kg (9,233 lb); MTOW 5,484kg (12,090 lb)
Performance: Maximum speed 740km/h (460 mph) at 6,095m (20,000 ft); operational ceiling 12,192m (40,000 ft); range 1,770km (1,100 miles)
Load: Four 20mm cannon

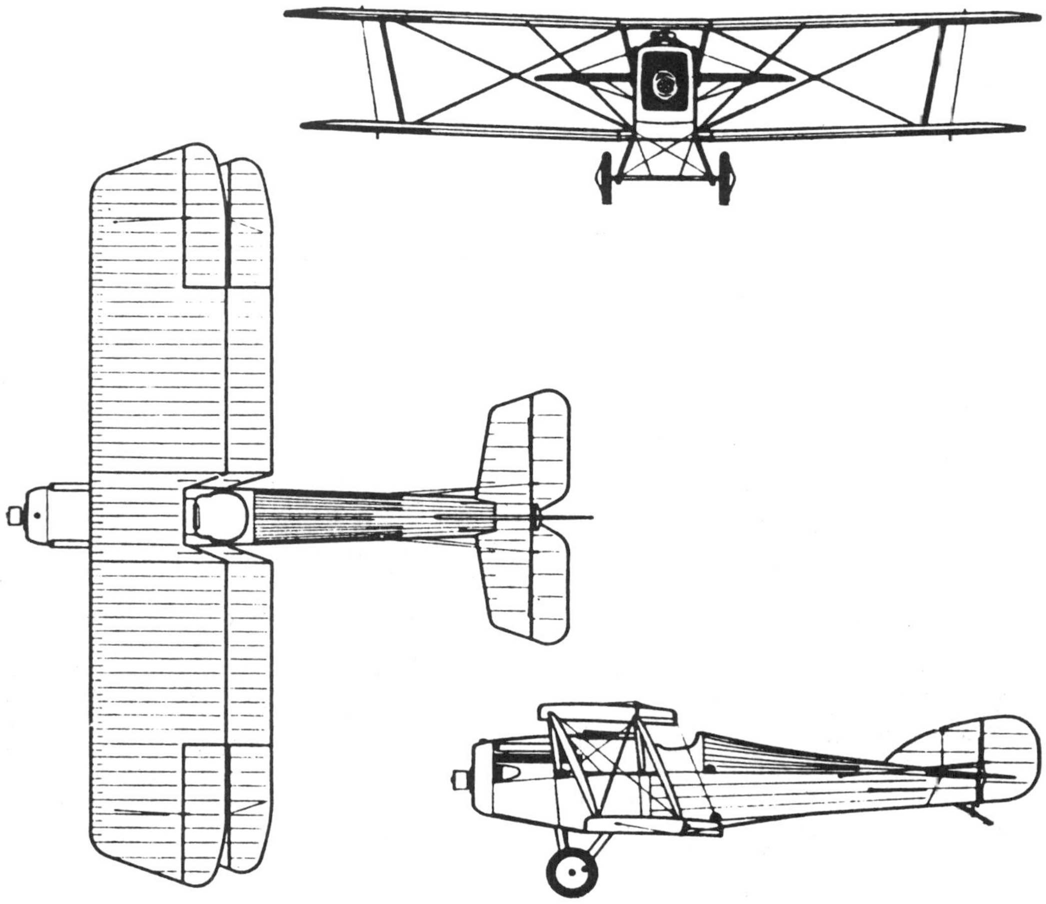

MARTINSYDE F3 and F4 BUZZARD
(U.K.)

The F3 was basically a single-seat derivative of the F2. two-seat prototype, and with a 213kW (285 hp) Rolls-Royce Falcon inline engine possessed excellent performance. Six pre-production F3s with the 205kW (275 hp) Falcon III paved the way for the definitive F4 Buzzard which located the pilot further aft and introduced a Hispano-Suiza engine. Large numbers were ordered, but with only 60 aircraft delivered by the end of the World War I the company was left with 20 surplus machines on its hands, which eventually went for export and conversion into civil aircraft.

MARTINSYDE F4 BUZZARD

Role: Fighter
Crew/Accommodation: One
Power Plant: One 300 hp Hispano-Suiza 8Fb water-cooled inline
Dimensions: Span 9.99m (32.79 ft); length 7.77m (25.48 ft); wing area 29.73m^2 (320 sq ft)
Weights: Empty 821kg (1,811 lb); MTOW 1,088kg (2,398 lb)
Performance: Maximum speed 213km/h (132.5 mph) at 4,572m (15,000 ft); operational ceiling 7,315m (24,000 ft); endurance 2 hours
Load: Two .303 inch machine guns

McDONNELL FD/FH PHANTOM
(U.S.A.)

This was the U.S. Navy's first carrierborne turbojet-powered single-seat fighter, and in 1942 the design task was entrusted to the small McDonnell company as the major aircraft manufacturers were fully committed with current programmes. The resulting XFD-1 prototype first flew in January 1945 as an unexceptional aeroplane, but an order was placed for 100 FD-1 production aircraft, the designation being changed to FH-1 before delivery was made in 1947. The aircraft remained in service less than two years.

McDONNELL FH-1 PHANTOM

Role: Naval carrierborne fighter
Crew/Accommodation: One
Power Plant: Two 726kgp (1,600 lb s.t.) Westinghouse J30-WE-20 turbojets
Dimensions: Span 12.42m (40.75 ft); length 11.35m (37.25 ft); wing area 25.6m^2 (276 sq ft)
Weights: Empty 2,792kg (6,156 lb); MTOW 4,323kg (9,531 lb)
Performance: Maximum speed 784km/h (487 mph) at sea level; operational ceiling 13,320m (43,700 ft); range 870km (540 miles) on internal fuel only
Load: Four .5 inch machine guns

McDONNELL F2H BANSHEE
(U.S.A.)

The Banshee was designed in 1945 as successor to the Phantom. The design had the same basic configuration as the Phantom, but was larger and more refined, and provided more fuel capacity. The XF2H-1 prototype flew in January 1947, and production of 792 aircraft followed. The main variants were the initial F2H-1 with 1,361kg (3,000 lb) thrust Westinghouse J34-WE-22 turbojets, the F2H-2 with greater fuel capacity and 1,474kg (3,250 lb) thrust J34-WE-34s, the F2H-2B fighter-bomber, the F2H-2N night-fighter, the F2H-2P photo-reconnaissance model, the F2H-3 naval carrierborne fighter, and the F2H-4 all-weather fighter with 1,633kg (3,600 lb) thrust J34-WE-38 engines.

McDONNELL F2H-3 BANSHEE

Role: Naval carrierborne fighter
Crew/Accommodation: One
Power Plant: Two, 1,474kgp (3,250 lb s.t.) Westinghouse J34-WE-34 turbojets
Dimensions: Span 12.73m (41.75 ft); length 14.68m (48.16 ft); wing area 27.3m^2 (294 sq ft)
Weights: Empty 5,980kg (13,183 lb); MTOW 11,437kg (25,214 lb)
Performance: Maximum speed 933km/h (580 mph) at sea level; operational ceiling 14,205m (46,600 ft); range 1,885km (1,170 miles) on internal fuel only
Load: Four 20mm cannon, plus up to 907kg (2,000 lb) of externally carried bombs

McDONNELL F3H DEMON
(U.S.A.)

Ordered in prototype form during September 1949, the F3H resulted from the U.S. Navy's appreciation of the fact that carrierborne fighters need not be inferior in any significant respect to land-based fighters. The design concept was somewhat different from that which had characterized the company's first two jet fighters, and introduced the layout followed in McDonnell's next few fighters with swept flying surfaces, a fuselage-mounted engine aspirated via long lateral inlets, and the tail mounted on what was essentially a boom above the jetpipe.

There were considerable problems and delays with the chosen Westinghouse J40 turbojet, so though the first XF3H-1 flew in August 1951 it was December 1953 before the F3H-1N began to enter service. This variant had the 4,944kg (10,900 lb) thrust J40-WE-22 version of this unreliable afterburning turbojet, and was seriously underpowered. Production was halted after the construction of only 58 F3H-1Ns, and the design was reworked to produce the F3H-2N with an 18 per cent larger wing and the 6,350kg (14,000 lb) thrust Allison J71-A-2; these 140 aircraft could carry four Sidewinder IR-homing air-to-air missiles, while the 80 F3H-2Ms were fitted for four Sparrow 1 radar-guided missiles. The last variant was the F3H-2 Sparrow-fitted strike fighter, of which 239 were produced before production ended in 1959.

McDONNELL F3H-2N/F-3C DEMON

Role: Naval carrierborne fighter
Crew/Accommodation: One
Power Plant: One 6,350kgp (14,000 lb s.t.) Allison J71-A-2 turbojet with reheat
Dimensions: Span 10.77m (35.33 ft); length 17.98m (59 ft); wing area 48.2m^2 (519 sq ft)
Weights: Empty 9,656kg (21,287 lb); MTOW 15,161kg (33,424 lb)
Performance: Maximum speed 1,170km/h (727 mph) at sea level; operational ceiling 13,000m (42,650 ft); range 1,900km (1,180 miles) with combat load
Load: Four 20mm cannon, plus four short-range, air-to-air missiles

McDONNELL F-101 VOODOO
(U.S.A.)

In 1946 McDonnell responded to a U.S. Army Air Force requirement for a long-range penetration or escort fighter with a design that first flew in XF-88 prototype form during February 1947. The validity of the concept was then questioned and the programme cancelled to save scarce financial resources. During the Korean War of the early 1950s, however, the U.S. Air Force found that its escorts were woefully deficient: those that had the range to escort the bombers lacked the performance to tangle with Soviet jet fighters, while those with the performance to tackle the opposing fighters lacked the range to serve as escorts. The XF-88 was thus revived in a modernized and upgraded form as the Voodoo. It was clearly beyond the capabilities of any fighter to escort the very long-ranged Convair B-36, however, and the F-101A became a tactical fighter. First flown in September 1954 with 6,804kg (15,000 lb) thrust J57-P-13 turbojets, the F-101A (77 aircraft) was followed by the RF-101A reconnaissance type (35), the F-101B two-seat long-range interceptor with 6,749kg (14,880 lb) thrust J57-P-55s (407), the TF-101B dual-control trainer (72), the F-101C single-seat tactical strike fighter (47) and the RF-101C reconnaissance type (166). The type was also operated by the Royal Canadian Air Force as the CF-101, and a number of converted models were evolved later.

McDONNELL F-101C VOODOO

Role: Long-range fighter
Crew/Accommodation: One
Power Plant: Two 6,804kgp (15,000 lb s.t.) Pratt & Whitney J57-P-13 turbojets with reheat
Dimensions: Span 12.09m (39.67 ft); length 21.14m (69.42 ft); wing area 34.19m^2 (368 sq ft)
Weights: Empty 11,794kg (26,00 lb); MTOW 21,940kg (48,366 lb)
Performance: Maximum speed 1,617km/h (873 knots) Mach 1.51 at 10,670m (35,000 ft); operational ceiling 15,330m (50,300 ft); radius 1,254km (218 naut. miles)
Load: Four 20mm cannon, plus up to 1,688kg (3,721 lb) of ordnance, including a nuclear weapon

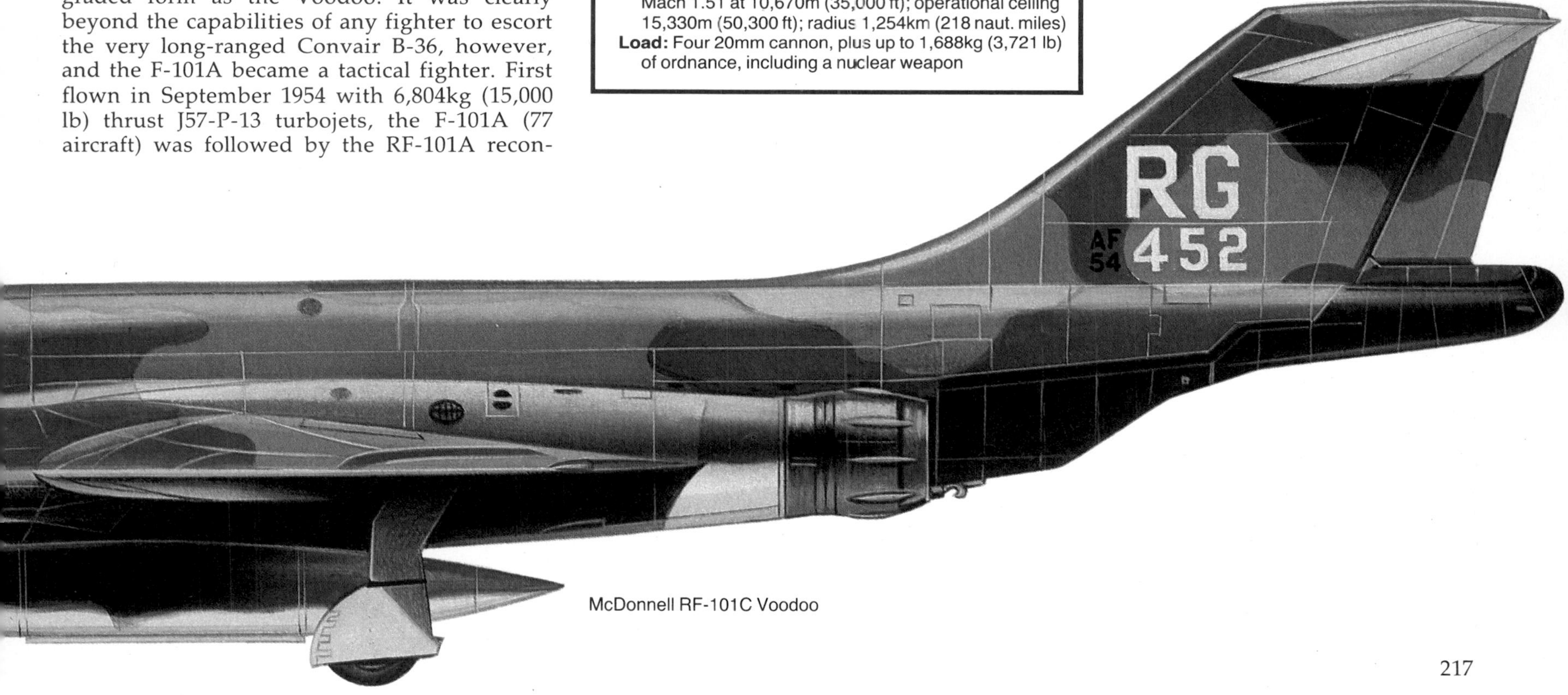

McDonnell RF-101C Voodoo

McDONNELL DOUGLAS F-4 PHANTOM II
(U.S.A.)

In October 1979 construction of the 5,057th Phantom II was completed, ending the West's largest warplane production programme since World War II. As may be imagined, the programme was devoted to a quite exceptional type that must be numbered in the five most important warplanes of all time for its combination of outright performance, electronic and weapon versatility, comparatively good agility and enormous strength. The type was planned initially as an all-weather attack aeroplane, but then adapted into an all-weather fleet-defence fighter and tactical fighter during the design process.

The XF4H-1 prototype flew in May 1958 with early examples of equally classic J79 afterburning turbojet. Production was then inaugurated for the F4H-1F (later F-4A) pre-production type with 7,326kg (16,150 lb) thrust J79-GE-2/2A engines, the F4H-1 (later F-4B) strike fighter with 7,711kg (17,000 lb) thrust J79-GE-8 engines, the RF-4B reconnaissance aeroplane for the U.S. Marine Corps, the F-4C (originally F-110A) attack fighter for the U.S. Air Force with 7,711kg (17,000 lb) thrust J79-GE-15 engines, the RF-4C U.S.A.F. tactical reconnaissance aeroplane, the F-4D version of the F-4C with mission electronics tailored to U.S.A.F. rather than U.S. Navy requirements, the F-4E major production version for the U.S.A.F with 8,119kg (17,900 lb) thrust J79-GE-17 engines, improved radar, leading-edge slats and an internal 20mm rotary-barrel cannon, the F-4F air superiority version of the F-4E for West Germany, the F-4J for the U.S. Navy with 8,119kg (17,900 lb) thrust J79-GE-10 engines, a revised wing and modified tail, the F-4K version of the F-4J for the Royal Navy with Rolls-Royce Spey turbofans, and the F-4M version of the F-4K for the Royal Air Force.

There have been several other versions produced by converting older airframes with more advanced electronics as well as other features, the most important being the similar F-4N and F-4S developments of the F-4B and F-4J for the U.S. Navy, the F-4G for the U.S.A.F.'s 'Wild Weasel' radar-suppression role, and the Super Phantom (or Phantom 2000) rebuild by Israel Aircraft Industries to modernize the F-4E with the latest electronics and an advanced cockpit.

McDonnell F.4G 'Wild Weasel'

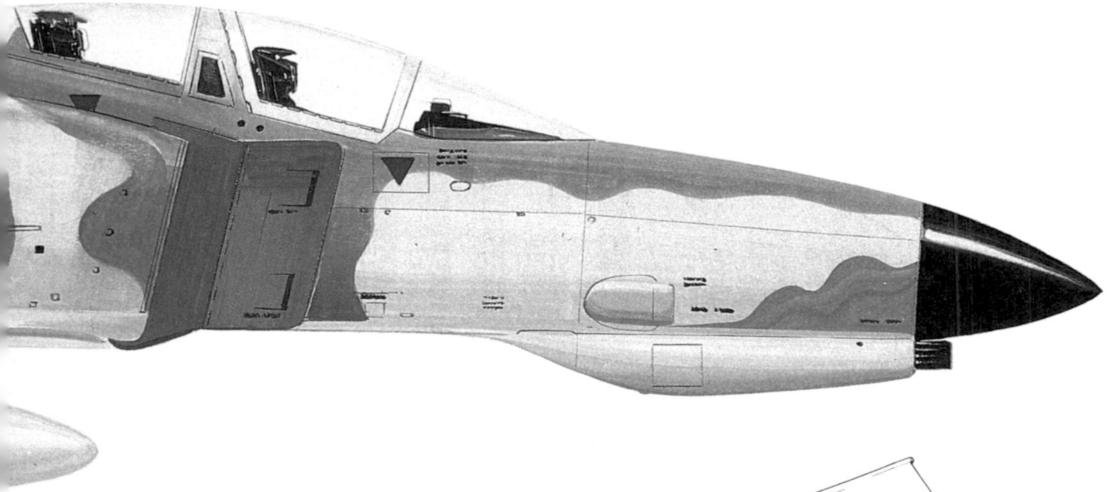

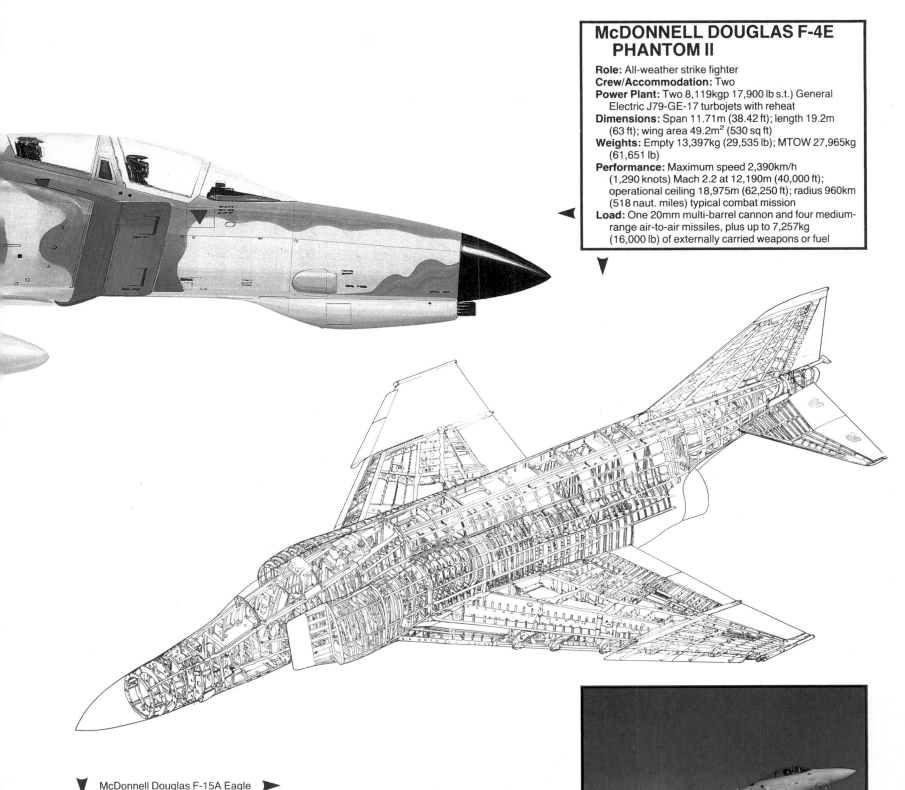

McDONNELL DOUGLAS F-4E PHANTOM II

Role: All-weather strike fighter
Crew/Accommodation: Two
Power Plant: Two 8,119kgp 17,900 lb s.t.) General Electric J79-GE-17 turbojets with reheat
Dimensions: Span 11.71m (38.42 ft); length 19.2m (63 ft); wing area 49.2m^2 (530 sq ft)
Weights: Empty 13,397kg (29,535 lb); MTOW 27,965kg (61,651 lb)
Performance: Maximum speed 2,390km/h (1,290 knots) Mach 2.2 at 12,190m (40,000 ft); operational ceiling 18,975m (62,250 ft); radius 960km (518 naut. miles) typical combat mission
Load: One 20mm multi-barrel cannon and four medium-range air-to-air missiles, plus up to 7,257kg (16,000 lb) of externally carried weapons or fuel

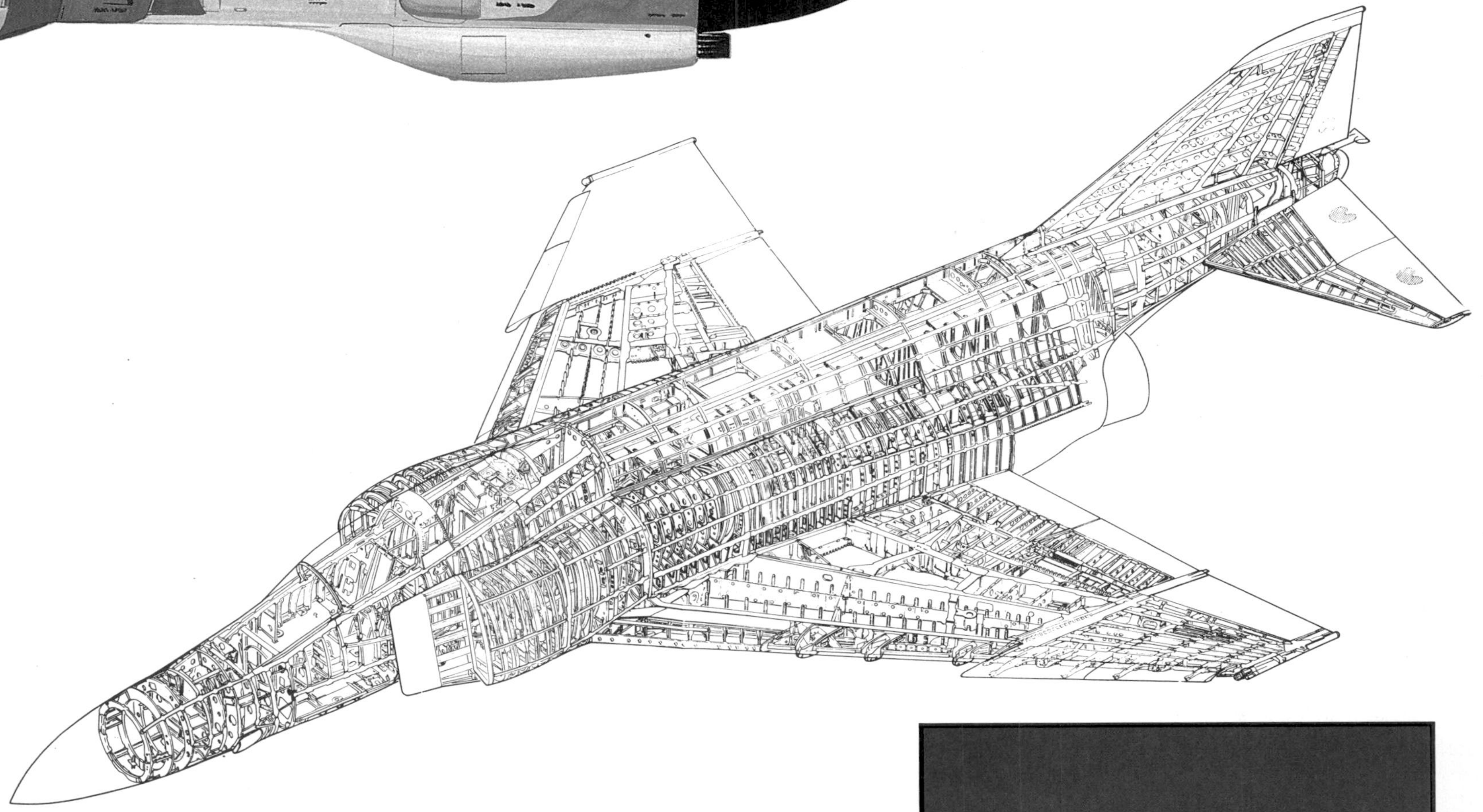

McDonnell Douglas F-15A Eagle

McDONNELL DOUGLAS F-15 EAGLE
(U.S.A.)

The F-15 was planned as the U.S.A.F.'s successor to the F-4 in the air-superiority role, and emerged for its first flight in July 1972 as a massive aeroplane with twin turbofans, sophisticated aerodynamics and advanced electronics. The type has since proved a first-class and versatile warplane in its initial F-15A single- and F-15B two-seat versions. In 1979 these initial versions were superseded in production by the F-15C and F-15D respectively, these having more advanced systems and provision for external carriage of the so-called FAST packs that provide significantly more fuel capacity and weapon-carriage capability for a negligible increase in drag. In 1989 the U.S.A.F. received its first F-15E two-seat interdiction aircraft derived from the F-15D, and developments currently in hand centre on a long-range strike version able to operate from short lengths of damaged runway with the aid of 2D vectoring nozzles and canard foreplanes controlled by an advanced fly-by-wire system.

McDONNELL DOUGLAS F-15E EAGLE

Role: All-weather strike fighter
Crew/Accommodation: Two
Power Plant: Two 10,637kgp 23,450 lb s.t.) Pratt & Whitney F100-PW-220 turbofans with reheat
Dimensions: Span 13.05m (42 ft); length 19.43m (63.75 ft); wing area 56.5m^2 (608 sq ft)
Weights: Empty 14,379kg (31,700 lb); MTOW 36,741kg (81,000 lb)
Performance: Maximum speed 2,698km/h (1,456 knots) Mach 2.54 at 12,192m (40,000 ft); operational ceiling 18,300m (60,000 ft); radius 1,420km (670 naut. miles) with 3,175kg (7,000 lb) of weapons
Load: Up to 10,659kg (23,500 lb) of weaponry, including one 20mm multi-barrel cannon mounted internally

McDonnell Douglas ATF-18A

McDONNELL DOUGLAS F/A-18 HORNET
(U.S.A.)

Serving with the U.S. Navy and Marine Corps, the dual-capability F/A-18 is one of the West's most important current carrierborne warplanes, and has secured useful export orders for land-based use. The type was derived from Northrop's losing contender in the U.S.A.F. lightweight fighter competition, enlarged and developed jointly by Northrop and McDonnell Douglas for production under the latter's leadership. The first aeroplane flew in November 1979, and the initial F/A-18A single- and F/A-18B two-seaters have been superseded in production by the F/A-18C and night-attack F/A-18D with more advanced electronics and provision for later weapon types.

McDONNELL DOUGLAS F/A-18A HORNET

Role: Naval carrierborne strike fighter
Crew/Accommodation: One
Power Plant: Two 7,257kgp (16,000 lb s.t.) General Electric F404-GE-400 turbofans with reheat
Dimensions: Span 11.4m (36.5 ft); length 17.1m (56 ft); wing area 37.16m^2 (400 sq ft)
Weights: Empty 12,700kg (28,000 lb); MTOW 25,401kg (56,000 lb)
Performance: Maximum speed 1,457km/h (786 knots) Mach 1.18 at sea level; operational ceiling 15,240m (50,000 ft); radius 740km (399 naut. miles) with missiles and internal fuel only
Load: One20mm multi-barrel cannon, two medium-range and two short-range air-to-air missiles, plus externally carried weapons that in total exceed 7,711kg (17,000 lb)

McDONNELL DOUGLAS/ BRITISH AEROSPACE AV-8B HARRIER II
(U.S.A./U.K.)

From the basic Harrier, McDonnell Douglas with BAe assistance, has developed the much improved Harrier II with a larger wing of supercritical section and single-piece graphite/epoxy construction. This improves wingborne lift, adds more hardpoints and increases internal fuel capacity while reducing structure weight. In combination with other lift-improvement modifications, zero-scarf vertoring nozzles, a more powerful Pegasus turbofan and a new cockpit, this has transformed the Harrier into a much more potent machine. The first prototype flew in November 1978, and all four Harrier II pre-production aircraft had flown by April 1979, and the type is now in production for the U.S. Marine Corps as the AV-8B and for the Royal Air Force as the Harrier GR.Mk 5. A night attack package is being retrofitted on Marine Corps AV-8Bs, and the adoption of a similar package for British aircraft will result in the Harrier GR.Mk 7 aircraft. The companies are presently examining the possibility of a radar-carrying development.

McDONNELL DOUGLAS AV-8B HARRIER II

Role: Naval shipborne V/STOL strike fighter
Crew/Accommodation: One
Power Plant: One 9,607kgp (21,108 lb s.t.) Pratt & Whitney/Rolls-Royce Pegasus 11 vectored thrust turbofan
Dimensions: Span 9.25m (30.3 ft); length 14.1m (46.3 ft); wing area 21.36m^2 (230 sq ft)
Weights: Empty 5,783kg (12,750 lb); MTOW 13,494kg (29,750 lb) with rolling take-off
Performance: Maximum speed 1,078km/h (582 knots) Mach 0.88 at sea level; operational ceiling 9,144+m (30,000+ ft); radius 1,114km (601 naut. miles) with 1,588kg (3,500 lb) or weapons
Load: Up to 4,137kg (9,200 lb) of ordnance, including air-to-ground missiles, air-to-air missiles and 25mm cannon pods

McDONNELL DOUGLAS HELICOPTERS H-6 CAYUSE, MODEL 500 and DEFENDER SERIES
(U.S.A.)

The model 369 was developed by Hughes as a prototype for the U.S. Army light observation helicopter competition of 1964, which it duly won. Very large production orders were placed for the OH-6A Cayuse service version, but then came problems with delivery rates and rising cost, and the programme was ended with delivery of just over 1,400 helicopters. The type was also produced in civil form as the Model 500 and improved Model 530, the latter developed after the purchase of Hughes by McDonnell Douglas in an almost bewildering number of export military versions.

McDONNELL DOUGLAS HELICOPTERS OH-6A CAYUSE

Role: Civil and military light utility helicopter
Crew/Accommodation: One, plus up to four passengers
Power Plant: One 250 shp Allison 250-C18 turboshaft
Dimensions: Span 9.24m (30.3 ft); rotor diameter 8.03m (26.33 ft)
Weights: Empty 452kg (995 lb); MTOW 1,080kg (2,400 lb)
Performance: Cruise speed 200km/h (125 mph) at 1,524m (5,000 ft); operational ceiling 4,800m (15,800 ft); range 611km (380 miles) with full payload
Load: Up to 353kg (780 lb)

McDONNELL DOUGLAS HELICOPTERS AH-64 APACHE
(U.S.A.)

The Apache was designed by Hughes, now McDonnell Douglas Helicopters, as a heavyweight battlefield and anti-tank helicopter to supplement the lighter Bell AH-1 HueyCobra series. The type is expensive to produce and difficult to maintain, but offers unexcelled electronic sophistication and firepower. The prototype first flew in September 1975, and AH-64A production helicopters have proved highly adaptable. The manufacturer has proposed an AH-64B version with more advanced missiles, a fly-by-light control system using optical fibres rather than electrical wiring, and voice-operated controls among other advanced features.

McDONNELL DOUGLAS HELICOPTERS AH-64A APACHE

Role: Armed and armoured attack helicopter
Crew/Accommodation: Two
Power Plant: Two 1,694 shp General Electric T700-GE-710 turboshafts
Dimensions: Overall length rotors turning 17.73m (58.17 ft); rotor diameter 14.63m (48 ft)
Weights: Empty 4,491kg (9,900 lb); MTOW 7,893kg (17,400 lb)
Performance: Cruise speed 288km/h (155 knots) at sea level; operational ceiling 6,401m (21,000 ft); range 579km (360 miles) with full warload
Load: One 30mm cannon, plus up to 469kg (1,034 lb) consisting of 16 Hellfire or 8 TOW anti-armour missiles

Messerschmitt Bf 109E

MESSERSCHMITT Bf 109
(Germany)

The Bf 109 was Germany's most important fighter of World War II in numerical terms, and bore the brunt of the air war until supplemented by the Focke-Wulf Fw 190 from 1941. The type went through a large number of production variants, and in common with other German aircraft was developed within these basic variants into a number of subvariants with factory- or field-installed modification packages. The type was designed in 1934 to provide the German Air Force with its first 'modern' fighter of all-metal stressed-skin construction with a low-set cantilever wing, retractable landing gear and enclosed cockpit. The first prototype flew in May 1935 with 518kW (695 hp) Rolls-Royce Kestrel inline engine, but the second had the 455kW (610 hp) Junkers Jumo 210A for which the aeroplane had been originally designed.

The overall production figure has not survived, but it is thought that some 35,000 aircraft were produced. The limited-number Bf 109A, B and C variants can be regarded mostly as pre-production and development variants with differing Jumo 210s and armament fits. With the Bf 109D the Daimler-Benz DB 600A inline engine was introduced, paving the way for the first large-scale production variant, the Bf 109E produced in variants up to the E-9 with the 820kW (1,100 hp) DB 601A. The Bf 109F introduced a more refined fuselage with reduced armament, and was powered by the DB 601E or N in variants up to the F-6.

The most important production model was the Bf 109G with the DB 605A and provision for cockpit pressurization in variants up to the G-16. Later in the war there appeared comparatively small numbers of the Bf 109H high-altitude fighter with increased span in variants up to the H-1, and the Bf 109K improved version of the Bf 109G with the DB 605 in variants up to the K-14.

MESSERSCHMITT Bf 109G-6

Role: Fighter
Crew/Accommodation: One
Power Plant: One 1,475 hp Daimler-Benz DB605A water-cooled in-line
Dimensions: Span 9.92m (32.55 ft); length 9.02m (29.59 ft); wing area 16.5m^2 (172.75 sq ft)
Weights: Empty 2,700kg (5,953 lb); MTOW 3,150kg (6,945 lb)
Performance: Maximum speed 623km/h (387 mph) at 7,000m (22,967 ft); operational ceiling 11,750m (38,551 ft); range 725km (450 miles)
Load: One 30mm cannon, two 20mm cannon and two 13mm machine guns, plus a 500kg (1,102 lb) bomb

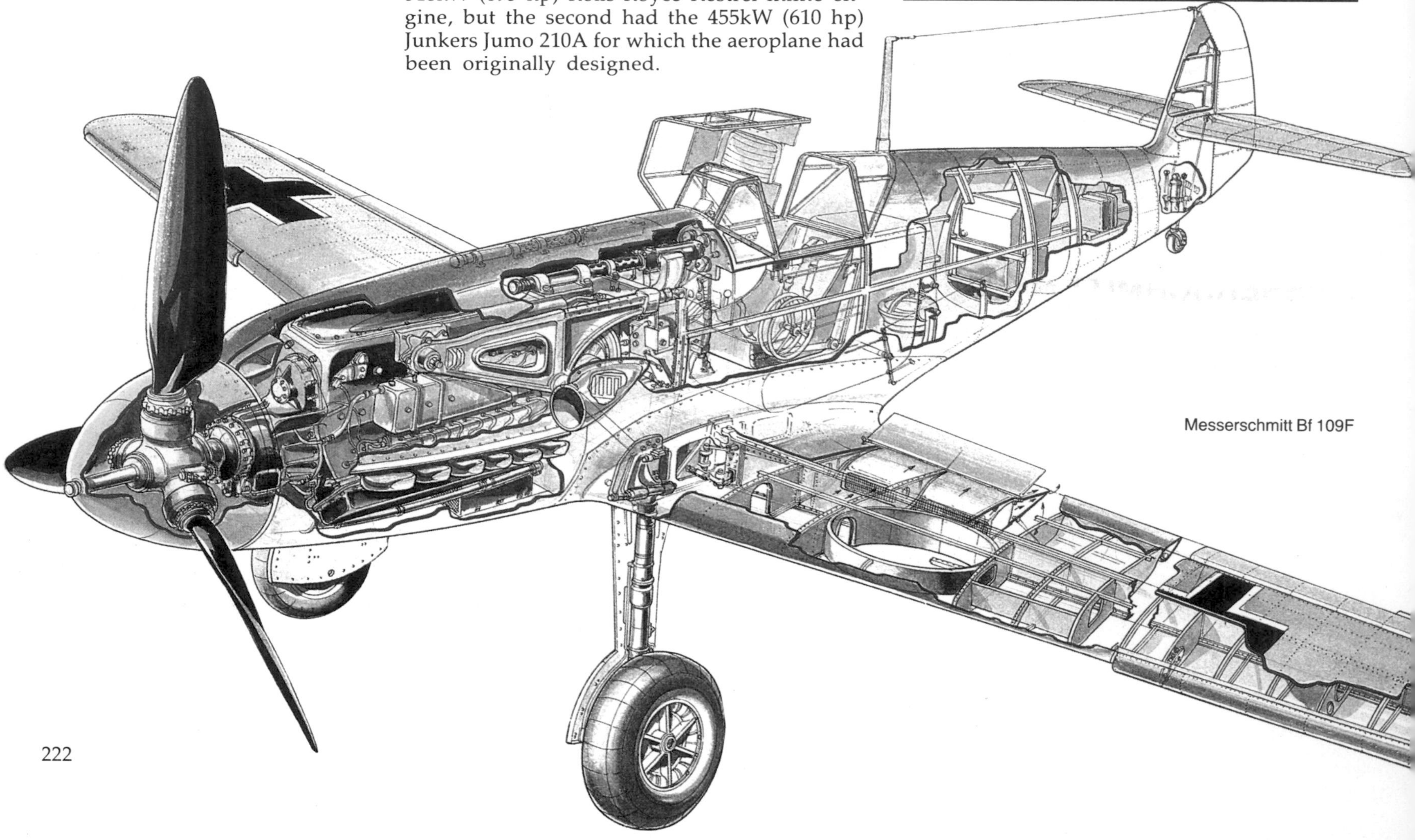

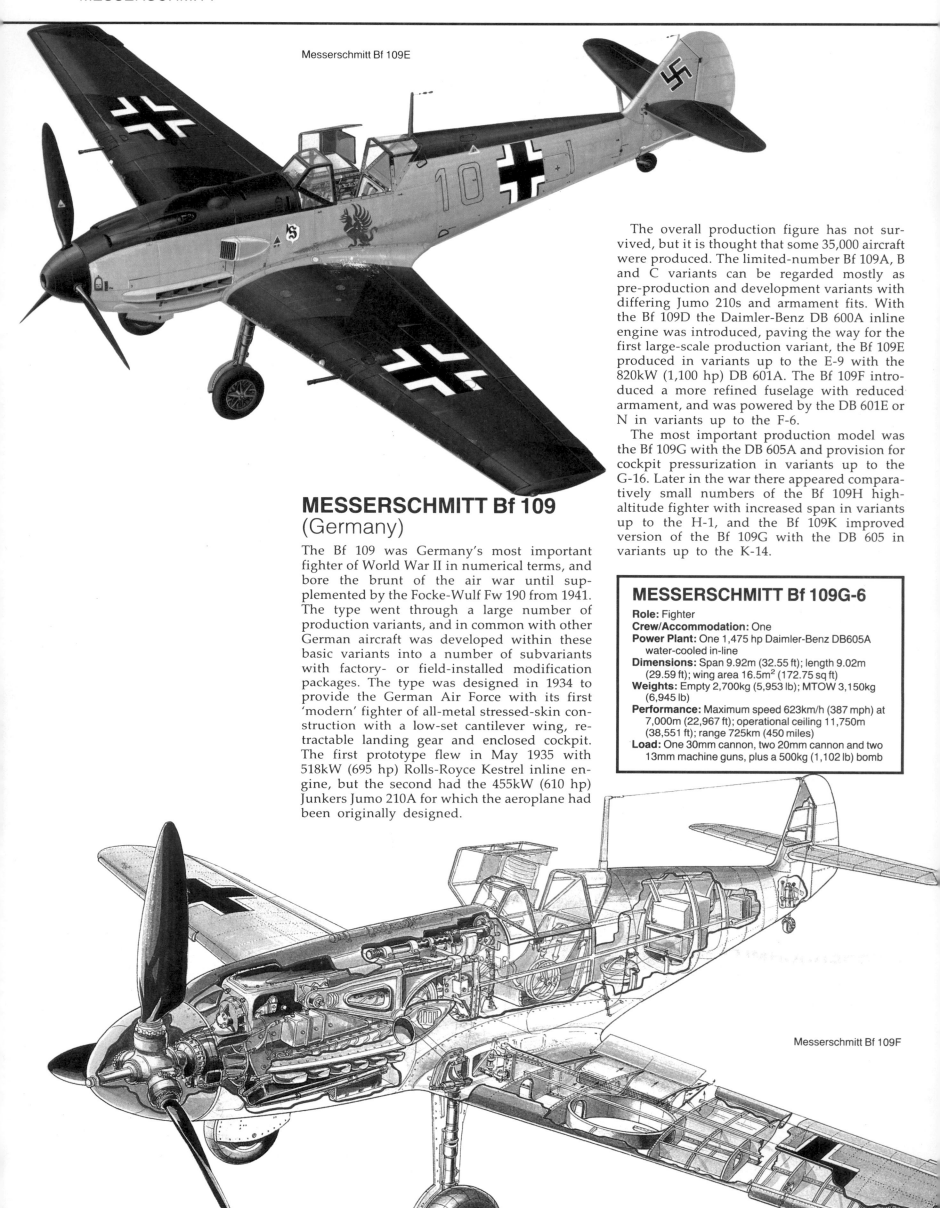

Messerschmitt Bf 109F

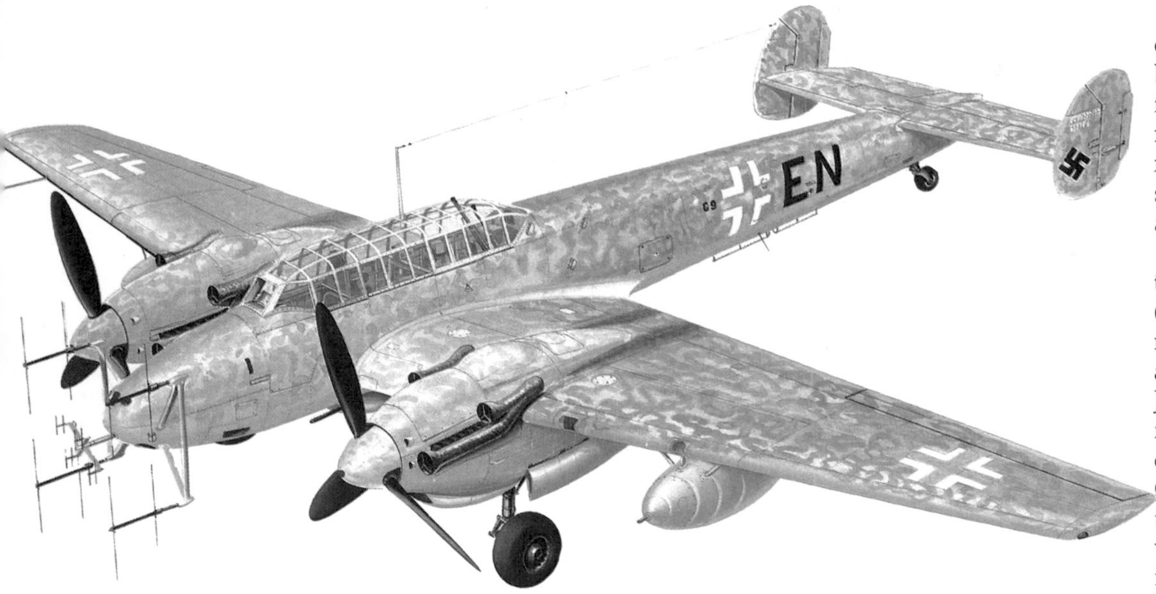

▲ MESSERSCHMITT Bf 110
(Germany)

The Bf 110 twin-engined fighter was first flown in May 1936 with two 679kW (910 hp) Daimler-Benz DB 600A inline engines. The type had been planned as a heavy fighter with potent nose armament, though secondary use as a high-speed fighter-bomber was also envisaged. The type was used initially as a heavy interceptor, proving quite successful, but was then pressed into operation as an escort, where its bulk and lack of manoeuvrability against opposing single-seat fighters left it at a severe disadvantage. After limited use in the fighter-bomber and reconnaissance roles the Bf 110 finally found its metier as a radar-equipped night fighter with increasingly heavy armament, and in this role was World War II's most successful aircraft in terms of numbers of aircraft destroyed.

About 6,050 Bf 110s were produced, and after the Bf 110A pre-production type with 507kW (680 hp) Junkers Jumo 210D inline engines, the most important variants were the aerodynamically refined Bf 110B reconnaissance fighter with Jumo 210s in variants up to the B-3, the Bf 110C fighter, fighter-bomber and reconnaissance aircraft with DB 601A inlines in variants up to the C-5, the Bf 110D long-range type in variants up to the D-3, the Bf 110E improved model in variants up to the E-3, the Bf 110F with DB 601F inline engines in variants up to the F-4, the Bf 110G night-fighter in variants up to the G-4, and the Bf 110H improved version in variants up to the H-4.

MESSERSCHMITT Bf 110C-4

Role: Long-range fighter
Crew/Accommodation: Two
Power Plant: Two 1,100 hp Daimler-Benz, DB601A water-cooled inline
Dimensions: Span 16.2m (53.15 ft); length 21.1m (39.67 ft); wing area 38.5m^2 (413 sq ft)
Weights: Empty 5,200kg (11,466 lb); MTOW 6,750kg (14,884 lb)
Performance: Maximum speed 560km/h (349 mph) at 7,000m (22,967 ft); operational ceiling 10,000m (32,811 ft); range 775km (481 miles) with full bombload
Load: Two 20mm cannon and five 7.9mm machine guns

▲

MESSERSCHMITT Me 163 KOMET (COMET)
(Germany)

The Me 163 was one of the most remarkable aircraft of World War II, a swept-wing tailless interceptor powered by a liquid-fuelled rocket. Powered prototypes were evaluated from the summer of 1941, the type revealing phenomenal climb and speed, but abysmally low powered endurance after lift-off from a jettisonable dolly. Landing was always traumatic, for the fuel dregs had a tendency to explode if the aeroplane made a heavy landing on its central skid. The 10 Me 163A pre-production aircraft were used as training gliders, but the 70 Me 163Bs were operational aircraft that saw limited combat use.

There were a number of developments including the Me 163S tandem-seat training glider, the Me 163C with a longer fuselage, and the Me 163D with a bubble canopy, a longer fuselage and tricycle landing gear.

MESSERSCHMITT Me 163C KOMET

Role: Interceptor
Crew/Accommodation: One
Power Plant: One 2,000kgp (4,409 lb s.t.) Walter 109-509C rocket
Dimensions: Span 9.8m (32.15 ft); length 7.035m (23.08 ft); wing area 20.7m^2 (223 sq ft)
Weights: Empty 2,500kg (5,512 lb); MTOW 5,125kg (11,299 lb)
Performance: Maximum speed 950km/h (590 mph) at 4,000m (13,123 ft); operational ceiling 16,000m (52,493 ft); range 125km (78 miles)
Load: Two 30mm cannon

MESSERSCHMITT Me 210 and Me 410 HORNISSE (HORNET)
(Germany)

The Me 210 was designed as the Bf 110's successor, and though resembling it in layout, it was more refined aerodynamically and powered by two 783kW (1,050 hp) Daimler-Benz DB 601A inline engines when first flown in September 1939. A distinctive feature was the pair of remote-control machine gun barbettes on the sides of the fuselage aft of the wing. Some 550 aircraft were built in Me 210 and Hungarian-built Me 210C variants, but the type was prone to severe stalling and spinning problems, and production ceased in spring 1942. Introduction of a longer rear fuselage and wing leading-edge slats cured the problems to create the Me 410, which was powered by the DB 603A inline engine and began to enter service in January 1943. Production amounted to 1,610 aircraft in Me 410A-1 to A-3 high-speed bomber, bomber destroyer and reconnaissance variants, and Me 410B-1 and B-3 equivalent versions with the DB 603G engine and further developed Me 410B-6 anti-shipping model.

MESSERSCHMITT Me 410A-1 HORNISSE

Role: Fighter bomber
Crew/Accommodation: Two
Power Plant: Two 1,720 hp Daimler-Benz DB 603A-1 water-cooled inlines
Dimensions: Span 16.4m (53.81 ft); length 12.4m (40.68 ft); wing area 36.2m² (389.65 sq ft)
Weights: Empty 6,150kg (13,558 lb); MTOW 10,650kg (23,479 lb)
Performance: Maximum speed 625km/h (388 mph) at 6,700m (21,982 ft); operational ceiling 10,000m (32,810 ft); range 2,330km (1,447 miles) with no bombs
Load: Two 20mm cannon, two 13mm and two 7.9mm machine guns, plus up to 1,000kg (2,205 lb) internally-stowed bombload

MESSERSCHMITT Me 262 SCHWALBE (SWALLOW)
(Germany)

This could have been the world's first operational jet fighter, but was enormously delayed by manoeuvrings within the German political and aircraft establishment. With its clean lines, tricycle landing gear, slightly swept wings and axial-flow turbojets, it was certainly the most advanced fighter to see service in World War II. Design started late in 1938, and the type first flew in April 1941 with a nose-mounted piston engine and tailwheel landing gear. The five prototypes were followed by 23 pre-production Me 262A-0s before the Me 262A-1 entered service as the first production variant; the -1a had four 30mm cannon and the -1b added 24 air-to-air unguided rockets. Total production was in the order of 1,100 aircraft, and later variants were the Me 262A-2 fighter-bomber, the Me 262A-5 reconnaissance fighter, the Me 262B-1a two-seat conversion trainer and the Me 262B-2 night-fighter. Many variants were under consideration or development at the end of the war.

MESSERSCHMITT Me 262A-1a SCHWALBE

Role: Fighter
Crew/Accommodation: One
Power Plant: Two 900kgp (1,984 lb s.t.) Junkers Jumo-004B turbojets
Dimensions: Span 12.5m (41.01 ft); length 10.605m (34.79 ft); wing area 21.68m² (233.3 sq ft)
Weights: Empty 4,000kg (8,820 lb); MTOW 6,775kg (14,938 lb)
Performance: Maximum speed 868km/h (536 mph) at 7,000m (22,800 ft); operational ceiling 11,000m (36,080 ft); range 845km (524 miles) at 6,000mm (19,685 ft) cruise altitude
Load: Four 30mm cannon

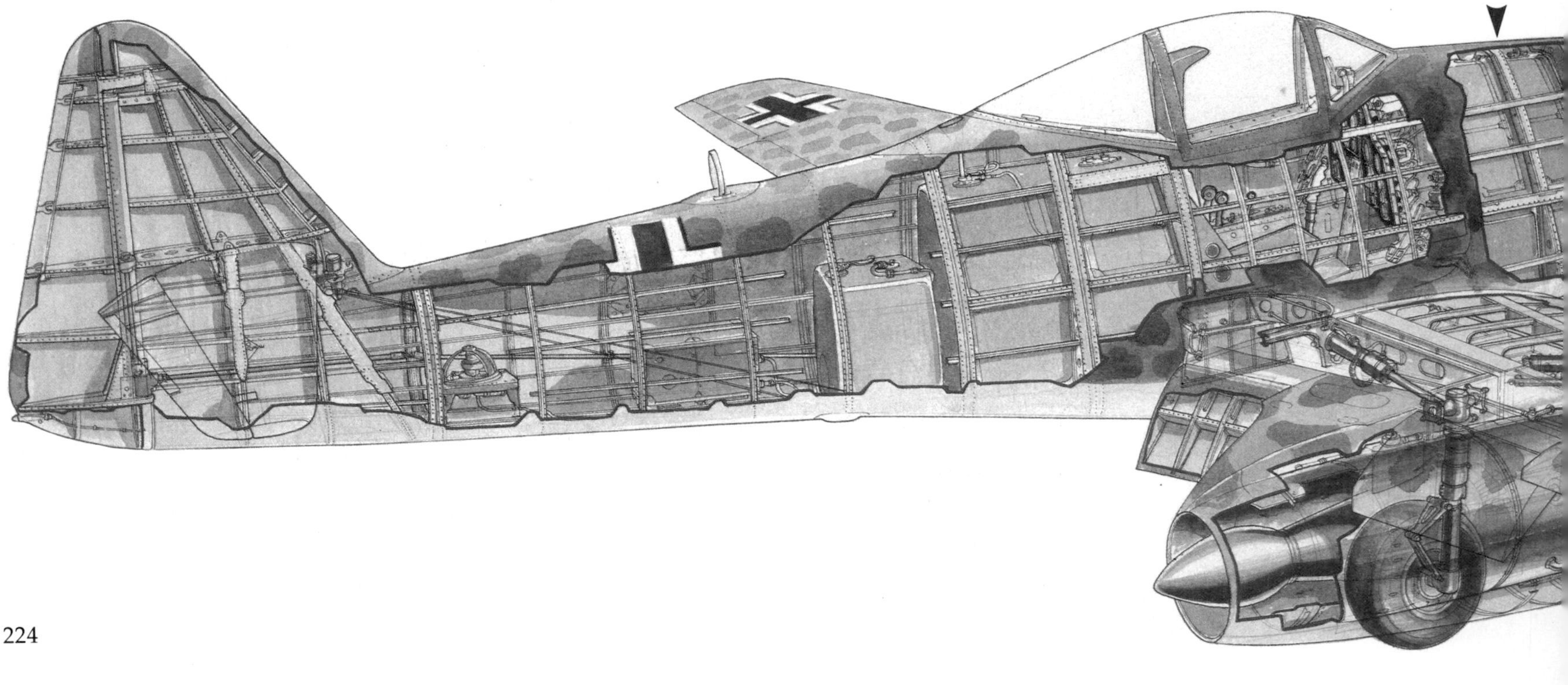

Messerschmitt Me321 Gigant

MESSERSCHMITT Me 321 and Me 323 GIGANT (GIANT)
(Germany)

The Me 321 was developed as a heavy transport glider able to accommodate light vehicles when towed by the Heinkel He 111Z tug (two He 111 twin-engined bombers joined by a common centre section fitted with a fifth engine). The type first flew in February 1941, and 200 aircraft were produced in Me 321A-1 unarmed and Me 321B-1 armed versions. The type also offered possibilities as a powered aeroplane, and two Me 321s were converted as prototypes with four and six Gnome-Rhone 14N radial engines, the latter arrangement being selected for the Me 323D production model. Some 211 were built in the baseline Me 323D and more heavily armed Me 323E models.

MESSERSCHMITT Me 323D-1 GIGANT

Role: Heavy military transport
Crew/Accommodation: Five, plus up to 130 troops
Power Plant: Six 1,140 hp Gnome Rhone 14N air-cooled radials, plus 500kgp (1,102 lb s.t.) Walter 109-500 rockets needed to assist in high weight take-offs
Dimensions: Span 55m (180.45 ft); length 28.15m (92.36 ft); wing area 300m^2 (3,228 sq ft)
Weights: Empty 27,330kg (60,252 lb); MTOW 43,000kg (94,799 lb)
Performance: Cruise speed 218km/h (136 mph) at sea level; operational ceiling 4,000m (13,123 ft); range 1,100km (683 miles) with full payload
Load: Up to fifteen 7.9mm machine guns, plus up to 13,000kg (28,660 lb) of cargo

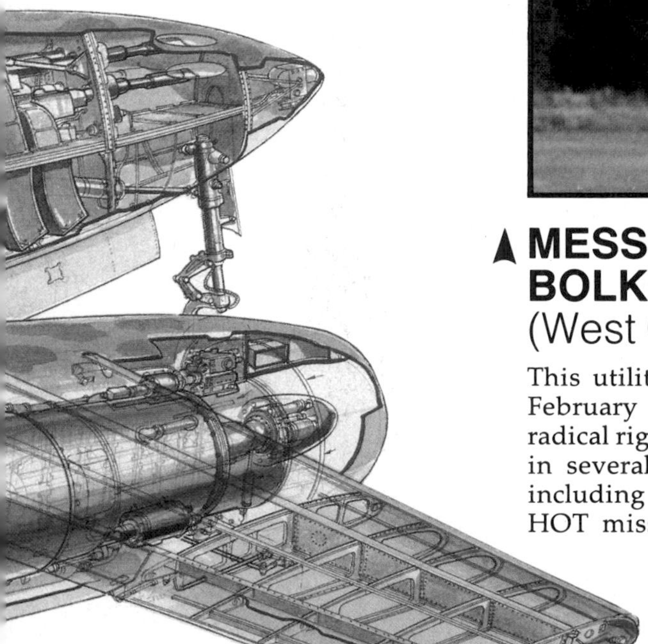

▲ MESSERSCHMITT-BOLKOW-BLOHM BO 105
(West Germany)

This utility light helicopter was first flown in February 1967, and was notable for its then-radical rigid rotor. The type has been developed in several civil and military forms, the latter including the PAH-1 anti-tank type with six HOT missiles.

MBB BO 105

Role: Light utility transport helicopter
Crew/Accommodation: One, plus up to four passengers
Power Plant: Two 420 shp Allison 250-C20B turboshafts
Dimensions: Span 11.86m (38.9 ft); rotor length 9.84m (32.3 ft)
Weights: Empty 1,388kg (3,060 lb); MTOW 2,400kg (5,291 lb)
Performance: Maximum speed 270km/h (145 knots) at sea level; operational ceiling 5,180m (17,000 ft); range 575km (310 naut. miles) with 200kg (441 lb) payload
Load: Up to 451kg (995 lb)

MIKOYAN-GUREVICH MiG-3 (U.S.S.R.)

As the starting point for a new interceptor, the MiG team produced I-65 and I-61 design concepts, the latter in variants with the Mikulin AM-35A and AM-37 inline engines. The I-61 was deemed superior and ordered in I-200 prototype form. This first flew in April 1940 and proved to have excellent speed with the AM-35A. the type was ordered into production as the MiG-1, but as range and strength were minimal, only 100 were delivered before the MiG-1 was superseded by the strengthened and aerodynamically refined MiG-3. Some 3,322 such aircraft were built, but these saw only limited use as they were completely outclassed by German fighters in the vital low/medium-altitude combat band. ▼

MIKOYAN-GUREVICH MiG-3

Role: Fighter
Crew/Accommodation: One
Power Plant: One 1,350 hp Mikulin AM-35A water-cooled in-line
Dimensions: Span 10.3m (33.79 ft); length 8.15m (26.74 ft); wing area 17.44m^2 (187.7 sq ft)
Weights: Empty 2,595kg (5,720 lb); MTOW 3,285kg (7,242 lb)
Performance: Maximum speed 640km/h (398 mph) at 7,000m (22,965 ft); operational ceiling 12,000m (39,370 ft); range 820km (510 miles) with full warload
Load: Three 12.7mm machine guns, plus up to 200kg (441 lb) of bombs or rockets

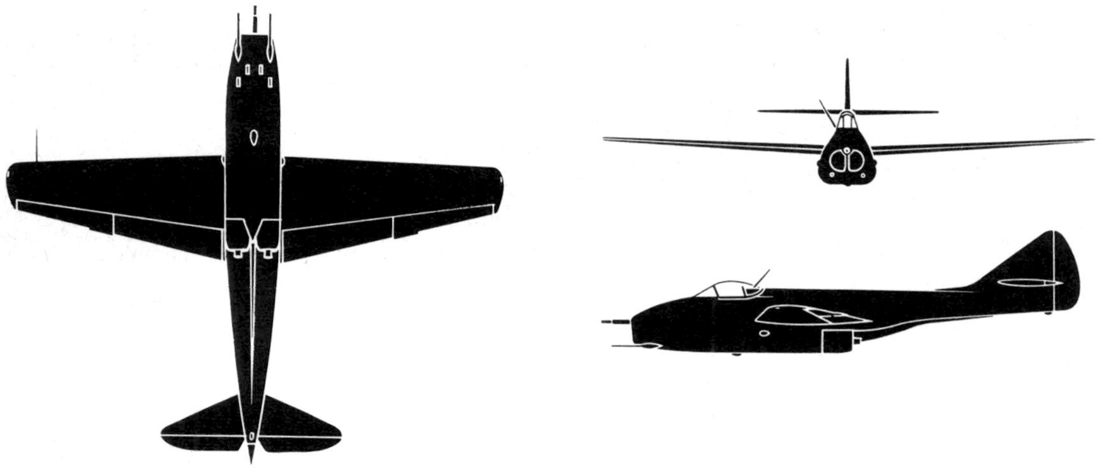

▲ MIKOYAN-GUREVICH MiG-9 (U.S.S.R.)

This was the production fighter that emerged from trials with the I-300 prototype, a type that drew heavily on German aerodynamic research in World War II and was powered by two examples of the RD-20 turbojet derived from the German BMW 003. The I-300 first flew in April 1946, and about 1,000 MiG-9s followed between late 1946 and 1948 to give the U.S.S.R. its first jet-powered MiG fighter.

MIKOYAN-GUREVICH MiG-9FR 'FARGO'

Role: Fighter
Crew/Accommodation: One
Power Plant: Two 1,000kgp (2,205 lb s.t) RD-21 turbojets
Dimensions: Span 10m (32.81 ft); length 9.75m (31.99 ft); wing area 18.2m^2 (195.9 sq ft)
Weights: Empty 3,570kg (7,870 lb); MTOW 5,070kg (11,177 lb)
Performance: Maximum speed 965km/h (521 knots) at 5,000m (16,404 ft); operational ceiling 13,000m (42,651 ft); range 800km (432 naut. miles)
Load: One 37mm and two 23mm cannon

MIKOYAN-GUREVICH MiG-15 and MiG-17
(U.S.S.R.)

The MiG-15 was the North American F-86 Sabre's main opponent in the Korean War, and was the production version of the I-310 prototype that first flew in late 1947. The MiG-15 was the U.S.S.R.'s first mass-production swept-wing fighter, its major variant being the improved MiG-15bis that could outclimb and out-turn the Sabre in most flight regimes. Many thousands of the series were produced, most of them as standard day fighters, but small numbers as MiG-15P all-weather fighters and Mig-15SB fighter-bombers. There was also an important MiG-15UTI tandem-seat advanced and conversion trainer, and licensed production was undertaken in Czechoslovakia and Poland of the S.102 and LIM variants.

The MiG-17 was the production version of the I-330 prototype developed to eliminate the MiG-15's tendency to snap-roll into an uncontrollable spin during a high-speed turn: a new wing of 45 degrees rather than 35 degrees sweep was introduced together with more power, a longer fuselage and a revised empennage. Several thousand aircraft were delivered from 1952 onwards in variants such as the MiG-17 day fighter, MiG-17F improved day fighter with the VK-1F afterburning engine, MiG-17PF limited all-weather fighter, MiG-17F improved day fighter, and MiG-17PFU missile-armed fighter. The type was also built in China, Czechoslovakia and Poland with the designations J-5, S.104 and LIM-5/6 respectively.

MIKOYAN-GUREVICH MiG-17PF 'FRESCO-D'

Role: Fighter
Crew/Accommodation: One
Power Plant: One 3,380kgp 7,452 lb s.t.) Klimov VK/1FA turbojet with reheat
Dimensions: Span 9.63m (31.59 ft); length 11.26m (36.94 ft); wing area 22.6m² (243.26 sq ft)
Weights: Empty 4,182kg (9,220 lb); MTOW 6,330kg (13,955 lb)
Performance: Maximum speed 1,074km/h (667 mph) at 4,000m (13,123 ft); operational ceiling 15,850m (52,001 ft); range 360km (224 miles) with full warload
Load: Three 23mm cannon, plus up to 500 kg (1,102 lb) of bombs or unguided rockets

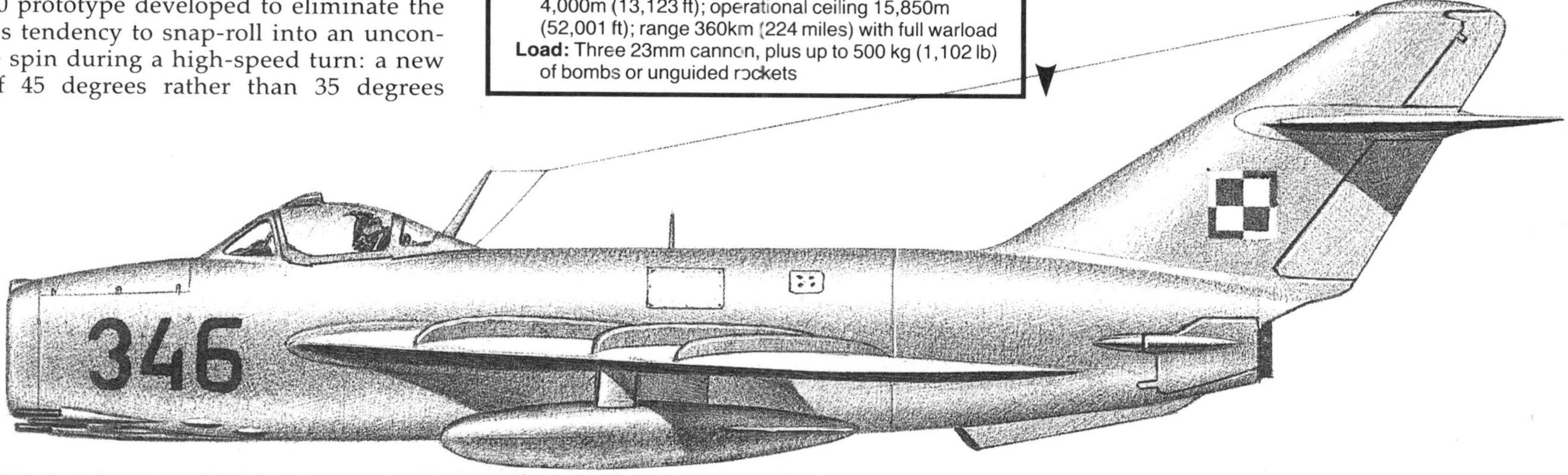

MIKOYAN-GUREVICH MiG-19
(U.S.S.R.)

▲ Just as the Mig-15 and North American F-86 Sabre were contemporaries, the MiG-19 vied with the North American F-100 Super Sabre for the honour of becoming the world's first genuinely supersonic fighter. There remains some doubt as to which fighter entered definitive service first, but there is no doubt that the Soviet fighter was an altogether more remarkable design with greater long-term potential. The first prototype flew in September 1953 with two Mikulin AM-5 afterburning turbojets and the type began to enter service as the MiG-19P. This suffered severe control difficulties at transonic speed, and was rapidly supplanted in production by the MiG-19S with a slab tailplane and Tumanskii RD-9B (renamed AM-5) turbojets. The MiG-19S entered service in mid-1955, and production of several thousands of aircraft produced the MiG-19SF improved fighter, the MiG-19PF all-weather fighter and the MiG-19PM missile-equipped fighter. Czech and Polish production produced the S.105 and LIM-7 versions, and China produces the excellent J-6 series of MiG-19 derivatives.

MIKOYAN-GUREVICH MiG-19S 'FARMER-C'

Role: Strike fighter
Crew/Accommodation: One
Power Plant: Two 3,250kgp (7,165 lb s.t.) Klimov RD-9B turbojets with reheat
Dimensions: Span 9m (29.53 ft); length 12.6m (41.34 ft); wing area 25m² (269.1 sq ft)
Weights: Empty 5,172kg (11,402 lb); MTOW 8,900kg (19,621 lb)
Performance: Maximum speed 1,450km/h (901 mph) Mach 1.33 at 10,000m (32,808 ft); operational ceiling 17,500m (57,415 ft); range 685km (426 miles) with full bombload
Load: Three 30mm cannon, plus up to two short-range air-to-air missiles or 500kg (1,102 lb) of bombs

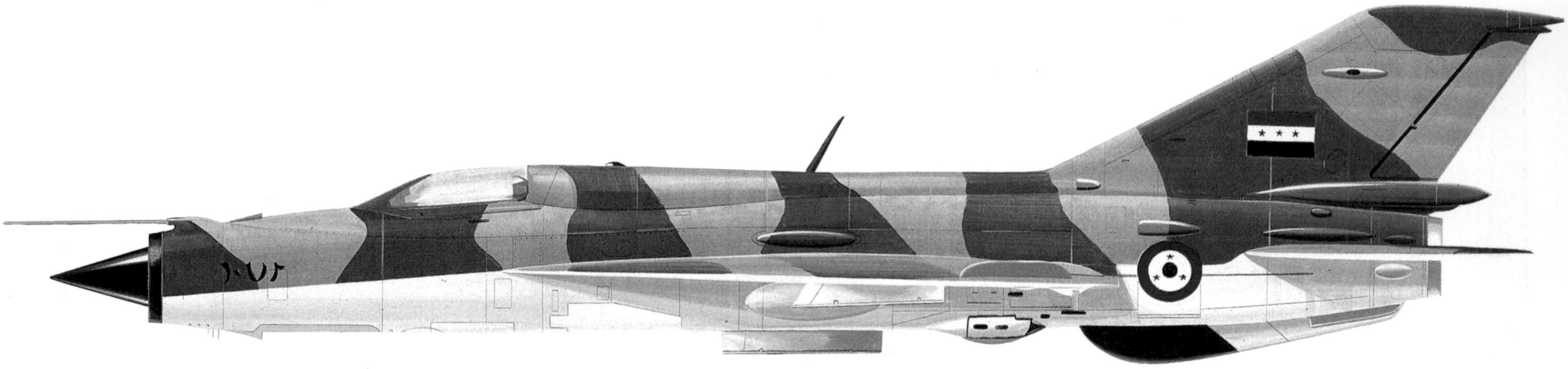

MIKOYAN-GUREVICH MiG-21
(U.S.S.R.)

The MiG-21 was designed after the U.S.S.R. had digested the implications of the Korean War to provide a short-range interceptor. The type was analogous to the Lockheed F-104 Starfighter in rationale, but was a radically different aeroplane based on a tailed delta configuration and light weight to ensure adequate performance on a relatively low-powered afterburning turbojet, the Tumanskii R-11. Differently configured Ye-2A and Ye-5 prototypes were flown in 1956, the latter paving the way for the definitive Ye-6 prototype that flew in 1957.

More than 11,000 examples of the MiG-21 were produced in variants such as the MiG-21 clear-weather interceptor, MiG-21PF limited all-weather fighter with search and track radar, MiG-21PFS fighter with blown flaps and provision for RATO units, MiG-21FL export version of the MiG-12PFS but without blown flaps or RATO provision, the MiG-21PFM improved version of the MiG-21PFS, MiG-21PFMA second-generation dual-role fighter with a larger dorsal hump and four rather than two underwing hardpoints, MiG-21M export version of the MiG-21PFMA, MiG-21R tactical reconnaissance version, MiG-21MF with the more powerful but lighter R-13-30 engine, MiG-21RF reconnaissance version of the MiG-21MF, MiG-21SMT aerodynamically refined version of the MiG-21MF with increased fuel and ECM capability, MiG-21bis third-generation multi-role fighter, and MiG-21bisF definitive third-generation fighter with a re-engineered airframe, updated electronics and R-25 engine. There have also been three MiG-21U conversion trainer versions.

MIKOYAN-GUREVICH MiG-21SMT 'FISHBED-K'

Role: Strike fighter
Crew/Accommodation: One
Power Plant: One 6,600kgp (14,550 lb s.t.) Tumansky R-13 turbojet with reheat
Dimensions: Span 7.15m (23.46 ft); length 13.46m (44.16 ft); wing area 23m^2 (247.57 sq ft)
Weights: Empty 5,450kg (12,015 lb); MTOW 7,750kg (17,085 lb)
Performance: Maximum speed 2,230km/h (1,386 mph) Mach 2.1 at 12,000m (39,370 ft); operational ceiling 18,000m (59,055 ft); radius 500km (311 miles) with full warload
Load: Two 23mm cannon, plus up to 1,000 kg (2,205 lb) of air-to-air missiles or bombs depending upon mission

MIKOYAN-GUREVICH MiG-23
(U.S.S.R.)

First flown in prototype form during 1966, the MiG-23 marked a radical development in the company's design philosophy by introducing variable-geometry wings to reconcile the apparently incompatible features of good field and cruise performance with high dash speed. The type was developed as an air-combat fighter with useful attack capability, and began to enter service in 1970. The type has been extensively developed in terms of its electronics and overall capabilities, the latest variant being that known in the West only by its NATO reporting designation 'Flogger-K'.

MIKOYAN-GUREVICH MiG-23S 'FLOGGER-B'

Role: All-weather fighter with variable-geometry wing
Crew/Accommodation: One
Power Plant: One 11,500kgp (25,350 lb s.t.) Tumansky R-29 B turbofan with reheat
Dimensions: Span spread 14.25m swept 8.17m, (spread 46.75 ft; swept 26.8 ft); length 18.15m (59.55 ft); wing area 27.26m^2 (293.42 sq ft)
Weights: Empty 8,000kg (17,637 lb); MTOW 20,100kg (44,312 lb)
Performance: Maximum speed 1,350km/h (839 mph) Mach 1.1 at 305m (1,001 ft); operational ceiling 18,000m (59,055 ft); range 850km (528 miles) with full missile load
Load: One 23mm cannon, plus two medium-range and two short-range air-to-air missiles

Mikoyan MiG-31 'Foxhound' ▼

Mikoyan MiG-25 'Foxbat' ▲

MIKOYAN-GUREVICH MiG-25 and MiG-31
(U.S.S.R.)

The MiG-25 was designed to provide the Soviets with an interceptor capable of dealing with the U.S.A.'s incredible North American B-70 Valkyrie Mach 3 high-altitude strategic bomber. But when the B-70 was cancelled the Soviets continued development of this very high-performance interceptor which first flew as the Ye-266 in 1964. The type is built largely of stainless steel with titanium leading edges to deal with friction-generated heat at Mach 3 dash speeds, but at such speeds is virtually incapable of manoeuvre.

The type entered service in the late 1960s, and variants are the 'Foxbat-A' interceptor with four air-to-air missiles, the 'Foxbat-B' operational-level reconnaissance aeroplane, the 'Foxbat-C' two-seat conversion trainer and the 'Foxbat-D' improved reconnaissance aeroplane. The designation 'Foxbat-E' has been given by NATO to 'Foxbat-A' interceptors converted to deal with low-level intruders such as penetration bombers and cruise missiles.

The MiG-31 is a development of the MiG-25 with a more effective look-down/shoot-down radar and missile system to destroy low-level attackers, and its maximum speed is a 'mere' Mach 2.4.

MIKOYAN-GUREVICH MiG-25 'FOXBAT-A'

Role: High speed high altitude interceptor
Crew/Accommodation: One
Power Plant: Two 12,250 kgp (27,010 lb s.t.) Tumansky R-31 turbojets with reheat
Dimensions: Span 13.95m (45.75 ft); length 23.82m (78.15 ft); wing area 56.83m^2 (611.7 sq ft)
Weights: Empty 20,000kg (44,100 lb); MTOW 37,425kg (82,500 lb)
Performance: Maximum speed 3,006km/h (1,622 knots) at 10,975m (36,000 ft); operational ceiling 24,385m (80,000 ft); range 1,450km (782 naut. miles)
Load: Four medium-range and two short-range air-to-air missiles

MIKOYAN-GUREVICH MiG-27
(U.S.S.R.)

The MiG-27 is the dedicated attack derivative of the MiG-23 with a revised forward fuselage offering heavy armour protection and fitted with terrain-avoidance radar. The MiG-27 has fixed air intakes and a simple on-off afterburner nozzle as it requires only transonic performance in its low-altitude role, but special target-acquisition and weapon-guidance equipment is installed. It has a multi-barrel cannon and additional hardpoints for the larger offensive load.

The two basic variants are the 'Flogger-D' and 'Flogger-J' which differ from each other only in detail. Export aircraft are the MiG-23 'Flogger-F' and 'Flogger-H' combining the airframe of the MiG-23 with the armoured nose of the MiG-27. ➤

MIKOYAN-GUREVICH MiG-27 'FLOGGER-J'

Role: Strike fighter with variable/geometry wing
Crew/Accommodation: One
Power Plant: One 11,500kgp 25,350 lb s.t.) Tumansky R-29-300 turbofan with reheat
Dimensions: Span spread 14.25m, swept 46.75 ft, spread 46.75 ft, swept 26.8; length 16.5m (54.13 ft); wing area 27.26m^2 (293.42 sq ft)
Weights: Empty 10,818kg (23,849 lb); MTOW 18,000kg (39,683 lb)
Performance: Maximum speed 1,102km/h (594 knots) Mach 0.95 at 305m (1,001 ft); operational ceiling 13,000+m (46,650+ ft); radius 390km (210 naut. miles) with full weapons load
Load: One 23mm cannon, plus up to 3,000kg (6,614 lb) of weapons

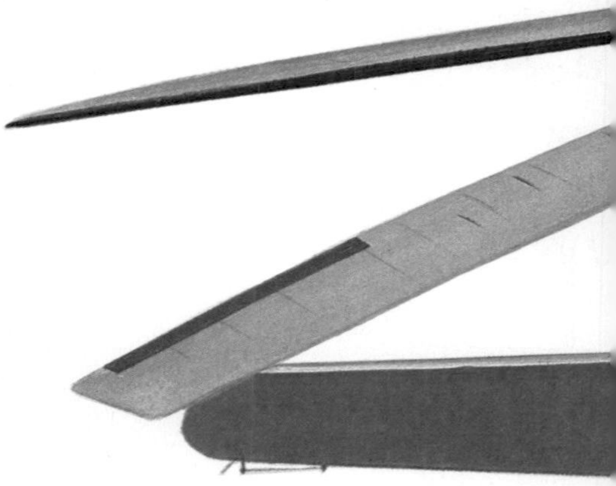

MIKOYAN-GUREVICH MiG-29 ▲
(U.S.S.R.)

Deployed in about 1986, the MiG-29 is the U.S.S.R.'s latest air-combat fighter, a singularly advanced type with blended aerodynamics, a modern engine and an advanced weapon system combining the inputs of several sensors in a good cockpit so that the pilot can extract maximum capability from his several air-to-air missile types. The fighter has been developed in single- and two-seat variants, and clearly possesses very significant development potential.

MIKOYAN-GUREVICH MiG-29 'FULCRUM-A'

Role: Fighter
Crew/Accommodation: One
Power Plant: Two 8,300kgp (18,300 lb s.t.) Tumansky R-33D turbofans with reheat
Dimensions: Span 11.36m (37.27 ft); length 17.32m (56.83 ft); wing area 35.2m^2 (378.9 sq ft)
Weights: Empty 8,175kg (18,025 lb); MTOW 18,000kg (39,700 lb)
Performance: Maximum speed 2,440km/h (1,320 knots) Mach 2.3 at 11,000m (36,090 ft); operational ceiling 17,000m (56,000 ft); range 2,100km (1,130 naut. miles)
Load: One 30mm cannon, plus two medium-range and four short-range air-to-air missiles

MIL Mi-4
(U.S.S.R.)

This large but essentially conventional helicopter first flew in May 1952, and production of about 3,500 piston-engined machines was completed by 1969. The type serves in a number of civil and generally unarmed military transport tasks, and has also been built in the Republic of China as the Zhi-5.

MIL Mi-4 'HOUND-A'

Role: Military troop/utiity transport helicopter
Crew/Accommodation: Three, plus up to 14 troops
Power Plant: One 1,700 hp Shvetsov ASh-82V air-cooled radial
Dimensions: Rotor diameter 21m (68.92 ft)
Weights: Empty 5,268kg (11,650 lb); MTOW 7,800kg (17,200 lb)
Performance: Cruise speed 160km/h (99 mph) at sea level; operational ceiling 4,200m (13,780 ft); range 595km (370 miles) with full payload
Load: Up to 1,600kg (3,525 lb)

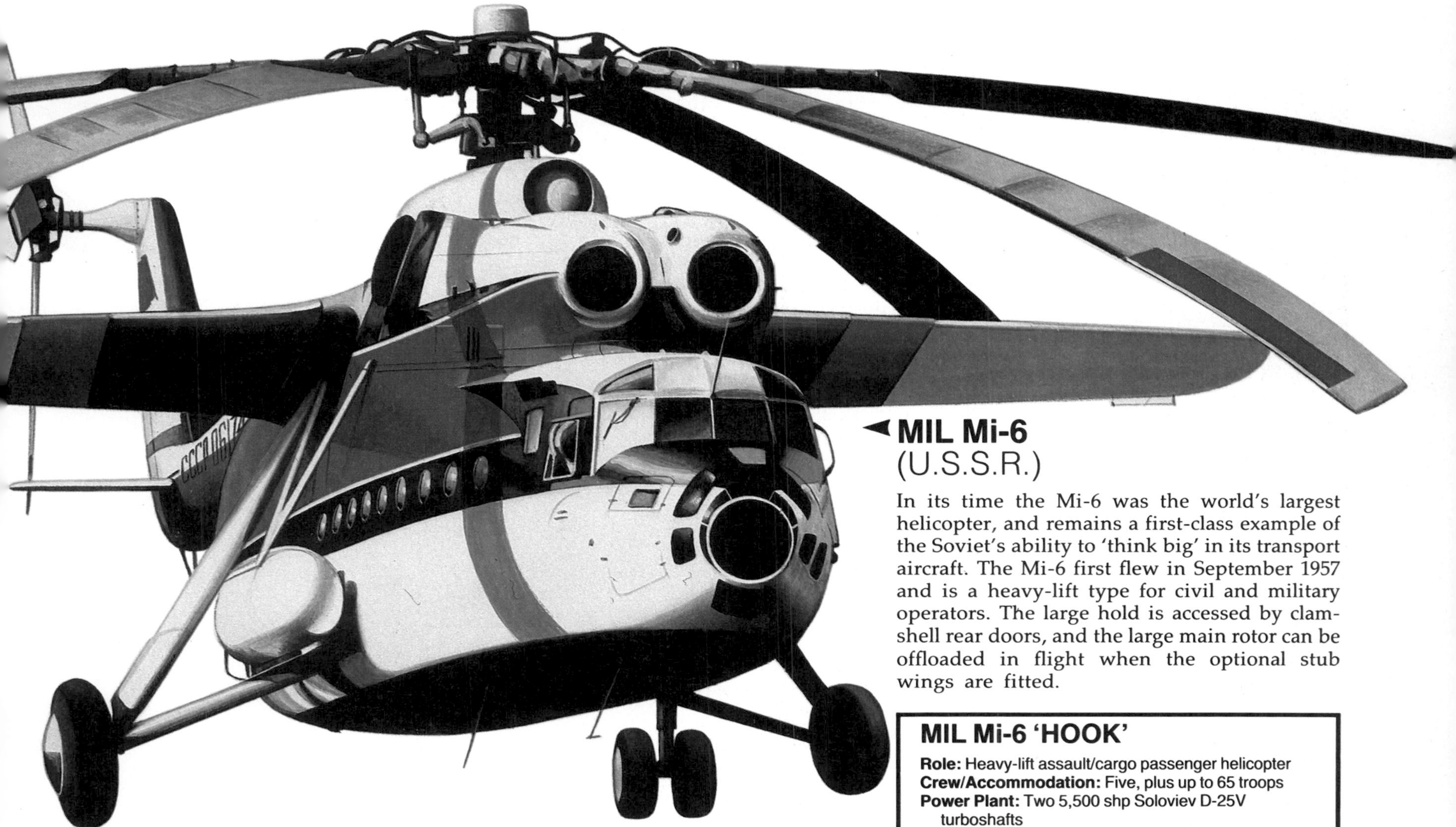

MIL Mi-6
(U.S.S.R.)

In its time the Mi-6 was the world's largest helicopter, and remains a first-class example of the Soviet's ability to 'think big' in its transport aircraft. The Mi-6 first flew in September 1957 and is a heavy-lift type for civil and military operators. The large hold is accessed by clamshell rear doors, and the large main rotor can be offloaded in flight when the optional stub wings are fitted.

MIL Mi-6 'HOOK'

Role: Heavy-lift assault/cargo passenger helicopter
Crew/Accommodation: Five, plus up to 65 troops
Power Plant: Two 5,500 shp Soloviev D-25V turboshafts
Dimensions: Overall length rotors turning 41.74m (136.94 ft); rotor diameter 35m (114.83 ft)
Weights: Empty 27,240kg (60,055 lb); MTOW 42,500kg (93,700 lb)
Performance: Cruise speed 250km/h (155 mph) at sea level; operational ceiling 4,500m (14,750 ft); range 620km (385 miles) with 8,000kg (17,637 lb) payload
Load: Up to 20,117kg (44,350 lb)

MIL Mi-8 and Mi-17
(U.S.S.R.)

The Mi-8 is in essence a turbine-powered version of the Mi-4, and though first flown with a single Soloviev turboshaft driving a four-blade main rotor was then revised with two Isotov turboshafts driving a five-blade main rotor. The greater lifting power of the Mi-8 permitted the incorporation of a cabin considerably larger than that of the Mi-4, and the type has found several civil applications. The Mi-8 is best known as a military type, however, in assault transport ('Hip-C'), armed support ('Hip-E'), anti-tank ('Hip-F') and electronic warfare ('Hip-D, G, J and K') versions.

The Mi-17 ('Hip-H') is a relatively simple upgrade of the Mi-8, with 1,417kW (1,900 shp) TV3-117MT turboshafts for better performance in 'hot-and-high' conditions.

MIL Mi-8 'HIP'

Role: Military and commercial passenger transport helicopter
Crew/Accommodation: Two/three, plus up to 28 passengers or 24 troops
Power Plant: Two 1,500 shp Isotov TB-2-117 turboshafts
Dimensions: Overall length rotors turning 25.28m (82.94 ft); rotor diameter 21.29m (69.85 ft)
Weights: Empty 7,161kg (15,787 lb); MTOW 12,000kg (26,455 lb)
Performance: Cruise speed 200km/h (124 mph) at sea level; operational ceiling 4,000m (13,125 ft); range 425km (264 miles) with 3,000kg (6,614 lb) payload
Load: Up to 4,000kg (8,820 lb)

MIL Mi-14
(U.S.S.R.)

The Mi-14 is a land-based anti-submarine helicopter based on the Mi-8, but with the dynamic system of the Mi-17. The type has a boat hull and stabilizing floats that also accommodate the retracted main landing gear units. The three variants are the 'Haze-A' anti-submarine, 'Haze-B' mine countermeasures and 'Haze-C' search and rescue helicopters.

MIL Mi-14 'HAZE-A'

Role: Land-based anti-submarine helicopter
Crew/Accommodation: Four/five
Power Plant: Two 1,900 shp Isotov TV-3-117 turboshafts
Dimensions: Overall length rotors turning 25.3m (83 ft); rotor diameter 21.29m (69.85 ft)
Weights: Empty – not available; MTOW 13,000kg (28,660 lb)
Performance: Cruise speed 200km/h (108 knots) at sea level
Load: not available

MIL Mi-24, Mi-25 and Mi-35
(U.S.S.R)

This series of battlefield, export and improved 'Hind' helicopters, was developed from the mid-1960s and introduced to service in 1974 to provide Soviet and their allied ground forces with a highly capable assault transport ('Hind-A, B and C') helicopter. It was then developed into a large battlefield support helicopter ('Hind-D, E and F') with tandem cockpits for the gunner and pilot, more extensive armament on the stub wings, and a 12.7mm (0.5 inch) four-barrel machine gun in a trainable under-nose barbette replaced in the 'Hind-F' by a 30mm twin barrel cannon on the starboard side of the fuselage. The battlefield support type retains the transport's hold, probably for reload weapons or, in emergencies, an anti-tank missile team.

MIL Mi-24 'HIND-D'

Role: Attack/assault helicopter
Crew/Accommodation: Two, plus up to eight troops
Power Plant: Two 2,200 shp Isotov TV-3-117 turboshafts
Dimensions: Length overall rotors turning 21.5m (70.54 ft); rotor diameter 16.76m (55 ft)
Weights: Empty 8,400kg (18,520 lb); MTOW 10,000kg (22,046 lb)
Performance: Maximum speed 273km/h (170 mph) at 1,000m (3,281 ft); operational ceiling 4,500m (14,750 ft); range 160km (99 miles) with full warload
Load: Up to 1,500 kg (3,300 lb)

MIL Mi-26
(U.S.S.R.)

The Mi-26 is the replacement for the Mi-6, and first flew in December 1977. The Mi-26 has a fuselage similar in configuration and size to that of the Mi-6, but a smaller-diameter eight-blade main rotor driven by twin turboshafts providing nearly twice as much power as those of the Mi-6. The result is good performance with a large payload.

MIL Mi-26 'HALO'

Role: Military heavy-lift assault helicopter
Crew/Accommodation: Five, plus up to 85 troops
Power Plant: Two 11,240 shp Lotarev D-136 turboshafts
Dimensions: Overall length rotors turning 40.025m (131.32 ft); rotor diameter 32m (105 ft)
Weights: Empty 28,200kg (62,170 lb); MTOW 56,000kg (123,450 lb)
Performance: Maximum speed 255km/h (137 knots) at sea level; operational ceiling 4,600m (15,100 ft); range 800 km (432 naut. miles) with 9,000kg (19,841 lb) payload
Load: Up to 20,000kg (44,090 lb)

MILES MAGISTER
(U.K.)

The Magister was the Royal Air Force's most important elementary trainer in the opening stages of World War II, and was also the service's first monoplane trainer. The type was derived from the Hawk Trainer, and deliveries of the type began in May 1937. Production lasted to 1941 and comprised 1,293 in the U.K. and another 100 licence-built in Turkey.

MILES MAGISTER

Role: Elementary trainer
Crew/Accommodation: Two
Power Plant: One 130 hp de Havilland Gipsy Major air-cooled inline
Dimensions: Span 10.31m (33.83 ft); length 7.51m (24.63 ft); wing area 16.35m² (176 sq ft)
Weights: Empty 583kg (1,286 lb); MTOW 862kg (1,900 lb)
Performance: Maximum speed 225km/h (140 mph) at sea level; operational ceiling 5,029m (16,500 ft); range 591km (367 miles)
Load: Up to 109kg (240 lb) including student pilot

Miles Master Mk.II

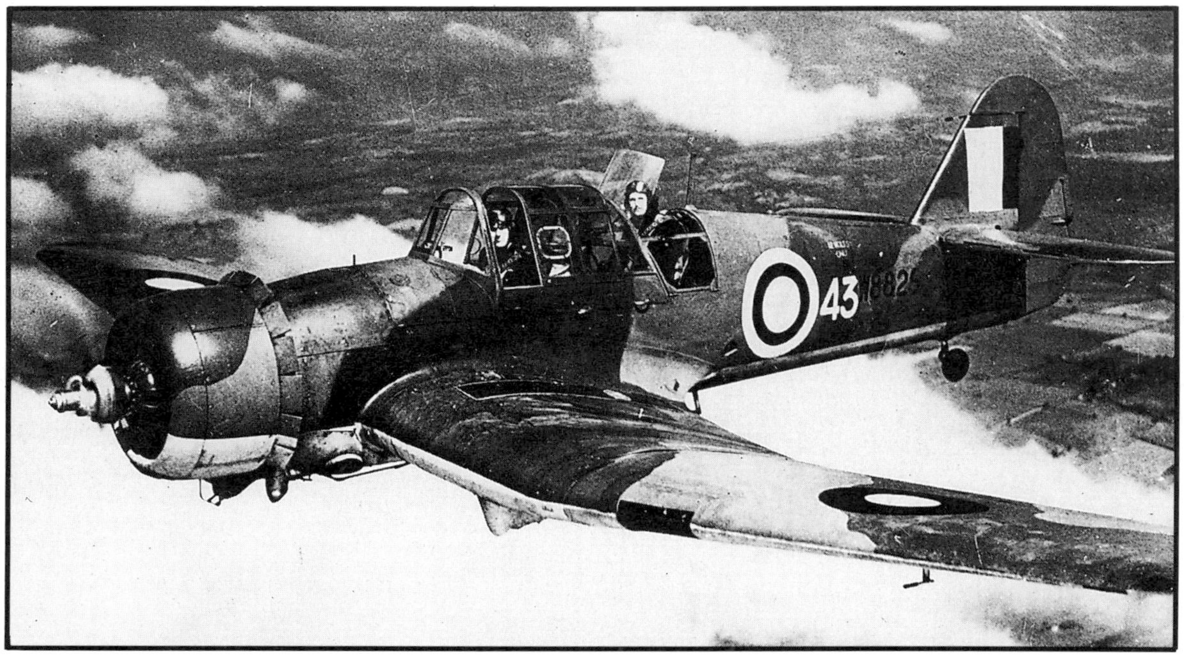

MILES MASTER
(U.K.)

The Master was the advanced trainer counterpart to the Magister with considerably more power and performance in an airframe with retractable landing gear and enclosed accommodation. The type was derived from the private-venture Kestrel prototype that flew in June 1937 with a 556kW (745 hp) Rolls-Royce Kestrel XVI inline engine. Production followed of 900 Master Mk Is with the 533kW (715 hp) Kestrel XXX, 1,747 Master Mk IIs with the 649kW (870 hp) Bristol Mercury XX radial engine, and 602 Master Mk IIIs with the 615kW (825 hp) Pratt & Whitney Wasp Junior radial engine. There were also 26 examples of the extemporized Master Fighter with six 7.7mm (0.303 inch) machine guns.

MILES MASTER Mk I

Role: Advanced trainer
Crew/Accommodation: Two
Power Plant: One 715 hp Rolls-Royce Kestrel XXX water-cooled inline
Dimensions: Span 11.89m (39 ft); length 9.27m (30.42 ft); wing area 30.19m² (325 sq ft)
Weights: Empty 1,982kg (4,370 lb); MTOW 2,528kg (5,573 lb)
Performance: Maximum speed 364km/h (226 mph) at 4,389m (14,400 ft); operational ceiling 8,230m (27,000 ft); range 779km (484 miles)
Load: Up to 201kg (444 lb) including student pilot.
Note: 24 Master Is were converted into single seat fighters with six .303 inch machine guns

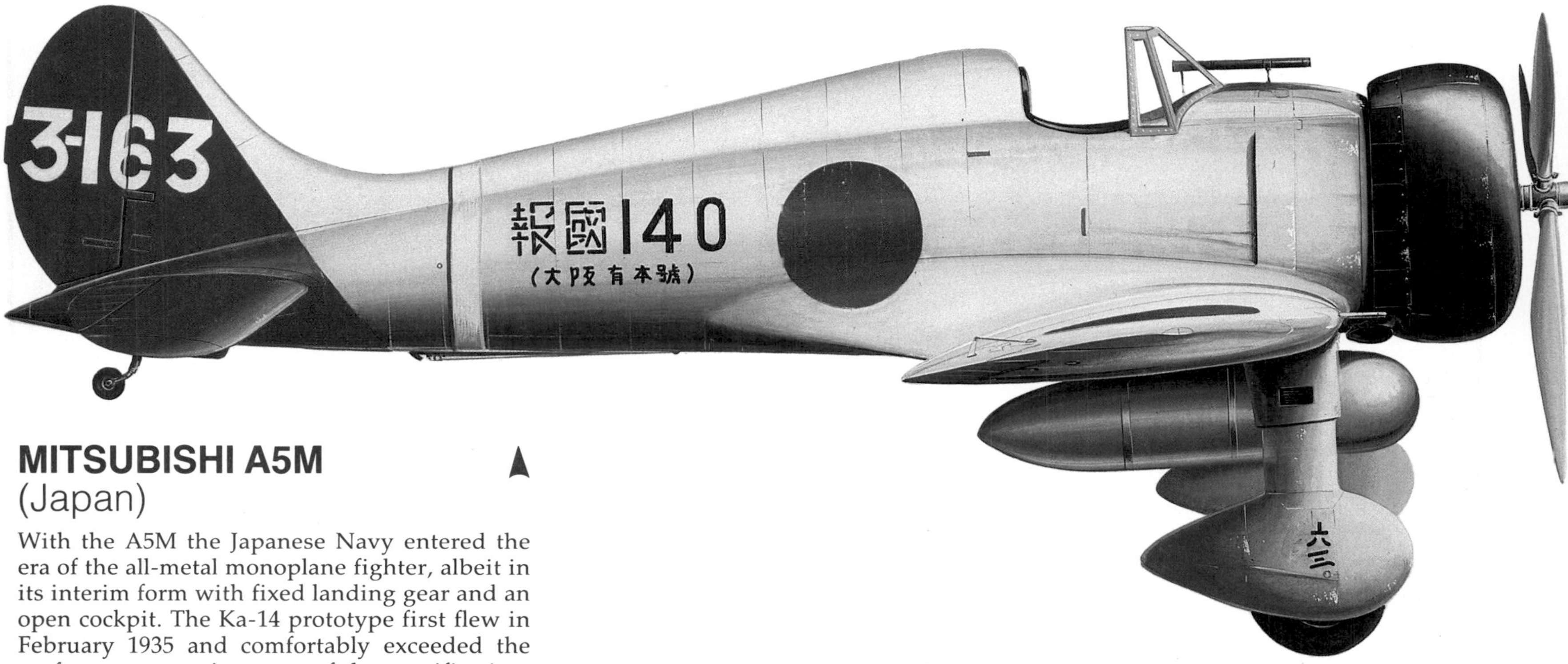

MITSUBISHI A5M

(Japan)

With the A5M the Japanese Navy entered the era of the all-metal monoplane fighter, albeit in its interim form with fixed landing gear and an open cockpit. The Ka-14 prototype first flew in February 1935 and comfortably exceeded the performance requirements of the specification, though handling left much to be desired. The type was therefore extensively modified before reappearing as the A5M1 production model with the 436kW (535 hp) Nakajima Kotobuki 2-KAI-1 radial engine. Production of the series totalled 788 from Mutsubishi, 264 from the Omura Naval Arsenal and 39 from Watanabe. Variants after the A5M1 were the A5M2a and b with the 455 and 477kW (610 and 640 hp) Kotobuki 2-KAI-3 and Kotobuki 3 radial engines respectively, the A5M4 with the 529kW (710 hp) Kotobuki 41 and the A5M4-K two-seat trainer. By 1942 surviving aircraft had been relegated to second-line duties.

MITSUBISHI A5M4 'CLAUDE'

Role: Naval carrierborne fighter
Crew/Accommodation: One
Power Plant: One 710 hp Nakajima Kotobuki 41 air-cooled radial
Dimensions: Span 11m (36.09 ft); length 7.57m (24.84 ft); wing area 17.8m^2 (191.6 sq ft)
Weights: Empty 1,216kg (2,681 lb); MTOW 1,671kg (3,684 lb)
Performance: Maximum speed 434km/h (270 mph) at 3,000m (9,840 ft); operational ceiling 9,800m (32,150 ft); range 1,200km (746 miles)
Load: Two 7.7mm machine guns, plus up to 60kg (132 lb) of bombs

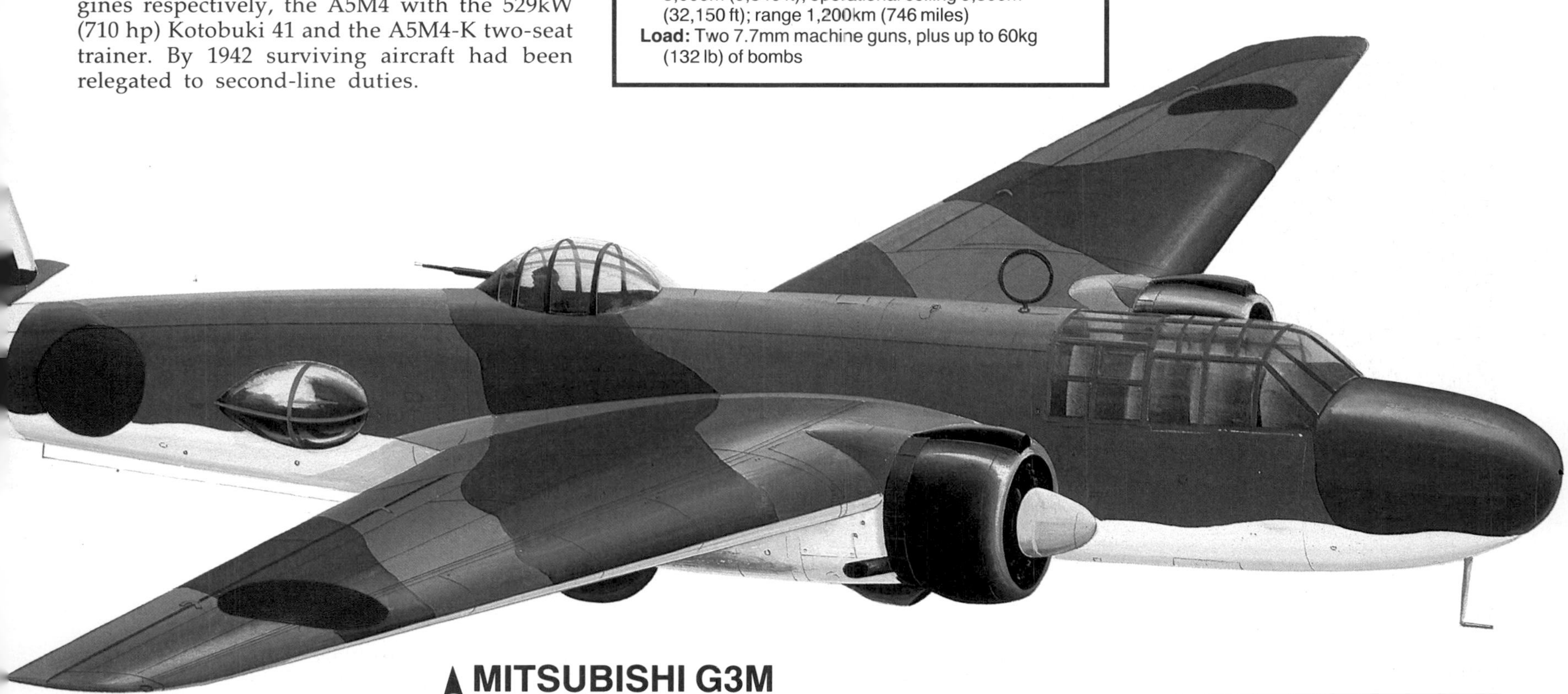

MITSUBISHI G3M

(Japan)

From its Ka-15 bomber/transport prototype of 1935, Mitsubishi evolved the G3M1 bomber with 679kW (910 hp) Mitsubishi Kinsei 3 radial engines. This was the Japanese Navy's most important bomber of the period up to 1942. Included in the overall production total of 1,048 were the prototypes, the G3M1, the G3M2 with 802kW (1,075 hp) Kinsei 42 or 43s and greater fuel capacity, and the G3M3 with 969kW (1,300 hp) Kinsei 51 engines. Some of the bombers were converted as transports with the designations G3M1-L, L3Y1 and L3Y2.

MITSUBISHI G3M2 'NELL'

Role: Bomber
Crew/Accommodation: Seven
Power Plant: Two 1,075 hp Mitsubishi Kinsei 42 air-cooled radials.
Dimensions: Span 25m (82.03 ft); length 16.45m (53.96 ft); wing area 75m^2 (807.3 sq ft)
Weights: Empty 5,194kg (11,446 lb); MTOW 10,300kg (22,708 lb)
Performance: Maximum speed 380km/h (236 mph) at 4,180m (13,715 ft); operational ceiling 9,130m (29,950 ft); range 1,450km (901 miles) with full warload
Load: One 20mm cannon and four 7.7mm machine guns, plus up to 1,000kg (2,205 lb) of bombs

MITSUBISHI Ki-21
(Japan)

The Ki-21 was one of the Japanese Army's most important bombers of World War II, and though classified by its operator as a heavy type was in Allied terms little more than a light bomber. The Ki-21-I prototype first flew with the Nakajima Ha-5 radial engine and proved to have excellent performance. The type entered service in 1938 as the Ki-21-Ia. Total production was 2,064 in three Ki-21-I subvariants and two Ki-21-II subvariants with the 1,119kW (1,500 hp) Mitsubishi Ha-101 radial engine for much improved performance.

MITSUBISHI Ki-21-II 'SALLY'

Role: Bomber
Crew/Accommodation: Seven
Power Plant: Two 1,500 hp Mitsubishi Ha101 air-cooled radials
Dimensions: Span 22.5m (73.82 ft); length 16m (52.49 ft); wing area 69.6m^2 (749.1 sq ft)
Weights: Empty 6,070kg (13,382 lb); MTOW 8,710kg (21,407 lb)
Performance: Maximum speed 478km/h (297 mph) at 4,000m (13,123 ft); operational ceiling 10,000m (32,808 ft); range 1,150km (715 miles) with full bombload
Load: One 12.7mm and five 7.7mm machine guns, plus up to 1,000kg (2,205 lb) of bombs

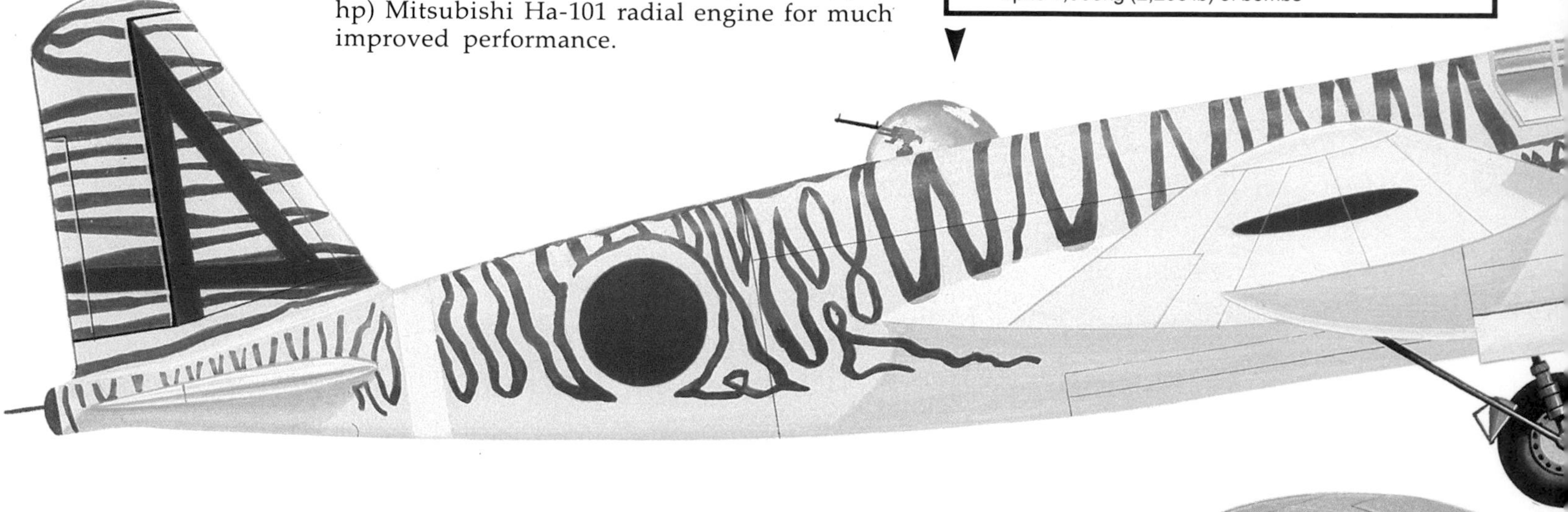

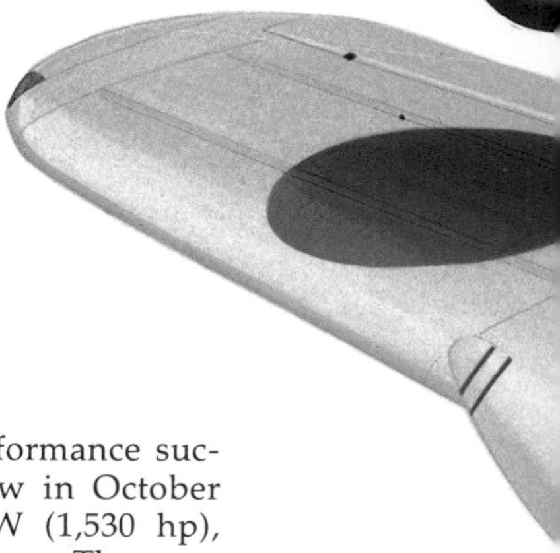

MITSUBISHI G4M
(Japan)

This was schemed as a high-performance successor to the G3M, and first flew in October 1939 as the G4M1 with 1,141kW (1,530 hp), Mitsubishi Kasei 11 radial engines. The type had the required performance, but was notably lacking in protection and such vital features as self-sealing fuel tanks. Production of the series totalled 2,446 and included the initial production G4M1, the G4M2 with 1,342kW (1,800 hp) Kasei 21s and progressively improved armament in several subvariants, the G4M2a with 1,361kW (1,825 hp) Kasei 25s and the final G4M3 with armour protection and the essential self-sealing fuel tanks.

MITSUBISHI G4M2a 'BETTY'

Role: Bomber
Crew/Accommodation: Seven
Power Plant: Two 1,825 hp Mitsubishi MK4T Kasei 25 air-cooled radials
Dimensions: Span 25m (82.02 ft); length 20m (65.62 ft); wing area 78.125m^2 (841 sq ft)
Weights: Empty 8,390kg (18,499 lb); MTOW 15,000kg (33,070 lb)
Performance: Maximum speed 438km/h (272 mph) at 4,600m (15,090 ft); operational ceiling 8,950m (29,365 ft); range 3,640km (2,262 miles) with full bombload
Load: Two 20mm cannon and four 7.7mm machine guns, plus up to 1,000kg (2,205 lb) of bombs, or one heavyweight torpedo

MITSUBISHI A6M REISEN (ZERO FIGHTER)
(Japan)

This will remain Japan's best known aeroplane of World War II, and was in its early days without doubt the finest carrierborne fighter anywhere in the world. The Zero was the first naval fighter able to deal on equal terms with the best of land-based fighters, and was notable for its heavy firepower combined with good performance, great range and considerable agility. Such a combination could only be achieved with a lightweight and virtually unprotected airframe, and this meant that from 1943 the Zero could not be developed effectively to maintain it as a competitive fighter. The A6M1 prototype first flew in April 1939 with a 582kW (780 hp) Mitsubishi MK2 Zuisei radial engine, and though performance and agility were generally excellent the type was somewhat slower than anticipated. The A6M2 prototype therefore introduced the 690kW (925 hp) Nakajima NK1C Sakae radial engine, and this was retained for the first production series, production of which amounted to 11,283 aircraft to the end of World War II.

Major variants after the A6M2 were the A6M3 with Sakae 21, the A6M5 with improved armament and armour in a, b and c subvariants, the A6M6 with the Sakae 31, and the A6M7 dive-bomber and fighter. There were also a number of experimental and development models as well as the A6M2-N floatplane fighter which was built by Nakajima.

▲

MITSUBISHI A6M5 'ZEKE'

Role: Naval carrierborne fighter
Crew/Accommodation: One
Power Plant: One 1,130 hp Nakajima Sakae 21 air-cooled radial
Dimensions: Span 11m (36.09 ft); length 9.09m (29.82 ft); wing area 21.3m² (229.3 sq ft)
Weights: Empty 1,894kg (4,176 lb); MTOW 2,952kg (6,508 lb)
Performance: Maximum speed 565km/h (351 mph) at 6,000m (19,685 ft); operational ceiling 11,740m (38,517 ft); range 1,570km (976 miles)
Load: Two 20mm cannon and two 7.7mm machine guns

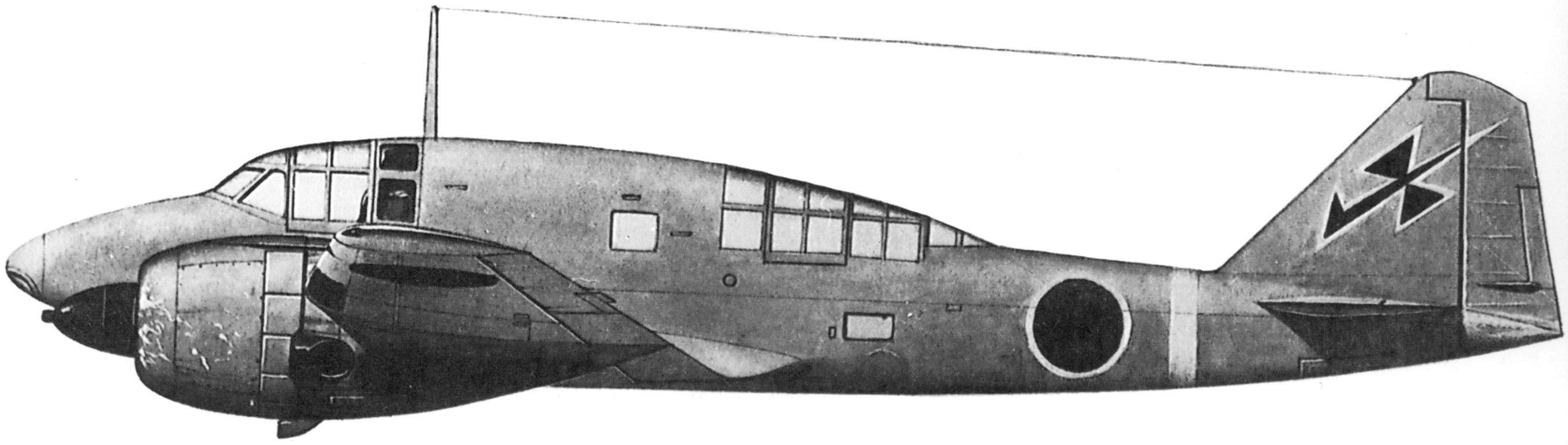

MITSUBISHI Ki-46
(Japan)

This was arguably the finest reconnaissance aeroplane of World War II, designed specifically for the strategic role from 1937. The Ki-46 prototype flew in November 1939 with two 671kW (900 hp) Mitsubishi Ha-21-I radial engines and the Ki-46-I production model was basically similar. The Ki-46-II introduced 805kW (1,080 hp) Mitsubishi Ha-102s for higher performance, and the Ki-46-III introduced an unstepped forward fuselage and 1,119kW (1,500 hp) Mitsubishi Ha-112s for performance that was improved so much that some aircraft were converted to Ki-46-III KAI interceptors and Ki-46-IIIB ground-attack aircraft. Total production was 1,742.

MITSUBISHI Ki-46-I 'DINAH'

Role: Long-range reconnaissance
Crew/Accommodation: Two
Power Plant: Two 900 hp Mitsubishi Ha-26-I air-cooled radials
Dimensions: Span 14.7m (48.23 ft); length 11m (36.09 ft); wing area 32m^2 (344.4 sq ft)
Weights: Empty 3,379kg (7,449 lb); MTOW 4,822kg (10,631 lb)
Performance: Maximum speed 540km/h (336 mph) at 4,070m (13,353 ft); operational ceiling 10,830m (35,531 ft); range 2,100km (1,305 miles)
Load: One 7.7mm machine gun

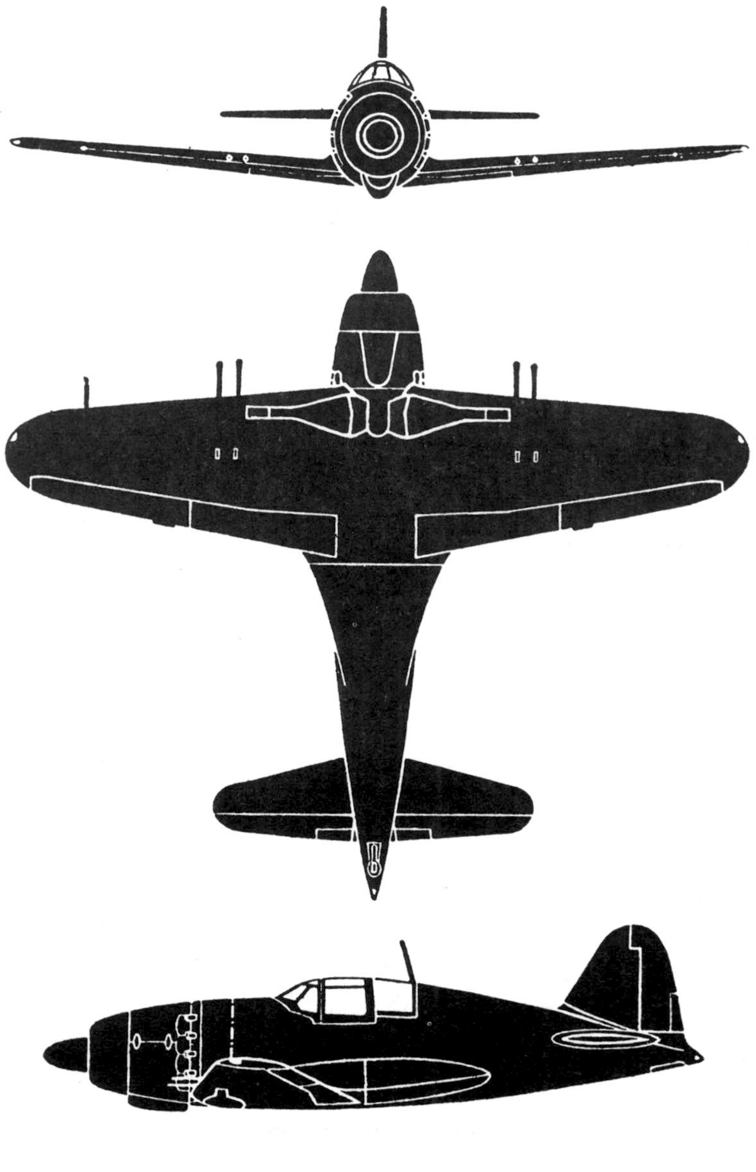

MITSUBISHI J2M RAIDEN (THUNDERBOLT)
(Japan)

The Raiden was designed from 1938 as an interceptor, and the first J2M1 prototype flew in March 1942. Development was protracted because of engine problems: Mitsubishi had been forced to use the 1,066kW (1,430 hp) Mitsubishi Kasei 13 radial engine, and the need to provide a fine entry led to the adoption of a long extension shaft inside a finely tapered nose. Prototype performance was inadequate, so the J2M2 adopted the 1,357kW (1,820 hp) Mitsubishi MK4R-A Kasei 23a radial engine to ensure the required performance. It was December 1943 before the J2M3 production model began to enter service and total production was 476. The J2M3, J2M3a, J2M5 and J2M5a variants were not particularly successful.

MITSUBISHI J2M5 RAIDEN 'JACK'

Role: Interceptor
Crew/Accommodation: One
Power Plant: One 1,820 hp Mitsubishi Kasei 26a air-cooled radial
Dimensions: Span 10.8m (35.43 ft); length 9.95m (32.64 ft); wing area 20.05m^2 (215.8 sq ft)
Weights: Empty 2,839kg (6,259 lb); MTOW 3,482kg (7,676 lb)
Performance: Maximum speed 615km/h (382 mph) at 6,585m (21,604 ft); operational ceiling 11,500m (37,730 ft); range 555km (345 miles)
Load: Four 20mm cannon

MITSUBISHI Ki-67 HIRYU (FLYING DRAGON)
(Japan)

This was designed from 1941 as a tactical heavy bomber, and first flew in December 1942 with 1,417kW (1,900 hp) Mitsubishi Ha-104 radial engines. Production was delayed as the Japanese Army considered a whole range of derivatives based on this high-performance basic aeroplane, but in December 1943 it decided on just the Ki-67-I level and torpedo bomber. Only 698 of these first-class aircraft were built, together with 22 examples of the Ki-109 derivative with a 75mm nose gun in the heavy fighter role.

MITSUBISHI Ki-67-I OTSU HIRYU 'PEGGY'

Role: Bomber
Crew/Accommodation: Eight
Power Plant: Two 1,900 hp Mitsubishi Ha-104 air-cooled radials
Dimensions: Span 22.5m (73.82 ft); length 18.7m (61.35 ft); wing area 65.85m² (708.8 sq ft)
Weights: Empty 8,649kg (19,068 lb); MTOW 13,765kg (30,347 lb)
Performance: Maximum speed 537km/h (334 mph) at 6,090m (19,980 ft); operational ceiling 9,470m (31,070 ft); range 2,800km (1,740 miles) with 500kg (1,102 lb) bombload
Load: One 20mm cannon and four 12.7mm machine guns, plus up to 1,030kg (2,359 lb) of ordnance, including one heavyweight torpedo

MITSUBISHI F-1 and T-2
(Japan)

Bearing a marked similarity in layout and power plant to the SEPECAT Jaguar, the T-2 first flew in July 1971 as the XT-2. The type has been produced as the T-2 advanced flying and T-2A weapons trainer to a total of 88 aircraft. The F-1 is a single-seat combat counterpart with the rear-cockpit volume used for mission electronics required in the ground-attack role. The prototypes flown in June 1975 were modified T-2s, and 71 aircraft were built.

MITSUBISHI F-1

Role: Strike fighter
Crew/Accommodation: One
Power Plant: Two 3,310kgp (7,300 lb s.t.) IHI-built Rolls-Royce/Turbomeca Adour TF40-IHI-801A turbofans with reheat
Dimensions: Span 7.88m (25.83 ft); length 17.85m (58.56 ft); wing area 21.18m² (227.97 sq ft)
Weights: Empty 6,358kg (14,017 lb); MTOW 13,700kg (30,203 lb)
Performance: Maximum speed 1,700km/h (917 knots) Mach 1.6 at 12,192m (40,000 ft); operational ceiling 00,000m (00,000 ft); range 352km (190 naut. miles) with 1,814kg (4,000 lb) warload
Load: One 20mm cannon, plus up to 3,629kg (8,000 lb) of weapons/fuel

MITSUBISHI DIAMOND I and DIAMOND II
(Japan)

First flown in August 1978 as the MU-300, the Diamond I was designed as a turbofan-powered 'bizjet' and secured comparatively small but useful production orders in its initial and more powerful Diamond IA versions. The basic design was then extensively developed aerodynamically and with greater power as the Diamond II before the whole package was sold to the American manufacturer Beech as the Beechjet 400, subsequently upgraded to Beechjet 400A standard.

MITSUBISHI DIAMOND II

Role: Executive transport
Crew/Accommodation: Two, plus up to seven passengers
Power Plant: Two 1,315kgp (2,900 lb s.t.) Pratt & Whitney Canada JT15D-5 turbofans
Dimensions: Span 13.25m (43.50 ft); length 14.75m (48.42 ft); wing area 22.43m² (241.4 sq ft)
Weights: Empty 4,120kg (9,080 lb); MTOW 7,100kg (15,660 lb)
Performance: Cruise speed 813km/h (439 knots) at 11,890m (39,000 ft); operational ceiling 12,500m (41,000 ft); range 2,930km (1,580 naut. miles) with 4 passengers
Load: Up to 975kg (2,150 lb)

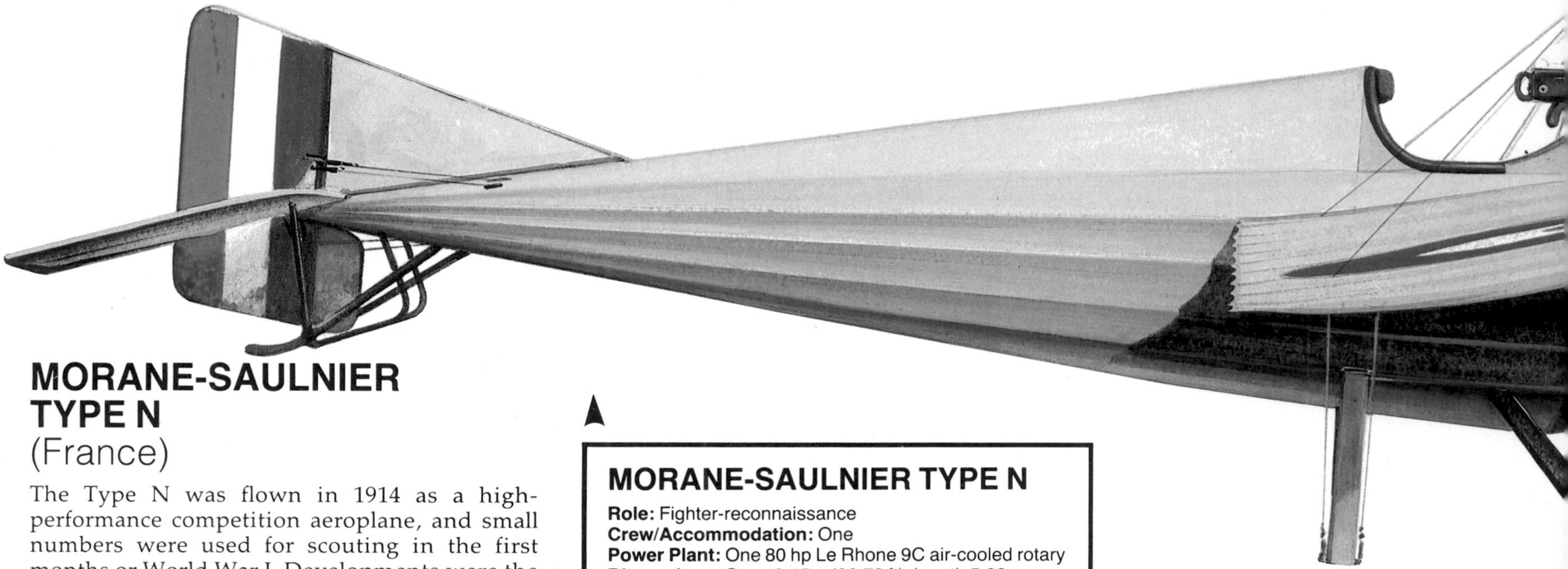

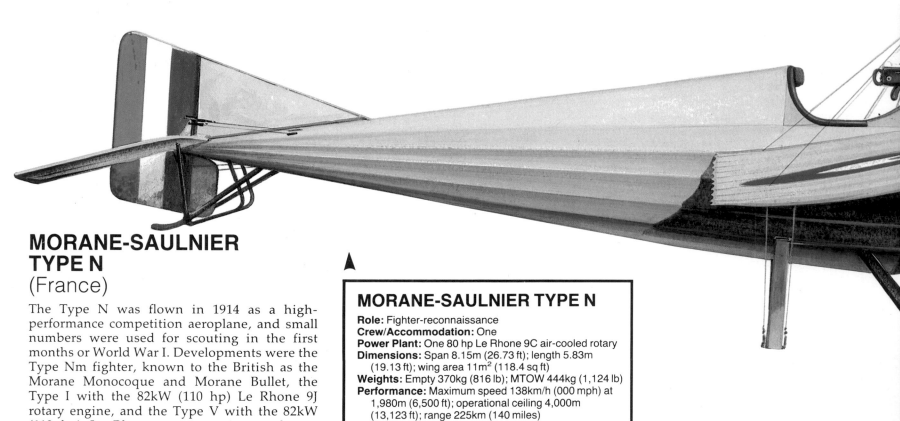

MORANE-SAULNIER TYPE N
(France)

The Type N was flown in 1914 as a high-performance competition aeroplane, and small numbers were used for scouting in the first months or World War I. Developments were the Type Nm fighter, known to the British as the Morane Monocoque and Morane Bullet, the Type I with the 82kW (110 hp) Le Rhone 9J rotary engine, and the Type V with the 82kW (110 hp) Le Rhone rotary engine, a deeper forward fuselage and wings of larger area.

MORANE-SAULNIER TYPE N

Role: Fighter-reconnaissance
Crew/Accommodation: One
Power Plant: One 80 hp Le Rhone 9C air-cooled rotary
Dimensions: Span 8.15m (26.73 ft); length 5.83m (19.13 ft); wing area 11m² (118.4 sq ft)
Weights: Empty 370kg (816 lb); MTOW 444kg (1,124 lb)
Performance: Maximum speed 138km/h (000 mph) at 1,980m (6,500 ft); operational ceiling 4,000m (13,123 ft); range 225km (140 miles)
Load: One .303 inch machine gun

MORANE-SAULNIER TYPE AI
(France)

The Type AI parasol-wing fighter was first flown in the summer of 1917 with a 112kW (150 hp) Gnome Monosoupape 9 rotary engine, and entered production as the MoS.27 and MoS.29 with single or twin 7.7mm (0.303 inch) machine guns. Some 1,210 aircraft were built, but they had only a short service career because of structural problems. Production was then centred on the MoS.39 single-seat advanced trainer with any of several lower-powered rotaries.

MORANE-SAULNIER TYPE AI

Role: Fighter
Crew/Accommodation: One
Power Plant: One 150 hp Gnome-Monosupape 9N rotary
Dimensions: Span 8.51m (27.92 ft); length 5.65m (18.54 ft); wing area 13.39m^2 (144.13 sq ft)
Weights: Empty 421kg (928 lb); MTOW 649kg (1,431 lb)
Performance: Maximum speed 225km/h (140 mph); operational ceiling 7,000m (22,965 ft); endurance 1 hour 45 min
Load: One 7.7mm machine gun

MORANE-SAULNIER MS.406
(France)

This was France's most numerous fighter at the beginning of World War II, but was essentially an interim 'modern' fighter and, therefore, already obsolescent. The type was derived from the limited-production MS.405 with the 641kW (860 hp) Hispano-Suiza HS 12Ygrs inline engine, and first flew in January 1939. About 1,403 aircraft of the MS.405/MS.406 family were completed, the family including the EFW D-3800 and D-3801 built in Switzerland. Aircraft in Finland were re-engined with the 820kW (1,100 hp) Klimov M-105P engine taken from downed or captured Soviet aircraft.

MORANE-SAULNIER MS. 406C1

Role: Fighter
Crew/Accommodation: One
Power Plant: One 860 hp Hispano-Suiza HS 12Ygvs water-cooled in-line
Dimensions: Span 10.62m (34.84 ft); length 8.17m (26.8 ft); wing area 16m^2 (172.2 sq ft)
Weights: Empty 1,872kg (4,127 lb); MTOW 2,471kg (5,448 lb)
Performance: Maximum speed 490km/h (304 mph) at 4,500m (14,764 ft); operational ceiling 9,400m (30,840 ft); range 750km (466 miles)
Load: One 20mm cannon and two 7.5mm machine guns

MYASISHCHEV M-4
(U.S.S.R.)

First flown late in 1953, the M-4 was the U.S.S.R.'s first four-jet strategic bomber. The type lacked the range demanded in its specification, but was still an enormous aerodynamic and structrual achievement. Production totalled about 200 'Bison-A' bombers, numbers of these being converted later into 'Bison-B' and 'Bison-C' maritime reconnaissance aircraft. Most surviving aircraft have been modified as inflight-refuelling tankers.

MYASISHCHEV M-4 'BISON-A'

Role: Long-range bomber/reconnaissance/tanker
Crew/Accommodation: Six
Power Plant: Four 8,700kgp (19,180 lb s.t.) Mikulin AM-3D turbojets
Dimensions: Span 50.48m (165.62 ft); length 47.20m (154.86 ft); wing area 300m^2 (3,229 sq ft)
Weights: MTOW 165,000kg (363,760 lb)
Performance: Maximum speed 998km/h (539 knots) at 12,000m (39,370 ft); operational ceiling 15,600m (51,181 ft); radius 5,600km (3,022 naut. miles) reconnaissance mission
Load: Eight 23mm cannon, plus up to 4,500kg (9,920 lb) of internally carried bombs

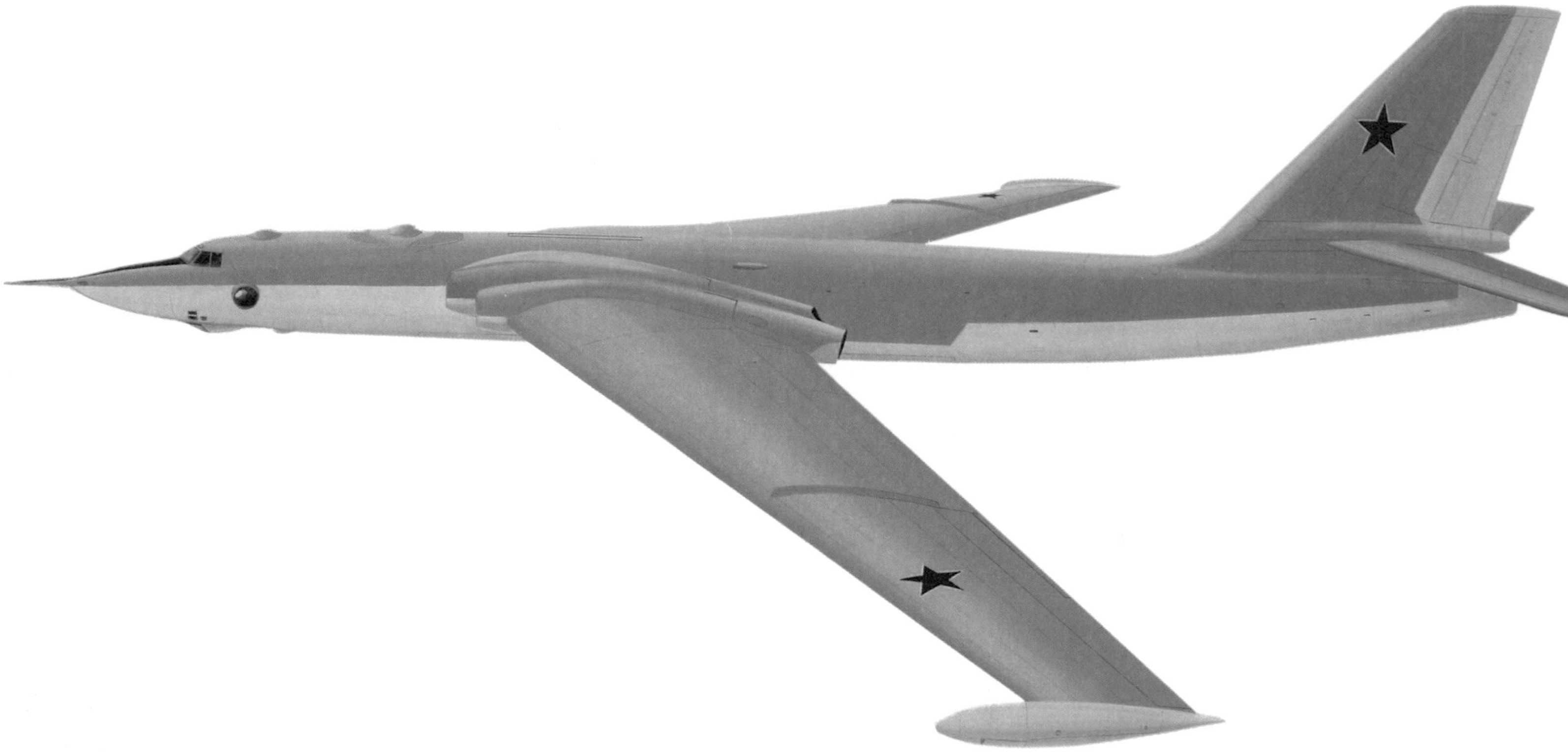

NAKAJIMA Ki-27
(Japan)

This was the Japanese Army's equivalent to the Navy's Mitsubishi A5M, and though it was an interim 'modern' fighter with fixed landing gear (selected because of its light weight) it had more advanced features such as flaps and an enclosed cockpit. The type was evolved from the company's private-venture Type PE design, and production between 1937 and 1942 totalled 3,399 aircraft in the original Ki-27 and modestly improved Ki-27b variants. The type was used operationally up to 1942, when its light structure and poor armament forced its relegation to second-line duties.

NAKAJIMA Ki-27 'NATE'

Role: Fighter
Crew/Accommodation: One
Power Plant: One 710 hp Nakajima Ha-1b air-cooled radial
Dimensions: Span 11.3m (37.07 ft); length 7.53m (24.7 ft); wing area 18.6m^2 (199.7 sq ft)
Weights: Empty 1,110kg (2,447 lb); MTOW 1,650kg (3,638 lb)
Performance: Maximum speed 460km/h (286 mph) at 3,500m (11,480 ft); operational ceiling 8,600m (28,215 ft); range 1,710m (1,060 miles)
Load: Two 7.7mm machine guns, plus up to 100kg (220 lb) of bombs

NAKAJIMA Ki-43 HAYABUSA (PEREGRINE FALCON)
(Japan)

Just as the A5M was superseded by the A6M in naval service, the Ki-27 was supplanted by the Ki-43 in army service, and this useful though lightly armed fighter was the mainstay of the Japanese Army for much of World War II. The Ki-43 prototype flew in January 1939 with a 727kW (975 hp) Nakajima Sakae radial engine but extensive development, including a larger wing, was necessary before the Ki-43-I entered production. Subvariants went from the -Ia with two 7.7mm (0.303 inch) machine guns via the -Ib with one 7.7mm and one 12.7mm (0.5 inch) guns to the -Ic with two 12.7mm guns.

Early operations revealed the type's lack of protection, and the Ki-43-IIa introduced armour and self-sealing tanks as well as the more powerful 858kW (1,150 hp) Ha-115 radial engine. The Ki-43-IIb had minor improvements, and the only other service variant in the 5,919-aircraft production total was the Ki-43-II KAI product-improved Ki-43-IIb. Prototype versions were the Ki-43-IIIa high-altitude fighter and the Ki-43-IIIb with two 20mm cannon and the 932kW (1,250 hp) Mitsubishi Ha-112.

NAKAJIMA Ki-43-II 'OSCAR'

Role: Interceptor
Crew/Accommodation: One
Power Plant: One 1,150 hp Nakajima Ha115 air-cooled radial
Dimensions: Span 11.44m (37.53 ft); length 8.92m (29.27 ft); wing area 31.4m² (230.3 sq ft)
Weights: Empty 1,975kg (4,354 lb); MTOW 2,642kg (5,825 lb)
Performance: Maximum speed 515km/h (320 mph) at 6,000m (19,685 ft); operational ceiling 11,215m (36,795 ft); range 1,200km (745 miles)
Load: Two 12.7mm machine guns

▼

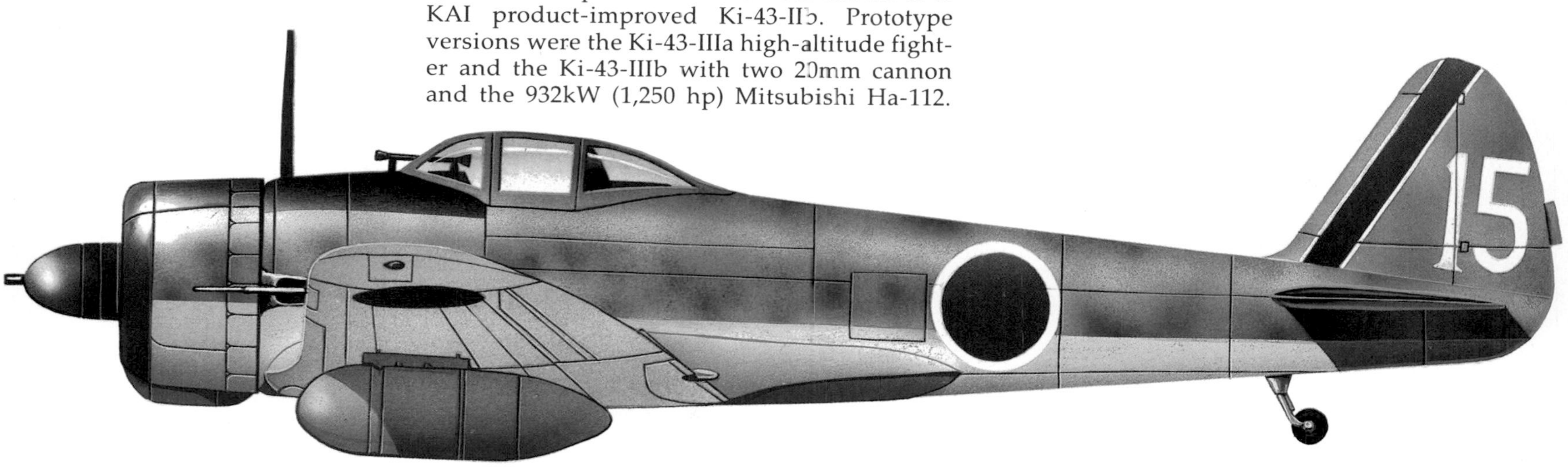

NAKAJIMA Ki-44 SHOKI (DEMON)
(Japan)

▲ The Ki-44 was designed at much the same time as the Ki-43 to provide the Japanese Army with a fast-climbing interceptor, and while resembling the Ki-43 in overall terms had a larger-diameter front fuselage to accommodate the 932kW (1,250 hp) Nakajima Ha-41 radial engine. The type first flew in August 1940, and was then ordered into production as the Ki-44-Ia despite the reservations expressed by service pilots about its high landing speed and relative lack of manoeuvrability. The Ki-44-Ia's two 12.7mm (0.5 inch) and two 7.7mm (0.303 inch) machine guns were replaced in the Ki-44-Ib and refined Ki-44-Ic by four 12.7mm guns. Then came the Ki-44-II and later Ki-44-III series with the 1,133 and 1,491kW (1,520 and 2,000 hp) Nakajima Ha-109 and Ha-145 radial engines respectively: the -IIa and -IIb had the same armament as the -Ia and -Ib, but the -IIc introduced four 20mm cannon or two 40mm cannon plus two 12.7mm machine guns. The -IIIa had four 20mm cannon, and the -IIIb two 37mm and two 20mm cannon. Total production up to late 1944 was 1,225.

NAKAJIMA Ki-44-IIb 'TOJO'

Role: Interceptor
Crew/Accommodation: One
Power Plant: One 1,520 hp Nakajima Ha-109 air-cooled radial
Dimensions: Span 9.45m (31 ft); length 8.75m (28.71 ft); wing area 15m² (161.46 sq ft)
Weights: Empty 2,106kg (4,643 lb); MTOW 2,770kg (6,107 lb)
Performance: Maximum speed 605km/h (375 mph) at 5,200m (17,060 ft); operational ceiling 11,200m (36,745 ft); range 1,690km (1,050 miles)
Load: Four 12.7mm machine guns

NAKAJIMA B5N
(Japan)

This was the Japanese Navy's most important level and torpedo bomber in the first half of World War II, and resulted from a 1935 requirement. For its period the type was aerodynamically and structurally advanced, but was underpowered with the 574kW (770 hp) Nakajima Hikari 3 radial engine. Production totalled 1,149, the original B5N1 being supplemented from 1939 by the B5N2 with the 746kW (1,000 hp) Nakajima Salae radial engine. The B5N2 was withdrawn from first-line service in 1944.

NAKAJIMA B5N1 'KATE'

Role: Naval carrierborne torpedo strike
Crew/Accommodation: Three
Power Plant: One 770 hp Nakajima Hikari 3 air-cooled radial
Dimensions: Span 15.52m (50.92 ft); length 10,34m (33.92 ft); wing area 37.7m^2 (405.8 sq ft)
Weights: Empty 2,107kg (4,645 lb); MTOW 3,650kg (8,047 lb)
Performance: Maximum speed 350km/h (217 mph) at 2,450m (8,038 ft); operational ceiling 8,420m (27,625 ft); range 1,100km (684 miles) with torpedo
Load: One 7.7mm machine gun, plus an 800kg (1,764 lb) torpedo

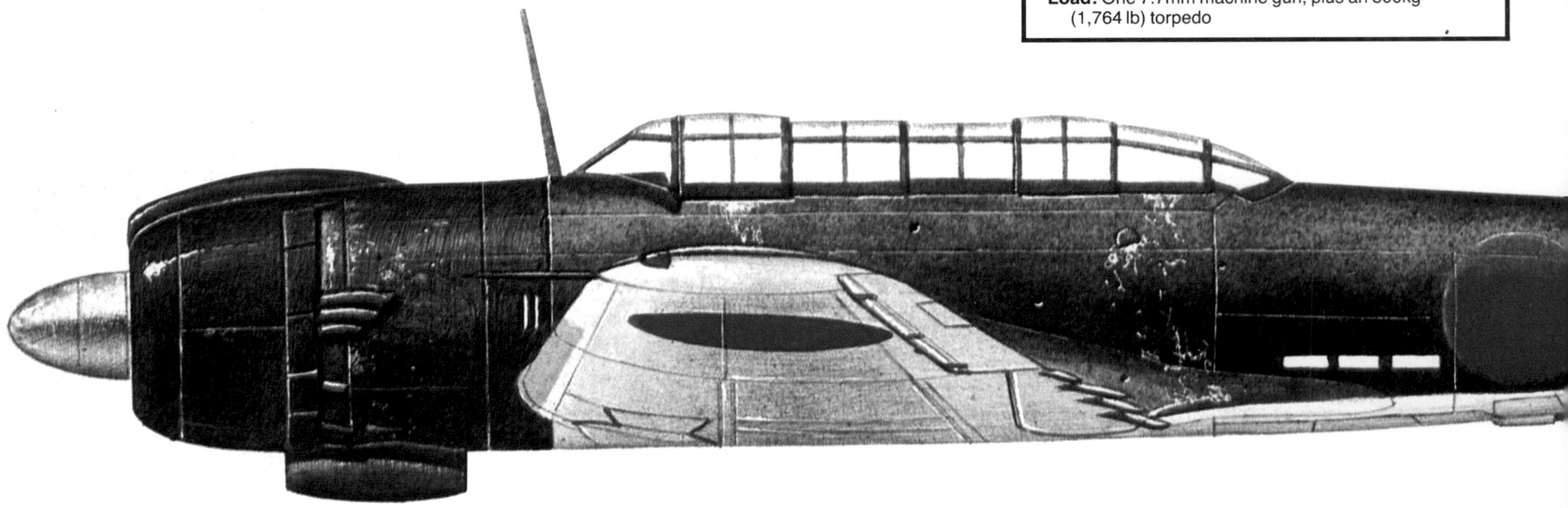

NAKAJIMA B6N TENZAN (HEAVENLY CLOUD)
(Japan)

The B6N was planned as successor to the B5N to a 1939 requirement, and used a basically similar airframe with a revised tail unit and a considerably more powerful engine, the 1,342kW (1,800 hp) Nakajima NK7A Mamoru. The prototype engine flew in 1941. Production aircraft were ordered with the designation B6N1, but after only 135 examples had been completed the company was instructed to stop making the Mamoru radial. Most of the 1,133 aircraft were therefore B6N2s with the 1,380kW (1,850 hp) Mitsubishi MK4T Kasei 25 radial engine. The sole variant of this model was the B6N2a with improved rearward-firing armament.

NAKAJIMA B6N2 'JILL'

Role: Naval carrierborne torpedo strike
Crew/Accommodation: Three
Power Plant: One 1,850 hp Mitsubishi MK4T Kasei 25 air-cooled radial
Dimensions: Span 14.89m (48.85 ft); length 10.87m (35.66 ft); wing area 37.2m^2 (400.4 sq ft)
Weights: Empty 3,010kg (6,636 lb); MTOW 5,650kg (12,456 lb)
Performance: Maximum speed 482km/h (300 mph) at 4,900m (16,075 ft); operational ceiling 9,040m (29,660 ft); range 1,745km (1,084 miles) with torpedo
Load: Two 7.7mm machine guns, plus an 800kg (1,764 lb) torpedo

NAKAJIMA Ki-49 DONRYU (STORM DRAGON)
(Japan)

This was designed as the Mitsubishi Ki-21's successor, and first flew in August 1939 with 708kW (950 hp) Nakajima Ha-5 KAI radial engines, changed in later prototypes to 932kW (1,250 hp) Nakajima Ha-41 engines of the type used in the Ki-49-I production version ordered in March 1941. For its bombload the Ki-49 was vastly overcrewed yet lacking in protection, and combat experience exposed both these factors as well as lack of performance. The Ki-49-IIa introduced 1,119kW (1,500 hp) Nakajima Ha-109 radial engines plus better protection and self-sealing fuel tanks; the Ki-49-IIb had better defensive armament. Performance was still poor, so the Ki-49-III used 1,805kW (2,240 hp) Nakajima Ha-117 radial engines. Only six of this variant (within an overall total of 819) had been produced before production was ended in December 1944.

NAKAJIMA Ki-49-IIa 'HELEN'

Role: Bomber
Crew/Accommodation: Eight
Power Plant: Two 1,500 Nakajima Ha-109 air-cooled radials
Dimensions: Span 20.3m (66.6 ft); length 16.2m (53.15 ft); wing area 67m^2 (721.1 sq ft)
Weights: Empty 6,530kg (14,396 lb); MTOW 10,680kg (23,545 lb)
Performance: Maximum speed 490km/h (305 mph) at 5,000m (16,400 ft); operational ceiling 8,160m (26,772 ft); range 2,400km (1,491 miles) with 250kg (551 lb) warload
Load: One 20mm cannon and five 7.7mm machine guns, plus up to 750kg (1,653 lb) of bombs

NAKAJIMA J1N GEKKO (MOONLIGHT)
(Japan)

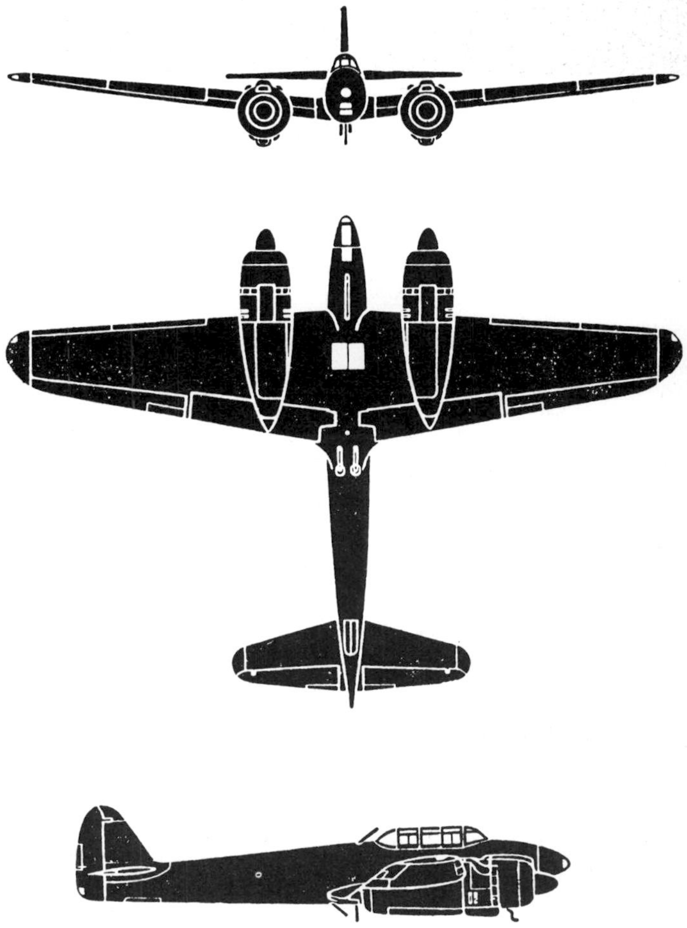

The J1N was developed as a long-range escort fighter to accompany naval bombers probing deep into China, and the J1N1 prototype flew in May 1941 with 843kW (1,130 hp) Nakajima Sakae 21/22 radial engines. Development was concentrated on giving the J1N a reconnaissance role, and it was in this task that the J1N1-C (later J1N1-R) entered service. With a rearward-firing 20mm cannon in place of the standard 13mm (0.51 inch) machine gun the type was designated J1N1-F. From 1943 the type also adapted to the night-fighter role with the designation J1N1-C KAI, while production night-fighters were called J1N1-S and received the name Gekko. The only other night-fighter variant was the J1N1-Sa with the rearward-firing cannon deleted. Production of all J1Ns totalled 479.

NAKAJIMA J1N1-S 'IRVING'

Role: Night fighter
Crew/Accommodation: Two
Power Plant: Two 1,130 hp Nakajima NKF1 Sakae 21 air-cooled radials
Dimensions: Span 16.98m (55.71 ft); length 12.18m (39.96 ft); wing area 40m^2 (430.56 sq ft)
Weights: Empty 4,852kg (10,697 lb); MTOW 7,527kg (16,594 lb)
Performance: Maximum speed 507km/h (316 mph) at 5,000m (16,404 ft); operational ceiling 9,320m (30,610 ft); range 2,550km (1,585 miles)
Load: Four 20mm cannon

NAKAJIMA Ki-84 HAYATE (GALE)
(Japan)

▲ The Hayate was planned as successor to the Hayabusa, but though it was an excellent fighter, it was too late to affect the outcome of the war for Japan when it began to enter service in the summer of 1944. Design was launched in 1942 and the Ki-84 prototypes were followed by no fewer than 83 service trials and 42 pre-production aircraft before production of the Ki-84-Ia began in late 1943 with an armament of two 20mm cannon and two 12.7mm (0.5 inch) machine guns Total construction was 3,514 aircraft including the Ki-84-Ib with four 20mm cannon, the Ki-84-Ic with two 30mm and two 20mm cannon, and the Ki-84-II that introduced a measure of wood into the structure to ease pressure on requirements for scarce light alloys.

NAKAJIMA Ki-84-Ia 'FRANK'

Role: Fighter
Crew/Accommodation: One
Power Plant: One 1,800 hp Nakajima Ha-45 air-cooled radial
Dimensions: Span 11.24m (36.88 ft); length 9.92m (32.55 ft); wing area 21m^2 (226.04 sq ft)
Weights: Empty 2,680kg (5,908 lb); MTOW 3,750kg (8,267 lb)
Performance: Maximum speed 624km/h (388 mph) at 6,000m (19,685 ft); operational ceiling 11,000m (36,090 ft); range 1,120km (696 miles) with full warload
Load: Two 20mm cannon and two 12.7mm machine guns, plus up to 500kg (1,102 lb) of bombs

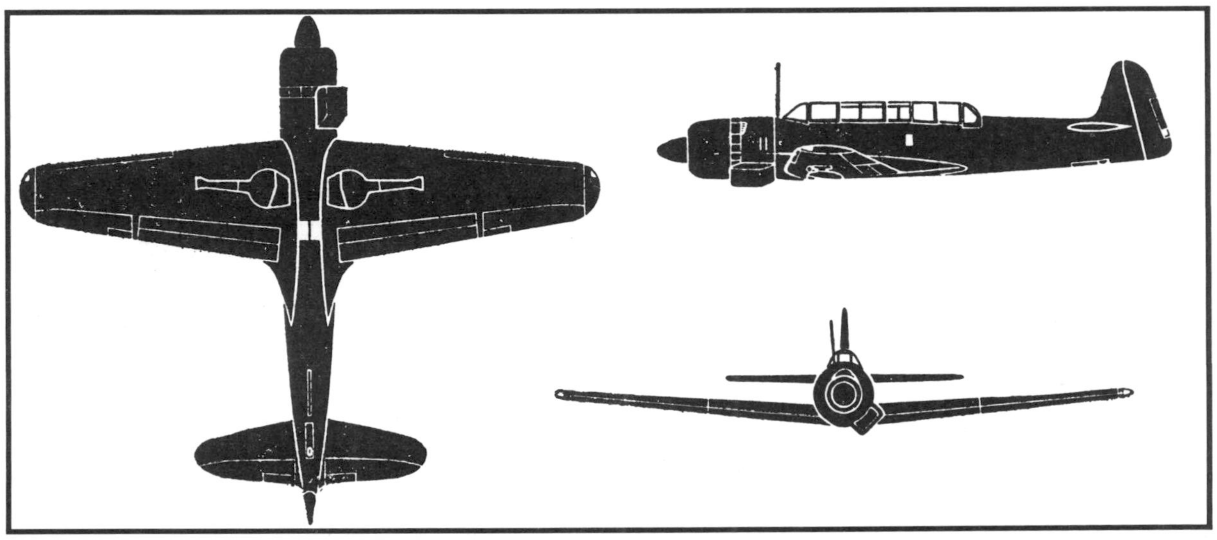

NAKAJIMA C6N SAIUN (PAINTED CLOUD)
(Japan)

The Saiun resulted from the Japanese Navy's appreciation early in the Pacific campaign of World War II that it lacked an effective carrier-borne long/range reconnaissance aeroplane. The resulting airframe was modelled conceptually on that of the B6N with modifications for the reconnaissance role, and first flew in May 1943 with a 1,358kW (1,820 hp) Nakajima Homare 11 radial engine. The production version was the C6N1 with 1,484kW (1,990 hp) Homare 21, and 463 aircraft were built. Some aircraft were later converted to C6N1-S night-fighters.

NAKAJIMA C6N1 'MYRT'

Role: Naval carrierborne long-range reconnaissance
Crew/Accommodation: Three
Power Plant: One 1,990 hp Nakajima NK9H Homare 21 air-cooled radial
Dimensions: Span 12.5m (41 ft); length 11m (36.09 ft); wing area 25.5m^2 (274.5 sq ft)
Weights: Empty 2,968kg (6,453 lb); MTOW 5,274kg (11,627 lb)
Performance: Maximum speed 609km/h (379 mph) at 6,000m (19,685 ft); operational ceiling 10,740m (35,236 ft); range 4,000km (2,486 miles) without external fuel
Load: One 7.9mm machine gun

NAVY/CURTISS NC
(U.S.A.)

Only 10 NCs were built, but these have a secure niche in the history of aviation as the first aircraft to cross the Atlantic, albeit in stages. The type was designed by the U.S. Navy and built Curtiss as an anti-submarine flying boat, the plans being completed in September 1917. The first four 'boats differed in detail and power plant arrangement from the last six, and were too late for World War I service as the first 'boat flew only in October 1918.

NAVY/CURTISS NC-4

Role: Long-range maritime patrol flying boat
Crew/Accommodation: Six
Power Plant: Four 420 hp Liberty 12 water-cooled inlines
Dimensions: Span 38.4m (126 ft); length 20.8m (68.25 ft); wing area 221.1m^2 (2,380 sq ft)
Weights: Empty 7,188kg (15,847 lb); MTOW 12,701kg (28,000 lb)
Performance: Cruise speed 122km/h (76 mph) at sea level; operational ceiling 1,372m (4,500 ft); range 2,366km (1,470 miles)
Load: Three .303 inch machine guns

NIEUPORT TYPE 11 BEBE (BABY) and TYPE 17
(France)

The Type 11 was the fighter largely instrumental for the defeat of 'Fokker Scourge' in 1916. Planned as a competition sesquiplane, the Type 11 was used as a scout from 1915 with the 60kW (80 hp) Le Rhone rotary engine, and was then turned into a fighter by the addition of a machine gun on the upper-wing centre section to fire over the propeller's swept disc. The Type 16 was a version with the 82kW (110 hp) Le Rhone for better performance. The Type 17 was a further expansion of the same design except with greater strength but the same 82kW Le Rhone. The Type 17bis introduced a 97kW (130 hp) Clerget rotary engine and a synchronized machine gun on the upper fuselage. Further improved variants were the Types 21 and 23.

NIEUPORT TYPE 11

Role: Fighter
Crew/Accommodation: One
Power Plant: One 80 hp Gnome or Le Rhone 9C air-cooled rotary
Dimensions: Span 7.55m (24.77 ft); length 5.8m (19.03 ft); wing area 13m^2 (140 sq ft)
Weights: Empty 344kg (759 lb); MTOW 550kg (1,213 lb)
Performance: Maximum speed 156km/h (97 mph) at sea level; operational ceiling 4,600m (15,090 ft); range 330km (205 miles) with full bombload
Load: One .303 inch machine gun

Nieuport 17C.1

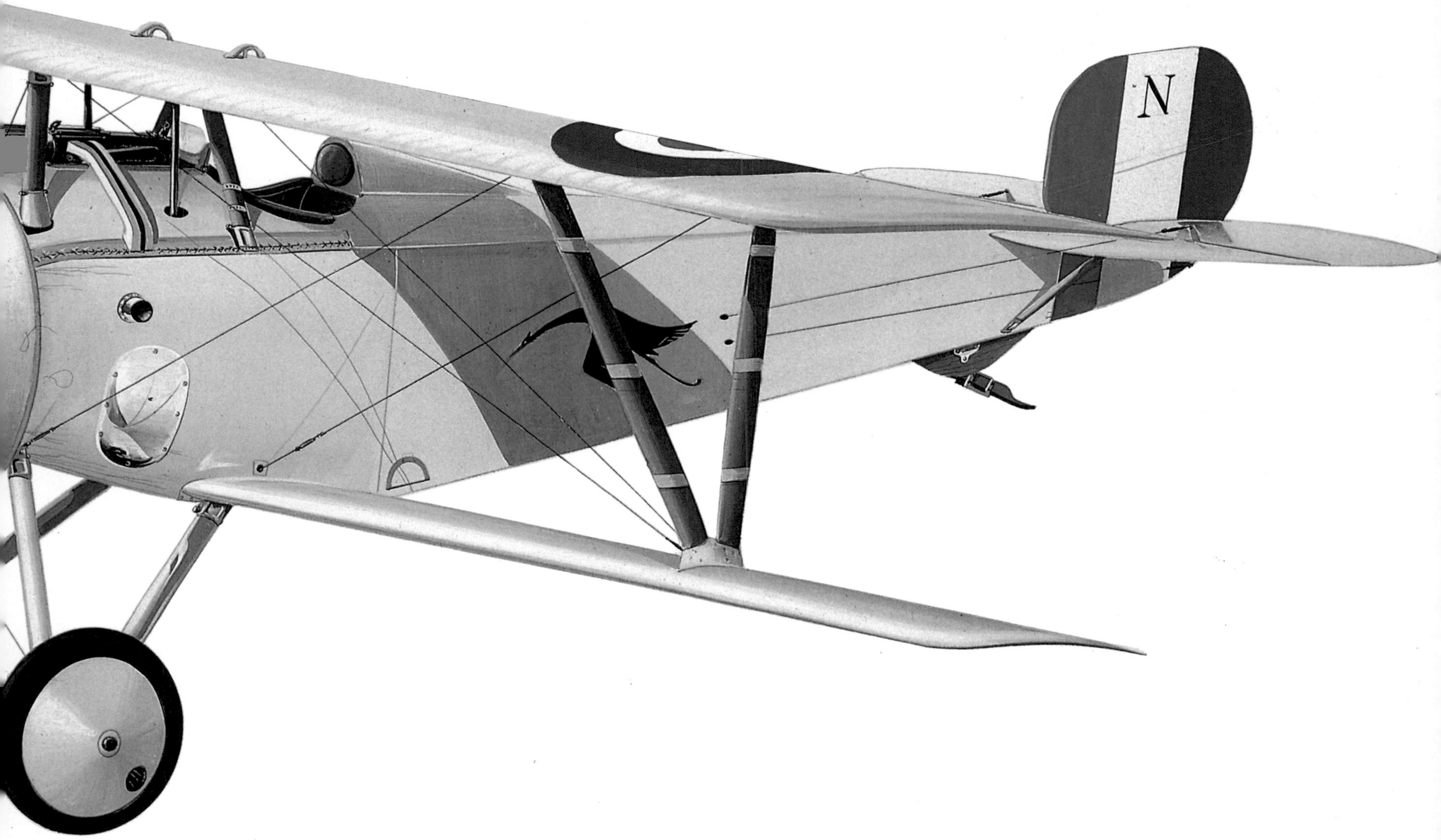

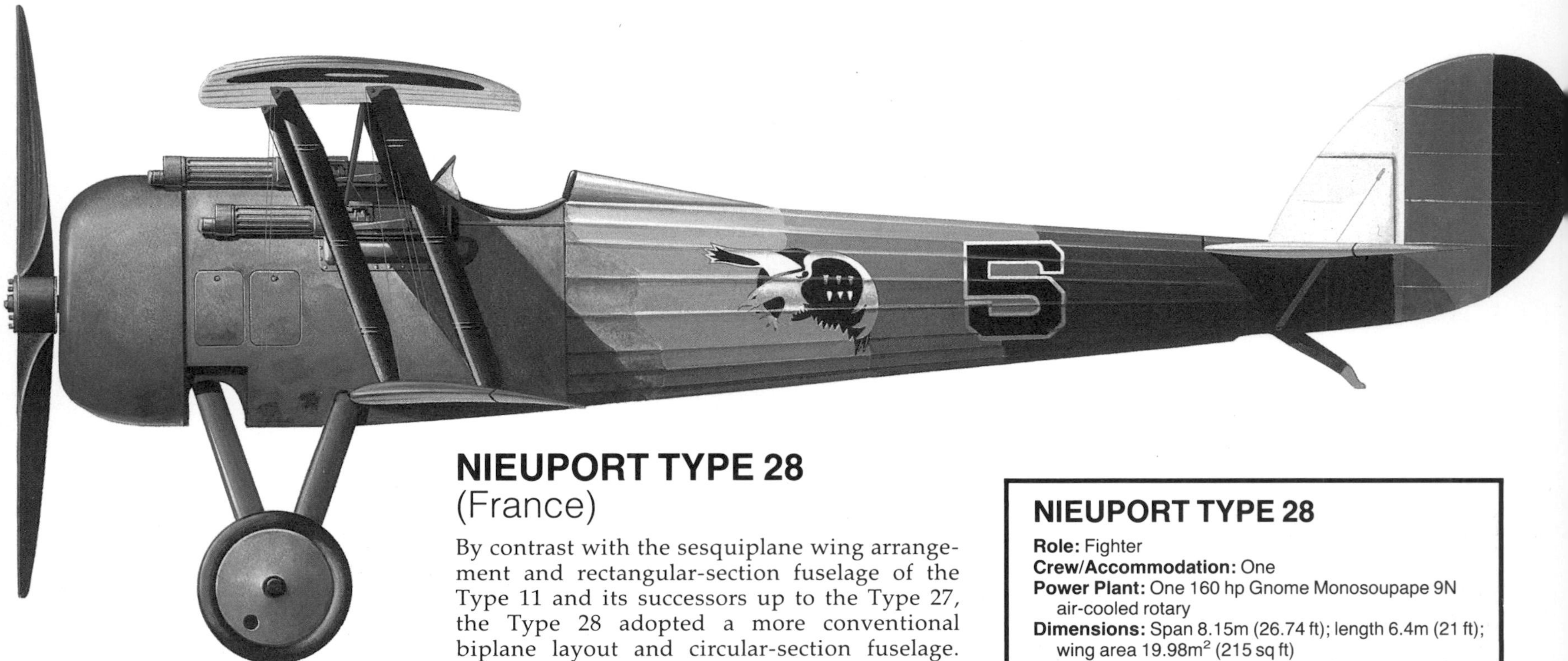

NIEUPORT TYPE 28
(France)

By contrast with the sesquiplane wing arrangement and rectangular-section fuselage of the Type 11 and its successors up to the Type 27, the Type 28 adopted a more conventional biplane layout and circular-section fuselage. The aeroplane first flew in June 1918 with a 119kW (160 hp) Gnome rotary engine, but was inferior in performance to the contemporary SPADs and thus used as a major fighter only by the Americans, who received 297 Type 28s.

NIEUPORT TYPE 28

Role: Fighter
Crew/Accommodation: One
Power Plant: One 160 hp Gnome Monosoupape 9N air-cooled rotary
Dimensions: Span 8.15m (26.74 ft); length 6.4m (21 ft); wing area 19.98m^2 (215 sq ft)
Weights: Empty 435kg (960 lb); MTOW 700kg (1,543 lb)
Performance: Maximum speed 213km/h (128 mph) at sea level; operational ceiling 5,300m (17,388 ft); range 250km (155 miles)
Load: Two .303 inch machine guns

NOORDUYN NORSEMAN
(Canada)

The Norseman was a remarkable bushplane whose design was inaugurated in 1934, and the Norseman I prototype flew in November 1935 with 313kW (420 hp) Wright R-975-E3 radial engine. The production model was the generally similar Norseman II, which was considered underpowered and supplanted by small numbers of the Norseman III and Norseman IV with the 336kW (450 hp) Pratt & Whitney Wasp SC and 410kW (550 hp) Wasp R-1340-AN-1. With improvements the Norseman IV became the Norseman V and, for the military in World War II, the Norseman VI. The Norseman VII with metal flying surfaces remained a post-war production, and total production up to the early 1950s was some 900.

NOORDUYN NORSEMAN Mk VI

Role: Ulility transport
Crew/Accommodation: One, plus up to nine passengers
Power Plant: One 550 hp Pratt & Whitney R-1340-AN-1 Wasp air-cooled radial
Dimensions: Span 15.7m (51.5 ft); length 9.75m (32 ft); wing area 30.19m^2 (325 sq ft)
Weights: Empty 2,125kg (4,680 lb); MTOW 3,360kg (7,400 lb)
Performance: Cruise speed 238km/h (148 mph) at sea level; operational ceiling 5,181m (17,000 ft); range 747km (464 miles) with full payload
Load: Up to 755kg (1,665 lb)

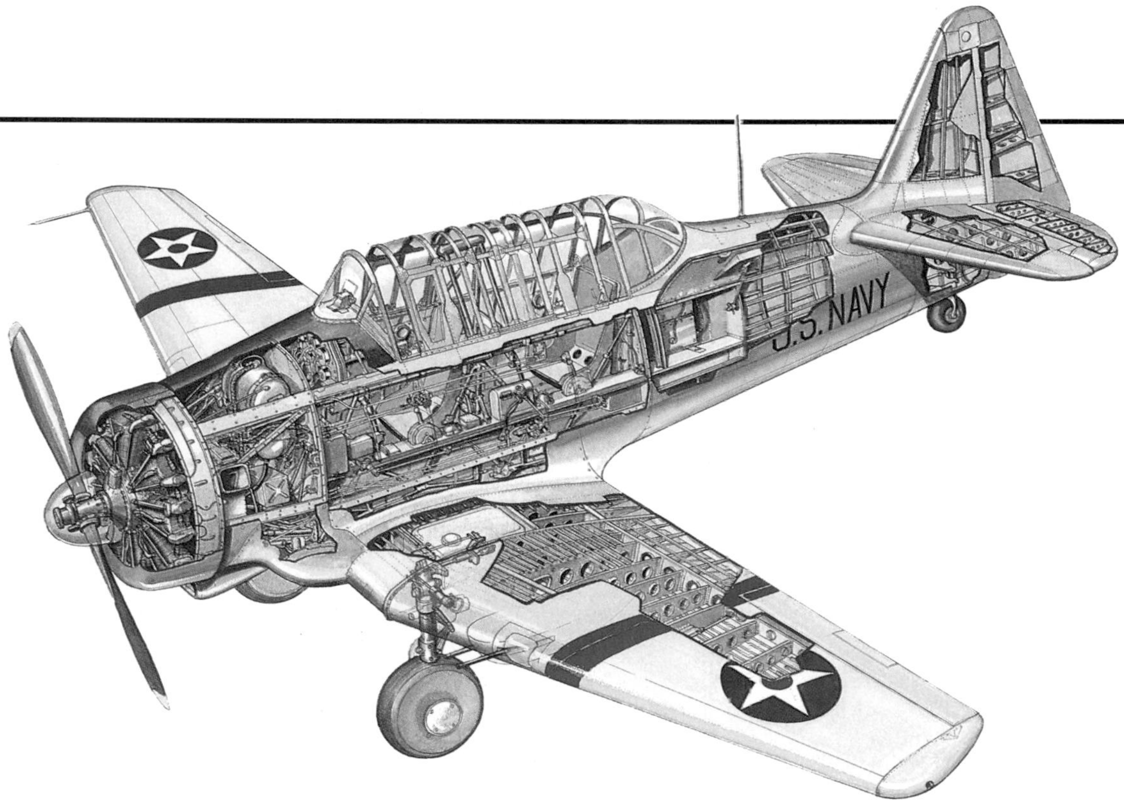

NORTH AMERICAN NA-18 (BT-9/14 and NJ) and NBA-26 (T-6, SNJ and HARVARD)
(U.S.A.)

This was the Western Alliance's most important trainer of World War II, and production was in the order of 17,000 aircraft. The series was pioneered by the NA-16 prototype that flew in April 1935 with fixed landing gear and a 298kW (400 hp) Wright R-975 radial engine. The type was then ordered by the U.S. Army and Navy and the BT-9 and NJ series, additional aircraft being produced for export to several countries including Canada, where the aeroplane was known as the Yale. The NA-18 was then developed further with retractable landing gear as a combat trainer, and this NA-26 was first ordered as the AT-6 (initially BC-1) and SNJ series for the U.S. Army and Navy respectively. British aircraft were designated Harvard. From 1949 some 2,086 aircraft were rebuilt with many modifications as T-6s.

NORTH AMERICAN T-6D TEXAN (RAF HARVARD)

Role: Advanced trainer
Crew/Accommodation: Two
Power Plant: One 550 hp Pratt & Whitney R-1340-AN1 Wasp air-cooled radial
Dimensions: Span 12.81m (42.02 ft); length 8.84m (28.99 ft); wing area 23.57m^2 (253.72 sq ft)
Weights: Empty 1,886kg (4,158 lb); MTOW 2,722kg (6,000 lb)
Performance: Maximum speed 330km/h (205 mph) at 1,524m (5,000 ft); operational ceiling 6,553m (21,500 ft); range 1,207km (750 miles)
Load: Two .303 inch machine guns

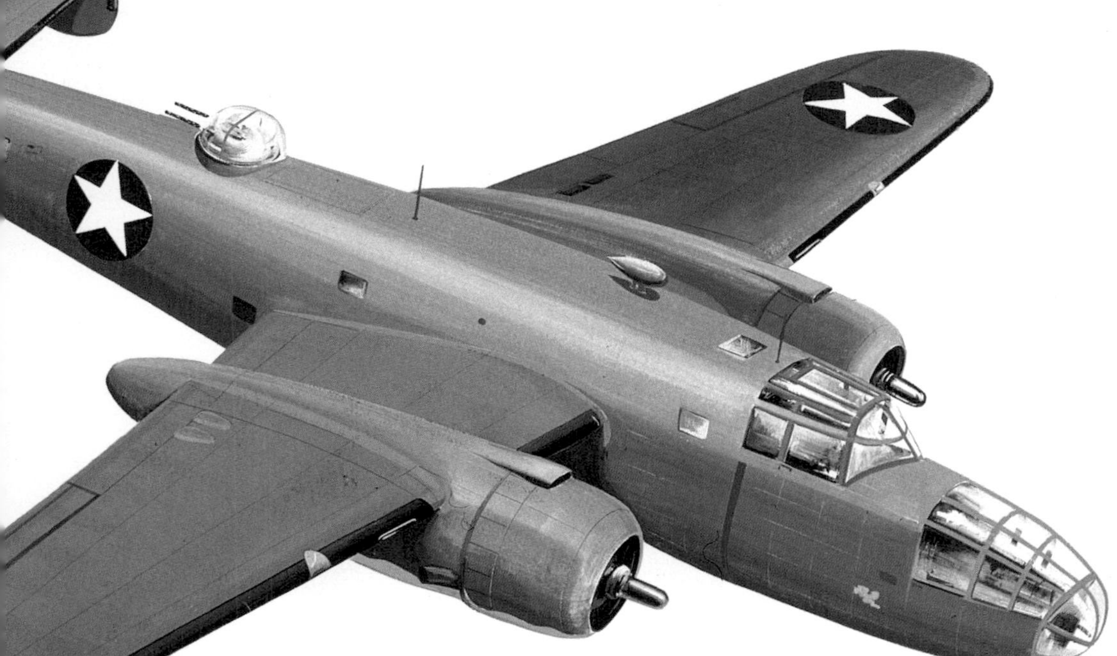

NORTH AMERICAN B-25C MITCHELL

Role: Medium bomber
Crew/Accommodation: Five/six
Power Plant: Two 1,700 hp Wright R-2600-19 Cyclone air-cooled radials
Dimensions: Span 20.6m (67.58 ft); length 16.13m (52.92 ft); wing area 56.67m^2 (610 sq ft)
Weights: Empty 9,208kg (20,300 lb); MTOW 15,422kg (34,000 lb)
Performance: Maximum speed 457km/h (284 mph) at 4,572m (15,000 ft); operational ceiling 6,401m (21,200 ft); range 2,414km (1,500 miles) with full bombload
Load: Four .5 inch machine guns, plus up to 1,361 kg (3,000 lb) of bombs

NORTH AMERICAN NA-62 (B-25 MITCHELL)
(U.S.A.)

The NA-40 was designed to meet a U.S. Army requirement for a twin-engined attack bomber, and emerged as a shoulder-wing monoplane with tricycle landing gear and 820kW (1,100 hp) Pratt & Whitney R-1830-S6C3-G radial engines for its first flight in January 1939. The engines were soon replaced by 969kW (1,300 hp) Wright GR-2600-A71 radial engines, and this basic power plant was retained throughout the rest of the type's 9,816-aircraft production run. Further development produced the NA-62 design with the wing lowered to the mid-position, the fuselage widened for side-by-side pilot seating, and power provided by 1,268kW (1,700) R-2600-9 engines.

The type entered production as the B-25, the first of 25 being flown in August 1940. Later variants were the B-25A (40) with armour and self-sealing fuel tanks, B-25B (120) with power-operated dorsal and ventral turrets, B-25D (2,290) improved aircraft, B-25G (405) with 75mm nose gun, B-25H (1,000) with the 75mm gun and between 14 and 18 12.7mm (0.5 inch machine guns), and B-25J (4,390 with R-2600-92 radial engines and 12 12.7mm machine guns). Other variants included in the total were the F-10 reconnaissance, the AT-25 and TB-25 trainers, and the PBJ U.S. Navy versions.

NORTH AMERICAN NA-73 (P-51 MUSTANG)
(U.S.A.)

The Mustang was perhaps one of the great fighters of World War II in terms of all-round performance and capability, and resulted from a British requirement of April 1940, which stipulated that the first prototype must be flown within 120 days. The first prototype flew in October of the same year with an 820kW (1,100 hp) Allison V-1710-F3R inline engine. Mustang production totalled 15,469.

The first variant was the Mustang Mk I reconnaissance fighter with an armament of four 12.7mm (0.5 inch) machine guns; two of this type were evaluated by the U.S. Army with the designation XP-51. The next variants were the Mustang Mk IA and equivalent P-51 with four 20mm cannon, and the Mustang Mk II and equivalent P-51A longer-range versions with more power and four machine guns. U.S. Army offshoots were the F-6 and F-6A reconnaissance aircraft and the A-36A dive-bomber and ground-attack aeroplane. Operations had shown, however, that tactical capability was hampered by the V-1710 engine, so the basic airframe was revised to take the Rolls-Royce Merlin built under licence in the U.S.A. by Packard as the V-1650.

Production versions were the Mustang Mk III with four machine guns, and the equivalent P-51B and P-51C with original and bubble canopies; there were also F-6C reconnaissance aircraft. The classic and most extensively built variant was the P-51D (British Mustang Mk IV) with a cutdown rear fuselage, a bubble canopy, six machine guns, greater power and more fuel; the F-6D was the reconnaissance version. The P-51D had the range to escort U.S. bombers on deep raids, and was the decisive fighter of World War II's second half. Later variants expanded on the theme of the P-51D: the P-51H was a lightened version, the P-51K was a similarly lightened variant with an Aeroproducts propeller, and the F-6K was the reconnaissance conversion of the P-51K. The type was also built under licence in Australia with designations running from Mustang Mk 20 to Mustang Mk 24.

North American P-51B ▲

North American P-51D in Swiss markings ▼

NORTH AMERICAN P-51D MUSTANG

Role: Day fighter
Crew/Accommodation: One
Power Plant: One 1,450 hp Packard/Rolls-Royce Merlin V-1650-7 water-cooled inline
Dimensions: Span 11.28m (37 ft); length 9.83m (32.25 ft); wing area 21.83m^2 (235 sq ft)
Weights: Empty 3,466kg (7,635 lb); MTOW 5,493kg (12,100 lb)
Performance: Maximum speed 703km/h (437 mph) at 7,625m (25,000 ft); operational ceiling 12,192m (40,000 ft); range 2,655km (1,650 miles) with maximum fuel
Load: Six .5 inch machine guns, plus up to 907 kg (2,000 lb) of externally carried bombs or fuel tanks

NORTH AMERICAN NA-134 and NA-140 (FJ FURY)
(U.S.A.)

The NA-64 carrierborne jet fighter prototype was produced in competition with the McDonnell FD Phantom and Vought F6U Pirate, and first flew in January 1945 with the 1,733kg (3,820 lb) thrust General Electric J35-GE-2 engine. A total of 200 aircraft was ordered, but only 30 FJ-1 fighters were delivered with the 1,814kg (4,000 lb) thrust J35-A-2, the same basic engine taken over and developed by Allison. The FJ designation did not die, however, for the U.S. Navy later took carrierborne equivalents of the swept-wing Sabre series: the FJ-2 was based on the F-86E with folding wings and 200 were delivered, the FJ-3 had a deeper fuselage for the Wright J65-W-2 or -4 turbojet and 538 were delivered, the FJ-4 (later F-1E) was a redesigned attack variant with J65-W-16A of which 152 were delivered. The FJ-4B (later AF-1E) was an attack aeroplane with a new airframe and 222 were delivered.

NORTH AMERICAN FJ-1 FURY

Role: Naval carrierborne fighter
Crew/Accommodation: One
Power Plant: One 1,814kgp (4,000 lb s.t.) Allison J35-A-2 turbojet
Dimensions: Span 11.63m (38.16 ft); length 10.49m (34.42 ft); wing area 20.53m^2 (221 sq ft)
Weights: Empty 4,011kg (8,843 lb); MTOW 7,076kg (15,600 lb)
Performance: Maximum speed 880km/h (547 mph) at 2,743m (9,000 ft); operational ceiling 9,754m (32,000 ft); range 2,414km (1,500 miles)
Load: Six .5 inch machine guns

NORTH AMERICAN NA-140 (F-86 SABRE)
(U.S.A.)

This was the most important air combat fighter fielded by the Americans in the Korean War, and resulted from a reworking of the XP-86 prototype design to incorporate swept wings after German research data had been captured in World War II's closing stages. The first prototype flew in October 1947 with a 1,701kg (3,750 lb) thrust General Electric TG-180 (later J35-GE-3) axial-flow turbojet. The type was then re-engined with the General Electric J47 turbojet to become the YP-86A, leading to the P-86A (later F-86A) production model with the 2,200kg (4,850 lb) thrust J47-GE-1 engine. These 554 aircraft with four J47 marks up to a thrust of 2,708kg (5,970 lb) were followed in chronologicgical order by the F-86E with slab a tailplane, the F-86F with a revised wing, the F86-D redesigned night and all-weather fighter, the F-86H fighter-bomber with the J73 engine, the F-86K simplified version of the F-86D, and the F-86L rebuilt version of the F-86D with a larger wing and updated electronics. The Sabre was also built in Australia as the CAC Sabre with the Rolls-Royce Avon turbojet, and in Canada as the Canadair Sabre with the Orenda turbojet.

NORTH AMERICAN F-86F SABRE ▲

Role: Day fighter
Crew/Accommodation: One
Power Plant: One 2,708 kgp (5,970 lb s.t.) General Electric J47-GE-27 turbojet
Dimensions: Span 11.3m (37.08 ft); length 11.43m (37.5 ft); wing area 26.76m^2 (288 sq ft)
Weights: Empty 4,967kg (10,950 lb); MTOW 7,711kg (17,000 lb)
Performance: Maximum speed 1,110km/h (690 mph) at sea level; operational ceiling 15,240m (50,000 ft); range 1,263km (785 miles) without external fuel
Load: Six .5 inch machine guns, plus up to 907 kg (2,000 lb) of bombs or fuel carried externally

▲ NORTH AMERICAN NA-180 (F-100 SUPER SABRE) (U.S.A.)

Vying with the Mikoyan-Gurevich MiG-19 for the distinction of having been the first supersonic fighter to have entered service, the Super Sabre was a radical revision of the Sabre concept with a wing swept at 45 degrees and an afterburning engine aspirated by a nose inlet in the much lengthened fuselage. The first YF-100 flew in April 1953 with the Pratt & Whitney J57 turbojet, and proved capable of exceeding Mach 1 without difficulty. The F-100A entered service in September 1954 with the 6,804kg (15,000 lb) thrust J57-P-7 or 7,257kg (16,000 lb) thrust J57-P-39. These 203 fighters were followed by 475 F-100C fighter-bombers with the J57-P-21 engine, inflight-refuelling capability and eight underwing hardpoints.

The version produced in largest numbers, however, was the F-100D attack fighter with the 7,711kg (17,000 lb) thrust J57-P-21A engine, trailing-edge flaps and provision for tactical nuclear weapons. These 1,274 aircraft were followed by 339 F-100F tandem-seat trainers. There were also a number of reconnaissance and trainer variants produced by converting aircraft of the production variants.

NORTH AMERICAN F-100D SUPER SABRE

Role: Strike fighter
Crew/Accommodation: One
Power Plant: One 7,711 kgp (17,000 lb s.t.) Pratt & Whitney J57-P-21A turbojet with reheat
Dimensions: Span 11.82m (38.78 ft); length 15.03m (49.31 ft); wing area 37.18m^2 (400.18 sq ft)
Weights: Empty 9,361kg (20,638 lb); MTOW 17,258kg (38,048 lb)
Performance: Maximum speed 1,492km/h (805 knots) Mach 1.39 at 10,670m (35,000 ft); operational ceiling 15,636m (51,300 ft); radius 448km (242 naut. miles) with 1,361 kg (3,000 lb) of weapons
Load: Four 20mm cannon, plus up to 3,402 kg (7,500 lb) of externally carried weapons, including tactical nuclear bombs

NORTH AMERICAN NA-247 (A-5 VIGILANTE) (U.S.A.) ▲

This was designed as a Mach 2 all-weather carrierborne strike aeroplane, and the YA3J-1 prototype first flew in August 1958 with 6,804kg (15,000 lb) thrust General Electric J79-GE-2 afterburning turbojets aspirated via the first variable-geometry inlets fitted on an operational type. The A3J-1 began to enter service in June 1961, and just over a year later was redesignated A-5A. These 57 aircraft were followed by the A-5B long-range version, of which only six were built, as a change in the U.S. Navy's strategic nuclear role led to the Vigilante's adaptation for the reconnaissance role. Thus 55 RA-5Cs were built, being complemented by 53 A-5As converted to the same configuration.

NORTH AMERICAN A-5A VIGILANTE

Role: Naval carrierborne nuclear bomber
Crew/Accommodation: Two
Power Plant: Two 7,324 kgp (16,150 lb s.t.) General Electric J79-GE-2/4/8 turbojets with reheat
Dimensions: Span 16.15m (53 ft); length 23.11m (75.83 ft); wing area 71.45m^2 (769 sq ft)
Weights: Empty 17,009kg (37,498 lb); MTOW 36,287kg (80,000 lb)
Performance: Maximum speed 2,229km/h (1,203 knots) Mach 2.1 at 12,192m (40,000 ft); operational ceiling 14,326m (47,000 ft); range 3,862km (2,084 naut. miles) with nuclear weapons
Load: One multi-megaton warhead class nuclear weapon

NORTHROP DELTA
(U.S.A.)

The Delta was a nine-seat transport developed by mating the flying surfaces of the earlier Gamma to a new fuselage. The type was denied an American airline role as the 1934 Air Commerce Act prohibited the movement of passengers at night or over rough country in a single-engined aeroplane. Three had been sold before the advent of the Act, another eight were executive transports and one was sold to AB Aerotransport in Sweden. Another 20 aircraft were built by Canadian Vickers with wheel or float landing gear.

NORTHROP DELTA 1A

Role: Passenger/executive transport
Crew/Accommodation: One, plus up to eight passengers
Power Plant: One 710 hp Wright SR-1820-F3 Cyclone air-cooled radial
Dimensions: Span 14.55m (47.75 ft); length 10.44m (34.25 ft); wing area 33.7m^2 (363 sq ft)
Weights: Empty 1,860kg (4,100 lb); MTOW 3,175kg (7,000 lb)
Performance: Cruise speed 301km/h (187 mph) at sea level; operational ceiling 6,095m (20,000 ft); range 2,495km (1,550 miles)
Load: Up to 708 kg (1,560 lb)

NORTHROP P-61 BLACK WIDOW
(U.S.A.)

This was the U.S.A.'s first purpose-designed night-fighter, a massive twin-boom machine with nose radar, nose-mounted cannon and a four-gun remotely controlled barbette above the rear of the central nacelle. The two XP-61 prototypes and 13 YP-31 service test aircraft were followed by 200 P-61A production aircraft that began to enter service in late 1943, the last 163 without the dorsal barbette. There followed 450 P-61Bs with provision for four drop tanks or 726kg (1,600 lb) bombs on underwing hard-points, and the last 250 of these again carried the barbette. The last production model was the P-61C with 2,088kW (2,800 hp) R-2800-73 radial engines; only 41 had been completed by the end of World War II, when further production was cancelled.

NORTHROP P-61A BLACK WIDOW

Role: Night fighter
Crew/Accommodation: Three
Power Plant: Two 2,000 hp Pratt & Whitney R-2800-65 Double Wasp air-cooled radials
Dimensions: Span 20.12m (66 ft); length 14.88m (48.83 ft); wing area 61.5m^2 (662 sq ft)
Weights: Empty 9,510kg (20,965 lb); MTOW 12,701kg (28,000 lb)
Performance: Maximum speed 594km/h (369 mph) at 6,096m (20,000 ft); operational ceiling 10,088m (33,100 ft); range 2,366km (1,470 miles)
Load: Four 20mm cannon and four .5 inch machine guns (on first 37 P-61As only). Interception directed by SCR-720 radar

NORTHROP F-89 SCORPION
(U.S.A.)

The F-89 was designed as replacement for the P-61 in the all-weather and night-fighting roles, and first flew during August 1958 in XF-89 prototype form. The type entered service as the F-89A with Allison J33-A-21 turbojets, later changed to J35-A-35 afterburning turbojets, and the last 30 of the 48 F-89As were delivered as F-89Bs with an autopilot. The 164 F-89Cs had a revised tail, and the 682 F-89Ds had a new Hughes fire-control system and, instead of the original six 20mm cannon, wingtip pods housing a total of 104 unguided air-to-air rockets. The final production variant was the F-89H, of which 156 were built with more power and the tip pods each revised for the carriage of 21 rockets and three Falcon air-to-air missiles. The F-89J designation was applied to F-89Ds modified to carry two Genie nuclear-tipped rockets and four Falcon missiles on underwing hardpoints.

NORTHROP F-89D SCORPION

Role: Night/all-weather fighter
Crew/Accommodation: Two
Power Plant: Two 3,266kgp (7,200 lb s.t.) Allison J35-A-35 turbojets with reheat
Dimensions: Span 18.41m (60.42 ft); length 16.41m (53.83 ft); wing area 52.21m^2 (562 sq ft)
Weights: Empty 11,428kg (25,194 lb); MTOW 21,219kg (46,780 lb)
Performance: Maximum speed 1,022km/h (635 mph) at 3,231m (10,600 ft); operational ceiling 15,000m (49,217 ft); range 2,205km (1,370 miles)
Load: One hundred and four 2.75 inch Mighty Mouse rockets

NORTHROP F-5 FREEDOM FIGHTER and TIGER II
(U.S.A.)

The F-5 Freedom Fighter was developed from Northrop's private-venture N-156 design as a supersonic fighter with the light weight, compact dimensions and simple avionics that would suit it to operation by nations requiring an essentially defensive type of low purchase and operating costs combined with relatively simple maintenance requirements. The type was developed in F-5A single-seat and F-5B two-seat variants, and first flew in July 1959 with 1,850kg (4,850 lb) thrust General Electric J85-GE-13 afterburning turbojets. Production of the F-5A and F-5B totalled 818 and 290 respectively in variously designated versions for a number of countries.

However, the mantle of the type had already been assumed by the more capable Tiger II produced in F-5E single-seat and F-5F two-seat forms with 2,268kg (5,000 lb) thrust J85-GE-21 engines for much improved payload and performance. Deliveries began in 1972, and large scale production followed before the line was closed in the mid-1980s. There are also RF-5A and RF-5E reconnaissance versions.

NORTHROP F-5E TIGER II

Role: Strike fighter
Crew/Accommodation: One
Power Plant: Two 2,268kgp (5,000 lb s.t.) General Electric J85-GE-21 turbojets with reheat
Dimensions: Span 8.13m (26.66 ft); length 14.68m (48.16 ft); wing area 17.3m^2 (186.2 sq ft)
Weights: Empty 4,392kg (9,683 lb); MTOW 11,195kg (24,680 lb)
Performance: Maximum speed 1,730km/h (934 knots) Mach 1.63 at 11,000m (36,090 ft); operational ceiling 15,790m (51,800 ft); radius 222km (138 miles) with full warload
Load: Two 20mm cannon, plus up 3,175kg (7,000 lb) of ordnance, including two short-range air-to-air missiles

NORTHROP T-38 TALON
(U.S.A.)

The Talon uses basically the same airframe/power plant combination as the F-5A Freedom Fighter, but is a supersonic advanced trainer. The first YT-38 prototype flew in December 1956, and production of 1,187 T-38As were built up to 1972. Some have been modified as AT-38A attack trainers.

NORTHROP T-38A TALON

Role: Advanced trainer
Crew/Accommodation: Two
Power Plant: Two 1,746kgp (3,850 lb s.t.) General Electric J85-GE-5 turbojets with reheat
Dimensions: Span 7.72m (25.33 ft); length 14.12m (46.33 ft); wing area 15.79m^2 (170 sq ft)
Weights: Empty 3,361kg (7,410 lb); MTOW 5,335kg (11,761 lb)
Performance: Maximum speed 1,483km/h (800 knots) Mach 1.33 at 10,973m (36,000 ft); operational ceiling 14,021m (46,000 ft); range 1,566km (845 naut. miles)
Load: None

A Canadian CF-5D

NORTHROP B-2
(U.S.A.)

Due to enter service in the early 1990s after a first flight late in 1989, the B-2 is an almost unbelievably advanced and expensive aeroplane of the flying wing type without vertical surfaces. The design and construction are optimized to reduce the airframe's radar reflectivity and curtail heat emission from the electronics and engines. The type is designed for the strategic role with a heavy load of conventional or nuclear weapons, and should operate in the electronically silent mode relying on inertial and satellite navigation systems.

NORTHROP B-2

Role: Long-range, low-detectability bomber
Crew/Accommodation: Two/three
Power Plant: Four 8,620kgp (19,000 lb s.t.) General Electric F118-GE-100 turbofans
Dimensions: Span 52.43m (172 ft); length 21.03m (69 ft)
Weights: MTOW 136,080kg (300,000 lb)
Performance: Maximum speed 1,012km/h (546 knots) at 15,240m (50,000 ft); range 11,110+km (6,000+naut. miles) unrefuelled
Load: Up to sixteen SRAM II, AGM-129 or B83 nuclear bombs stowed internally and carried on rotary dispensers within three weapons bays.
Note: data on empty weight and operational ceiling remain classified

PANAVIA TORNADO
(Italy/U.K./West Germany)

Currently one of the NATO alliance's premier front-line aircraft types, the Tornado was planned from the late 1960s as a multi-role combat aeroplane able to operate from and into short or damaged runways for long-range interdiction missions at high speed and very low level. The keys to the mission are a variable-geometry design with wings carrying extensive leading- and trailing-edge high-lift devices, compact turbofan engines fitted with thrust reversers, and an extremely advanced sensor and electronic suite.

The first of nine Tornado prototypes flew in August 1974, and after a protracted development the first of 1,000+ Tornados for the U.K., Germany and Italy and for export began to enter service in 1980. The three variants are the Tornado IDS baseline interdiction and strike aeroplane, the Tornado ADV air-defence fighter with revised weapons and radar in a longer fuselage, and the Tornado ECR electronic combat and reconnaissance type. From the early 1990s the type is to be updated with more modern electronics suited to the carriage and delivery of the most advanced 'smart' weapons.

PANAVIA TORNADO GR. Mk 1

Role: All-weather, low-level strike reconnaissance
Crew/Accommodation: Two
Power Plant: Two 7,257kgp (16,000 lb s.t.) Turbo-Union BB199 Mk 101 turbofans with reheat
Dimensions: Span spread 13.9m, swept 8.6m (spread 45.6 ft swept 28.2 ft); length 16.7m (54.8 ft); wing area 30m^2 (322.9 sq ft)
Weights: Empty 10,433kg (23,000 lb); MTOW 26,490kg (58,400 lb)
Performance: Maximum speed 1,483km/h (800 knots) Mach 1.2 at 152m (500 ft); operational ceiling 15,240+m (50,000+ ft); range 1,062km (573 naut. miles) unrefuelled with 1,657kg (3,652 lb) warload
Load: Two 27mm cannon, plus up to 7,257 kg (16,000 lb) of externally carried weaponry or fuel

ROYAL SAUDI AIR FORCE

PERCIVAL P.6 MEW GULL ▲
(U.K.)

First flown in March 1934, the P.2 prototype was of angular and somewhat austere appearance that gave little hint of the beautiful P.6 Mew Gull derivatives it would sire. Altogether, four examples of the P.6 were to be built and these subsequently dominated the British air racing scene during the three years prior to September 1939 and the outset of war. Of exceptionally well-proportioned shape, the P.6 Mew Gulls were constantly in the headlines, frequently being flown by the aircraft's designer/pilot Captain Edgar Percival in such events as the annually held King's Cup air race.

PERCIVAL P.6 MEW GULL

Role: Racer
Crew/Accommodation: One
Power Plant: One 205 hp De Havilland Gipsy Six Series II air-cooled inline
Dimensions: Span 7.54m (24.75 ft); length 6.88m (21.92 ft); wing area 8.18m^2 (88 sq ft)
Weights: Empty 562kg (1,240 lb); MTOW 1,066kg (2,350 lb)
Performance: Maximum speed 398km/h (247 mph) at sea level; range 3,219km (2,000 miles) with 3,861l (85 Imp gal) tankage
Note: figures are for Alex Henshaw's modified G-AEXF as configured for February 1939, his record-breaking England-Cape Town return flight

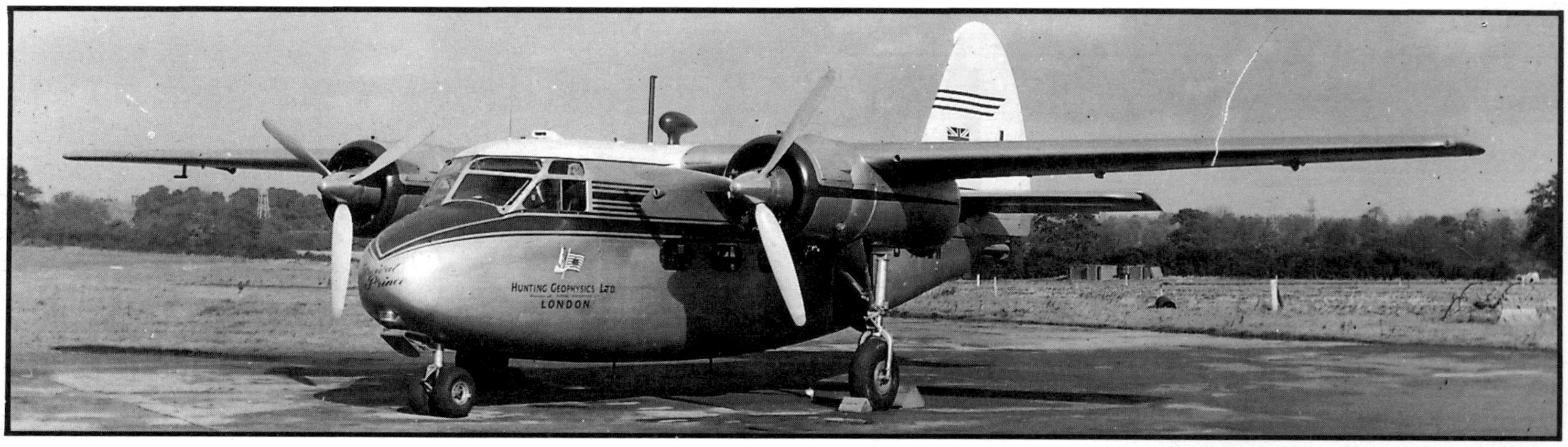

PERCIVAL P.50 PRINCE and P.66 PEMBROKE ▲
(U.K.)

First flown in May 1947, the P.48 Merganser was a trim five-seat transport with retractable tricycle landing gear and two 221kW (296 hp) de Havilland Gipsy Queen 51 inline engines. Production of the engine was then halted, and Percival revised the type as the larger P.50 Prince with 388kW (520 hp) Alvis Leonides 501/4 radial engines. This flew in May 1948 and was followed by 24 production aircraft with engines in the 388/418kW (520/560 hp) power range. Further development produced the P.66 Pembroke suitable for the military communications role, and over 100 were produced in basic and Sea Prince naval forms.

PERCIVAL PRINCE

Role: Short-range passenger/cargo transport
Crew/Accommodation: Two, plus up to 12 passengers
Power Plant: Two 520 hp Alvis Leonides 501/4 air-cooled radials
Dimensions: Span 17.3m (56 ft); length 14.12m (46.33 ft); wing area 33.9m^2 (365 sq ft)
Weights: Empty 3,340kg (7,364 lb); MTOW 7,500kg (10,650 lb)
Performance: Cruise speed 288km/h (179 mph) at 1,525m (5,000 ft); operational ceiling 7,160m (23,500 ft); range 925km (575 miles) with maximum payload
Load: Up to 910kg (2,018 lb)

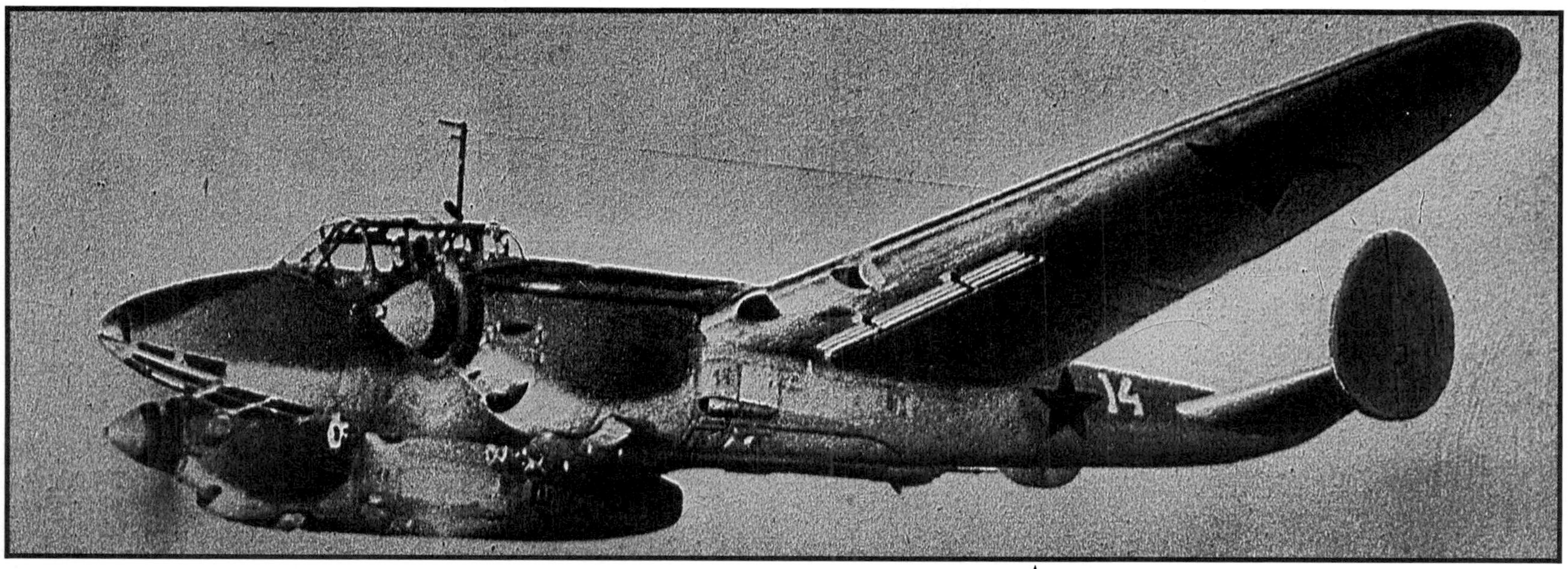

PETLYAKOV Pe-2
(U.S.S.R.)

This was one of the U.S.S.R.'s most important tactical aircraft of World War II, and resulted from the VI-10 high altitude fighter prototype with 783kW (1,050 hp) Klimov M-105 inline engines. The planned role was then changed to dive-bombing, resulting in the PB-100 design with a crew of three rather than two, dive-brakes and other modifications. The type entered service as the Pe-2 during November 1940, and eventually 11,427 were built by early 1945 in variants intended for the bombing, reconnaissance, bomber-destroyer, night-fighter and conversion trainer roles.

PETLYAKOV Pe-2

Role: Dive bomber
Crew/Accommodation: Three/four
Power Plant: Two 1,100 hp Klimov M-105R water-cooled inlines
Dimensions: Span 17.16m (56.23 ft); length 12.66m (41.54 ft); wing area 40.5m^2 (435.9 sq ft)
Weights: Empty 5,876kg (12,954 lb); MTOW 8,496kg (18,730 lb)
Performance: Maximum speed 540km/h (336 mph) at 5,000m (16,404 ft); operational ceiling 8,800m (28,871 ft); range 1,500km (932 miles)
Load: One 12.7mm and two 7.62mm machine guns, plus up to 1,200kg (2,646 lb) of bombs

PETLYAKOV Pe-8
(U.S.S.R.)

This was the U.S.S.R.'s only advanced heavy bomber of World War II, but was built to the extent of only 81 aircraft by the end of production in October 1941. The type first flew as the ANT-42 in December 1936. In its developed form, epitomized by the second prototype, the most distinctive features were a piston-engined supercharger in the fuselage to provide air to all four 820kW (1,100 hp) Mikulin AM-100 inline engines, and a machine gun turret in the rear end of the inboard engine nacelles. The production aircraft had more power and were initially designated TB-7, and from 1942 at least 48 were re-engined with 1,268kW (1,700 hp) Shvetsov ASh-82FN radial engines.

PETLYAKOV Pe-8 (TB-7)

Role: Long-range bomber
Crew/Accommodation: Ten
Power Plant: Four 1,350 hp Mikulin AM-35A water-cooled inlines
Dimensions: Span 39.94m (131.04 ft); length 22.47m (73.72 ft); wing area 190m^2 (2,045 sq ft)
Weights: Empty 17,825kg (39,297 lb); MTOW 33,325kg (73,469 lb)
Performance: Maximum speed 438km/h (272 mph) at 7,600m (24,934 ft); operational ceiling 7,000m (22,966 ft); range 4,700km (2,920 miles)
Load: Two 20mm cannon, two 12.7mm and 7.62mm machine guns, plus up to 4,000kg (8,818 lb) of internally carried bombs

PFALZ D III (Germany)

The D III was an elegant fighter whose capabilities have been overshadowed by the success of the Albatros V-strutters. The type was designed after Pfalz had undertaken licence-construction of LFG fighters and bore a similarity to these machines when it appeared in the summer of 1917. The D III entered service late in 1917 with the 119kW (160 hp) Mercedes D.III inline engine. Included in the 600 or so production aircraft were numbers of the D IIIa variant with the 134kW (180 hp) Mercedes D.IIIa engine.

PFALZ DIIIa

Role: Fighter
Crew/Accommodation: One
Power Plant: One 180 hp Mercedes DIIIa water-cooled inline
Dimensions: Span 9.4m (30.84 ft); length 6.95m (22.8 ft); wing area 22.17m^2 (237.75 sq ft)
Weights: Empty 695kg (1,532 lb); MTOW 933kg (2,056 lb)
Performance: Maximum speed 165km/h (102.5 mph) at 3,050m (10,000 ft); operational ceiling 5,500+m (18,045 ft); endurance 2.5 hours
Load: Two 7.9mm machine guns

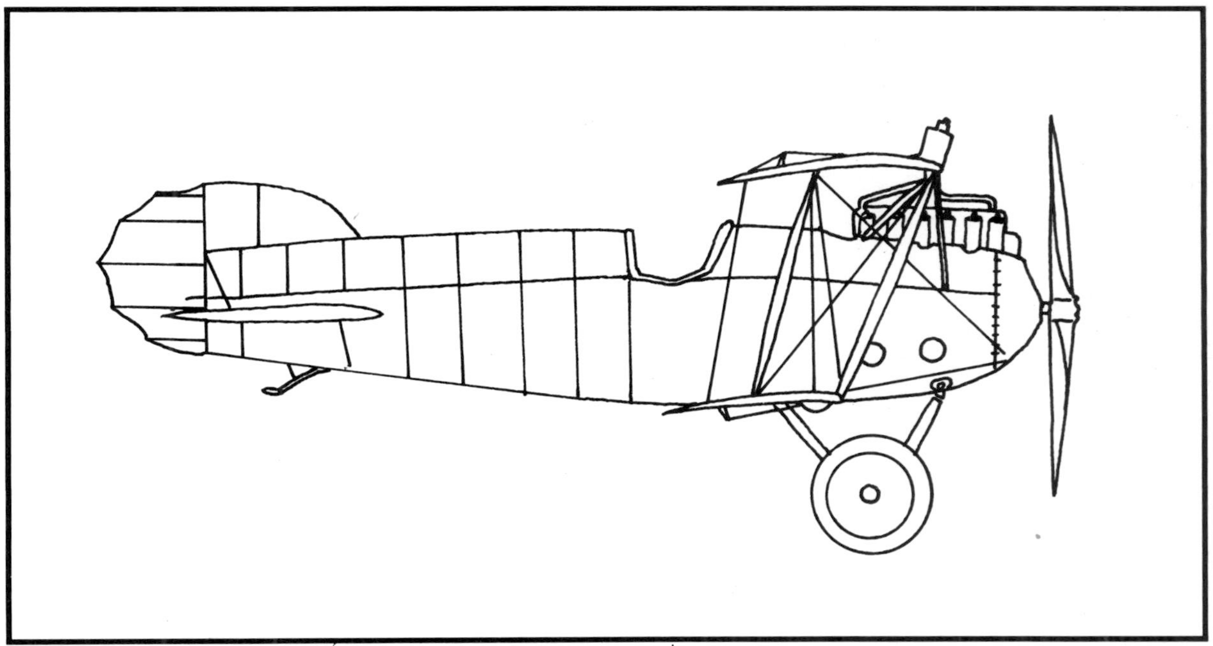

PHOENIX D I to D III (Austria-Hungary)

Phoenix licence-built Hansa-Brandenburg aircraft before turning to its own fighter design, therefore, it is little surprising that the D I which appeared in mid-1917 resembled the German type strongly. In fact the Austro-Hungarian aeroplane can be regarded as a development of the Hansa-Brandenburg D I. The type entered production with the 149kW (200 hp) Hiero inline engine, but was rapidly developed into the D II with balanced elevators and upper-wing ailerons. Further refinement led to the D III with a 172kW (230 hp) Hiero inline engine. Total production of the series was 158 up to November 1918.

PHOENIX D III

Role: Fighter
Crew/Accommodation: One
Power Plant: One 230 hp Hiero water-cooled inline
Dimensions: Span 9.88m (32.42 ft); length 9.14m (30 ft)
Weights: Empty 681kg (1,501 lb); MTOW 831kg (1,831 lb)
Performance: Maximum speed 200km/h (125 mph) at sea level; endurance 3 hours
Load: Two 7.9mm machine guns

PIAGGIO P.108
(Italy)

This was Italy's only heavy bomber of World War II, a workmanlike but unexceptional type notable only for its double-stepped nose contour. The aeroplane was developed from the P.50 II and first flew in 1939. The type was planned in anti-ship, bomber, airliner and military transport versions, but the only type to enter more than limited production was the P.108B bomber. There were a few P.108C airliners, these and some of the P.108Bs later being converted into military transports. Total production was 163.

PIAGGIO P.108B

Role: Long-range heavy bomber
Crew/Accommodation: Seven
Power Plant: Four 1,500 hp Piaggio P.XII RC35 air-cooled radials
Dimensions: Span 32m (104.99 ft); length 22.29m (73.13 ft); wing area 135m^2 (1,453.1 sq ft)
Weights: Empty 17,325kg (38,195 lb); MTOW 29,885kg (65,885 lb)
Performance: Maximum speed 430km/h (267 mph) at 4,200m (13,780 ft); operational ceiling 8,500m (27,890 ft); range 3,520km (2,187 miles) with 2,000kg (4,409 lb) warload
Load: Six 12.7mm machine guns, plus up to 3,500kg (7,716 lb) of bombs

PIAGGIO P.166
(Italy)

First flown in November 1957, the P.166 was an Italian attempt to break into the business transport market, and its most distinctive features were the high wing and pusher engines inherited from the P.136 amphibian design. The P.166 enjoyed only limited civil success, but was built in more substantial numbers for the military in a number of forms.

PIAGGIO P.166C

Role: Executive/passenger transport
Crew/Accommodation: One, plus up to 12 passengers
Power Plant: Two 380 hp Lycoming IGSO-540-A1C air-cooled flat-opposed
Dimensions: Span 14.33m (47.01 ft); length 11.89m (39.01 ft); wing area 26.56m^2 (285.9 sq ft)
Weights: Empty 2,640kg (5,820 lb); MTOW 3,950kg (8,708 lb)
Performance: Cruise speed 337km/h (209 mph) at 4,572m (15,000 ft); operational ceiling 8,230m (27,000 ft); range 1,170km (727 miles)
Load: Up to 1,000kg (2,205 lb)

PIAGGIO P.180 AVANTI
(Italy)

Due to enter service in 1989, the Avanti is typical of the modern breed of fast executive transports. It embodies advanced aerodynamics (in this instance a canard configuration), a comparatively high proportion of advanced materials in its structure, and twin pusher turboprops. The whole package offers not only high performance but also modest purchase and operating costs together with straight forward maintenance.

PIAGGIO P.180 AVANTI

Role: Executive transport
Crew/Accommodation: Two, plus up to eight passengers
Power Plant: Two 850 hp Pratt & Whitney of Canada PT6A-66 turboprops
Dimensions: Span 13.84m (45.42 ft); length 14.17m (46.48 ft); wing area 15.76m^2 (169.64 sq ft)
Weights: Empty 3,130kg (6,900 lb); MTOW 4,767kg (10,510 lb)
Performance: Maximum speed 740km/h (400 knots) at 8,230m (27,000 ft); operational ceiling 12,495m (41,000 ft); range 3,335km (1,800 naut. miles) with 4 passengers
Load: Up to 907kg (2,000 lb)

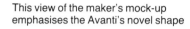

This view of the maker's mock-up emphasises the Avanti's novel shape

PILATUS PC-6 PORTER and TURBO-PORTER
(Switzerland)

With sales of over 500 aircraft achieved, the PC-6 has been one of the major successes of Switzerland's small aerospace industry. The type was designed as a STOL utility transport and first flew in May 1959 with the 254kW (340 hp) Avco Lycoming GSO-480/B1A6 piston engine. Sales were made of this initial version and also of the PC-6/350 with the 261kW (350 hp) IGO-540-A1A. Further capability was added in the Turbo-Porter by the adoption of a turboprop to produce the PC-6/A with the Turbomeca Astazou, the PC-6/N with the Pratt & Whitney Canada PT6A and the PC-6/C with the Garrett TPE331.

PILATUS PC-6/B2-H4 TURBO-PORTER

Role: Short take-off and landing utility transport
Crew/Accommodation: One, plus up to ten passengers
Power Plant: One 680 hp Pratt & Whitney Canada PT6A-27 turboprop
Dimensions: Span 15.87m (52.07 ft); length 11m (36.09 ft); wing area 30.15m^2 (324.5 sq ft)
Weights: Empty 1,270kg (2,800 lb); MTOW 2,800kg (6,173 lb)
Performance: Cruise speed 213km/h (132 mph) at 3,050m (10,000 ft); operational ceiling 7,620m (25,000 ft); range 730km (453 miles) with full payload
Load: Up to 1,130kg (2,491 lb)

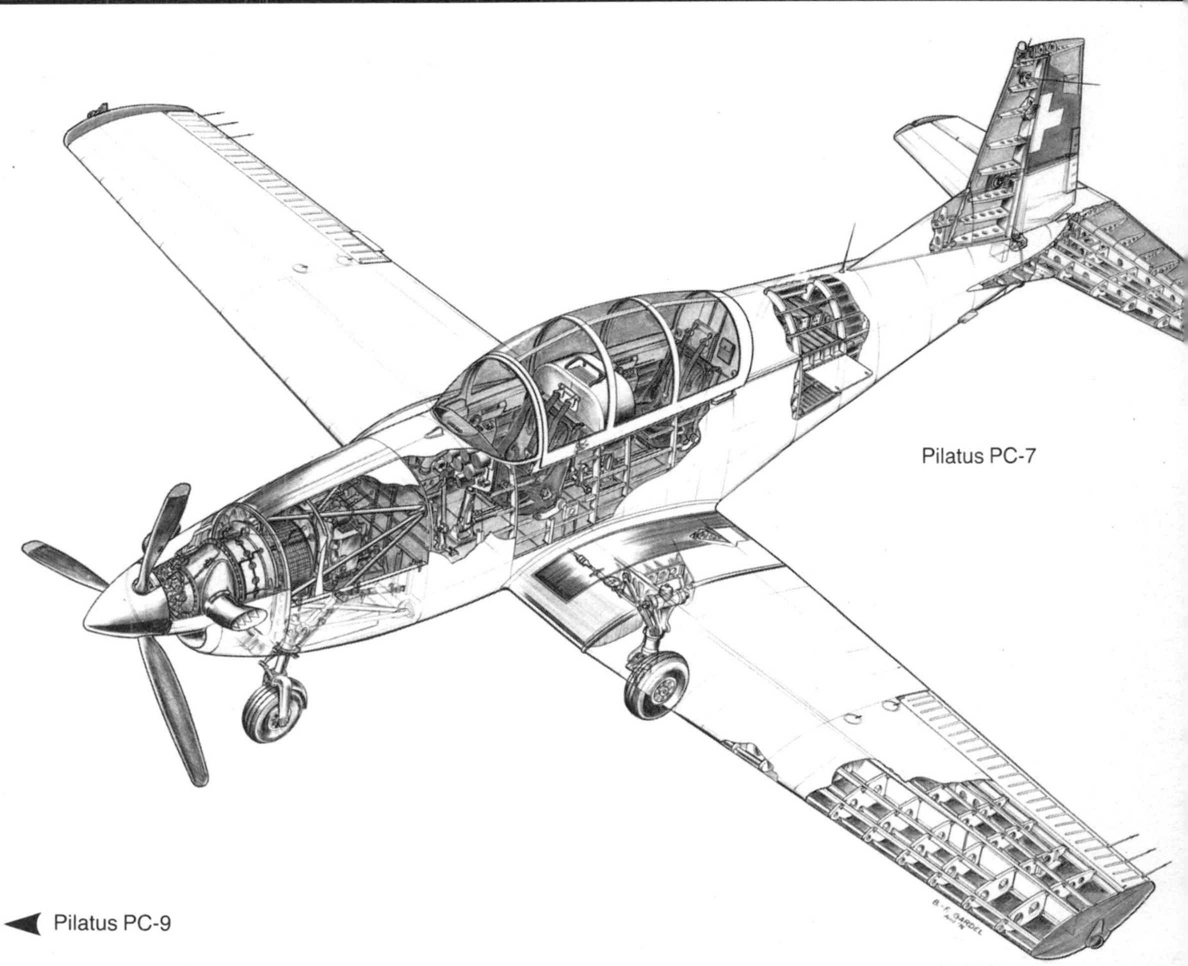

Pilatus PC-7

◄ Pilatus PC-9

PILATUS PC-7 TURBO-TRAINER and PC-9
(Switzerland)

The PC-7 is a turboprop-powered development of the P-3 piston-engined trainer, and first flew in April 1966 as the P-3B with the 485kW (650 shp) Pratt & Whitney Canada PT6A-25. The first PC-7 production aeroplane followed only in August 1978 as most military operators were content with piston- or turbojet-powered trainers. Further development produced the PC-9 which first flew in July 1984 with a 857kW (1,150 shp) PT6A-62, reduced-span wing and other structural changes for much enhanced performance.

PILATUS PC-7 TURBO-TRAINER

Role: Light strike/trainer
Crew/Accommodation: Two
Power Plant: One 550 shp Pratt & Whitney Canada PT6A-25A turboprop
Dimensions: Span 10.4m (34.1 ft); length 9.78m (32.1 ft); wing area 16.6m^2 (178.7 sq ft)
Weights: Empty 1,130kg (2,932 lb); MTOW 2,700kg (5,950 lb)
Performance: Maximum speed 411km/h (222 knots) at sea level; operational ceiling 9,755m (32,000 ft); range 1,350km (730 naut. miles)
Load: Up to 800kg (1,791 lb) externally carried

PIPER J-3 CUB and L-4 GRASSHOPPER
(U.S.A.)

First flown in September 1930 as the Taylor Cub, this was Piper's first production aeroplane after the Taylor brothers had been bought out by W.T.Piper in 1937. The Cub is a classic braced high-wing monoplane with any of several flat-four piston engines. Some 14,125 Cubs were built in several series, and another 5,703 military liaison and observation aircraft expanded this number with the original designation O-59 that was soon changed to L-4. Such has been the abiding popularity of the type that it was reinstated in production during 1988 with a number of refined features that fail to obscure the essentially simple nature of the basic aeroplane.

PIPER L-4 GRASSHOPPER

Role: Observation/communications
Crew/Accommodation: Two
Power Plant: One 65 hp Continental 0-170-3 air-cooled flat-opposed
Dimensions: Span 10.74m (35.25 ft); length 6.71m (22 ft); wing area 16.63m^2 (179 sq ft)
Weights: Empty 331kg (730 lb); MTOW 553kg (1,220 lb)
Performance: Maximum speed 137km/h (85 mph) at sea level; operational ceiling 2,835m (9,300 ft); range 306km (190 miles)
Load: Up to 86kg (190 lb)

PIPER PA-18 SUPER CUB and L-18/L-21 SERIES
(U.S.A.)

The Super Cub was a refined development of the Cub with engines in the 67/112kW (90/150 hp) range. The type first flew in November 1949 and remained in production up to 1981, being built to the extent of 6,650 aircraft. The type was also used by the military as the L-18 with the 71kW (90 hp) Continental C90-8F, or L-21 with the 93kW (125 hp) Avco Lycoming O-290-11 engine: military production was 838.

PIPER PA-18 SUPER CUB

Role: Tourer/trainer
Crew/Accommodation: Two/one, plus one passenger
Power Plant: One 90 hp Continental C90 air-cooled flat-opposed
Dimensions: Span 10.72m (35.21 ft); length 6.82m (22.38 ft); wing area 16.58m^2 (178.5 sq ft)
Weights: Empty 360kg (790 lb); MTOW 680kg (1,500 lb)
Performance: Cruise speed 160km/h (100 mph) at sea level; operational ceiling 4,120m (13,500 ft); range 491km (305 miles)
Load: Up to 84kg (185 lb)

PIPER PA-22 TRI-PACER (U.S.A.)

The Tri-Pacer of 1951 was in essence the four-seat PA-20 Pacer revised with tricycle rather than tailwheel landing gear and with a 112kW (150 hp) Avco Lycoming O-320 rather than O-290 engine. The type was taken out of production in the early 1960s after the construction of 7,668 aircraft.

PIPER PA-22-160 TRI-PACER

Role: Tourer
Crew/Accommodation: One, plus up to three passengers
Power Plant: One 150 hp Lycoming O-320 air-cooled flat-opposed
Dimensions: Span 8.92m (29.27 ft); length 6.28m (20.6 ft); wing area 13.7m^2 (147.5 sq ft)
Weights: Empty 503kg (1,110 lb); MTOW 907kg (2,000 lb)
Performance: Cruise speed 216km/h (134 mph) at 2,134m (7,000 ft); operational ceiling 5,029m (16,500 ft); range 863km (536 miles) with three passengers
Load: Up to 231kg (510 lb)

PIPER PA-23 APACHE (U.S.A.)

First flown in March 1952, the PA-23 Twin-Stinson was a low-wing monoplane with a retractable tricycle landing gear and a tail unit terminating in endplate vertical surface. By the time the PA-23 reached the production line it had been recast as the PA-23 Apache using a conventional tail with a single vertical surface. The original Apache 150 was built to the extent of 2,166 aircraft with 112kW (150 hp) Avco Lycoming O-320 engines, and was followed by 816 Apache 160s with 119kW (160 hp) O-320-B engines, and 119 five/six-seat Apache 235s with a swept fin and equipped with 175kW (235 hp) O-540-B1A5 engines.

PIPER PA-23 APACHE 235

Role: Light executive transport/air taxi
Crew/Accommodation: One, plus up to four passengers
Power Plant: Two 235 hp Lycoming 0-540-B1A5 air-cooled flat-opposed
Dimensions: Span 11.33m (37.17 ft); length 8.41m (27.59 ft); wing area 19.23^2 (207 sq ft)
Weights: Empty 1,240kg (2,735 lb); MTOW 2,177kg (4,800 lb)
Performance: Cruise speed 307km/h (191 mph) at 2,135m (7,000 ft); operational ceiling 5,240m (17,200 ft); range 1,905km (1,185 miles) with four passengers
Load: Up to 336kg (740 lb)

PIPER PA-23 AZTEC
(U.S.A.)

The Aztec was developed to restore flagging sales in the Apache series, and was a more advanced six-seat aeroplane. The type entered production as the PA-23-235 in 1959 and remained available up to 1982 in a series of steadily more capable forms, the ultimate version being the PA-23T-350 Turbo Aztec F with 186 kW (250 hp) Avco Lycoming T10-540-C1A turbocharged engines.

PIPER PA-23 AZTEC F

Role: Light executive transport/air taxi
Crew/Accommodation: One, plus up to five passengers
Power Plant: Two 250 hp Lycoming T10-540-C1A air-cooled flat-opposed
Dimensions: Span 11.4m (37.3 ft); length 9.5m (31.2 ft); wing area 14.25m^2 (207 sq ft)
Weights: Empty 1,383kg (3,049 lb); MTOW 2,360kg (5,200 lb)
Performance: Cruise speed 331km/h (206 mph) at 1,170m (3,850 ft); operational ceiling 5,368m (17,600 ft); range 1,546km (835 miles) with full payload
Load: Up to 556kg (1,225 lb)

PIPER PA-24 COMANCHE
(U.S.A.)

The first of the Comanche series flew in May 1956 as the PA-24 with a retractable tricycle landing gear and a 134kW (180 hp) Avco Lycoming O-360-A1A engine. The type entered production as the PA-24-180, and production of the series totalled 4,708 by the time construction ended in 1973. The main variants, other than the PA-24-180, were the PA-24-250 and PA-24T-260 in which the last part of the designation indicates engine horsepower.

PIPER PA-24-250 COMANCHE

Role: Tourer
Crew/Accommodation: One, plus up to three passengers
Power Plant: One 250 hp Lycoming 0-540 air-cooled flat-opposed
Dimensions: Span 10.97m (36 ft); length 7.57m (24.84 ft); wing area 16.53m^2 (178 sq ft)
Weights: Empty 766kg (1,690 lb); MTOW 1,315kg (2,900 lb)
Performance: Cruise speed 290km/h (180 mph) at 2,440m (8,000 ft); operational ceiling 6,100m (20,000 ft); range 1,770km (1,100 miles) with three passengers
Load: Up to 252 kg (555 lb)

PIPER PA-31 NAVAJO
(U.S.A.)

When first flown in September 1964 as a prototype business aeroplane, the PA-31 Inca was the largest aeroplane then produced by Piper. The type had been renamed Navajo by the time production aircraft were being delivered in 1967. The basic Navajo remained in production up to 1972 in basic, turbocharged PA-31 Turbo Navajo, and PA-31P Pressurized Navajo variants, and was then succeeded by the Navajo Chieftain with a lengthened fuselage. Variants of this improved model are the PA-31-325 Turbo Navajo C/R with opposite rotating engines, the PA-31-325 Navajo C/R, the PA-31-350 Chieftain and the PA-31P-350 Mojave.

PIPER PA-31 NAVAJO

Role: Light executive transport/air taxi
Crew/Accommodation: Two, plus up to six passengers
Power Plant: Two 300 hp Lycoming 10-540-M air-cooled flat opposed
Dimensions: Span 12.4m (40.67 ft); length 9.9m (32.48 ft); wing area 21.3m^2 (229 sq ft)
Weights: Empty 1,632kg (3,603 lb); MTOW 2,809kg (6,200 lb)
Performance: Cruise speed 338km/h (210 mph) at 1,951m (6,400 ft); operational ceiling 6,248m (20,500 ft); range 2,414km (1,500 miles) with full payload
Load: Up to 662kg (1,460 lb)

PIPER PA-28 CHEROKEE (U.S.A.)

First flown in January 1960, the Cherokee and its successors have been a remarkable success story for Piper. The type has gone through a large number of developments and variants with engine horsepower indicated by the numerical suffix appended to the basic designation. Thus, the initial PA-28-150 with fixed tricycle landing gear was followed in chronological sequence by the PA-28-160, PA-28-180, PA-28-235 and PA-28-140. An upgraded series was the Cherokee Arrow with retractable landing gear in the form of the PA-28-180R and PA-28-200R. The first series was then redesignated the PA-28-140 becoming Cherokee Flite Liner and de luxe Cherokee Cruiser 2 Plus 2, the PA-28-180 becoming the Cherokee Challenger with slightly lengthened fuselage and wings, and the PA-28-235 becoming the Cherokee Charger.

In 1974 further changes in name were made: the Cherokee 2 Plus 2 became the Cherokee Cruiser, the Cherokee Challenger became the Cherokee Archer, and the Cherokee Charger became the Cherokee Pathfinder. A new member of the family introduced at this time was the PA-28-151 Cherokee Warrior based on the Cherokee Archer with a new and longer-span wing. The Cherokee Cruiser and Cherokee Pathfinder went out of production in 1977, the year in which the PA-28-236 Dakota was introduced. In the following year there appeared the PA-28-201T Turbo Dakota that went out of production in 1980 to leave in production aircraft now designated PA-28-161 Warrior II, PA-28-181 Archer II and PA-28RT-201T Turbo Arrow IV.

PIPER PA 28-161 CHEROKEE WARRIOR II

Role: Tourer
Crew/Accommodation: One, plus up to three passengers
Power Plant: One 150 hp Lycoming 0-320-E3D air-cooled flat-opposed
Dimensions: Span 10.65m (35 ft); length 7.2m (23.8 ft); wing area 15.8m^2 (170 sq ft)
Weights: Empty 590kg (1,301 lb); MTOW 1,065kg (2,325 lb)
Performance: Cruise speed 213km/h (133 mph) at 2,438m (8,000 ft); operational ceiling 3,930m (12,700 ft); range 1,660km (720 miles) with full payload
Load: Up to 342kg (775 lb)

PIPER PA-31 and PA-42 CHEYENNE
(U.S.A.)

First flown in October 1973 as the PA-31T, the Cheyenne was based on the pressurized Navajo but introduced two 462kW (620 shp) Pratt & Whitney Canada PT6A-28 turboprops for higher performance. This was redesignated Cheyenne II when the PA-31T-1 Cheyenne I was introduced with lower-powered PT6A-11 turboprops. A lengthened version of the Cheyenne II is the PA-31T Cheyenne IIXL. Further development of the same basic aeroplane resulted in the PA-42 Cheyenne III with a longer-span wing, a longer fuselage, more powerful PT6A-41 turboprops and a T-tail. The same basic airframe is used in the Cheyenne IV with Garrett TPE331-14A/B turboprops.

PIPER PA-31T CHEYENNE

Role: Executive transport
Crew/Accommodation: Two, plus up to six passengers
Power Plant: Two 620 shp Pratt & Whitney Canada PT6A-28 turboprops
Dimensions: Span 13.01m (42.68 ft); length 10.57m (34.68 ft); wing area 21.3m^2 (229 sq ft)
Weights: Empty 2,209kg (4,870 lb); MTOW 4,082kg (9,000 lb)
Performance: Cruise speed 393km/h (244 mph) at 7,620m (25,000 ft); operational ceiling 8,840m (29,000 ft); range 2,500km (1,555 miles) with six passengers
Load: Up to 685kg (1,510 lb) excluding the two crew members

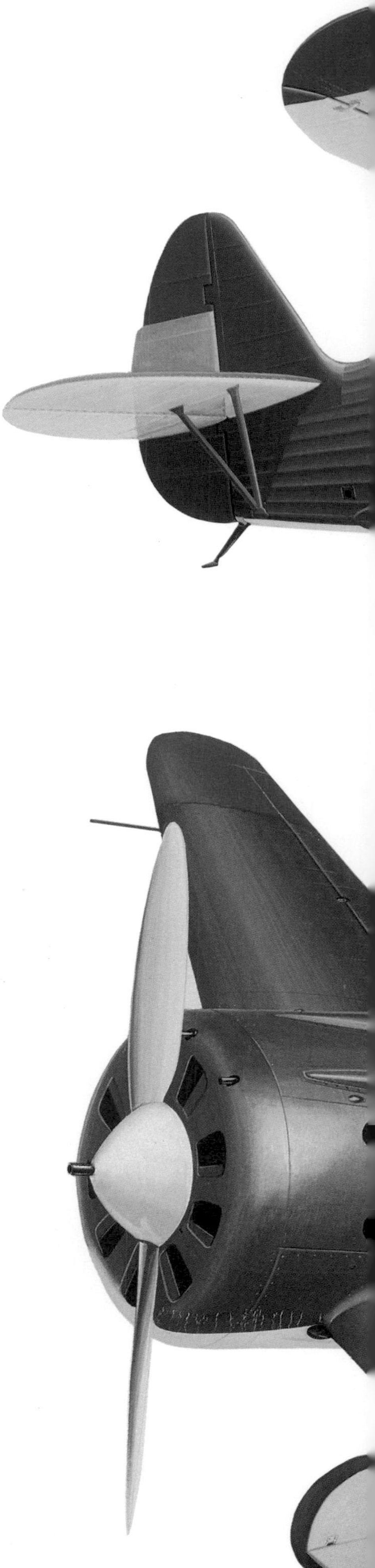

PITTS SPECIALS
(U.S.A.)

First flown in 1947 and then developed in single-seat S-1 and two-seat S-2 forms for factory- and home-built construction, the Pitts Specials are still among the world's most agile aircraft.

PITTS S-1 SPECIAL

Role: Aerobatic sportsplane
Crew/Accommodation: One
Power Plant: One 180 hp Lycoming 10-360-B4A air-cooled flat-opposed
Dimensions: Span 5.28m (17.33 ft); length 4.72m (15.5 ft); wing area 9.15m^2 (98.5 sq ft)
Weights: Empty 326kg (720 lb); MTOW 521kg (1,150 lb)
Performance: Maximum speed 285km/h (177 mph) at sea level; operational ceiling 6,795m (22,300 ft); range 507km (315 miles)
Load: None

POLIKARPOV I-15
(U.S.S.R.)

The I-15 was the U.S.S.R.'s last major biplane fighter, and a classic example of the genre with its trim plank-type interplane struts, cantilever main landing gear units and compact overall dimensions. The type first flew in October 1933 as the TsKB-3 prototype developed from the I-5 fighter. This first machine was powered by an imported radial engine, the 529kW (710 hp) Wright SGR-1820-F3 Cyclone. Initial production I-15s however, had the 358kW (480 hp) M-22 radial engine and thereby suffered considerably in performance by comparison with the prototype. These first aircraft had just two 7.62mm (0.303 inch) machine guns, but later aircraft had SGR-1820-F3 engines plus a Hamilton Standard two-pitch propeller (59 aircraft) and with the more powerful M-25 radial engine plus a two-pitch propeller had four guns of the same calibre. These 733 aircraft were followed by 2,408 examples of the I-15bis (otherwise the I-152) with the M-25V radial engine in a long-chord cowling and a straight upper-wing centre section rather than the I-15's gulled arrangement in which the inner section of the upper wings were angled down to meet the fuselage.

The final development was the I-153 which reverted to the gulled upper wing but introduced retractable landing gear. A total of 3,437 of this model was built in a number of subvariants with increasingly heavy armament.

POLIKARPOV I-153

Role: Fighter
Crew/Accommodation: One
Power Plant: One 1,000 hp Shvetsov M-62R air-cooled radial
Dimensions: Span 10m (32.81 ft); length 6.3m (20.67 ft); wing area 22.1m^2 (237.9 sq ft)
Weights: Empty 1,452kg (3,201 lb); MTOW 2,110kg (4,652 lb)
Performance: Maximum speed 443km/h (275 mph) at 5,000m (16,404 ft); operational ceiling 10,700m (35,105 ft); range 880km (547 miles)
Load: Four 7.62mm machine guns, plus up to 100kg (220 lb) of bombs or rocket projectiles

POLIKARPOV I-16
(U.S.S.R.)

This was the world's first low-wing monoplane fighter with retractable landing gear, and first flew as the TsKB-12 in December 1933 with a 358kW (480 hp) M-22 radial engine. The TsKB-12bis with an imported 529kW (710 hp) Wright SR-1820-F3 radial engine flew two months later and offered markedly better performance. The handling qualities of both variants were distinctly tricky, largely because of the very short fuselage, but the fighter was ordered into production as the I-16 Type 1 with the M-22.

Total production was 7,005 in variants with progressively more power and armament: the I-16 Type 4 used the imported Wright Cyclone, the I-16 Type 5 had the 522kW (700 hp) M-25 licensed Cyclone, the I-16 Type 6 had the 544kW (730 hp) M-25A, the I-16 Type 10 had the 559kW (750 hp) M-25V and four rather than two 7.62mm (0.312 inch) machine guns, the I-16 Type 17 was strengthened and had 20mm cannon in place of the two wing machine guns plus provision for six 8.2mm (0.320 inches) machine guns, the I-16 Type 18 had the 686kW (920 hp) M-62 radial engine and four machine guns, the I-16 Type 24 had the 746kW (1,000 hp) M-62 or 820kW (1,100 hp) M-63 radial engine, strengthened wings and four machine guns and the I-16 Types 28 and 30 that were reinstated in production during the dismal days of 1941 and 1942 had the M-63 radial engine.

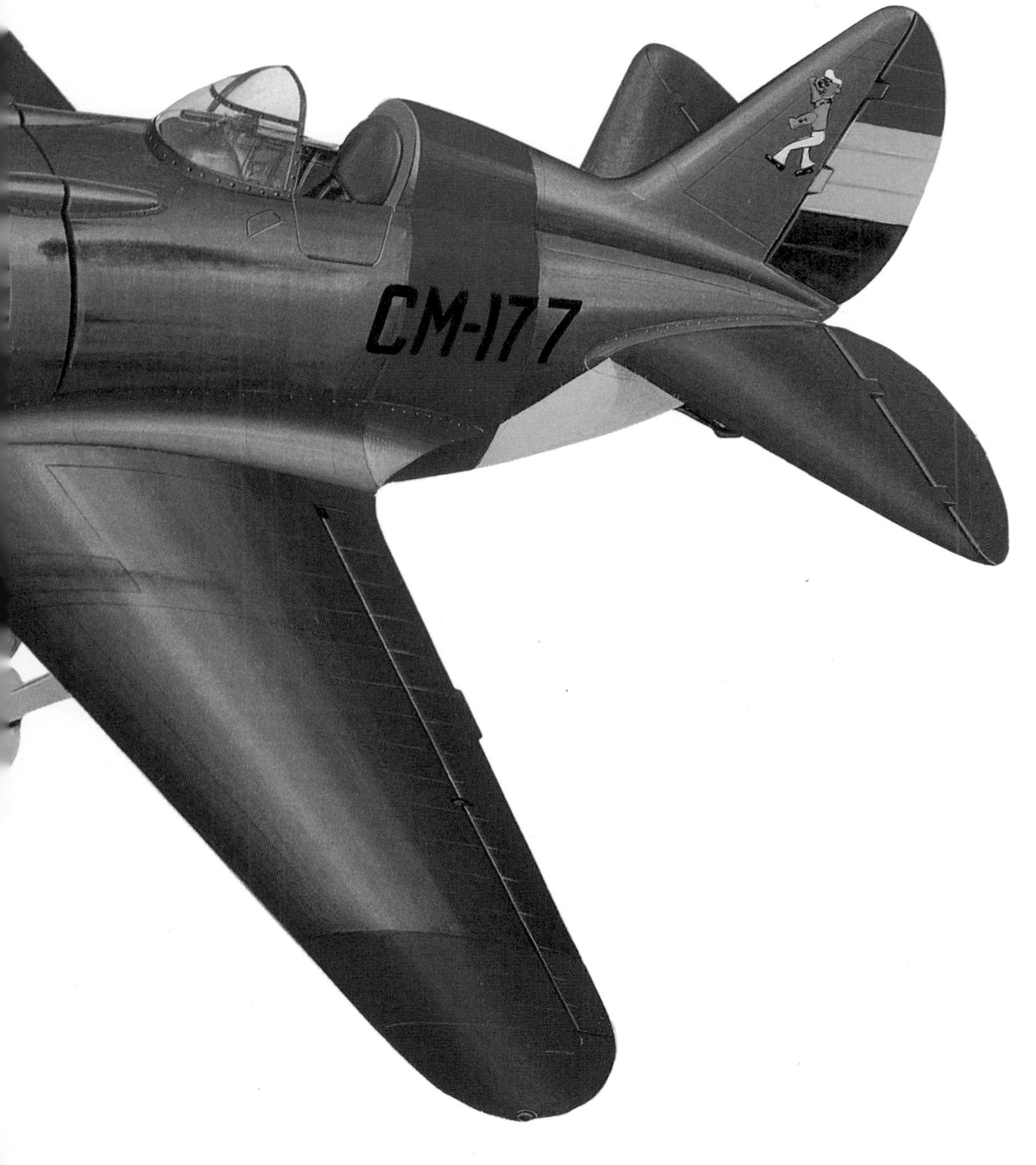

POLIKARPOV I-16 TYPE 24

Role: Fighter
Crew/Accommodation: One
Power Plant: One 1,000 hp Shvetsov M-62 air-cooled radial
Dimensions: Span 9m (29.53 ft); length 6.13m (20.11 ft); wing area 14.54m^2 (156.5 sq ft)
Weights: Empty 1,475kg (3,313 lb); MTOW 2,050kg (4,519 lb)
Performance: Maximum speed 525km/h (326 mph) at sea level; operational ceiling 9,000m (29,528 ft); range 700km (435 miles)
Load: Two 20mm cannon and two 7.62mm machine guns, plus six rocket projectiles

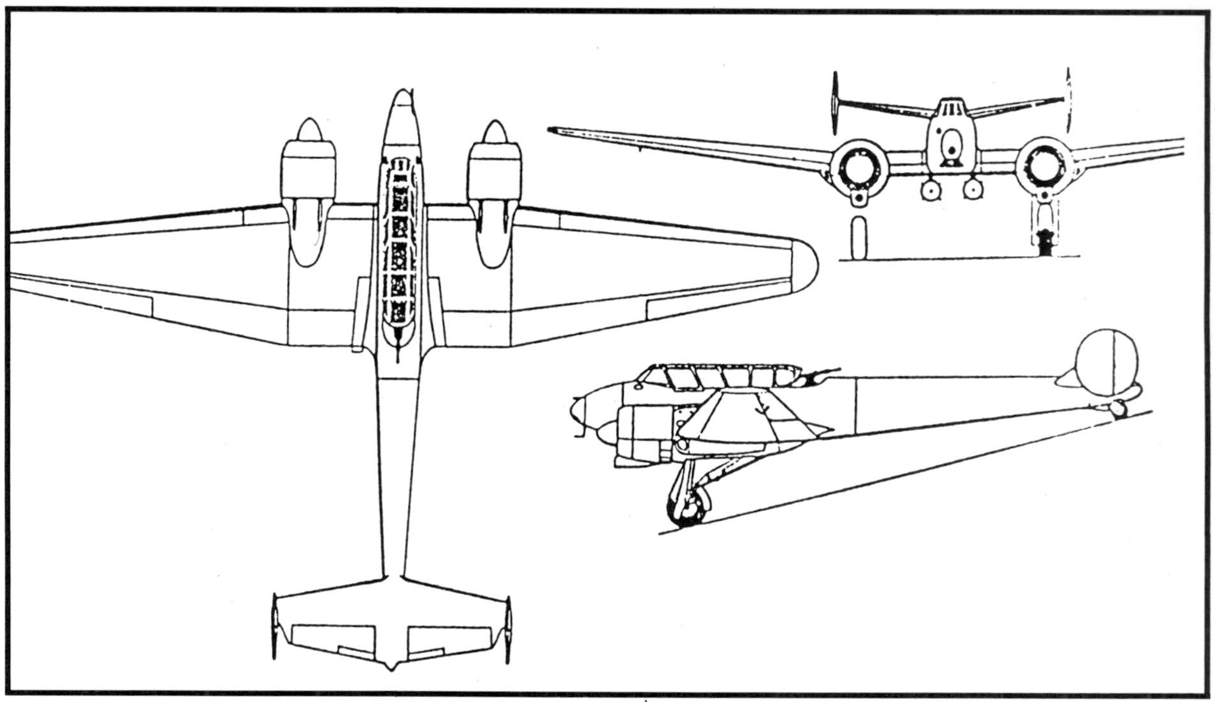

POTEZ 63
(France)

This was planned as a multi-role type able to serve the French Air Force in the two-seat interceptor and escort fighter, two-seat night-fighter and three-seat heavy command fighter roles. The Potez 630.01 prototype flew in March 1937 with Gnome-Rhone 14 Mars radial engines and offered the promise of success in several other roles. Large numbers were therefore ordered, but only 1,360 completed, which was perhaps just as well as the type proved a failure in its more powerful Potez 630 fighter, Potez 633 day bomber, Potez 637 army co-operation and Potez 63.11 reconnaissance forms.

POTEZ 63.11A3

Role: Close air support/reconnaissance
Crew/Accommodation: Three
Power Plant: Two 690 hp Gnome-Rhone 14M 4/5 air-cooled radials
Dimensions: Span 16m (52.49 ft); length 11m (36.09 ft); wing area 32.7m^2 (352 sq ft)
Weights: Empty 3,135kg (6,911 lb); MTOW 4,530kg (9,987 lb)
Performance: Maximum speed 424km/h (263 mph) at 5,000m (16,405 ft); operational ceiling 8,500m (27,885 ft); range 1,500km (932 miles)
Load: Three 7.5mm machine guns, plus up to 200kg (441 lb) of bombs

PZL P.7, P.11 and P.24
(Poland)

The common factors in these Polish fighters were all-metal construction and the Pulawski gull wing with parallel bracing struts. The P.6 and P.7 prototypes flew in August and October 1930 with the 336kW (450 hp) Bristol Jupiter VI and 362kW (485 hp) Jupiter VII respectively. The P.7 was selected for the P.7a production type, of which 150 were delivered. The P.11 was developed to reduce the obstruction to the pilot's fields of vision imposed by the large-diameter Jupiter by adopting the smaller diameter Bristol Mercury radial engine. Prototypes flew in August and December 1931 with the 384kW (515 hp) French-built Jupiter IXAsb and 395kW (530 hp) Mercury IVA. Poland took Mercury-engined aircraft as the P.11a (30 aircraft) and P.11c (175 aircraft with a lower engined position), and Romania took 50 P.11b fighters with the 373kW (500 hp) Gnome-Rhone 9K Mistral.

The final P.24 series was developed with the Gnome-Rhone 14K radial to avoid restrictions on exporting Polish fighters with imported Bristol engines and about 300 aircraft were produced for Bulgaria, Greece, Romania and Turkey.

PZL P.7A

Role: Fighter
Crew/Accommodation: One
Power Plant: One 485 hp Skoda-built Bristol Jupiter VIIF air-cooled radial
Dimensions: Span 10.57m (34.68 ft); length 6.98m (22.9 ft); wing area 17.9m^2 (192.7 sq ft)
Weights: Empty 1,090kg (2,403 lb); MTOW 1,476kg (3,254 lb)
Performance: Maximum speed 327km/h (203 mph) at 4,000m (13,125 ft); operational ceiling 8,150m (26,739 ft); range 600km (373 miles)
Load: Two 7.7mm machine guns

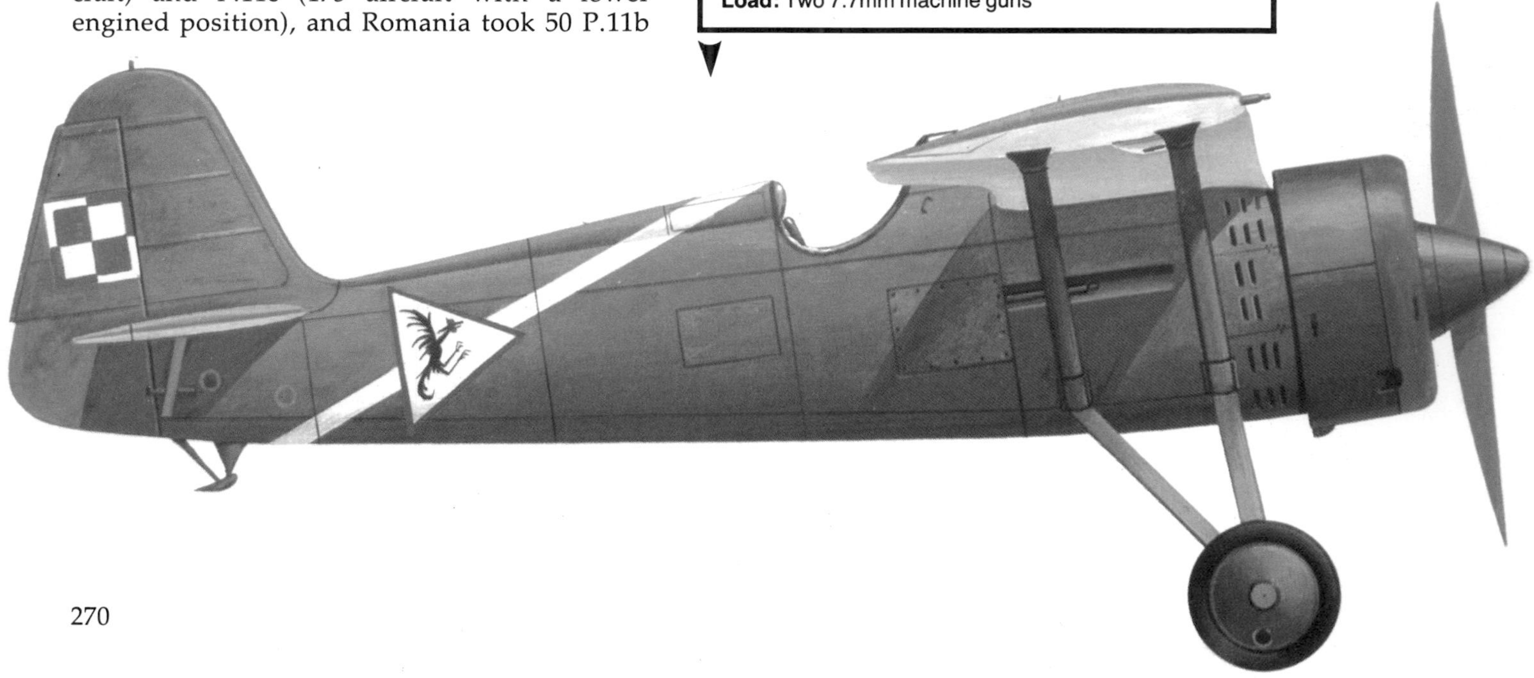

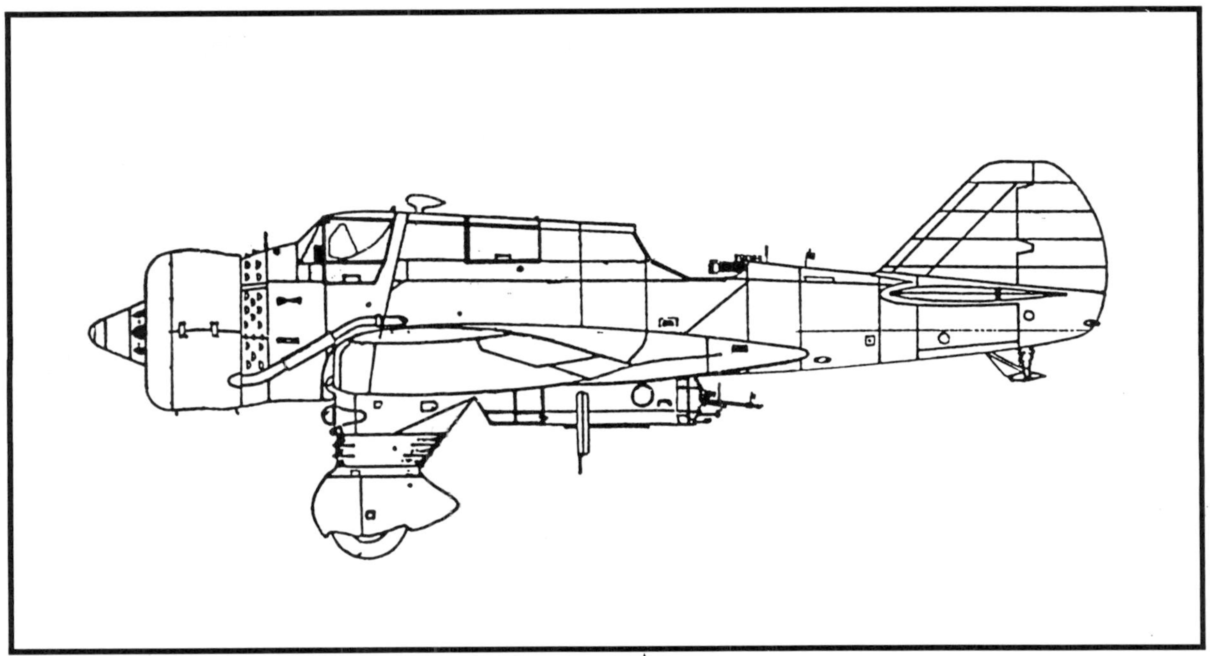

PZL P.23 KARAS (CRUCIAN CARP)
(Poland)

This was Poland's standard army co-operation aeroplane in September 1939, and could trace its origins to the PZL P.13 light transport project of 1931. This design was heavily adapted to create the P.23 with its long crew enclosure on the top of the fuselage and the only marginally smaller gondola on the fuselage's underside to provide observation and ventral defence positions. The type first flew in August 1934, and the two production variants were the Karas A with the 433kW (580 hp) Bristol Pegasus II radial engine (40 training aircraft) and the Karas B with the 507kW (680 hp) Pegasus VIII (210 operational aircraft). A similar type was ordered by Bulgaria as the P.43: the order comprised 12 P.43A aircraft with the 694kW (930 hp) Gnome-Rhone 14Kfs and 42 P-43B aircraft with the 731kW (980 hp) 14N-01.

PZL P.23B KARAS B

Role: Bomber/reconnaissance
Crew/Accommodation: Two
Power Plant: One 680 hp PZL-built Bristol Pegasus VIII air-cooled radial
Dimensions: Span 13.95m (45.77 ft); length 9.68m (31.77 ft); wing area 26.8m^2 (288.5 sq ft)
Weights: Empty 1,928kg (4,250 lb); MTOW 3,526kg (7,773 lb)
Performance: Maximum speed 319km/h (198 mph) at 3,650m (11,975 ft); operational ceiling 8,000m (26,246 ft); range 600km (410 miles) with full payload
Load: Three 7.6mm machine guns, plus up to 700kg (1,543 lb) of externally carried bombs

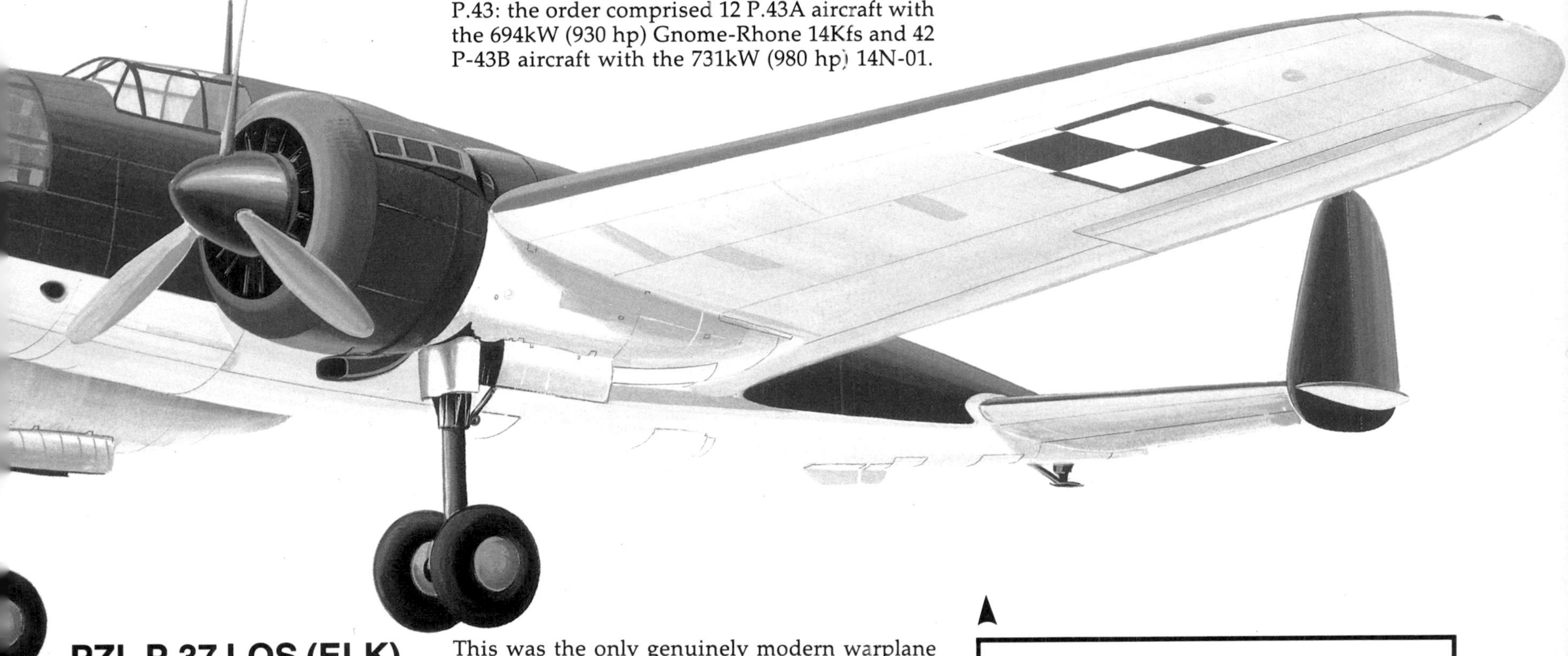

PZL P.37 LOS (ELK)
(Poland)

This was the only genuinely modern warplane in full-scale Polish service during 1939, and was a bomber derivative of the proposed P.30 transport. The P.31 prototype first flew in June 1936 with 651kW (873 hp) Bristol Pegasus XII radial engines of the type specified for the Los A initial production variant. Of these the first ten had a single vertical tail surface and the last 20 had twin endplate surfaces. Several export variants were produced with Gnome-Rhone radial engines, and Poland also ordered a total of ten Los B bombers with 690kW (950 hp) Pegasus XX radial engines, but only about 70 had been delivered by the time of Poland's surrender to the Germans in the first campaign of World War II.

PZL P.37 LOS B

Role: Medium bomber
Crew/Accommodation: Three
Power Plant: Two 950 hp PZL-built Bristol Pegasus XX air-cooled radials
Dimensions: Span 17.93m (58.83 ft); length 12.92m (42.39 ft); wing area 53.51m^2 (576 sq ft)
Weights: Empty 4,280kg (9,436 lb); MTOW 8,900kg (19,621 lb)
Performance: Maximum speed 445km/h (276 mph) at 3,400m (11,155 ft); operational ceiling 9,250m (30,350 ft); range 1,500km (932 miles) with maximum bombload
Load: Three 7.7mm machine guns, plus up to 2,500kg (5,678 lb) of bombs

Reggiane Re 2001

REGGIANE Re.2000 SERIES
(Italy)

Bearing a striking resemblance to the U.S.A.'s P-43 Lancer series, the Re.2000 was developed from 1937 as a compact fighter using a large radial engine. The first fruit was the Re.2000 Falco I (Falcon I) that first flew in 1938 with the 649kW (870 hp) Piaggio P.XI RC 40 engine. Only limited interest was expressed by the Italian forces, of which the navy ordered 36 examples (12 Re.2000 Serie II aircraft stressed for catapult launches, and 24 Re.2000 Serie III long-range aircraft); the type was exported to Hungary and Sweden as the Hejja (Hawk) and J 20 respectively. The Re.2001 Falco II used the 876kW (1,175 hp) Daimler-Benz DB 601A-1 or licence-built Alfa-Romeo RA.1000 RC 41-1a Monsone: production of 252 included 100 Re.2001 Serie I, II and III fighters with armament variations, and Re.2001 Serie IV fighter-bombers, and 150 Re.2001 CN night-fighters. The 50 Re.2002 Ariete (Ram) fighter-bombers used the Re.2001 airframe mated to the 876kW (1,175 hp) Piaggio P.XIX RC 45 radial engine, and the ultimate Re.2005 Sagittario (Archer) used the 1,100kW (1,475 hp) Alfa-Romeo RA.1050 RC 58 Tifone radial engine in the 37 aircraft which were produced.

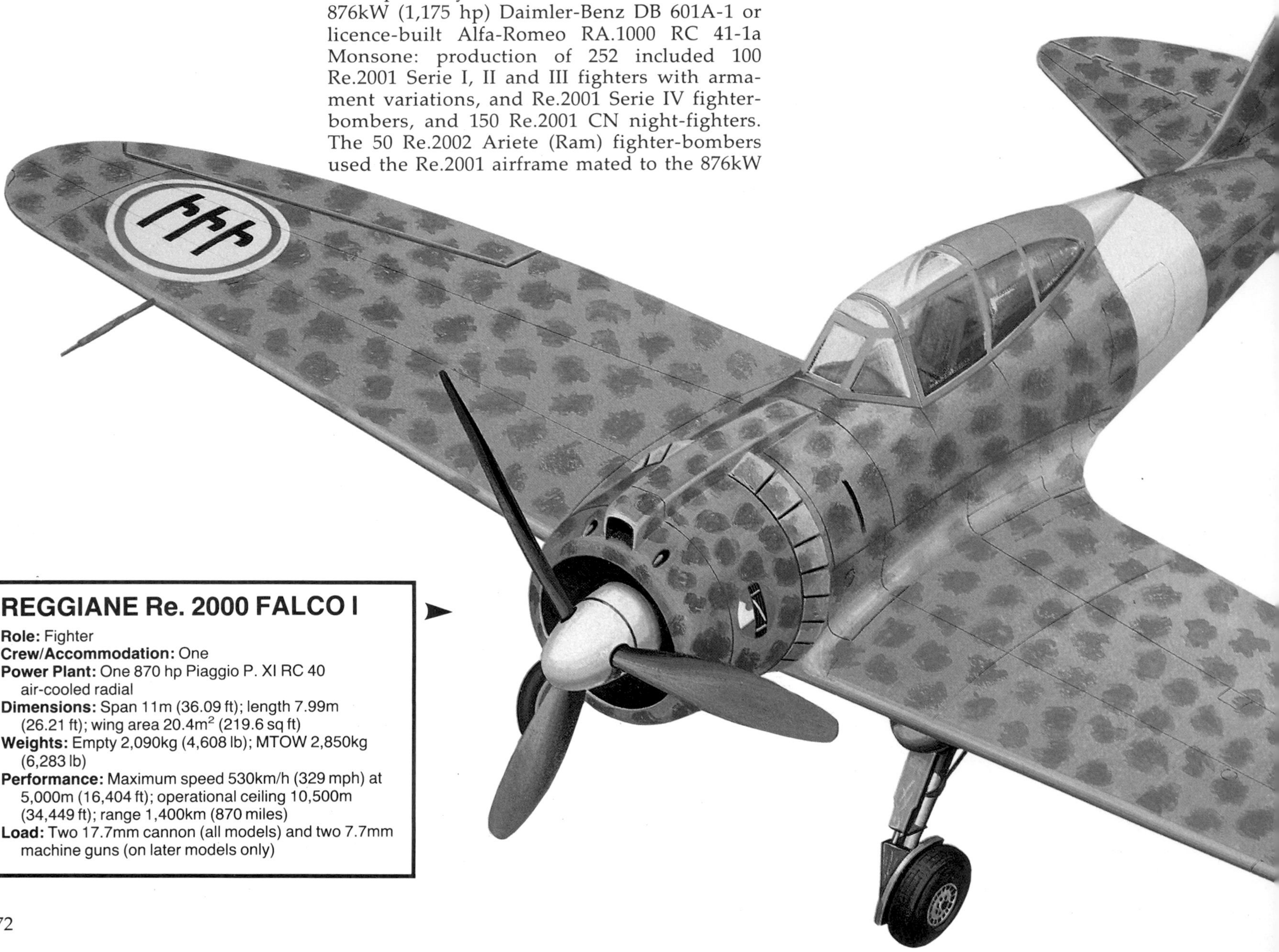

REGGIANE Re. 2000 FALCO I

Role: Fighter
Crew/Accommodation: One
Power Plant: One 870 hp Piaggio P. XI RC 40 air-cooled radial
Dimensions: Span 11m (36.09 ft); length 7.99m (26.21 ft); wing area 20.4m^2 (219.6 sq ft)
Weights: Empty 2,090kg (4,608 lb); MTOW 2,850kg (6,283 lb)
Performance: Maximum speed 530km/h (329 mph) at 5,000m (16,404 ft); operational ceiling 10,500m (34,449 ft); range 1,400km (870 miles)
Load: Two 17.7mm cannon (all models) and two 7.7mm machine guns (on later models only)

A rare photograph of a Swedish-operated Reggiane Re 2000

A line-up of Italian Re 2000s

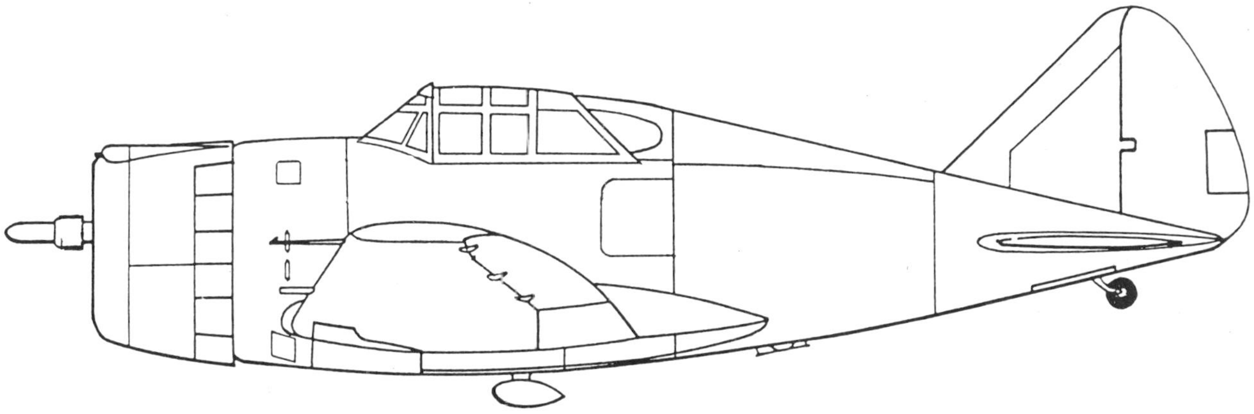

REPUBLIC P-43 LANCER
(U.S.A.)

The P-43 was an interim fighter developed from the P-35 with a turbocharged engine. The XP-41 was converted from a P-35 and served as a prototype, further refinement producing the P-43 of which 272 were built between 1940 and 1941 in the form of 54 P-43s with the R-1830-47 radial engine, 80 P-43As with the R-1830-49 radial engine and 125 P-43A-1s with the R-1830-57. The aircraft were later redesignated RP-43, RP-43A and RP-43A-1, 152 being modified with a camera installation as 150 P-43Bs and two P-43Cs.

REPUBLIC P-43A-1 LANCER

Role: Fighter
Crew/Accommodation: One
Power Plant: One 1,200 hp Pratt & Whitney R-1830-57 Twin Wasp air-cooled radial
Dimensions: Span 10.97m (36 ft); length 8.69m (28.5 ft); wing area 21.65m^2 (223 sq ft)
Weights: Empty 2,720kg (5,996 lb); MTOW 3,846kg (8,480 lb)
Performance: Maximum speed 573km/h (356 mph) at 6,096m (20,000 ft); operational ceiling 10,973m (36,000 ft); range 1,046km (650 miles) with a 91 kg (200 lb) bomb
Load: Four .5 inch machine guns, plus 91kg (200 lb) of bombs

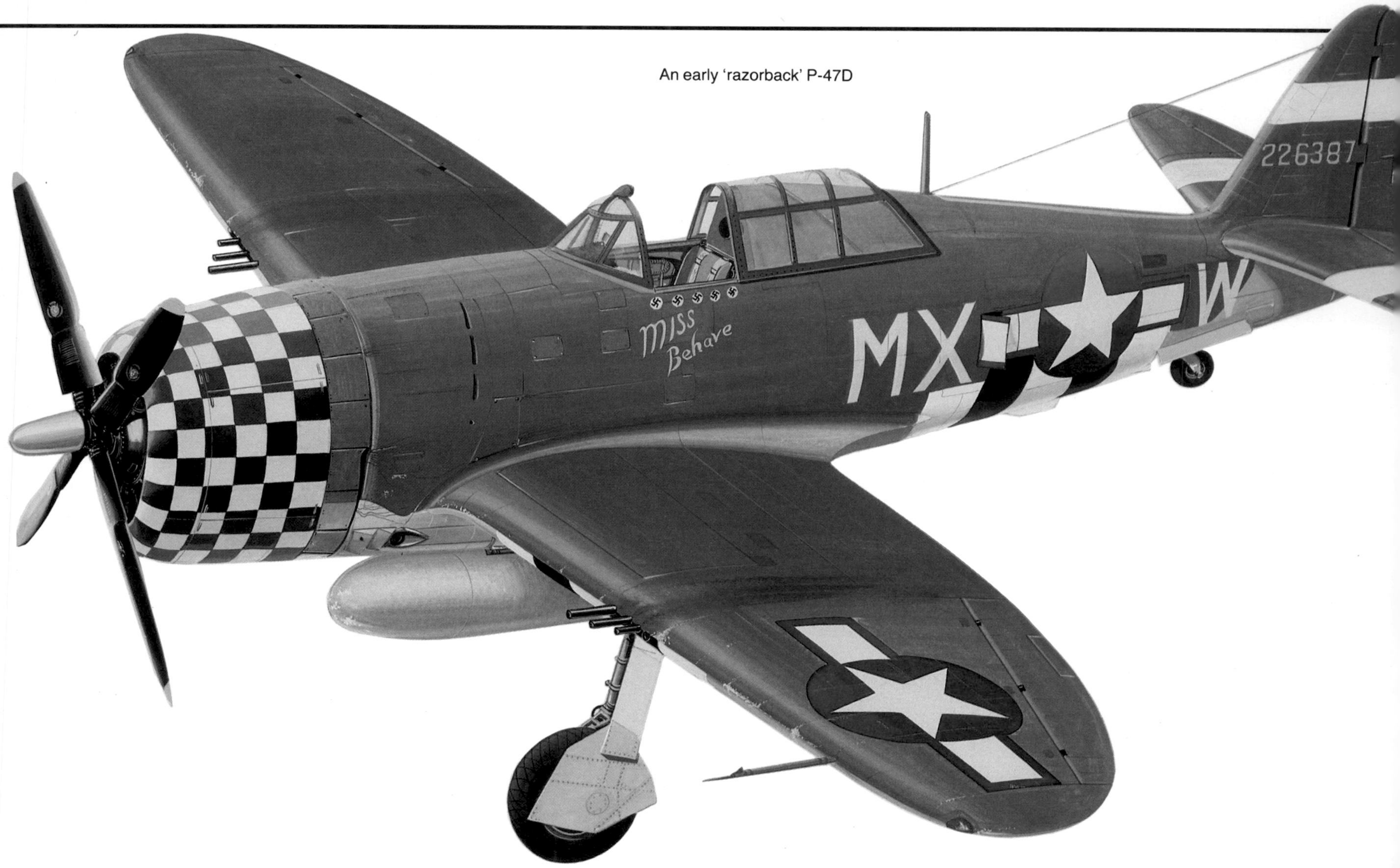

An early 'razorback' P-47D

REPUBLIC P-47 THUNDERBOLT
(U.S.A.)

The massive fuselage of this superb heavyweight fighter was dictated by the use of a large turbocharger, which had to be located in the rear of the fuselage for balance reasons, and therefore required extensive lengths of wide-diameter ducting. The type was clearly related to Republic's early portly-fuselage fighters, but was marked by very high performance, heavy firepower and great strength. The XP-47B prototype flew in May 1941 with the 1,380kW (1,850 hp) XR-2800 radial engine, later revised to develop 1,491kW (2,000 hp). This formed the basis of the 171 P-47B production aircraft with the R-2800-21 radial engine, and the 602 P-47Cs with a longer forward fuselage for the same engine or, in later examples, the 1,715kW (2,300 hp) R-2800-59.

The P-47D was the main production model, 12,602 being built with the 1,715kW 2,300 hp R-2800-21W or 1,890kW (2,535 hp) R-2800-59W water-injected radial engines; early aircraft had the original 'razorback' canopy/rear fuselage installation, but later machines introduced a 360 degree vision bubble canopy and a cut-down rear fuselage. P-47G was the designation given to 354 Wright-built P-47Ds, and the only other production models were the 130 P-47M 'sprinters', with the 2,088kW (2,800 hp) R-2800-57(C) radial engine, and the 1,816 P-47N long-range aircraft with a strengthened and longer wing plus the 2,088kW (2,800 hp) R-2800-77. The Thunderbolt was never an effective close-in fighter, but excelled in the high-speed dive-and-zoom attacks useful in long-range escort. The type was also a potent fighter-bomber.

REPUBLIC P-47C THUNDERBOLT

Role: Fighter
Crew/Accommodation: One
Power Plant: One 2,000 hp Pratt & Whitney R-2800-21 Double Wasp air-cooled radial
Dimensions: Span 12.42m (40.75 ft); length 10.99m (36.08 ft); wing area 27.87m^2 (300 sq ft)
Weights: Empty 4,491kg (9,900 lb); MTOW 6,770kg (14,925 lb)
Performance: Maximum speed 697km/h (433 mph) at 9,144m (30,000 ft); operational ceiling 12,802m (42,000 ft); range 772km (480 miles) with a 227kg (500 lb) bomb
Load: Eight .5 inch machine guns, plus up to 227kg (500 lb) of bombs

REPUBLIC F-84 THUNDERJET, THUNDERSTREAK and THUNDERFLASH
(U.S.A.)

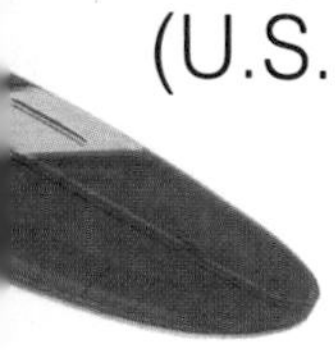

Republic's first turbojet-powered fighter-bomber was the Thunderjet, a straight-winged type that first flew in February 1946 as the XP-84 with the 1,701kg (3,750 lb) thrust General Electric J35-GE-7 engine. The 25 YP-84 service trial aircraft switched to the 1,814kg (4,000 lb) thrust Allison J35-A-15, the type chosen for the 226 P-84B initial production aircraft. The 191 P-84C (later F-84C) aircraft had the similarly rated J35-A-13C but a revised electrical system, and the 154 F-84Ds had the 2,268kg (5,000 lb) thrust J35-A-17D engine, revised landing gear and thicker-skinned wings.

Korean War experience resulted in the F-84E, of which 843 were built, with a lengthend fuselage, enlarged cockpit and improved systems. The F-84G was similar but powered by the 2,540kg (5,600 lb) thrust J35-A-29, and the 3,025 of this variant were able to deliver nuclear weapons in the tactical strike role.

The basic design was then revised as the Thunderstreak to incorporate swept flying surfaces and the more powerful Wright J65 turbojet for higher performance. Some 2,713 such F-84Fs were built, the first 375 with the J65-W-1 and the others with the more powerful 3,275kg (7,220 lb) thrust J65-W-3.

The final development was the RF-84F Thunderflash reconnaissance variant with the 3,538kg (7,800 lb) thrust J65W-7 aspirated via wing root inlets to leave the nose clear for the camera installation. There were a number of experimental and development variants, the most interesting of these being the GRF-84F (later RF-84K), which was designed to be carried by the huge Convair B-36 strategic bomber for aerial launch and recovery.

REPUBLIC F-84B THUNDERJET

Role: Fighter bomber
Crew/Accommodation: One
Power Plant: One 1,814kgp (4,000 lb s.t.) Allison J35-A-15 turbojet
Dimensions: Span 11.1m (36.42 ft); length 11.41m (37.42 ft); wing area 24.15^2 (260 sq ft)
Weights: Empty 4,326kg (9,538 lb); MTOW 8,931kg (19,689 lb)
Performance: Maximum speed 945km/h (587 mph) at 1,219m (4,000 ft); operational ceiling 12,421m (40,750 ft); range 2,063km (1,282 miles)
Load: Six .5 inch machine guns and thirty-two 5 inch rocket projectiles

The centre photograph is of a Portuguese-operated Thunderjet while those above and below depict the F-84F Thunderstreak

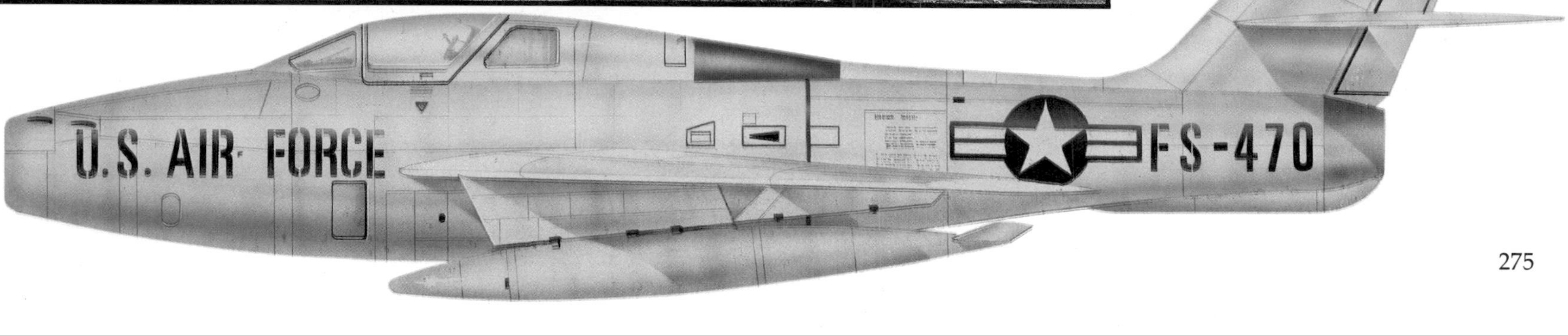

REPUBLIC F-105 THUNDERCHIEF
(U.S.A.)

The Thunderchief was the final major type to come from Republic before its merger with the Fairchild organization, and epitomized its manufacturer's reputation for massive tactical aircraft. The type was schemed as a strike fighter, and the YF-105A prototype flew in October 1965 with a 6,804kg (15,000 lb) thrust Pratt & Whitney J57-P-25 afterburning turbojet. Production began with 71 F-105B aircraft modelled on the YF-105B prototype that introduced an area-ruled fuselage and forward-swept inlets in the wing roots for a different engine, the 7,484kg (16,000 lb) thrust Pratt & Whitney J75-P-3. The major variant was the F-105D, of which 610 were built, with the 11,113kg (24,500 lb) J75-P-19W turbojet. The final version was the F-105F tandem two-seat conversion trainer, and of 86 aircraft 60 were later converted to F-105G 'Wild Weasel' aircraft in the defence-suppression role.

REPUBLIC F-105D THUNDERCHIEF

Role: Fighter
Crew/Accommodation: One
Power Plant: One 11,113kgp (24,500 lb s.t.) Pratt & Whitney J75-P-15W turbojet with reheat
Dimensions: Span 10.65m (34.94 ft); length 19.58m (64.25 ft); wing area 35.76m² (385 sq ft)
Weights: Empty 12,474kg (27,500 lb); MTOW 23,834kg (52,546 lb)
Performance: Maximum speed 2,369km/h (1,279 knots) Mach 2.23 at 11,000m (36,090 ft); operational ceiling 12,802m (42,000 ft); radius 1,152km (662 naut. miles)
Load: One 20mm multi-barrel cannon, plus up to 6,350kg (14,000 lb) of weapons/fuel

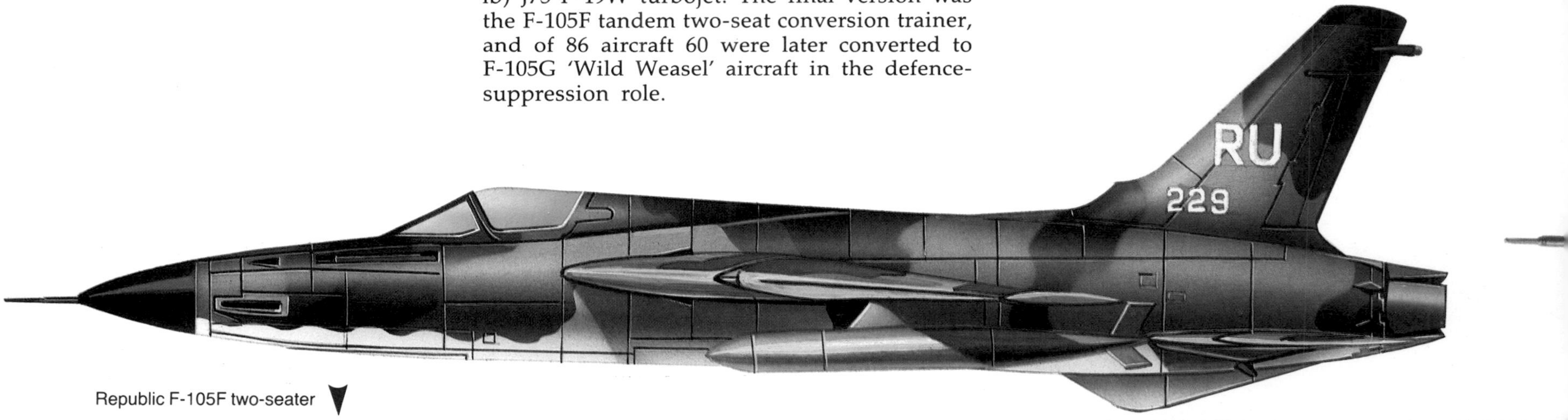

Republic F-105F two-seater

ROBIN LIGHTPLANES
(France)

The Robin company started life with a number of wooden lightplanes modelled on the original Jodel series. Typical types were the three-seat DR.220 series with fixed tailwheel landing gear, and the four-seat Regent series with fixed tricycle landing gear. The HR.100 two-seat and HR.200 four-seat series introduced metal construction and included versions with retractable tricycle landing gear. The R.1000, R.2000 and R.3000 switched to a conventionally dihedralled wing in place of the earlier type's Jodel-derived practice of a flat inner panel and upturned outer panel on each side.

ROBIN DR.400/120 DAUPHIN

Role: Light plane
Crew/Accommodation: One, plus up to three passengers
Power Plant: One 112 hp Lycoming 0-235-L2A air-cooled flat-opposed
Dimensions: Span 8.72m (28.61 ft); length 6.96m (22.83 ft); wing area 13.60m^2 (146.4 sq ft)
Weights: Empty 530kg (1,169 lb); MTOW 900kg (1,984 lb)
Performance: Cruise speed 215km/h (133 mph) at 2,250m (7,400 ft); operational ceiling 3,650m (12,000 ft); range 860km (534 miles) with three passengers
Load: Up to 213kg (470 lb)

ROCKWELL OV-10 BRONCO
(U.S.A.)

Developed as the North American NA-300 before the company became the aerospace core of the Rockwell Corporation, the OV-10 first flew in July 1965. It was a STOL light attack and reconnaissance aeroplane, with 447kW (600 hp) Garrett T76-G-6/8 turboprops at the front of the booms that supported the tail to leave the rear of the central nacelle clear to facilitate the loading of casualties or other loads. The OV-10A initial production model had more power and a 3.05m (10 ft) increase in span, and totals for the U.S. Marines and Air Force were 114 and 157 respectively. Other variants were produced for export, and the designation OV-10D is used for OV-10As modified to night observation aircraft with improved sensors and armament.

ROCKWELL OV-10A BRONCO

Role: Battlefield observation/artillery direction/utility
Crew/Accommodation: Two, plus up to six troops
Power Plant: Two 715 shp Garret AiResearch T76-G-10/12 turboprops
Dimensions: Span 12.19m (40 ft); length 12.12m (39.75 ft); wing area 27.03m^2 (291.0 sq ft)
Weights: Empty 3,260kg (7,190 lb); MTOW 6,550kg (14,444 lb)
Performance: Maximum speed 452km/h (280 mph) at sea level; operational ceiling 7,925m (26,000 ft); endurance 3.85 hours
Load: Four .303 inch machine guns, plus up to 1,451 kg (3,200 lb) of ordnance/fuel and/or sensors

ROCKWELL B-1B
(U.S.A.)

This supersonic penetration bomber was derived from the Mach 2 B-1A strategic bomber cancelled by President Carter in 1977. The B-1A first flew in December 1974 as a state-of-the-art type with variable-geometry wings and a fully variable engine inlet/nozzle combination. The lower-altitude and lower-performance B-1B had modified electronics, a strengthened airframe and landing gear, and plain inlets on revised engine nacelles. Production of 100 aircraft has been completed; these suffered an unfortunate early career because of fuel problems and inadequate electronic countermeasures.

ROCKWELL B-1B

Role: Long-range low-level variable-geometry stand-off bomber
Crew/Accommodation: Four
Power Plant: Four 13,608 kg (30,750 lb s.t.) General Electric F101-GE-102 turbofans with reheat
Dimensions: Span spread 41.66m, swept 23.83m, (spread 136.68 ft, swept 78.23 ft); length 44.43m (145.76 ft); wing area 181.2m^2 (1,950 sq ft)
Weights: Empty 87,090kg (192,000 lb); MTOW 213,367kg (477,000 lb)
Performance: Maximum speed 978km/h (529 knots) Mach 0.8 at 61m (200 ft); operational ceiling 15,240+m (50,000+ ft); range 10,378m (5,600 naut. miles) with 34,020 kg (75,000 lb) internal warload
Load: Up to 56,699 kg (125,000 lb) of weapons, including up to 24 short-range attack missiles carried on rotary launcher within the aircraft's three internal weapons bays

ROYAL AIRCRAFT FACTORY B.E.2 (U.K.)

This was the most important reconnaissance aeroplane available to the British in the first half of World War I. Though in many ways an admirable machine, it suffered for its good inherent stability; this had been designed into the type to make piloting easier, but proved a severe handicap when the type was attacked by enemy aircraft. The type first flew early in 1912 with a 52kW (70 hp) Renault inline engine, and production amounted to more than 3,200 aircraft. Major variants were the B.E.2a with wing-warping lateral control, the B.E.2b whose later examples introduced ailerons for lateral control, the B.E.2c with the 66kW (90 hp) RAF 1a inline engine and a defensive machine gun, the improved B.E.2d and the B.E.2e major production model.

ROYAL AIRCRAFT FACTORY B.E.2c

Role: Reconnaissance and artillery direct
Crew/Accommodation: Two
Power Plant: One 90 hp RAF1a air-cooled inline
Dimensions: Span 11.23m (36.83 ft); length 8.31m (27.25 ft); wing area 36.79m^2 (396 sq ft)
Weights: Empty 621kg (1,370 lb); MTOW 972kg (2,142 lb)
Performance: Maximum speed 116km/h (72 mph) at 1,981m (6,500 ft); operational ceiling 3,048m (10,000 ft); endurance 3.25 hours
Load: One/four .303 inch machine guns

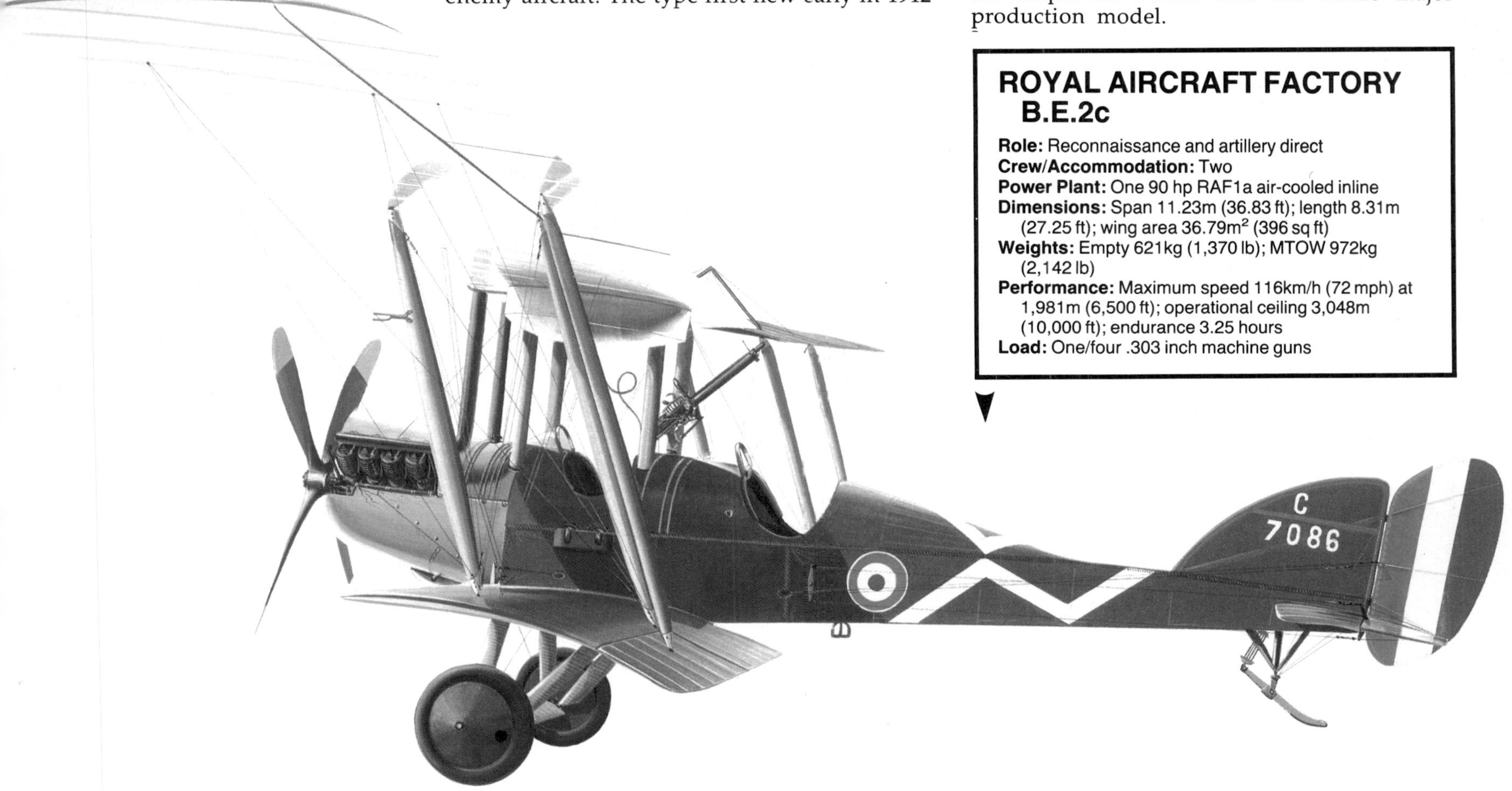

ROYAL AIRCRAFT FACTORY F.E.2 (U.K.)

The F.E.2 was developed as a pusher biplane fighter to allow the installation of forward-firing armament in the period at the beginning of World War I before the development of a synchronizer gear. The first production model was the F.E.2a with a 75kW (100 hp) Green inline engine, resulting in performance so poor that the first variant to reach squadrons in France late in 1915 was the F.E.2b with the 89kW (120 hp) Beardmore inline engine. The 1,939 F.E.2b machines were followed by about 250 F.E.2d night-flying aircraft with the positions of the crew reversed (pilot at the front instead of the rear) and the 186kW (250 hp) Rolls-Royce Eagle.

ROYAL AIRCRAFT FACTORY F.E.2b

Role: Fighter-reconnaissance
Crew/Accommodation: Two
Power Plant: One 120 hp Beardmore water-cooled inline
Dimensions: Span 14.55m (47.75 ft); length 9.83m (32.25 ft); wing area 45.9m^2 (494 sq ft)
Weights: Empty 935kg (2,061 lb); MTOW 1,378kg (3,037 lb)
Performance: Maximum speed 147km/h (91.5 mph) at sea level; operational ceiling 3,353m (11,000 ft); endurance 3.5 hours
Load: One/two .303 inch machine guns

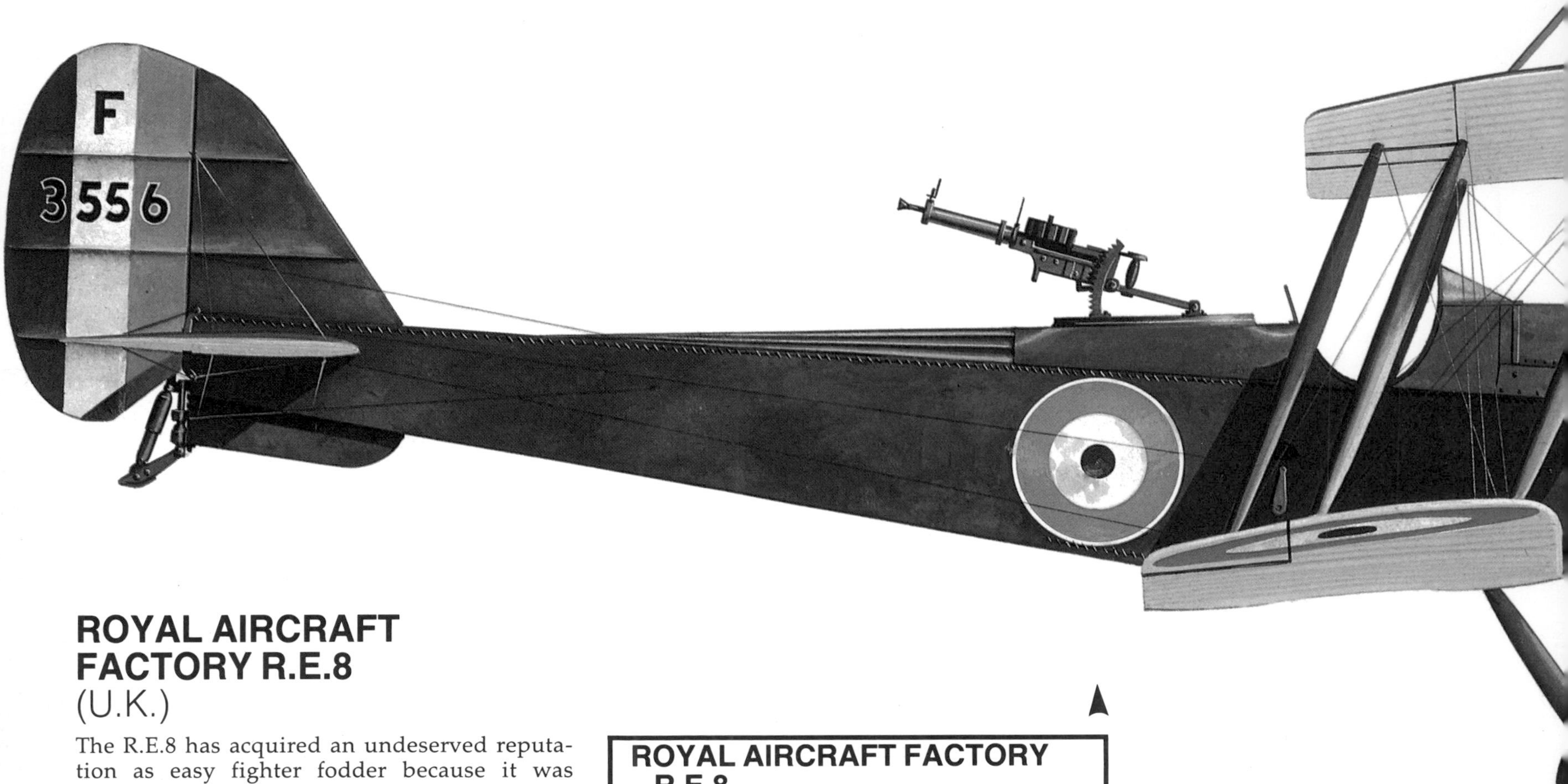

ROYAL AIRCRAFT FACTORY R.E.8
(U.K.)

The R.E.8 has acquired an undeserved reputation as easy fighter fodder because it was similar to the B.E.2 in being inherently stable. In fact it provided the British with a good artillery observation and reconnaissance aeroplane from early 1917. Production of the single variant was 4,077.

ROYAL AIRCRAFT FACTORY R.E.8

Role: Reconnaissance
Crew/Accommodation: Two
Power Plant: One 140 hp RAF4a
Dimensions: Span 14.5m (47.58 ft); length 8.5m (27.88 ft); wing area 35.07m^2 (377.5 sq ft)
Weights: Empty 818kg (1,803 lb); MTOW 1,301kg (2,869 lb)
Performance: Maximum speed 158km/h (98 mph) at 1,981m (6,500 ft); operational ceiling 3,353m (11,000 ft); endurance 2.25 hours
Load: Two/three .303 inch machine guns, plus up to 102kg (224 lb) of bombs

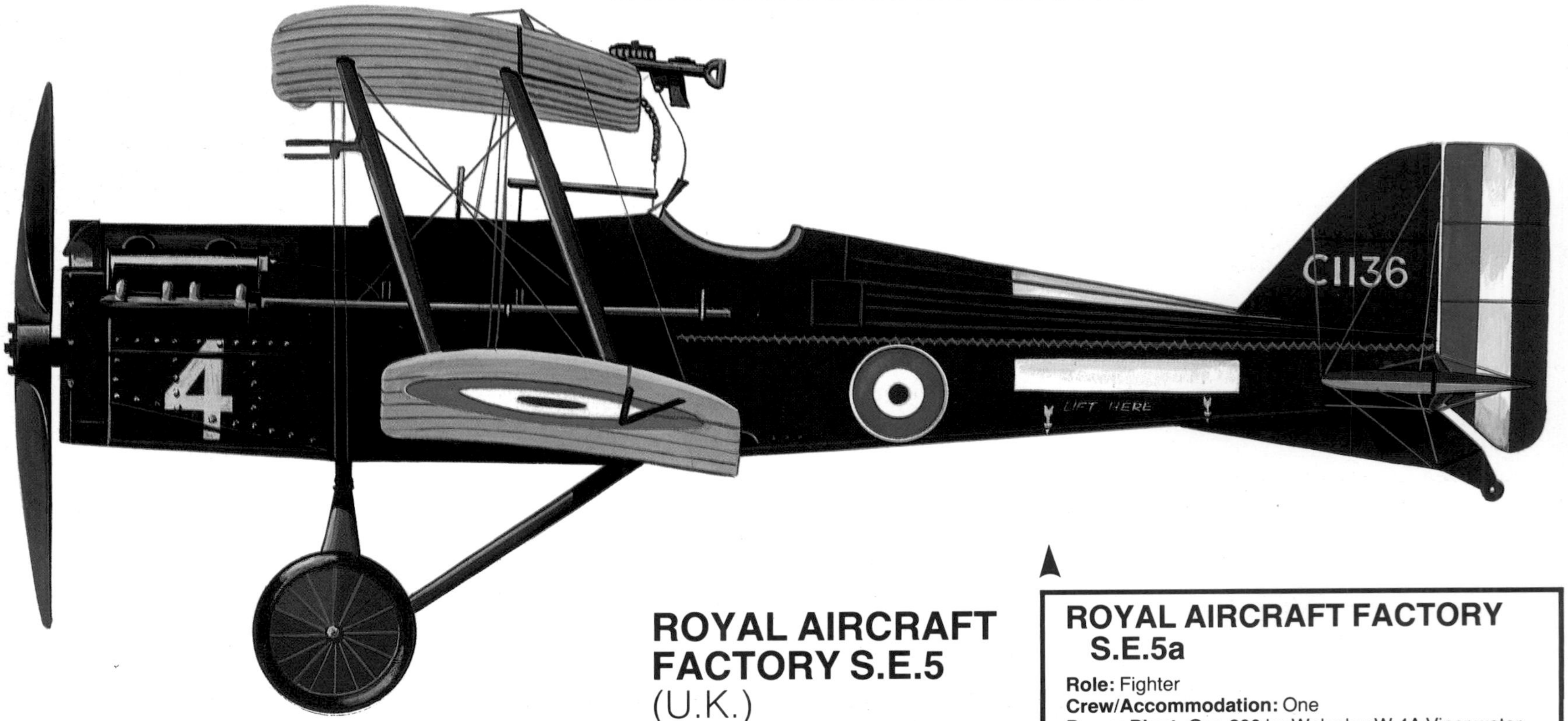

ROYAL AIRCRAFT FACTORY S.E.5
(U.K.)

The best aeroplane to be designed by the Royal Aircraft Factory, the S.E.5 was an exceptionally steady gun platform and without sacrifice of structural strength possessed good performance and agility. Powered by a 112kW (150 hp) Hispano-Suiza inline engine, the type began to enter service in April 1917 and was soon joined by the superlative S.E.5a version with a 149kW (200 hp) engine. Total production was 5,205.

ROYAL AIRCRAFT FACTORY S.E.5a

Role: Fighter
Crew/Accommodation: One
Power Plant: One 200 hp Wolseley W.4A Viper water-cooled inline
Dimensions: Span 8.12m (26.63 ft); length 6.38m (20.92 ft); wing area 22.84m^2 (245.8 sq ft)
Weights: Empty 635kg (1,399 lb); MTOW 880kg (1,940 lb)
Performance: Maximum speed 222km/h (138 mph) at sea level; operational ceiling 5,182m (17,000 ft); endurance 2.5 hours
Load: Two .303 inch machine guns, plus up to 45kg (100 lb) of bombs

RUMPLER C SERIES (Germany)

Rumpler was one of the main builders of general-purpose and reconnaissance aircraft for the German air arm in World War I, and its initial offering in the armed reconnaissance field was the C I with a 119kW (160 hp) Mercedes D.III inline engine, exchanged for a 134kW (180 hp) Argus As.III in the C Ia. The type entered service in early 1916, and production totalled about 300. Next came the C IV developed via the C III with cleaner lines and the 164kW (220 hp) Benz Bz.IV inline engine. Less than 100 were produced before the advent late in 1917 of the basically similar C VII with the 179kW (240 hp) Maybach Mb.IV inline engine.

RUMPLER C VIII

Role: Operational trainer
Crew/Accommodation: Two
Power Plant: One 180 hp Argus As III water-cooled inline
Dimensions: Span 12.18m (39.96 ft); length 8.02m (26.31 ft); wing area 36.2m^2 (391 sq ft)
Weights: Empty 874kg (1,923 lb); MTOW 1,374kg (3,023 lb)
Performance: Maximum speed 140km/h (87.5 mph) at sea level; operational ceiling 6,000m (19,685 ft); endurance 4 hours
Load: None

RUTAN VARI-VIGGEN (U.S.A.)

First flown in February 1972, the Vari-Viggen is an excellent example of the design philosophy of Burt Rutan, a far-sighted designer who favours new materials and unusual design concepts to secure maximum performance from comparatively low-powered engines. The type has been extensively produced by homebuilders, and the same basic design concept is carried to greater lengths in the high aspect ratio Vari-Eze.

RUTAN VARI-VIGGEN

Role: Lightplane
Crew/Accommodation: One, plus one passenger
Power Plant: One 150 hp Lycoming 0-320-A2A air-cooled flat-opposed
Dimensions: Span 5.79m (19 ft); length 5.79m (19.75 ft); wing area 11.06m^2 (119 sq ft)
Weights: Empty 431kg (950 lb); MTOW 771kg (1,700 lb)
Performance: Maximum speed 262km/h (163 mph) at sea level; operational ceiling 4,265m (14,000 ft); range 643km (40 miles) with passenger
Load: Up to 45kg (100 lb) as well as passenger

The Long-ez shown is a follow-on to the Vari-Viggan, using Rutan's characteristic canard layout

RYAN NYP
(U.S.A.)

The most celebrated aeroplane designed by Ryan, the single NYP (better known as the 'Spirit of St. Louis') was the aeroplane in which Charles Lindbergh achieved the first solo non-stop crossing of the Atlantic when he flew from New York to Paris in 1927.

RYAN NYP

Role: Long-range aircraft
Crew/Accommodation: One
Power Plant: One 220 hp Wright J5C Whirlwind air-cooled radial
Dimensions: Span 14.02m (46 ft); length 8.46m (27.75 ft); wing area 29.64m^2 (319 sq ft)
Weights: Empty 975kg (2,150 lb); MTOW 2,411kg (5,315 lb)
Performance: Maximum speed 169km/h (105 mph) at sea level; operational ceiling 4,877m (16,000 ft); range 6,775km (4,210 miles)
Load: Up to 1,359kg (2,995 lb) of fuel and lubricant.
Note: the actual distance covered by Lindbergh in his May 1927 New York to Paris flight was 5,816km (3,614 miles)

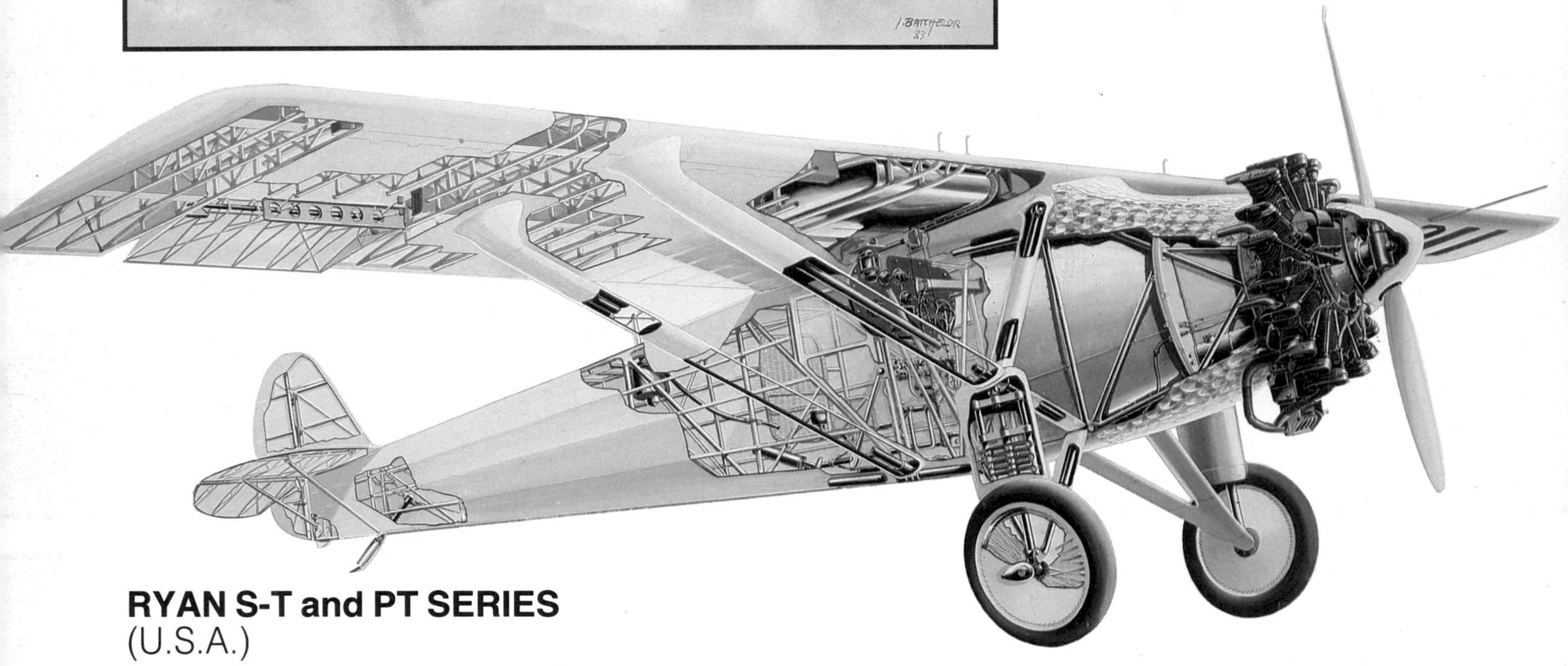

RYAN S-T and PT SERIES
(U.S.A.)

This series of wire-braced low-wing monoplanes with fixed landing gear, was one of the most important primary trainers used by the U.S. forces during World War II. The series began with the S-T of 1933 with 71kW (95 hp) Menasco B-4 Pirate inline engine. Five S-Ts were followed by 71 S-T-As and 11 S-T-A Specials, but at this point the U.S. forces became interested, one XPT-16 and 15 YPT-16s being followed by 30 PT-20s, 10 PT-21s and 1,023 PT-22 Recruits for the U.S. Army, as well as 100 NR-1 Recruits for the U.S. Navy. There were also several export variants.

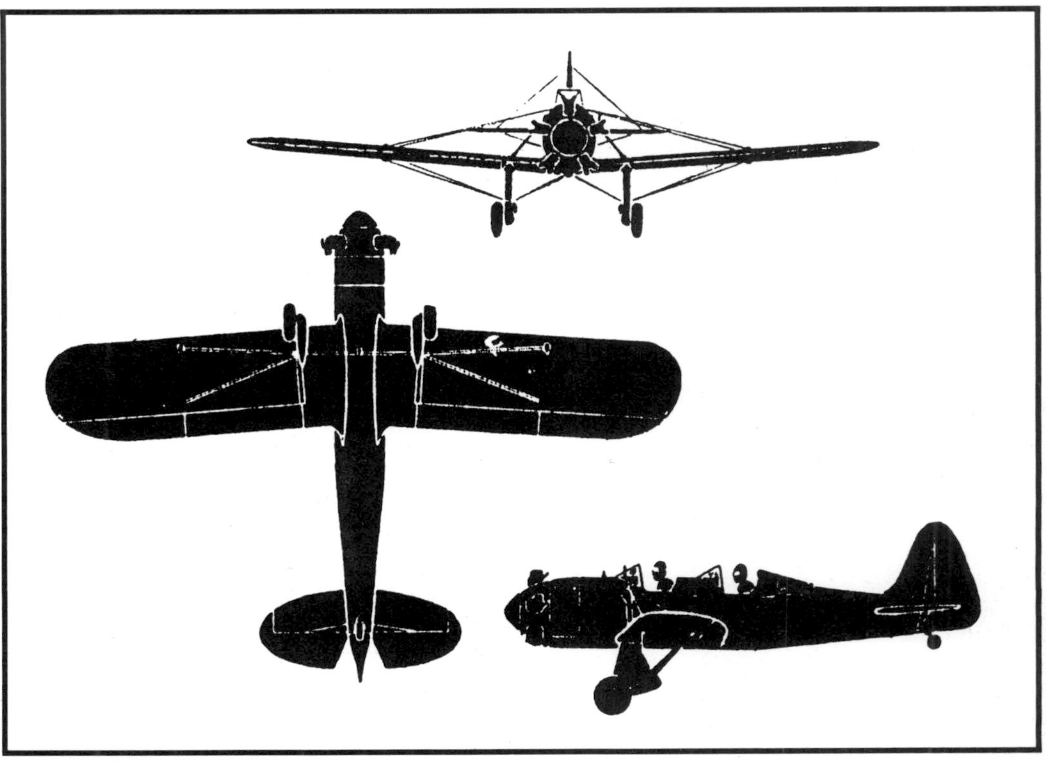

RYAN YPT-16

Role: Elementary trainer
Crew/Accommodation: Two
Power Plant: One 125 hp Menasco C-4/L-365 air-cooled inline
Dimensions: Span 9.14m (30 ft); length 6.55m (21.50 ft); wing area 11.52m^2 (124 sq ft)
Weights: Empty 499kg (1,100 lb); MTOW 726kg (1,600 lb)
Performance: Maximum speed 206km/h (128 mph) at sea level; operational ceiling 4,572m (15,000 ft); range 563km (350 miles)
Load: Up to 102kg (224 lb) including student pilot

SAAB 21 (Sweden)

This was Sweden's first indigenously designed fighter, the concept being developed from 1941 when Swedish Air Force was equipped with a miscellany of obsolescent imported aircraft. A radical approach was selected: a twin boom layout allowed the installation of a piston engine and pusher propeller in the rear of the central nacelle in such a way that this combination could be replaced by a turbojet once such an engine became available. Other advanced features were tricycle landing gear and an ejector seat. The first Saab 21A flew in July 1943 with a Daimler-Benz DB 605 inline engine, and the type entered service from late 1945 as the J 21A-1 and improved J 21A-2 fighters and as the A 21A attack aeroplane. Production totalled 299 aircraft. The Saab 21R prototype with the de Havilland Goblin centrifugal-flow turbojet first flew in March 1947, but production of J 21R fighters totalled only 60 after problems had delayed the development programme to the point at which it was virtually overtaken by that of the Saab 29. The aircraft were later converted to A 21R attack fighters.

SAAB 21A/J 21A-1

Role: Fighter
Crew/Accommodation: One
Power Plant: One 1,475 hp SFA-built Daimler-Benz DB 605B water-cooled in-line
Dimensions: Span 11.6m (38.08 ft); length 10.45m (34.25 ft); wing area 22.2m^2 (238.9 sq ft)
Weights: Empty 3,250kg (7,165 lb); MTOW 4,150kg (9,149 lb)
Performance: Maximum speed 640km/h (398 mph) at 6,000m (19,685 ft); operational ceiling 10,420m (34,186 ft); range 735km (457 miles)
Load: One 20mm cannon and four 13mm machine guns

SAAB 32 LANSEN (LANCE) (Sweden)

First flown in November 1953, the Saab 32 was a remarkable achievement for the small Swedish aircraft and aero engine industries, which developed a capable swept-wing airframe and an afterburning version of the Rolls-Royce Avon turbojet. Some 450 Lansens were produced between 1953 and 1960, the primary variants being the A 32A attack aeroplane with the 4,500kg (9,920 lb) Avon 100, the J 32B all-weather and night-fighter with the 6,900kg (15,212 lb) thrust Avon 200, and the S 32C reconnaissance aeroplane with cameras and radar surveillance equipment.

SAAB 32/A 32A LANSEN

Role: All-weather strike
Crew/Accommodation: Two
Power Plant: One 4,500kgp (9,920 lb s.t.) SFA-built Rolls-Royce Avon 100 turbojet with reheat
Dimensions: Span 13m (42.65 ft); length 14.65m (48.06 ft); wing area 37.4m^2 (402.6 sq ft)
Weights: Empty 7,440kg (16,402 lb); MTOW 13,000kg (28,660 lb)
Performance: Maximum speed 1,102km/h (595 knots) at sea level; operational ceiling 15,000m (49,213 ft); range 1,380km (744 naut. miles) with full warload
Load: Four 20mm cannon, plus up to 1,000kg (2,205 lb) of bombs or air-to-surface missiles

SAAB 35 DRAKEN (DRAGON)
(Sweden)

An even more remarkable achievement than the Saab 32, the Saab 35 was designed as a supersonic fighter using a double-delta planform for a large lifting area and fuel capacity combined with minimum profile drag. The first prototype flew in October 1955. The J 35A initial production variant began to enter service in 1958, offering outright performance comparable to that of the slightly later and heavier English Electric Lightning, but only one afterburning Rolls-Royce Avon turbojet.

Production totalled 525 in variants such as the J 35A fighter with the 7,000 kg (14,432 lb) thrust RM6B, the J 35B improved fighter with collision-course radar and a data-link system, the SK 35C tandem-seat operational trainer, the J 35D with the 7,830 kg (17,262 lb) thrust RM6C and more advanced electronics, the S 35E tactical reconnaissance aeroplane, and the J 35F with more advanced radar and Hughes Falcon air-to-air missiles. The type was also exported as the Saab 35X, and surviving J 35Fs have been upgraded to J 35J standard for service into the 1990s.

SAAB 35/J 35 DRAKEN

Role: Interceptor/strike/reconnaissance
Crew/Accommodation: One
Power Plant: One 7,830kgp (17,262 lb s.t.) Flygmotor-built Rolls-Royce Avon RM6C turbojet with reheat
Dimensions: Span 9.4m (30.83 ft); length 15.4m (50.33 ft); wing area 50m^2 (538.2 sq ft)
Weights: Empty (not available); MTOW 16,000kg (35,274 lb)
Performance: Maximum/Cruise speed 2,150km/h (1,160 mph) Mach 2.023 at 11,000m (36,090 ft); operational ceiling 18,300m (60,039 ft); range 1,149km (620 naut. miles) with 2,000 lb warload
Load: Two 30mm cannon, plus up to 4,082kg (9,000 lb) of bombs

SAAB 37 VIGGEN (THUNDERBOLT)
(Sweden)

With the Saab 37, first flown in February 1967, Sweden produced a true multi-role fighter with a thrust-reversible afterburning turbofan and a canard layout for true STOL capability using short lengths of road. Production was about 330, and the varaints have been the AJ 37 attack aeroplane with the 17,789kg (25,990 lb) thrust RM8A, the SF 37 overland reconnaissance aeroplane, the SH 37 overwater reconnaissance aeroplane, the SK 35 tandem two-seat operational trainer, and the JA 37 interceptor.

SAAB 37/JA 37 VIGGEN

Role: Interceptor
Crew/Accommodation: One
Power Plant: One 12,750kgp (28,109 lb s.t.) Volvo Flygmotor RM8B turbofan with reheat
Dimensions: Span 10.6m (34.78 ft); length 16.4m (53.81 ft); wing area 52.2m^2 (561.9 sq ft)
Weights: Empty 12,200kg (26,455 lb); MTOW 22,500kg (49,604 lb)
Performance: Maximum speed 2,231km/h (1,204 knots) Mach 2.10 at 11,000m (36,090 ft); operational ceiling 18,300m (60,039 ft); radius 483km (260 naut. miles) with 1,360kg warload
Load: One 30mm cannon, plus up to 6,000kg (13,277 lb) of externally-carried weapons/fuel, including two medium range and four short-range air-to-air missiles

SAAB 39 GRIPEN (GRIFFON)
(Sweden)

The tendency of Swedish warplanes to grow larger and heavier has been reversed with the Saab 39, a multi-role fighter, attack and reconnaissance aeroplane. This first flew in early 1989 as an advanced type using canard aerodynamics, a large measure of composite construction and a fly-by-wire control system for maximum performance and agility with a heavy warload. There have been technical problems and cost overruns, so the future of the 300+ production series may be in jeopardy.

SAAB 39/JA5 39 GRIPEN

Role: Interceptor/strike fighter/reconnaissance
Crew/Accommodation: One
Power Plant: One 8,210kgp (18,100 lb s.t.) Volvo-built General Electric F404J turbofan with reheat
Dimensions: Span 8m (26.25 ft); length 14.1m (46.25 ft)
Weights: MTOW 8,000kg (17,635 lb)
Performance: Maximum speed 1,471km/h (794 knots) Mach 1.20 at sea level
Load: One 27mm cannon, plus four medium-range and two short-range air-to-air missiles

SAVOIA-MARCHETTI S.55
(Italy)

This fascinating flying boat took part in a number of important mass flights in the 1930s. The type first flew in the summer of 1924 with a large wing as the structural core that supported twin pod-and-boom hulls whose tubular booms ended in a wide tailplane.

Total production included 90 military S.55s with 380kW (510 hp) Isotta-Fraschini Asso 500 inline engines, eight S.55C airliners, 23 S.55P airliners with greater accommodation, 16 S-55A long-range 'boats with 418kW (560 hp) Fiat A.22R engines, seven S.55M with a largely metal structure and Asso 500 engines, 32 S 55 Scafo Allargato 'boats with larger hulls and Asso 500 engines, 42 S.55 Scafo Allargatissimo 'boats with further enlarged hulls and Asso 500 engines, and 25 S.55X 'boats for the mass flights with 656kW (880 hp) Asso 750 engines.

SAVOIA-MARCHETTI S.55A

Role: Bomber/transport flying boat
Crew/Accommodation: Three, plus up to 18 troops
Power Plant: Three 560 hp Fiat A.22R water-cooled inlines
Dimensions: Span 24m (78.75 ft); length 16m (52.5 ft); wing area 92m^2 (990.27 sq ft)
Weights: Empty 4,385kg (9,667 lb); MTOW 7,300kg (16,093 lb)
Performance: Maximum speed 207km/h (129 mph) at sea level; operational ceiling 6,100m (13,450 ft); range 595km (370 miles)
Load: Up to 4,000kg (8,818 lb)

SAVOIA-MARCHETTI SM.79 SPARVIERO (SPARROWHAWK)
(Italy)

This was Italy's most important bomber of World War II, and in its specialist anti-ship version the best torpedo bomber of the war. The type was evolved from the company's earlier tri-motor types and first flew in late 1934. Production totalled about 1,370 for Italy and for export, and the primary variants were the SM.79-I bomber with 582kW (780 hp) Alfa-Romeo 126 RC 34 radial engines, the SM 79-II torpedo bomber with 746kW (1,000 hp) Piaggio P.XI RC 40 or 768kW (1,030 hp) Fiat A.80 RC 41 radial engines, the SM 79-III improved version of the SM.79-II without the ventral gondola, and the SM.79B twin-engined export version of the SM.79-I.

SAVOIA-MARCHETTI SM.79 SPARVIERO

Role: Bomber
Crew/Accommodation: Four
Power Plant: Three 780 hp Alfa Romeo 126 RC34 air-cooled radials
Dimensions: Span 21m (69.55 ft); length 15.62m (51.25 ft); wing area 61.7m^2 (664.2 sq ft)
Weights: Empty 6,800kg (14,991 lb); MTOW 10,500kg (23,148 lb)
Performance: Maximum speed 430km/h (267 mph) at 4,000m (13,125 ft); operational ceiling 6,500m (21,325 ft); range 1,900km (1,180 miles) with full bombload
Load: Three 12.7mm and two 7.7mm machine guns, plus up to 1,250kg (2,756 lb) of bombs or one torpedo

SAVOIA-MARCHETTI SM.82 CANGURU (KANGAROO)
(Italy)

This was a development of the SM.75 airliner and transport optimized for the military transport role with fixed instead of retractable landing gear. It first flew in 1935 and production of the sole service variant totalled 535.

SAVOIA-MARCHETTI SM.82 CANGURU

Role: Heavy bomber/transport
Crew/Accommodation: Four (bomber) or three, plus up to 40 troops (transport)
Power Plant: Three 950 hp Alfa Romeo 128 RC21 air-cooled radials
Dimensions: Span 29.68m (97.38 ft); length 22.90m (75.13 ft); wing area 118.6m^2 (1,276 sq ft)
Weights: Empty 10,550kg (23,260 lb); MTOW 17,820kg (39,286 lb)
Performance: Maximum speed 370km/h (230 mph) at 3,600m (8,211 ft); operational ceiling 6,000m (19,685 ft); range 3,000km (1,864 miles) with full bombload
Load: One 12.7mm and five 7.7mm machine guns, plus up to 4,000kg (8,820 lb) of bombs or cargo

This Twin Pioneer carries special geophysical survey sensors in its 'customised' wingtip housings.

SCOTTISH AVIATION TWIN PIONEER
(U.K.)

This was developed after the success of the single-engined Pioneer to provide a heavier lift capabaility without any sacrifice of STOL performance, The type first flew in June 1955 with 403kW (540 hp) Alvis Leonides 503/8 radial engines. Production totalled 87 aircraft in Twin Pioneer Series 1, 2 and 3 variants with 418kW (560 hp) Leonides 514/8, 447kW (600 hp) Pratt & Whitney R-1340-S1H1 and 564kW (640 hp) Leonides 531 engines respectively.

SCOTTISH AVIATION TWIN PIONEER SERIES 3

Role: Short take-off and landing utility transport
Crew/Accommodation: Two, plus up to 19 passengers
Power Plant: Two 640 hp Alvis Leonides 531/8B air-cooled radials
Dimensions: Span 23.32m (76.5 ft); length 13.79m (45.25 ft); wing area 62.25m^2 (670 sq ft)
Weights: Empty 4,564kg (10,062 lb); MTOW 6,622kg (14,600 lb)
Performance: Maximum speed 216km/h (134 mph) at 1,524m (5,000 ft); operational ceiling 6,477m (21,250 ft); range 927km (576 miles) with full payload
Load: Up to 1,134kg (2,500 lb)

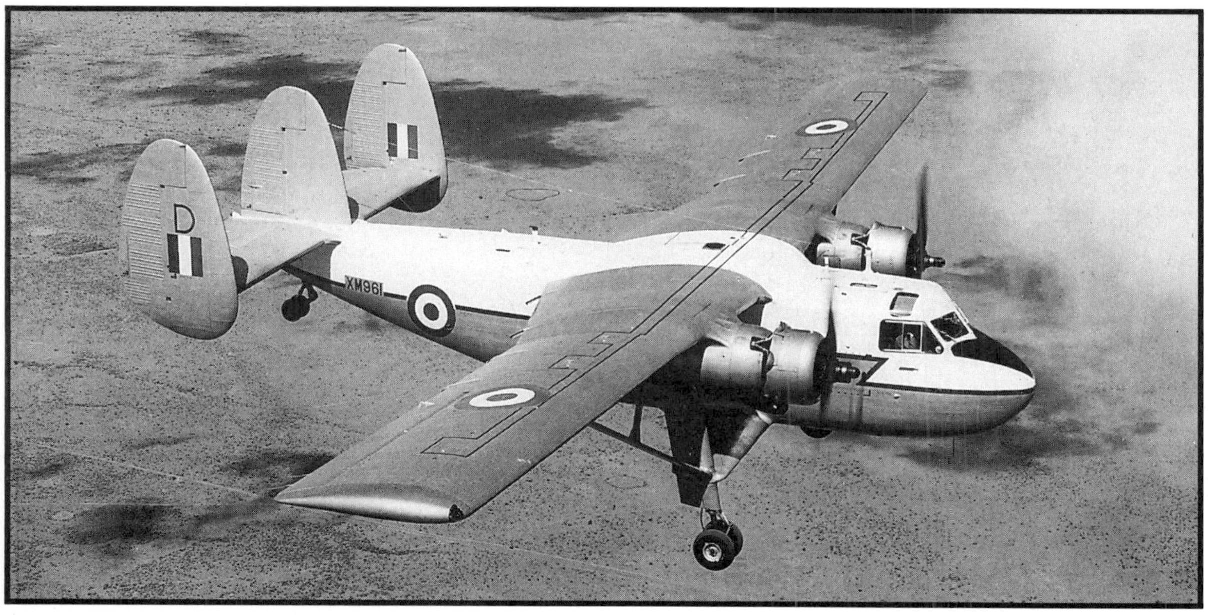

An R.A.F. Jaguar GR.1 lifting off.

A French Jaguar E two-seater.

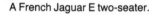

SEPECAT JAGUAR
(France/U.K.)

Developed by British and French interests as a supersonic trainer on the basis of the Breguet Br.121 design, and then produced by the specially formed SEPECAT consortium (comprising BAC and Dassault-Breguet) with engines from a comparable international grouping of Rolls-Royce and Turbomeca, the Jaguar first flew in September 1968. Such were the capabilities of the type that the major production variants were the Jaguar A and S attack aircraft (160 and 165 aircraft respectively for the French and British air forces) plus only a few Jaguar E and Jaguar B trainers (40 and 38 aircraft respectively for the French and British air forces). There has also been the Jaguar international for the export market, which has absorbed just over another 120 aircraft.

SEPECAT JAGUAR INTERNATIONAL

Role: Low-level strike fighter
Crew/Accommodation: One
Power Plant: Two 3,811kgp (8,400 lb s.t.) Rolls-Royce/Turbomeca Adour Mk 811 turbofans with reheat
Dimensions: Span 8.69m (28.5 ft); length 16.8m (55.1 ft); wing area 24.18m^2 (280.3 sq ft)
Weights: Empty 7,000kg (15,432 lb); MTOW 15,422kg (34,000 lb)
Performance: Maximum speed 1,350km/h (729 knots) Mach 1.1 at sea level; operational ceiling (not available); radius 852km (460 naut. miles) with 3,629kg (8,000 lb) warload
Load: Two 30mm cannon, plus up to 4,762 kg (10,500 lb) of weapons, including bombs, rockets or air-to-surface missiles, plus two short-range air-to-air missiles

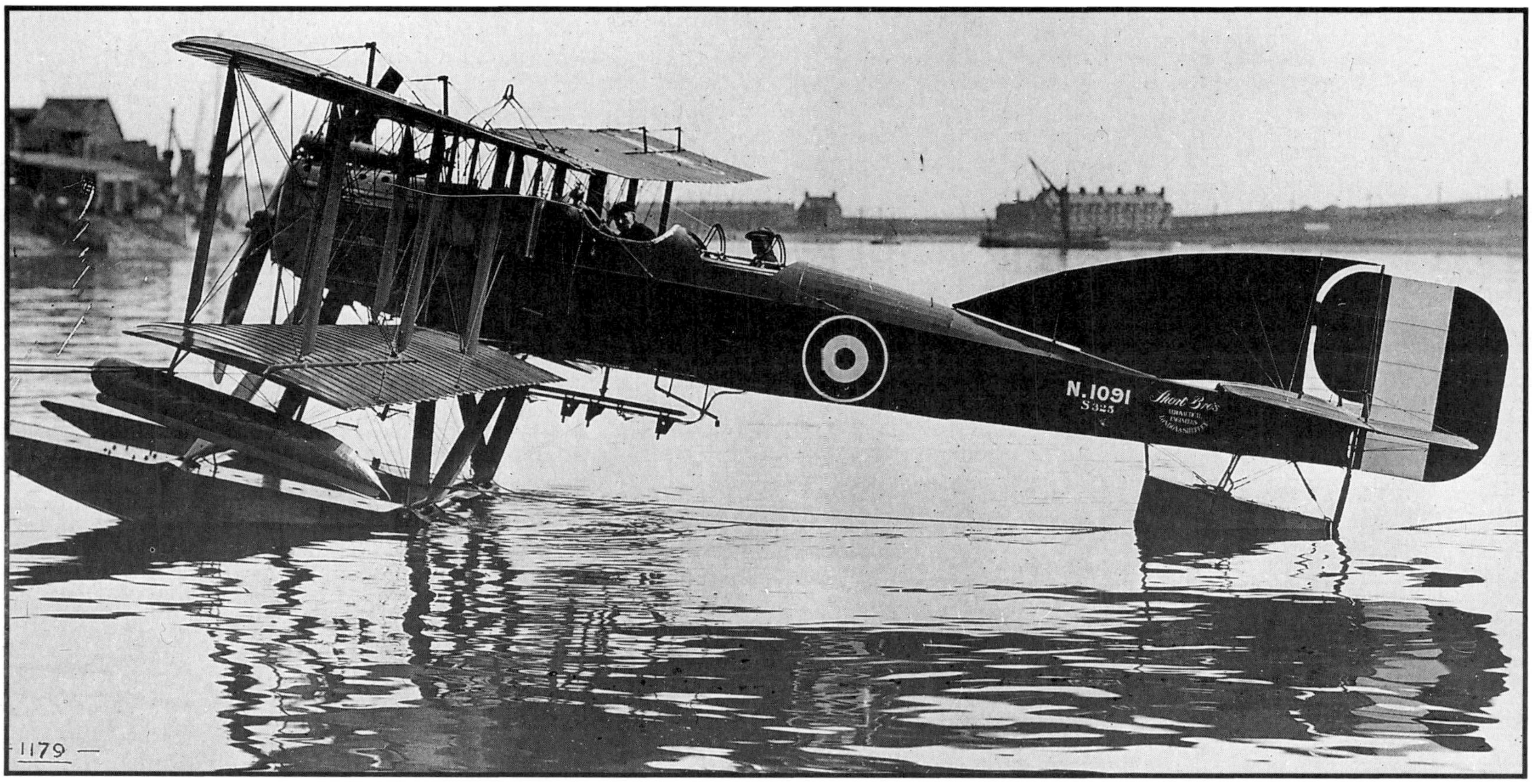

SHORT 184
(U.K.)

This was a pioneering torpedo bomber, a floatplane type that first flew in 1914 and was then built to the extent of 900 and more examples by the parent company and nine subcontractors. The Short 184 was flown with a variety of engine types, and served in just about every theatre in which British forces were involved during World War I.

SHORT 184

Role: Torpedo bomber/reconnaissance
Crew/Accommodation: Two
Power Plant: One 260 hp Sunbeam inline
Dimensions: Span 19.35m (63.5 ft); length 12.37m (40.58 ft); wing area 63.92m^2 (688 sq ft)
Weights: Empty 1,680kg (3,703 lb); MTOW 2,433kg (5,363 lb)
Performance: Maximum speed 142.4km/h (88.5 mph) at 610m (2,000 ft); operational ceiling 2,743m (9,000 ft); endurance 2.75 hours
Load: One .303 inch machine gun, plus one 14 inch torpedo

SHORT S.8 CALCUTTA and RANGOON
(U.K.)

The Calcutta was designed as a flying boat airliner suitable for use on Imperial Airways' routes to the Far East and Africa, and first flew in 1928. Five 'boats were built for Imperial Airways with a sixth going to France, where the type was built under licence as the Breguet Bre.521 Bizerte. The type also possessed military potential, and six basically similar aircraft were produced with armament under the name Rangoon.

SHORT S.8 CALCUTTA

Role: Passenger transport flying boat
Crew/Accommodation: Two and one cabin crew, plus up to 15 passengers
Power Plant: Three 540 hp Bristol Jupiter IX air-cooled radials
Dimensions: Span 28.35m (93 ft); length 20.12m (66 ft); wing area 169.56m^2 (1,825 sq ft)
Weights: Empty 6,280kg (13,845 lb); MTOW 10,206kg (22,500 lb)
Performance: Maximum speed 156km/h (97 mph) at 610m (2,000 ft); operational ceiling 4,115m (13,500 ft); range 1,046km (650 miles)
Load: Up to 1,395kg (3,075 lb)

SHORT S.23, S.30 and S.33 EMPIRE SERIES (U.K.)

The top picture shows the first of the Short S.30s about to 'unstick', while, in the lower view, an S.23 is seen over the Medway.

This series of superb flying boats started with an Imperial Airways' requirement for a type able to carry passengers and mail with complete reliability and safety but at higher speeds and over longer ranges than possible with the airline's previous generation of biplane 'boats. The series is generally known as the Empire type, and the first of the S.23 or 'C' class flew in July 1936 with 686kW (920 hp) Bristol Pegasus XC radial engines, and production totalled 31.

Further development towards producing a 'boat with transatlantic capability resulted in the S.30 with 664kW (890 hp) Bristol Perseus XIIC radial engines and greater fuel capacity to double the range of the S.23; production totalled nine. The ultimate expression of the series' design philosophy was the considerably larger S.33 with 1,029kW (1,380 hp) Bristol Hercules; only two had been completed before World War II, the third being completed during the war when most of the 'boats were impressed into military service before the survivors were returned to British Airways for important over-water routes.

SHORT S.23/EMPIRE C CLASS

Role: Passenger transport flying boat
Crew/Accommodation: Five, plus up to 24 passengers
Power Plant: Four 920 hp Bristol Pegasus XC air-cooled radials
Dimensions: Span 34.75m (114 ft); length 26.82m (88 ft); wing area 139.36m^2 (1,500 sq ft)
Weights: Empty 10,886kg (24,000 lb); MTOW 18,371kg (40,500 lb)
Performance: Cruise speed 266km/h (165 mph) at sea level; operational ceiling 2,377m (7,800 ft); range 1,223km (760 miles)
Load: Up to 3,300kg (7,275 lb)

SHORT S.25 SUNDERLAND and SANDRINGHAM
(U.K.)

This was the U.K.'s premier maritime reconnaissance flying boat of World War II, and derived ultimately from the S.23 'boats. The prototype flew in October 1937 with 753kW (1,010 hp) Bristol Pegasus XXII radial engines, and production of this variant totalled 90 before it was overtaken on the line by the Sunderland Mk II with 794kW (1,065 hp) Pegasus XVIIIs and a power-operated dorsal turret. These 43 'boats were in turn succeeded by the Sunderland Mk III, of which 456 were built with a revised hull and surface-search radar. The Sunderland Mk IV was developed for Pacific operations and became the Seaford, and the last Sunderland variant was the Mk V, of which 150 were built with 895kW (1,200 hp) Pratt & Whitney R-1830-90B radial engines.

Sunderlands were also used for civil transport, the first of 24 Sunderland Mk IIIs being handed over to British Airways in March 1943. The 'boats were later brought up to more comfortable standard as Hythes, and were then revised as the Sandringham with R-1830-92 radial engines and neat aerodynamic fairings over the original nose and tail turret positions.

SHORT SUNDERLAND Mk V

Role: Anti-submarine/maritime patrol
Crew/Accommodation: Seven
Power Plant: Four 1,200 hp Pratt & Whitney R-1830-90B Twin Wasp air-cooled radials
Dimensions: Span 34.39m (112.77 ft); length 26m (85.33 ft); wing area 156.6m^2 (1,687 sq ft)
Weights: Empty 16,783kg (37,000 lb); MTOW 27,250kg (60,000 lb)
Performance: Maximum speed 343km/h (213 mph) at 1,525m (5,000 ft); operational ceiling 5,455m (17,900 ft); range 4,300km (2,690 miles) with maximum fuel
Load: Six .5 inch machine guns and eight .303 inch machine guns, plus up to 908kg (2,000 lb) of bombs/depth charges

SHORT S.29 STIRLING
(U.K.)

This was one of the U.K.'s trio of four-engined heavy bombers in World War II, and the least successful of the three in its designed role: the specification limited the span to a maximum that could be accommodated in current hangars and thus lacked adequate altitude performance. The first prototype flew in May 1939 with 1,025kW (1,375 hp) Bristol Hercules II radial engines, and production followed of the Stirling Mk I with 1,189kW (1,595 hp) Hercules XIs (712 aircraft) and the improved Stirling Mk III with 1.230kW (1,650 hp) Hercules XVIs (1,047 aircraft). Stirling Mk IIIs were then converted as Stirling Mk IV glider tugs, these 450 aircraft being complemented by a contingent of 150 Stirling Mk V transports.

SHORT S.29 STIRLING Mk III

Role: Heavy night bomber
Crew/Accommodation: Eight
Power Plant: Four 1,650 hp Bristol Hercules XVI air-cooled radials
Dimensions: Span 30.2m (99.08 ft); length 26.59m (87.25 ft); wing area 135.6m^2 (1,460 sq ft)
Weights: Empty 21,200kg (46,900 lb); MTOW 31,790kg (70,000 lb)
Performance: Maximum speed 434km/h (270 mph) at 4,420m (14,500 ft); operational ceiling 5,181m (17,000 ft); range 950km (590 miles) with full warload
Load: Eight .303 inch machine guns, plus up to 6,350kg (14,000 lb) internally-carried bombload

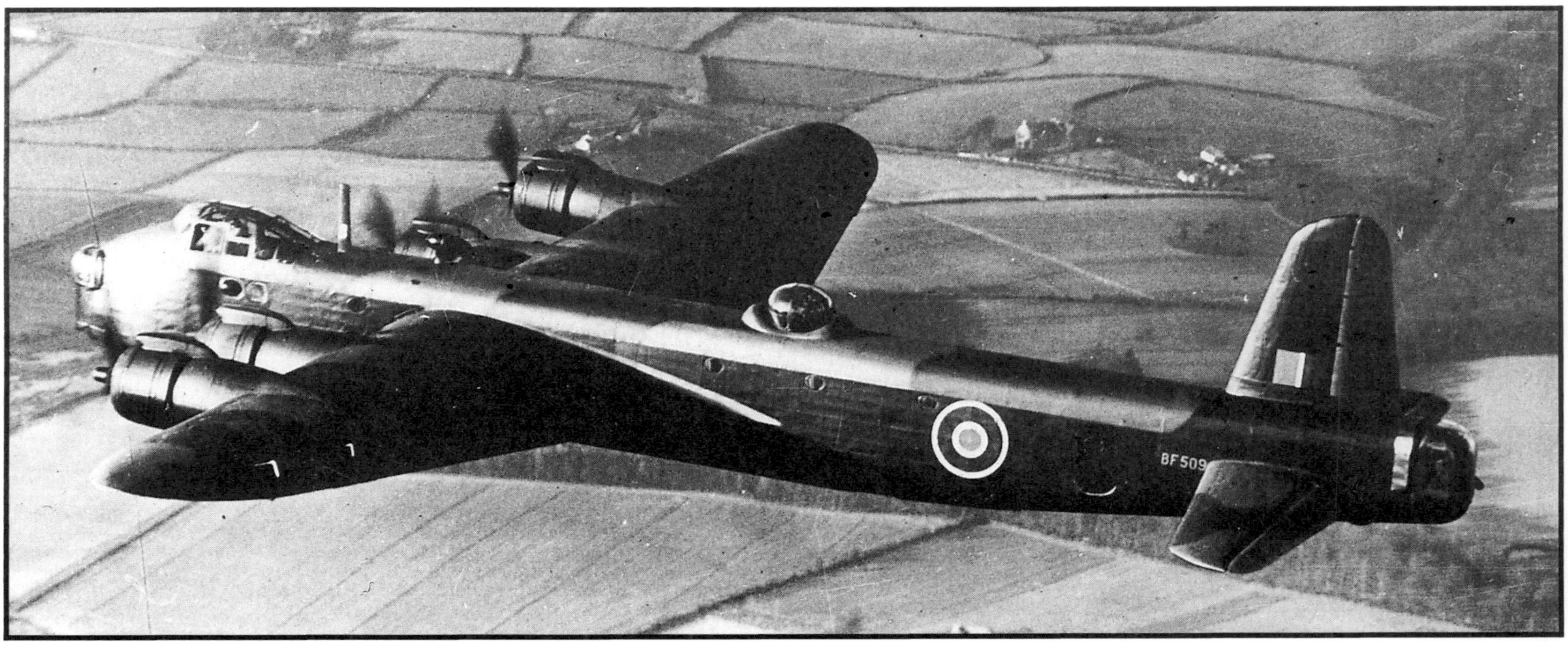

SHORTS 330 and C-23 SHERPA
(U.K.)

The Shorts 330 is essentially a stretched and refined Skyvan with retractable landing gear and more power. The type first flew in August 1974 as the SD3-30 with 862kW (1,156 shp) Pratt & Whitney Canada PT6A-45B turboprops, more powerful engines being used on the Shorts 330-200 current version, which has a military counterpart as the Shorts 330 Utility Tactical Transport. This last has been ordered by the U.S. Air Force as the C-23A Sherpa.

SHORTS C-23A SHERPA

Role: Short-range freight transport
Crew/Accommodation: Two and one load master/crew chief
Power Plant: Two 1,156 shp Pratt & Whitney of Canada PT6A-45B turboprops
Dimensions: Span 22.78m (74.75 ft); length 17.69m (58 ft); wing area 42.1m^2 (453 sq ft)
Weights: Empty 6,440kg (14,200 lb); MTOW 10,387kg (22,900 lb)
Performance: Maximum speed 352km/h (190 knots) at 1,830m (6,000 ft); operational ceiling 3,930+m (12,900 ft); range 370km (200 naut. miles) with full payload
Load: Up to 3,175kg (7,000 lb), including bulky cargo

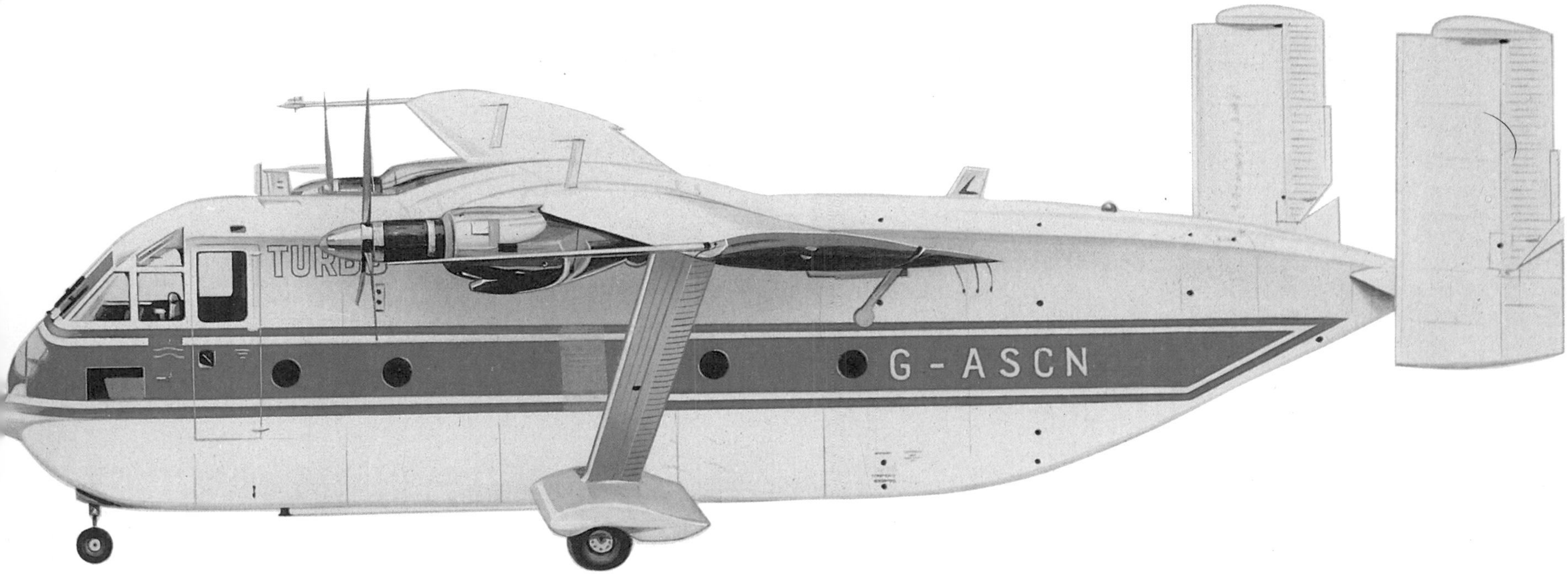

SHORTS SC.7 SKYVAN
(U.K.)

This STOL utility transport was inaugurated in 1959 and first flew in January 1963 with 291kW (390 hp) Continental GTS10-520 piston engines. The Skyvan Series 1A switched to 388kW (520 shp) Turbomeca Astazou IIA turboprops, and the Skyvan Series 2 production model had 544kW (730 shp) Astazou XIIs, replaced by 533kW (715 shp) Garrett TPE331-2-201A in the Skyvan Series 3. The production total of slightly more than 150 aircraft included the Skyvan 3M military variant.

SHORTS SC.7 SKYVAN SERIES 3

Role: Short take-off and landing freight/passenger transport
Crew/Accommodation: Two, plus up to 19 passengers
Power Plant: Two 715 shp Garrett AiResearch TPE 331-2-201A turboshafts
Dimensions: Span 19.79m (64.92 ft); length 12.45m (40.08 ft); wing area 34.66m^2 (373 sq ft)
Weights: Empty 3,306kg (7,289 lb); MTOW 5,670kg (12,500 lb)
Performance: Cruise speed 278km/h (150 knots) at 3,048m (10,000 ft); operational ceiling 6,401m (21,000 ft); range 311km (168 naut. miles)
Load: Up to 2,085kg (4,600 lb)

SIKORSKY ILYA MUROMETS
(Russia)

First flown in May 1913, the Russkii Vitiaz designed by Igor Sikorsky for the Russo-Baltic Waggon Works was the world's first four-engined aeroplane. The power plant comprised 75kW (100 hp) Argus inline engines, and the type was intended as a transporrt. The aeroplane was the basis for the Ilya Muromets heavy bomber that first flew in January 1914 and was operated by the Russians in World War I. Between 70 and 80 of this epoch-making type were delivered with a number of power plant combinations.

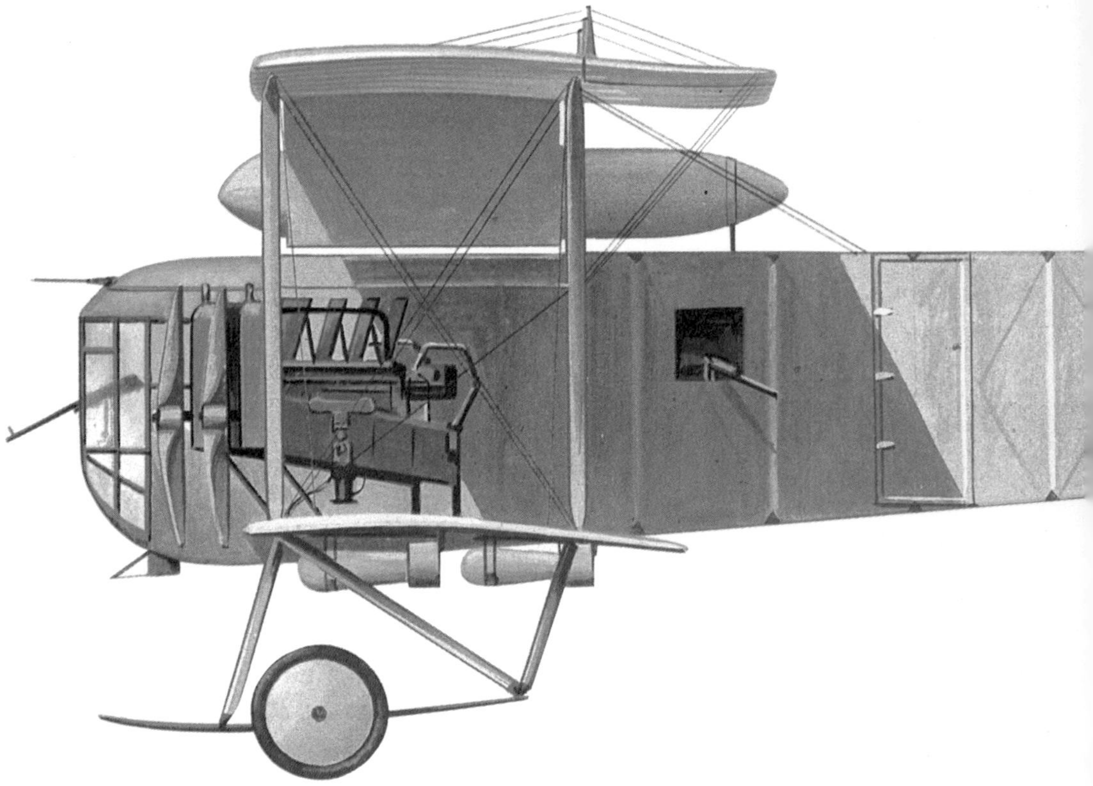

SIKORSKY ILYA MUROMETS TYPE YE (E)

Role: Long-range heavy bomber
Crew/Accommodation: Eight
Power Plant: Four 220 hp Renault water-cooled inlines
Dimensions: Span 30.4m (99.74 ft); length 18.5m (60.75 ft); wing area 190m^2 (2,045 sq ft)
Weights: Empty 4,200kg (9,260 lb); MTOW 6,100kg (13,448 lb)
Performance: Maximum speed 137km/h (85 mph) at sea level; operational ceiling 4,000m (13,123 ft); range 540km (336 miles) with full warload
Load: Eight 7.62mm machine guns, plus up to around 1,100kg (2,425 lb) of bombs

SIKORSKY S-38
(U.S.A.)

After World War I Sikorsky emigrated to the U.S.A., where his enthusiasm was drawn initially to the seaplane, especially in its amphibian form. The most important of these early aircraft was the S-38, a development of the S-36 amphibian of the late 1920s. The type had two 317kW (425 hp) Pratt & Whitney R-1340 Wasp radial engines, and began to enter service with Pan American as an eight-seat airliner in November 1931.

SIKORSKY S-38A

Role: Short-range passenger transport flying boat amphibian
Crew/Accommodation: Two, plus up to nine passengers
Power Plant: Two 425 hp Pratt & Whitney R-1340 Wasp air-cooled radials
Dimensions: Span 21.84m (71.66 ft); length 12.27m (40.25 ft); wing area 66.89m^2 (720 sq ft)
Weights: Empty 2,722kg (6,000 lb); MTOW 4,754kg (10,480 lb)
Performance: Cruise speed 169km/h (105 mph) at sea level; operational ceiling 4,877m (16,000 ft); range 483km (300 miles) with full payload
Load: Up to 1,270kg (2,800 lb)

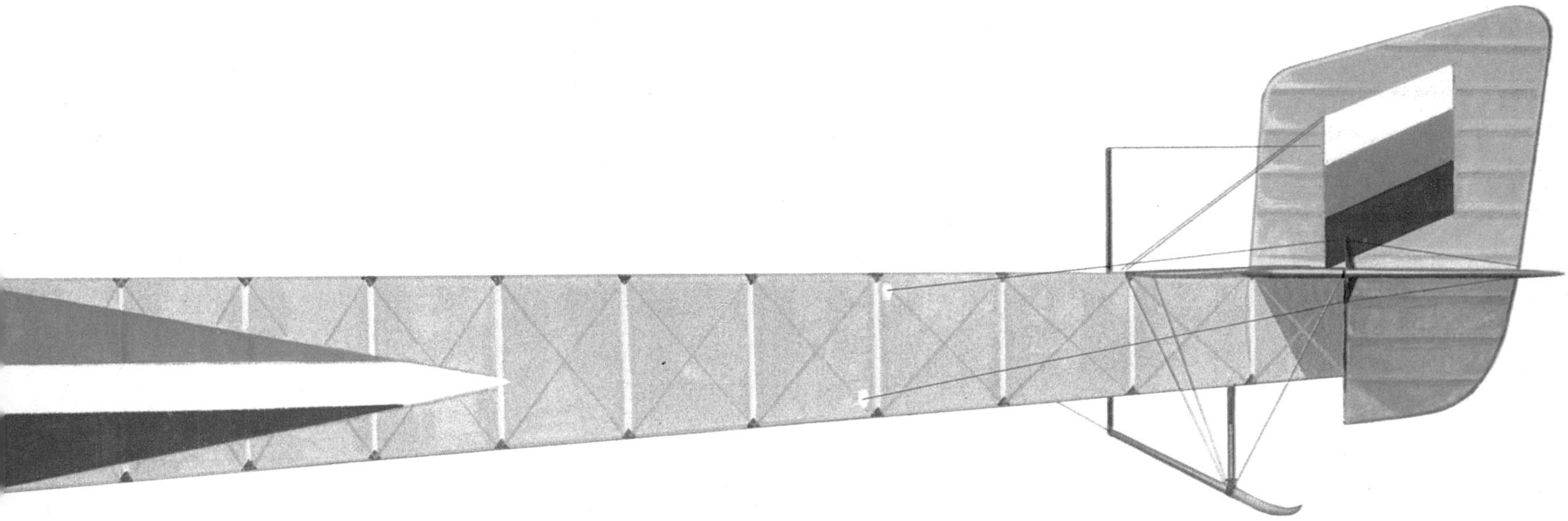

SIKORSKY S-42
(U.S.A.)

From the S-40 amphibian designed for use on Pan American's routes across the Caribbean and to South America, Sikorsky developed the larger and more powerful S-42. This was a parasol-winged flying boat intended for Pan American's planned transatlantic service, but used mainly for South American and transpacific routes (including pioneering flights across the South Pacific to New Zealand from August 1934.

Total production was 10 'boats including three S-42s with the 522kW (700 hp) Pratt & Whitney Hornet S5D1G, three S-42A 'boats with 559kW (750 hp) Hornet S1EGs and longer-span wings, and four S-42B 'boats with further refinements and Hamilton Standard constant-speed propellers.

SIKORSKY S-42B

Role: Intermediate/short-range passenger transport flying boat
Crew/Accommodation: Four and two cabin crew, plus up to 32 passengers
Power Plant: Four 800 hp Pratt & Whitney R-1690 Hornet air-cooled radials
Dimensions: Span 35.97m (118.33 ft); length 20.93m (68.66 ft); wing area 124.5m^2 (1,340 sq ft)
Weights: Empty 9,491kg (20,924 lb); MTOW 19,504kg (43,000 lb)
Performance: Cruise speed 225km/h (140 mph) at 610m (2,000 ft); operational ceiling 4,878m (16,000 ft); range 1,207km (750 miles) with full payload
Load: Up to 3,626kg (7,995 lb)

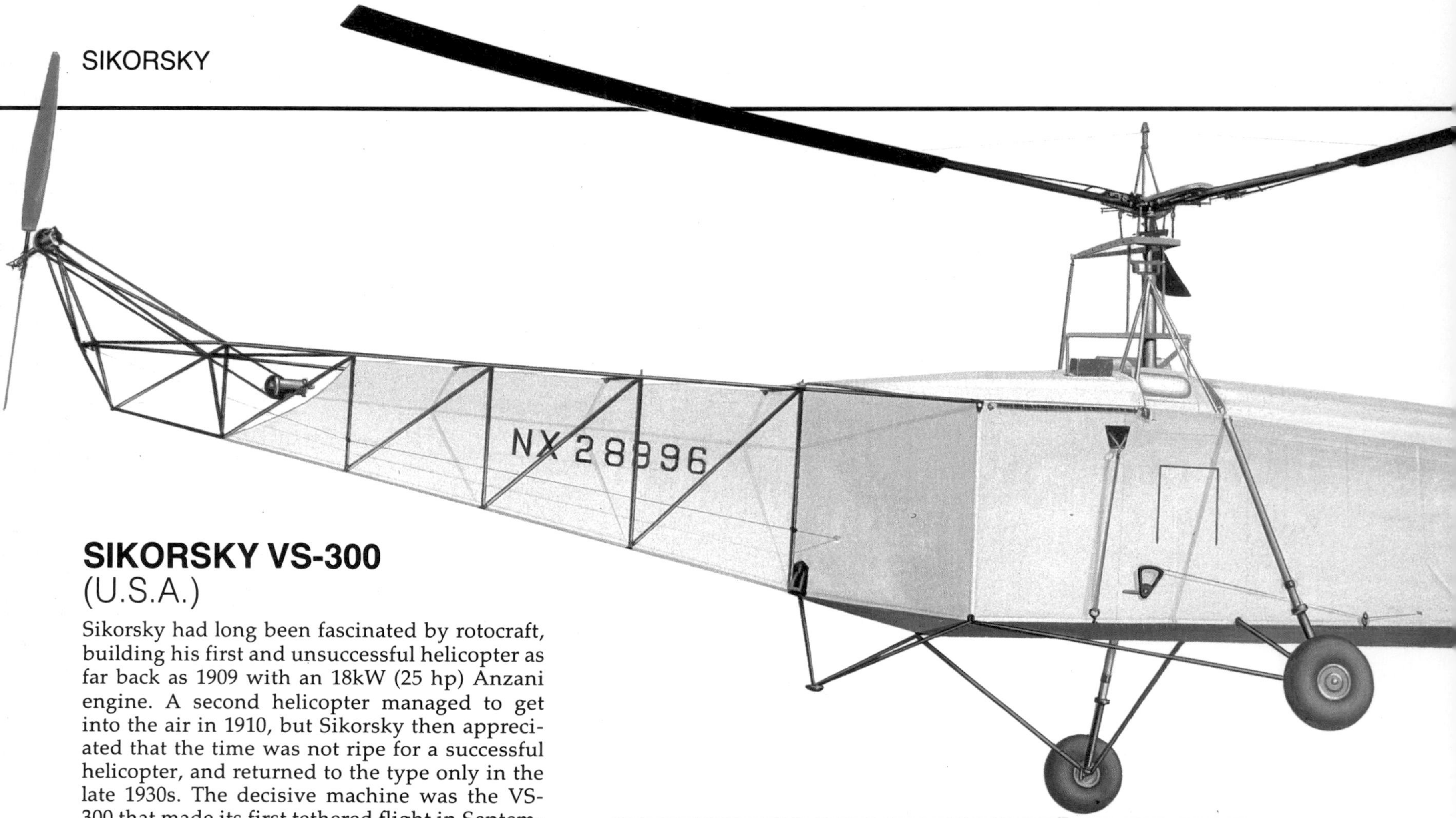

SIKORSKY VS-300
(U.S.A.)

Sikorsky had long been fascinated by rotocraft, building his first and unsuccessful helicopter as far back as 1909 with an 18kW (25 hp) Anzani engine. A second helicopter managed to get into the air in 1910, but Sikorsky then appreciated that the time was not ripe for a successful helicopter, and returned to the type only in the late 1930s. The decisive machine was the VS-300 that made its first tethered flight in September 1939. This prototype went through a large number of forms as Sikorsky tested various lift and control factors, and without any doubt may be judged to be the world's first successful single-rotor helicopter.

SIKORSKY VS-300

Role: Developmental helicopter
Crew/Accommodation: One
Power Plant: One 90 hp Franklin 4AC-199-E3 air-cooled flat-opposed
Dimensions: Overall length rotors turning 12.13m (39.8 ft); rotor diameter 9.14m (30 ft)
Weights: Empty 422kg (930 lb); MTOW 522kg (1,150 lb)
Performance: Cruise speed 97km/h (60 mph) at sea level; range 121km (75 miles)
Load: None

SIKORSKY R-4 and R-6
(U.S.A.)

This was developed from the VS-300 as the VS-316A, and may therefore be regarded as the world's first practical single-rotor helicopter. The VS-316A was in essence a refined and two-seat development of the VS-300 with a fabric-covered fuselage, and first flew in January 1942 with a 123kW (165 hp) Warner R-500 radial engine. To provide an evaluation batch the U.S. Army ordered 30 similar helicopters as three YR-4A and 26 YR-4B machines with the 134kW (180 hp) Warner R-550 driving a larger main rotor. Further production resulted in the R-4B, of which 100 were built with the 149kW (200 hp) R-550-3.

The type was also operated by the U.S. Navy with the designation HNS-1, and by the RAF as the Hoverfly Mk I. Installation of the R-4s dynamic system in a more streamlined fuselage with a semi-monocoque tail resulted in the VS-316B, which entered production as the R-6A; production totalled 193 helicopters with the 168kW (225 hp) Avco Lycoming O-435 engine; U.S. Navy and British examples were designated HOS-1 and Hoverfly Mk II.

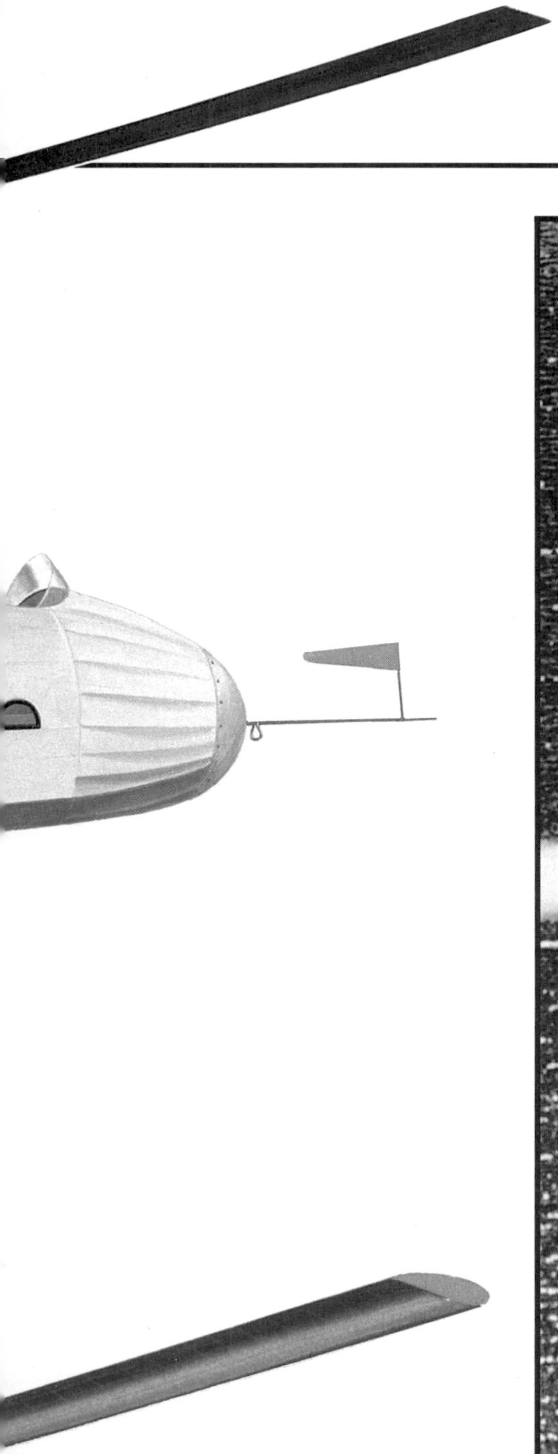

SIKORSKY S-51 and R-5
(U.S.A.)

Produced in parallel with the VS-316 series was the VS-337 tandem two-seater with a new rotor driven by the 336kW (450 hp) Pratt & Whitney R-985-AN-5 radial engine. This first flew in August 1943 and production totalled 53 in models such as the YR-5A and litter-carrying R-5A. The U.S. Navy used the type as the HO3S-1, and larger production was undertaken of the S-51 civil model, which reached a total of 379 including helicopters for the military. The British licence-built version of the S-51 was the Westland Dragonfly.

SIKORSKY R-4B

Role: Light utility/communications helicopter
Crew/Accommodation: One, plus one passenger/observer
Power Plant: One 200 hp Warner R-550-3 air-cooled radial
Dimensions: Overall length rotors turning 13.33m (43.75 ft); rotor diameter 11.58m (38 ft)
Weights: Empty 916kg (2,020 lb); MTOW 1,315kg (2,900 lb)
Performance: Maximum speed 121km/h (75 mph) at sea level; operational ceiling 2,438m (8,000 ft); range 209km (130 miles) with observer
Load: Up to 213kg (470 lb) including second crew member/passenger

SIKORSKY S-51

Role: Search and rescue/light utility helicopter
Crew/Accommodation: One/two, plus up to three passengers
Power Plant: One 450 hp Pratt & Whitney R-985-AN-5 Wasp Junior air-cooled radial
Dimensions: Overall length rotors turning 17.58m (57.67 ft); rotor diameter 14.94m (49 ft)
Weights: Empty 1,720kg (3,788 lb); MTOW 2,263kg (4,985 lb)
Performance: Maximum speed 166km/h (103 mph) at sea level; operational ceiling 4,389m (14,400 ft); range 467km (290 miles) with 104kg (230 lb) payload
Load: Up to 431kg (950 lb)

SIKORSKY S-55
(U.S.A.)

Development of the S-55 was launched in 1948 to meet a U.S. Air Force utility helicopter requirement. The YH-19 prototype flew in November 1949 with a 410kW (550 hp) Pratt & Whitney R-1340-57 radial engine mounted in the lower nose and driving the main rotor by means of a long shaft running diagonally up through the cockpit to the gearbox under the main rotor. Production of 50 H-19A helicopters followed with 447kW (600 hp) R-1340-57s.

Later variants were 270 H-19Bs with the 522kW (700 hp) Wright R-1300-3 radial engines, together with comparable versions for the U.S. Army as the H-19C and H-19D. The U.S. Navy took similar machines with the designations HO4S-1 and HO4S-3, while U.S. Marine Corps assault transports were the HRS-1 and HRS-2. The type was also built under licence in the U.K. as the Westland Whirlwind with the Alvis Leonides radial piston engine or Bristol Siddeley Gnome turboshaft.

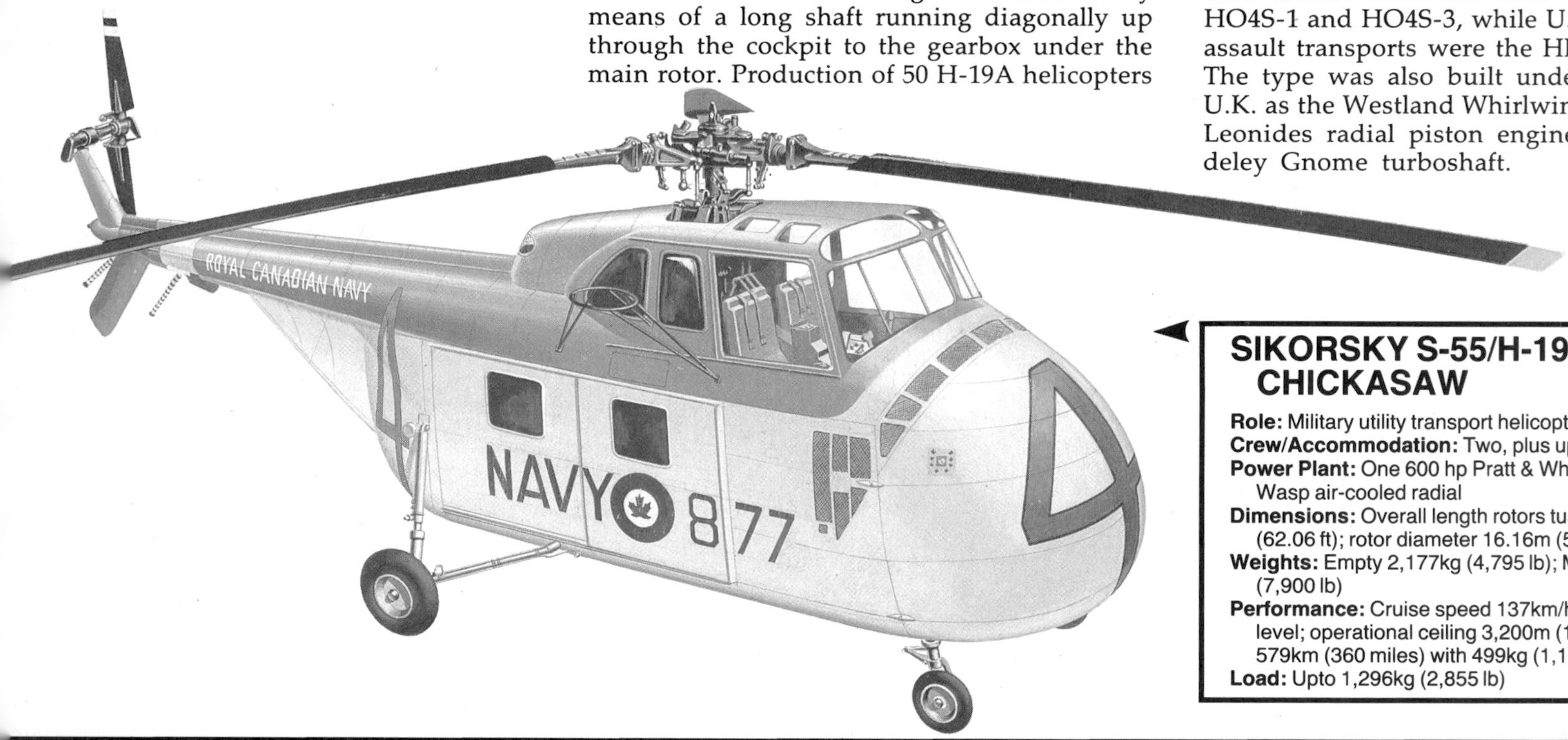

SIKORSKY S-55/H-19A CHICKASAW

Role: Military utility transport helicopter
Crew/Accommodation: Two, plus up to ten troops
Power Plant: One 600 hp Pratt & Whitney R-1340-57 Wasp air-cooled radial
Dimensions: Overall length rotors turning 18.91m (62.06 ft); rotor diameter 16.16m (53 ft)
Weights: Empty 2,177kg (4,795 lb); MTOW 3,587kg (7,900 lb)
Performance: Cruise speed 137km/h (85 mph) at sea level; operational ceiling 3,200m (10,500 ft); range 579km (360 miles) with 499kg (1,100 lb) payload
Load: Upto 1,296kg (2,855 lb)

SIKORSKY S-61
(U.S.A.)

The HSS had been compelled by its payload limitations to operate as a submarine hunter or a submarine killer, requiring two-helicopter teams for the anti-submarine role. The S-61 designed to overcome this limitation by adopting turboshaft power to create a helicopter with sufficient payload to carry sensors and weapons. The type first flew in March 1959 as the XHHS-2 with two General Electric T58 turboshafts, and was then produced as the HSS-2 (SH-3A from 1962) Sea King with 932kW (1,250 shp) T58-GE-8B engines. Further development resulted in the SH-3D with 1,044kW (1,400 shp) T58-GE-10 engines, conversions of earlier helicopters then produced the SH-3G utility and SH-3H multi-role and missile-warning type.

The Sea King anti-submarine helicopter has also been produced under licence in Italy by Agusta and in Japan by Mitsubishi. The Naval S-61A was complemented by the air force's S-61R with a revised fuselage allowing the incorporation of a rear ramp. This was produced as the CH-3 utility type in B, C and E variants. The E variant was also developed as a combat search-and-rescue type with the designation HH-3E; the U.S. Coast Guard used an unarmed version for ordinary search and rescue as the HH-3F Pelican. The S-61 was also developed for civil use as the non-amphibious S-61L and amphibious S-61N variants.

SIKORSKY S-61/SH-3G SEA KING

Role: Naval shipborne anti-submarine helicopter
Crew/Accommodation: Four
Power Plant: Two 1,400 shp General Electric T58-GE-10 turboshafts
Dimensions: Overall length rotors turning 22.15m (72.66 ft); rotor length 18.85m (62 ft)
Weights: Empty 5,382kg (11,865 lb); MTOW 9,299kg (20,500 lb)
Performance: Maximum speed 254km/h (137 knots) at sea level; operational ceiling 4,481m (14,700 ft); range 300km (161 naut. miles) with full warload
Load: Up to 544kg (1,200 lb)

SIKORSKY S-58
(U.S.A.)

The S-58 was designed to replace the HO4S in naval service, the object being the creation of an anti-submarine helicopter with the payload for an extended patrol with useful sensor or weapon load. The XHSS-1 prototype first flew in March 1954 with a 1,137kW (1,525 hp) Wright R-1820 radial engine. Total production was 1,820 helicopters in naval HSS Seabat, Marine HUS Seahorse and Army H-34 Choctaw variants with steadily improved features. The type was also built for civil use, and in the U.K. the S-58 was built under licence as the Westland Wessex with one Napier Gazelle or two Bristol Siddeley Gnome turboshafts.

SIKORSKY S-58/H.34A CHOCTAW

Role: Military assault/transport helicopter
Crew/Accommodation: Two, plus up to 18 troops
Power Plant: One 1,525 hp Wright R-1820-84 Cyclone air-cooled radial
Dimensions: Overall length rotors turning 22.68m (74.4 ft); rotor diameter 17.07m (56 ft)
Weights: Empty 3,461kg (7,630 lb); MTOW 6,038kg (13,300 lb)
Performance: Cruise speed 158km/h (98 mph) at sea level; operational ceiling 2,896m (9,500 ft); range 293km (182 miles) with full payload
Load: Up to 1,878kg (4,140 lb)

SIKORSKY S-65
(U.S.A.)

The S-65 and its successors are the West's largest helicopters, and the series first flew in prototype form during October 1964 to meet a U.S. Marine Corps requirement for an assault and logistic transport helicopter. This initial CH-53 Sea Stallion is a twin-engined machine powered in several variants by steadily more powerful General Electric T64 turboshafts. Other variants are the HH-53 'Jolly Green Giant' combat search and rescue helicopter and the RH/MH-53 minesweeping helicopter. The type has been upgraded considerably in the minesweeping role to produce the MH-53E Super Stallion which is equipped with three 3,266kW (4,380 shp) T64-GE-416 turboshafts driving an improved rotor.

SIKORSKY S-65A/MH-53E SUPER STALLION

Role: Big-shipborne heavyweight assault helicopter
Crew/Accommodation: Three, plus up to 55 troops
Power Plant: Three 4,380 shp General Electric T64-GE-416 turboshafts
Dimensions: Overall length rotors turning 30.48m (99.5 ft); rotor diameter 24.08m (79 ft)
Weights: Empty 15,071kg (33,226 lb); MTOW 33,339kg (73,500 lb) with externally-slung load
Performance: Maximum speed 315km/h (170 knots) at sea level; operational ceiling 3,780m (12,400 ft); range 93km (50 naut. miles) with full payload
Load: Up to 14,515kg (32,000 lb)

SIKORSKY S-70
(U.S.A.)

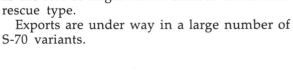

The S-70 series was developed to provide the U.S. Army with a tactical helicopter to replace the Bell UH-1 series, and first flew in October 1974. The type had been ordered in large numbers for the U.S. Army as the UH-60 Black Hawk, with subvariants such as the EH-60 electronic warfare type, and by the U.S. Navy as the SH-60 Seahawk in the medium anti-submarine (SH-60B) and carrier battle group protection (SH-70F) roles, and by the U.S. Air Force as the HH-60 Night Hawk combat search and rescue type.

Exports are under way in a large number of S-70 variants.

A U.S. Air Force operated HH-60F Night Hawk

SIKORSKY S-70/SH-60B SEAHAWK

Role: Naval shipborne anti-submarine/ship helicopter
Crew/Accommodation: Three
Power Plant: Two 1,713 shp General Electric T700-GE-401 turboshafts
Dimensions: Overall length rotors turning 19.76m (64.83 ft); rotor diameter 16.36m (53.67 ft)
Weights: Empty 6,204kg (13,678 lb); MTOW 9,183kg (20,244 lb) for anti-submarine mission
Performance: Maximum speed 234km/h (126 knots) at 1,524m (5,000 ft); operational ceiling 3,048m (10,000 ft); endurance 4.75 hours with reserves
Load: Up to approximately 590kg (1,300 lb)

SIKORSKY S-76 SPIRIT
(U.S.A.)

This aerodynamically refined helicopter was developed as a high-performance type with a twin-turboshaft power plant so that it could be used for the support of the offshore resources-exploitation industry. The type flew in March 1977, and has sold well in its initial and improved S-76 Mk II forms. The type has also been sold in H-76 Eagle form for military use.

SIKORSKY S-76/H-76 EAGLE

Role: Military and civil utility helicopter
Crew/Accommodation: Two, plus up to ten troops
Power Plant: Two 960 shp Pratt & Whitney Canada PT6B-36 turboshafts
Dimensions: Overall length rotors turning 16m (52.50 ft); rotor diameter 13.41m (44 ft)
Weights: Empty 2,525kg (5,610 lb); MTOW 4,761kg (10,300 lb)
Performance: Maximum speed 287km/h (155 knots) at sea level; operational ceiling 4,875m (16,000 ft); range 578km (313 naut. miles) with 998kg (2,200 lb) payload
Load: Up to 1,497kg (3,300 lb) externally slung

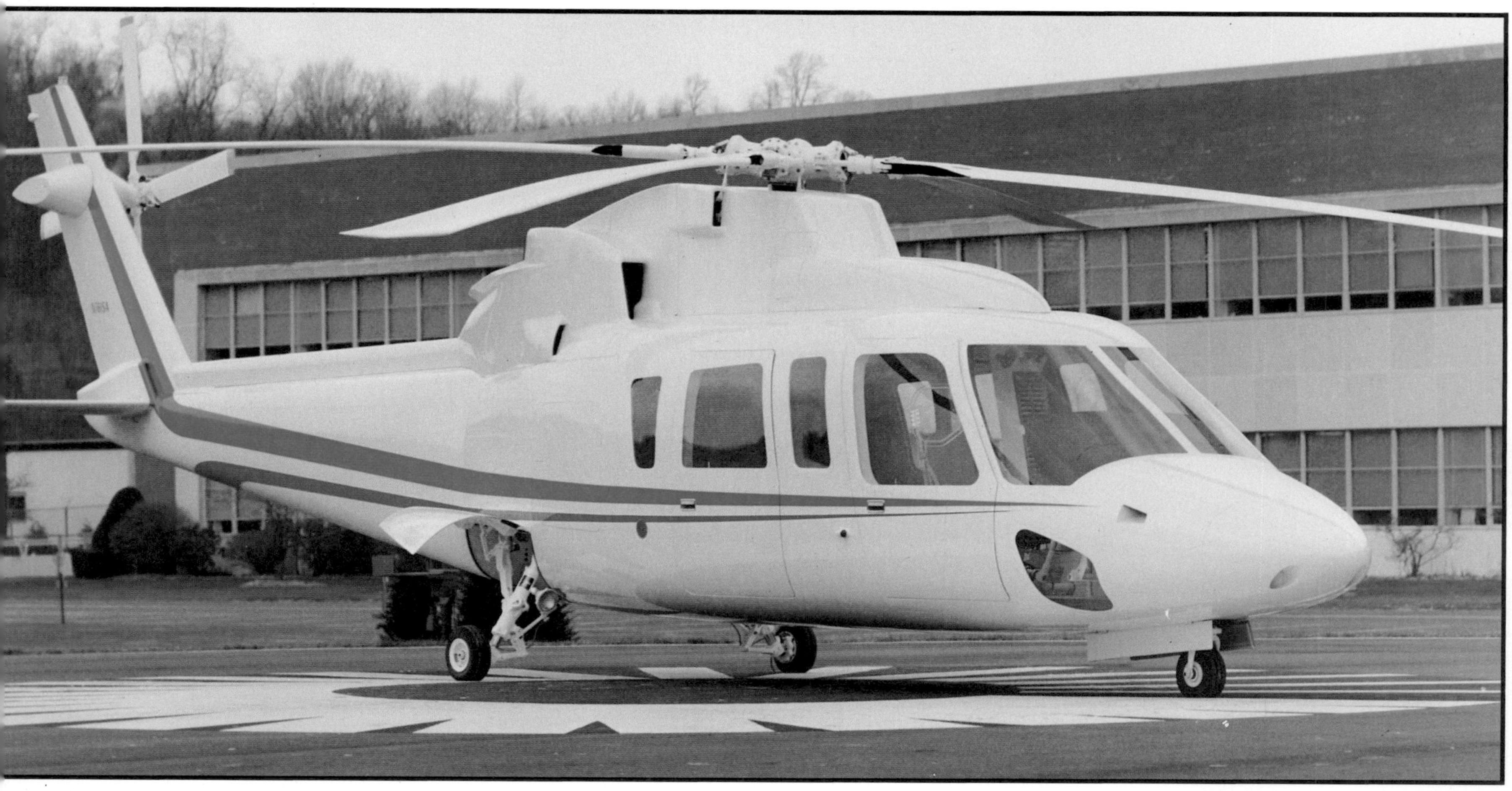

SNCA SUD-EST S.E.210 CARAVELLE
(France)

This pioneering aeroplane, the world's third jet-powered airliner to enter service, introduced the concept of engines pod-mounted on the sides of the rear fuselage to reduce cabin noise and also leave the wings free for their primary task of generating lift. The type resulted from a 1951 specification and first flew in May 1955 with 4,536kg (10,000 lb) thrust Rolls-Royce Avon RA.26 turbojets and a forward fuselage/cockpit essentially identical with that of the de Havilland Comet. Production totalled 280 in nine lengthened and thus more capacious models that culminated in the Caravelle 12 with 6,577kg (14,500 lb) thrust Pratt & Whitney JT8D-9 turbofans.

Sud Aviation Caravelle I

SNCA NORD 2500 NORATLAS
(France)

First flown in September 1949, the Noratlas was designed as a military transport whose configuration with twin booms projecting aft from the engines allowed both straight-in ground loading/unloading, and the aerial despatch of paratroops and paradropped equipment. The Nord 2500 prototype had 1,212kW (1,625 hp) Gnome-Rhone 14R radial engines, but the Nord 2501 production model switched to 1,521kW (2,040 hp) SNECMA-built Bristol Hercules 739 radial engines Production of 425 aircraft in several variants lasted to 1961.

NORD N.2501 NORATLAS

Role: Military cargo/paratroop transport
Crew/Accommodation: Four/five, plus up to 45 troops, or 36 paratroops
Power Plant: Two 2,040 hp SNECMA-built Bristol Hercules 739 air-cooled radials
Dimensions: Span 32.48m (106.56 ft); length 21.95m (72.01 ft); wing area 101m^2 (1,087.1 sq ft)
Weights: Empty 13,200kg (29,327 lb); MTOW 22,000kg (48,502 lb)
Performance: Maximum speed 322km/h (200 mph) at 3,050m (10,000 ft); operational ceiling 7,100m (23,300 ft); range 1,500km (932 miles)
Load: Up to 6,000kg (13,227 lb)

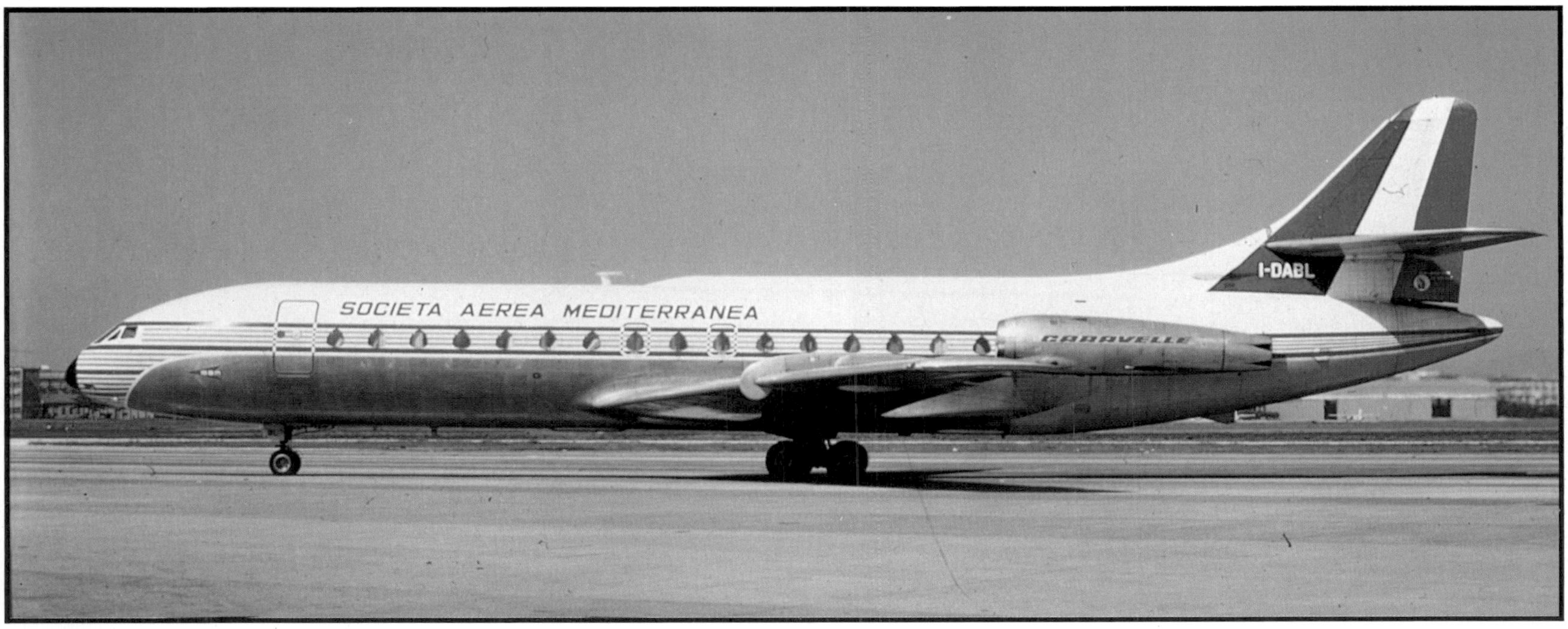

Sud Aviation Caravelle III

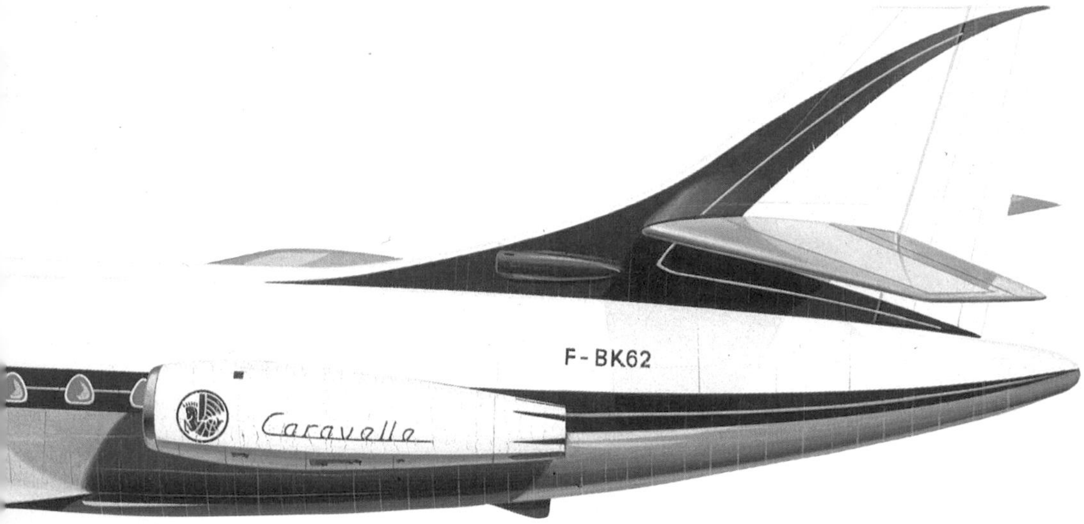

AEROSPATIALE (SUD AVIATION) CARAVELLE 12

Role: Short-range passenger transport
Crew/Accommodation: Two, four cabin crew, plus up to 140 passengers
Power Plant: Two 6,577kgp (14,500 lb s.t.) Pratt & Whitney JT8D-9 turbofans
Dimensions: Span 34.30m (112.5 ft); length 36.24m (118.75 ft); wing area 146.7m^2 (1,579 sq ft)
Weights: Empty 31,800kg (70,107 lb); MTOW 56,699kg (125,000 lb)
Performance: Maximum speed 785km/h (424 knots) at 7,620m (25,000 ft); operational ceiling 12,192m (40,000+ ft); range 1,870km (1,162 miles) with full payload
Load: Up to 13,200kg (29,101 lb)

SNCA SUD-OUEST S.O.4050 VAUTOUR (VULTURE)
(France)

The Vautour first flew in April 1949 in the form of the S.O.4000 prototype, though the design of this machine was extensively modified to create the S.O.4050 that first flew during October 1952. The three prototypes were flown in the form of a two-seat all-weather fighter with 2,400kg (5,291 lb) thrust SNECMA Atar 101B turbojets, a single-seat attack aeroplane with 2,820kg (6,217 lb) thrust Atar 101D turbojets, and a two-seat strike bomber with Armstrong Siddeley Sapphire turbojets. Production of all three types was ordered with the Atar 101E turbojet, and amounted to 30 Vautour II-A attack aircraft, 40 Vautour II-B bombers and 70 Vautour II-N night-fighters.

SUD S.O.4050 VAUTOUR II N

Role: All-weather/night fighter
Crew/Accommodation: Two
Power Plant: Two 3,300kgp (7,275 lb s.t.) SNECMA Atar 101E-3 turbojets
Dimensions: Span 15.1m (49.54 ft); length 16.5m (54.13 ft); wing area 45.3m^2 (487.6 sq ft)
Weights: Empty 9,880kg (21,782 lb); MTOW 17,000kg (37,479 lb)
Performance: Maximum speed 958km/h (595 mph) at 12,200m (40,026 ft); operational ceiling 14,000m (45,932 ft); range 2,750km (1,709 miles) with maximum fuel
Load: Four 20mm cannon

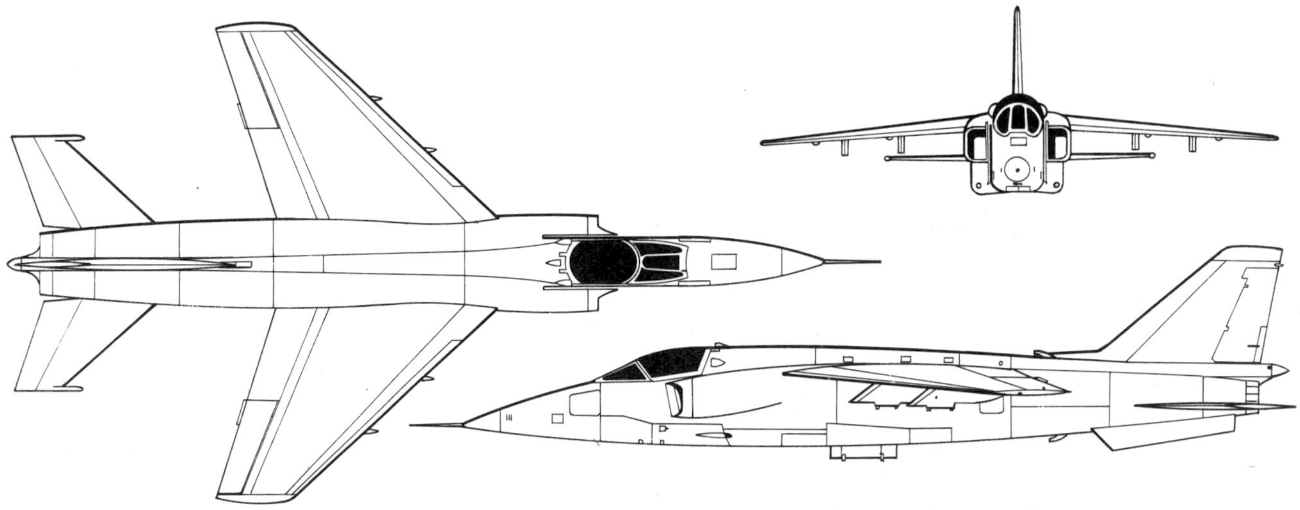

SOKO ORAO (EAGLE) and IAR-93
(Romania/Yugoslavia)

This simple attack fighter was developed jointly by Romania and Yugoslavia as the IAR-93 and Orao respectively. The type first flew in October 1974, and production of identically dimensioned single- and two-seaters has centred on the IAR-93A and Orao-A with 1,814kg (4,000 lb) thrust Rolls-Royce Viper Mk 632-41R non-afterburning turbojets, and the IAT-9B and Orao-B with 2,268kg (5,000 lb) thrust Viper Mk 633-47 afterburning turbojets. Both these engine variants are being built under licence by the Turbomecanica company.

SOKO ORAO-B

Role: Light strike fighter
Crew/Accommodation: One
Power Plant: Two 1,760kgp (5,000 lb s.t.) Rolls-Royce Viper 633-41 turbojets with reheat
Dimensions: Span 9.62m (31.56 ft); length 14.9m (48.88 ft); wing area 26m^2 (279.9 sq ft)
Weights: Empty 5,750kg (12,676 lb); MTOW 11,250kg (24,800 lb)
Performance: Maximum speed 1,160km/h (626 knots) at sea level; operational ceiling 13,500m (44,300 ft); radius 450km (242 naut. miles) with 1,000 kg (2,205 lb) bombload
Load: Up to 1,640 kg (3,615 lb) of externally carried ordnance

SOPWITH TABLOID
(U.K.)

The Tabloid was a side-by-side two-seater designed for competition flying, and first flew in 1913. The type's high performance appealed to the military, who were in need of a fast scout, and production of 36 such landplanes for the Royal Flying Corps and Royal Naval Air Service was followed by the building of 160 Tabloid floatplanes.

SOPWITH TABLOID

Role: Military scout
Crew/Accommodation One
Power Plant: One 80 hp Gnome air-cooled rotary
Dimensions: Span 7.77m (25.5 ft); length 6.2m (20.33 ft); wing area 31.71m^2 (341.3 sq ft)
Weights: Empty 331kg (730 lb); MTOW 508kg (1,120 lb)
Performance: Maximum speed 148km/h (92 mph) at sea level; operational ceiling 2,652m (8,700 ft); endurance 3.5 hours
Load: Some carried one .303 inch machine gun

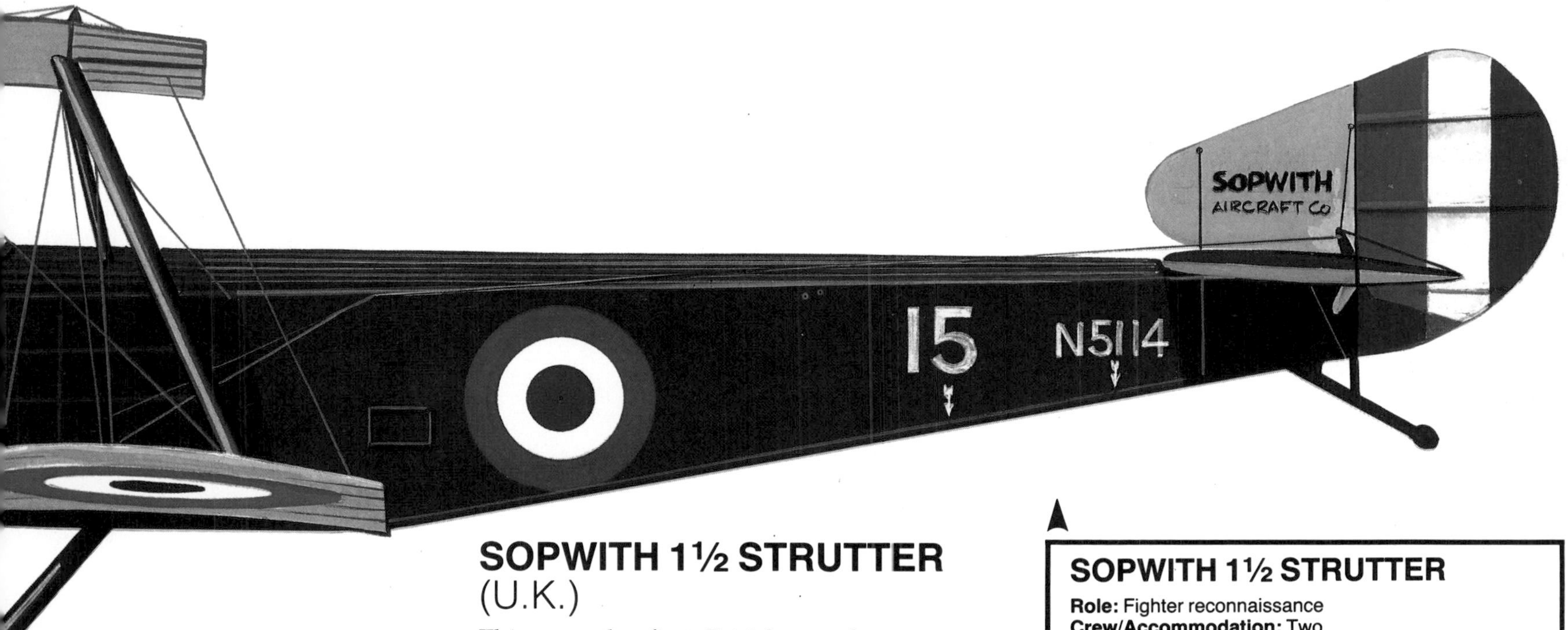

SOPWITH 1½ STRUTTER
(U.K.)

This was the first British warplane to enter service with a gun synchronizing mechanism, and proved very successful. The first prototype flew late in 1915 with an 82kW (110 hp) Clerget rotary engine, and was produced to the extent of 1,513 aircraft for the Royal Flying Corps as the Sopwith Two-Seater and for the Royal Navy Flying Service as the Sopwith Type 9700. Both variants were produced as two-seat fighters with the 97kW (130 hp) Clerget or single-seat bombers with the 82kW (110 hp) (Clerget), and it is believed that perhaps 4,500 similar aircraft were built under licence in France.

SOPWITH 1½ STRUTTER

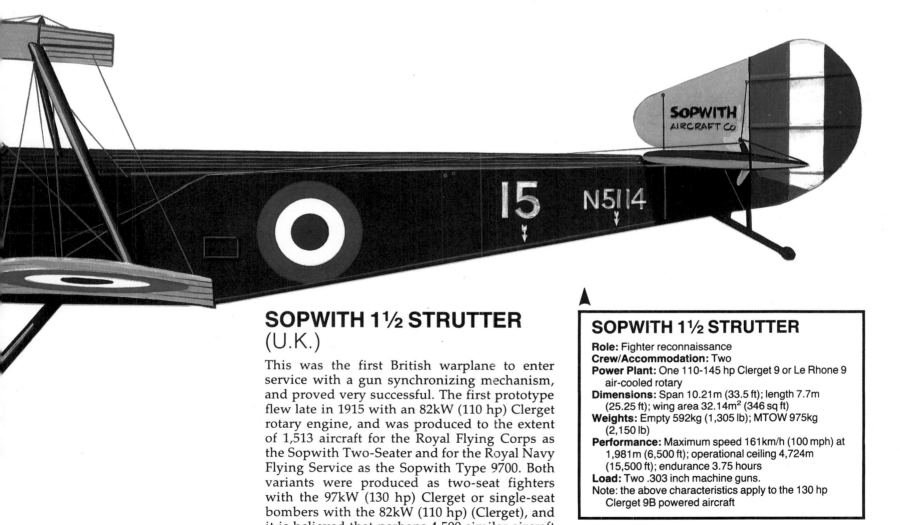

Role: Fighter reconnaissance
Crew/Accommodation: Two
Power Plant: One 110-145 hp Clerget 9 or Le Rhone 9 air-cooled rotary
Dimensions: Span 10.21m (33.5 ft); length 7.7m (25.25 ft); wing area 32.14m^2 (346 sq ft)
Weights: Empty 592kg (1,305 lb); MTOW 975kg (2,150 lb)
Performance: Maximum speed 161km/h (100 mph) at 1,981m (6,500 ft); operational ceiling 4,724m (15,500 ft); endurance 3.75 hours
Load: Two .303 inch machine guns.
Note: the above characteristics apply to the 130 hp Clerget 9B powered aircraft

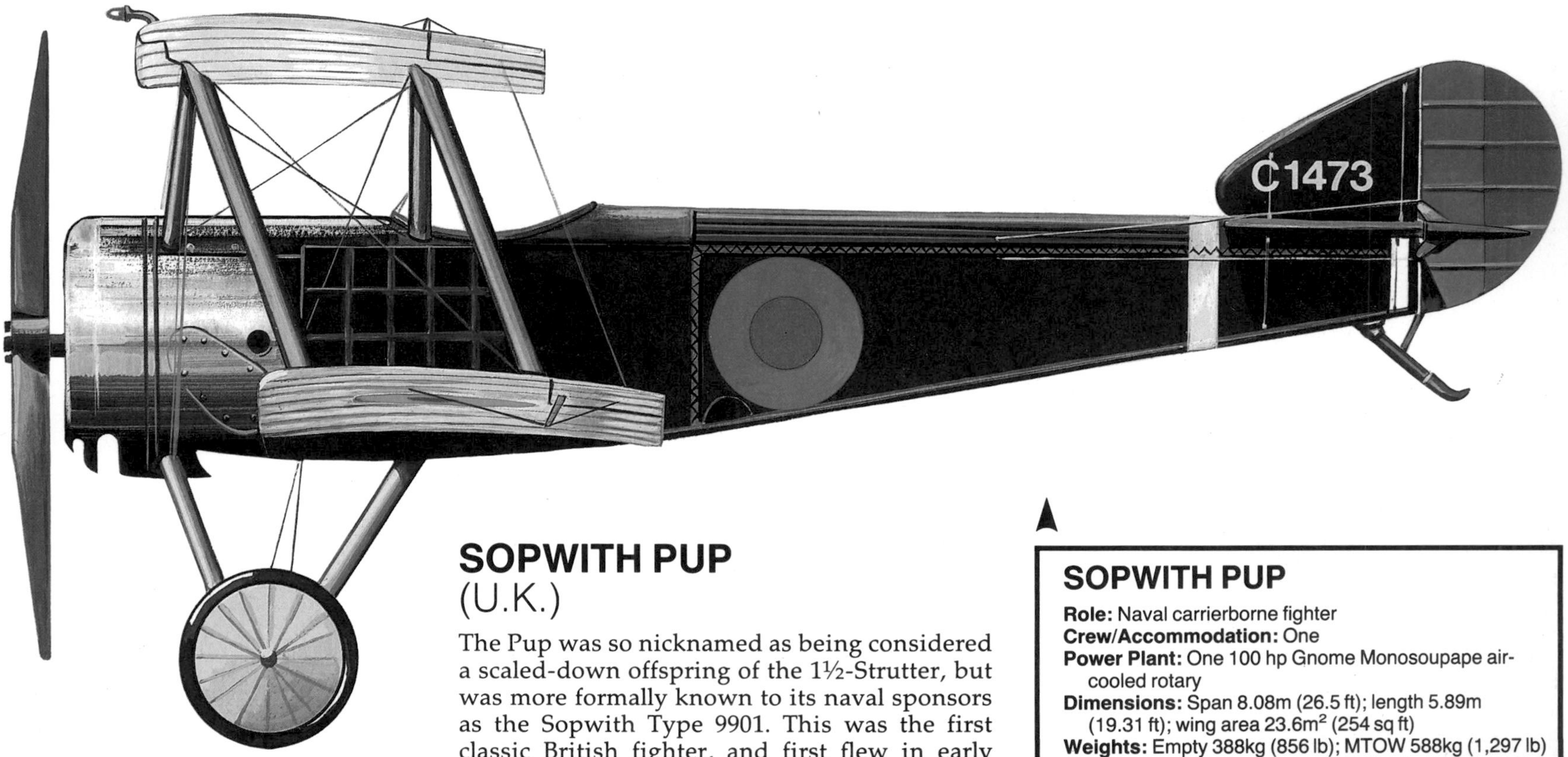

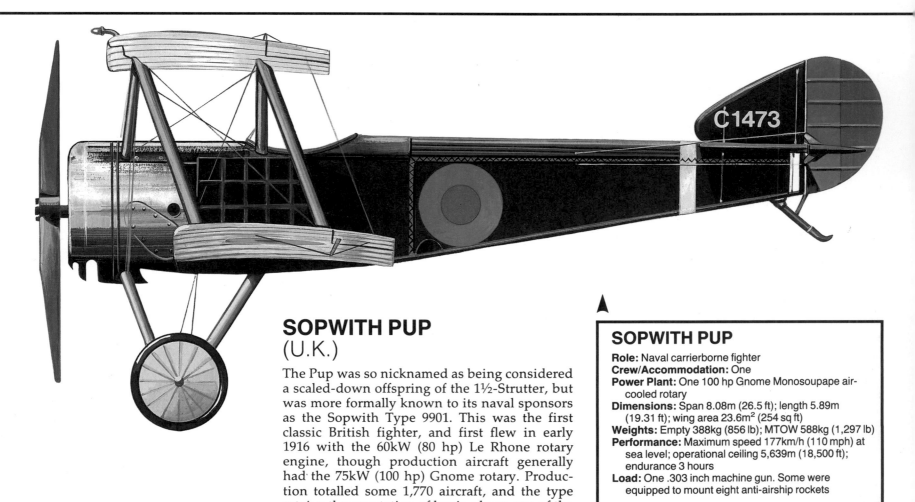

SOPWITH PUP
(U.K.)

The Pup was so nicknamed as being considered a scaled-down offspring of the 1½-Strutter, but was more formally known to its naval sponsors as the Sopwith Type 9901. This was the first classic British fighter, and first flew in early 1916 with the 60kW (80 hp) Le Rhone rotary engine, though production aircraft generally had the 75kW (100 hp) Gnome rotary. Production totalled some 1,770 aircraft, and the type retains the reputation of having been one of the more agile yet viceless biplane fighters ever to have entered service.

SOPWITH PUP

Role: Naval carrierborne fighter
Crew/Accommodation: One
Power Plant: One 100 hp Gnome Monosoupape air-cooled rotary
Dimensions: Span 8.08m (26.5 ft); length 5.89m (19.31 ft); wing area 23.6m^2 (254 sq ft)
Weights: Empty 388kg (856 lb); MTOW 588kg (1,297 lb)
Performance: Maximum speed 177km/h (110 mph) at sea level; operational ceiling 5,639m (18,500 ft); endurance 3 hours
Load: One .303 inch machine gun. Some were equipped to mount eight anti-airship rockets

SOPWITH CAMEL
(U.K.)

The Camel was clearly an evolution of the Pup's design concept, but had all its major masses (engine, fuel/lubricant, guns/ammunition and pilot) located in the forward 2.1m (7 ft) of the fuselage, on and around the centre of gravity to offer the least inertial resistance to agility. The type was therefore, supremely manoeuvrable: the torque of the powerful rotary meant that a three-quarter turn to the right could be achieved as swiftly as a quarter turn to the left, but this also meant that the type could easily stall and enter a tight spin when not flown with adequate care.

The type was more formally known to its naval sponsors as the Sopwith Biplane F.1., the nickname deriving from the humped fuselage over the breeches of the two machine guns. Production of 5,490 aircraft made this the most important British fighter of later 1917 and 1918, and a specialist derivative was the 2F.1 for shipboard use with folding wings.

SOPWITH F.1 CAMEL

Role: Fighter
Crew/Accommodation: One
Power Plant: One 130 hp Clerget 9B air-cooled rotary
Dimensions: Span/rotor diameter 8.53m (28 ft); length 5.71m (18.75 ft); wing area 21.5m^2 (231 sq ft)
Weights: Empty 436kg (962 lb); MTOW 672kg (1,482 lb)
Performance: Maximum speed 168km/h (104.5 mph) at 3,048m (10,000 ft); operational ceiling 5,486m (18,000 ft); endurance 2.5 hours
Load: Two .303 inch machine guns

SOPWITH TRIPLANE
(U.K.)

The Triplane was the fuselage and empennage of the Pup combined with a more powerful rotary engine and a narrow-chord triplane wing cellule to provide good agility, rate of climb and fields of vision. The type began to enter British naval service early in 1917 and soon acquired an awesome reputation with German pilots, who clamoured for their own triplane fighter. Total production was only 140.

SOPWITH TRIPLANE

Role: Fighter
Crew/Accommodation: One
Power Plant: One 130 hp Clerget air-cooled rotary (110 hp Clerget or Le Rhone in early aircraft)
Dimensions: Span 8.08m (26.5 ft); length 5.94m (19.5 ft); wing area 21.46m^2 (231 sq ft)
Weights: Empty 451kg (993 lb); MTOW 642kg (1,415 lb)
Performance: Maximum speed 187km/h (116 mph) at 1,829m (6,000 ft); operational ceiling 6,096m (20,000 ft); endurance 2.75 hours
Load: One or two .303 inch machine guns

SPAD S.7 and S.13
(France)

These two closely related aircraft were France's best fighters of World War I, combining performance and structural strength without too great a sacrifice of agility. The result was an excellent gun platform comparable with the S.E.5a in British service. The S.7's prototype was in effect the S.5 that flew towards the end of 1915, and the first S.7. flew early in 1916 with 112kW (150 hp) Hispano-Suiza 8Aa inline engines and a single synchronized machine gun. Delivery of the essentially similar first production series began in September 1916, being followed by an improved model with the 134kW (180 hp) HS 8Ac engine and wings of slightly increased span. About 5,000 aircraft were built, this number including about 300 of S.12 fighters with the 164kW (220 hp) HS 8Bec engine between whose cylinder banks nestled a 37mm cannon, Further development produced the S.13 that first flew in April 1917 for service from May of the same year. This has two guns, rather than one, more power and a number of aerodynamic refinements. Production reached a total of 8,472.

SPAD S.13

Role: Fighter
Crew/Accommodation: One
Power Plant: One 220 hp Hispano-Suiza 8Bec water-cooled inline
Dimensions: Span 8m (26.3 ft); length 6.2m (20.33 ft); wing area 21.1m^2 (227.1 sq ft)
Weights: Empty 565kg (1,245 lb); MTOW 820kg (1,807 lb)
Performance: Maximum speed 222km/h (138 mph) at sea level; operational ceiling 5,400m (17,717 ft); range 402km (250 miles)
Load: Two .303 inch machine guns

STINSON SR RELIANT
(U.S.A.)

The SR Reliant was introduced in 1933 and became one of the classic braced high-wing monoplanes typical of U.S. light transport manufacture in the 1930s. The type went through variants up to the SR-10 with a wide assortment of radial engines, and production of civil aircraft ended only when the U.S.A. entered World War II, when large numbers of Reliants were impressed as C-81s of various types. Wartime production amounted to some 500 aircraft delivered to the U.K. as AT-19 advanced trainer and liaison aircraft.

STINSON SR-9BD RELIANT

Role: Tourer
Crew/Accommodation: One, plus up to four passengers
Power Plant: One 245 hp Lycoming R-680-D6 air-cooled radial
Dimensions: Span 12.75m (41.83 ft); length 8.51m (27.92 ft); wing area 24.02m^2 (258.5 sq ft)
Weights: Empty 1,148kg (2,530 lb); MTOW 1,678kg (3,700 lb)
Performance: Cruise speed 229km/h (143 mph) at 1,951m (6,400 ft); operational ceiling 4,023m (13,200 ft); range 640km (400 miles) with three passengers
Load: Up to 328 kg (724 lb)

STINSON VOYAGER and L-5 SENTINEL
(U.S.A.)

The Voyager was Stinson's first lightplane and was modelled conceptually on the company's earlier high-wing transports. The type entered service in 1939 and was built in modest numbers with several flat-four piston engines. Total production was more than 5,700, most of these being built in World War II as L-5 Sentinel liaison and observation aircraft in variants up to the L-5G.

STINSON L-5 SENTINEL

Role: Observation/communications
Crew/Accommodation: Two, plus one passenger
Power Plant: One 190 hp Lycoming 0-435-1 air-cooled flat-opposed
Dimensions: Span 10.36m (34 ft); length 7.35m (24.1 ft); wing area 14.4m^2 (155 sq ft)
Weights: Empty 668kg (1,472 lb); MTOW 979kg (2,158 lb)
Performance: Cruise speed 180km/h (112 mph) at sea level; operational ceiling 4,816m (15,800 ft); range 676km (420 miles)
Load: Up to 100kg (220 lb) including passenger

SUKHOI, Su-7, Su-17, Su-20 and Su-22 SERIES
(U.S.S.R.)

The Su-7 has a superb reputation as a ground-attack fighter able to absorb virtually any amount of combat damage yet still deliver its ordnance with great accuracy. On the other side of the coin, however, the type has such short range on internal fuel that at least two hard-points have to be used for drop tanks rather than ordnance if the type is to have a useful tactical radius. Various prototypes flew in the mid-1950s, and the Su-7 was ordered into production during 1958. The type was developed in steadily improved Su-7 variants, and was then transformed into the far more potent Su-17 with variable-geometry outer wing panels. The Su-7IG prototype of 1966 confirmed that field performance and range were markedly improved and since then the type had been extensively built in Su-17, Su-20 and Su-22 forms for use by the Soviet Warsaw Pact and its allies.

SUKHOI Su-7BMk 'FITTER A'

Role: Strike-fighter
Crew/Accommodation: One
Power Plant: One 9,600kgp (21,164 lb s.t.) Lyulka AL-7F-1 turbojet with reheat
Dimensions: Span 8.77m (28.77 ft); length 16.8m (55.12 ft); wing area 34.5m^2 (371.4 sq ft)
Weights: Empty 8,616kg (18,995 lb); MTOW 13,500kg (29,762 lb)
Performance: Maximum speed 1,160km/h (720 mph) Mach 0.95 at 305m (1,000 ft); operational ceiling 13,000+m (42,650 ft); radius 460km (285 miles) with 1,500kg (3,307 lb) warload
Load: Two 30mm cannon, plus up to 2,500kg (5,512 lb) of weapons/fuel carried externally

SUKHOI Su-11 and Su-15
(U.S.S.R.)

The Su-11 all-weather fighter that began to enter service in the early 1960s was a development of the Su-9 with a lengthened fuselage accommadating more powerful radar for use with AA-3 air-to-air missiles. Production of the Su-9 and Su-11 totalled about 2,000 aircraft, the Su-11 share including the Su-11U two-seat conversion trainer. Further development of the same basic airframe produced the Su-15 with two Tumanskii R-13 afterburning turbojets replacing the single 10,000kg (22,046 lb) thrust Lyulka AL-7F-I afterurning turbojet of the Su-11. These engines are aspirated via inlets in the wing roots, leaving the forward fuselage exclusively to the radar. Production of about 1,500 Su-15s has been completed , and the type was developed in variants up to the 'Flagon-F'.

SUKHOI Su-15 'FLAGON F'

Role: All-weather/night interceptor
Crew/Accommodation: One
Power Plant: Two 7,200kgp (15,875 lb s.t.) Tumansky R-13F-300 turbojets with reheat
Dimensions: Span 10.5m (34.45 ft); length 22m (72.18 ft)
Weights: Empty 13,130kg (28,947 lb); MTOW 18,144kg (40,000 lb)
Performance: Maximum speed 2,444km/h (1,519 mph) Mach 2.3 at 11,000m (36,090 ft); operational ceiling 18,800m (61,680 ft); range 997km (620 miles) with full warload
Load: Four medium range air-to-air missiles

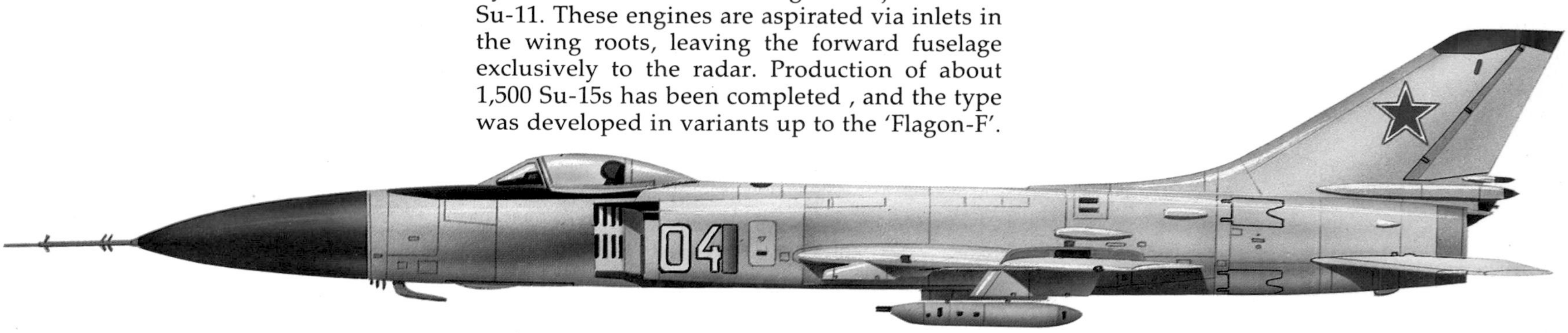

SUKHOI Su-24
(U.S.S.R.)

This was the first variable-geometry aeroplane designed as such in the U.S.S.R., and is a highly capable long-range interdictor in the same basic class as the Americans' General Dynamics F-111. The type flew in prototype form during 1969 or 1970, and production continues in several increasingly sophisticated variants that include the baseline 'Fencer-A', the improved 'Fencer-B', the electronically upgraded 'Fencer-C' with more powerful engines, the 'Fencer-D' version of the 'Fencer-C' with an inflight-refuelling probe, the 'Fencer-E' electronic warfare aeroplane and the 'Fencer-F' reconnaissance aeroplane.

SUKHOI Su-24 'FENCER D'

Role: Long-range variable-geometry strike
Crew/Accommodation: Two
Power Plant: Two 11,500kgp (25,350 lb s.t.) Tumansky R-29B turbofans with reheat
Dimensions: Span spread 17.25m swept 10.3m; spread 56.5 ft, (swept 33.75 ft); length 22.00m (72.17 ft)
Weights: Empty 19,000kg (41,888 lb); MTOW 40,800kg (89,948 lb)
Performance: Maximum speed 1,471km/h (914 mph) Mach 1.2 at sea level; operational ceiling 13,500+m (44,291+ ft); radius 1,500km (930 miles) with 3,000kg (6,614 lb) warload
Load: One 30mm multi-barrel cannon, plus up to 6,000kg (13,230 lb) of ordnance/fuel, including air-to-surface stand-off ranged missiles

SUKHOI Su-25
(U.S.S.R.)

This subsonic battlefield close-support aeroplane was produced as counterpart to the Americans' Fairchild Republic A-10A Thunderbolt II, but is a somewhat more refined type in its aerodynamics and, therefore, possesses usefully higher performance. The type was evaluated in pre-production form in Afghanistan during the early 1980s, and the lessons of this evaluation were used in the development of the current full-production type, which is produced in single- and two-seat forms.

SUKHOI Su-25 'FROGFOOT'

Role: Strike/close air support
Crew/Accommodation: One
Power Plant: Two 5,100kgp (11,240 lb s.t.) Tumansky R-13-1300 turbofans
Dimensions: Span 14.2m (46.59 ft); length 15.2m (49.87 ft); wing area 37.6m^2 (404.7 sq ft)
Weights: Empty 9,500kg (20,944 lb); MTOW 20,000kg (44,092 lb)
Performance: Maximum speed 877km/h (473 knots) at 3,000m (9,843 ft); operational ceiling 9,200+m (30,184+ ft); radius 546km (295 naut. miles) with full warload
Load: One 30mm cannon, plus up to 4,000kg (8,818 lb) of weapons/fuel carried externally

SUKHOI Su-27
(U.S.S.R.)

The Su-27 is the U.S.S.R.'s latest air-superiority fighter, an advanced type comparable with the Americans' McDonnell Douglas F-15 Eagle. The type has two powerful turbofan engines, a blended fuselage/wing design and twin vertical tail surfaces. It is also notable for the considerably higher level of electronic sophistication than was apparent in Soviet fighters of the preceding generation. The 'Flanker' is produced in single- and two-seat forms, the latter probably intended more for the combat role than conversion training.

SUKHOI Su-27 'FLANKER A'

Role: Interceptor
Crew/Accommodation: One
Power Plant: Two 13,608kgp (30,000 lb.s.t.) Tumansky R-32 turbofans with reheat
Dimensions: Span 14.5m (47.57 ft); length 21m (68.9 ft); wing area 71m^2 (764.2 sq ft)
Weights: Empty 15,400kg (33,951 lb); MTOW 28,400kg (62,611 lb)
Performance: Maximum speed 2,125+km/h (1,147+ knots) Mach 2.0+ at 11,000m (36,090 ft); operational ceiling 16,400+m (53,806+ ft); range 1,500km (809 naut. miles) with full missile warload
Load: One 30mm multi-barrel cannon, plus six long-range and four short-range air-to-air missiles

SUPERMARINE SOUTHAMPTON
(U.K.)

The Southampton maritime reconnaissance flying boat was a development of the Swan airliner, and first flew in March 1925. The first six 'boats were Southampton Mk Is with a wooden hull, but the other 62 were Southampton Mk IIs with a metal hull that was not only lighter and more durable than the wooden type, but also eliminated the water soakage problem that added some 181kg (400 lb) to the Southampton Mk I's weight.

SUPERMARINE SOUTHAMPTON Mk II

Role: Maritime reconaissance flying boat
Crew/Accommodation: Four
Power Plant: Two 500 hp Napier Lion XA water-cooled inlines
Dimensions: Span 22.86m (75 ft); length 15.15m (49.71 ft); wing area 134.5m^2 (1,448 sq ft)
Weights: Empty 4,398kg (9,697 lb); MTOW 6,895kg (15,200 lb)
Performance: Maximum speed 153km/h (95 mph) at sea level; operational ceiling 2,469m (8,100 ft); range 876km (544 miles)
Load: None

SUPERMARINE WALRUS
(U.K.)

The Walrus was a biplane amphibian with a pusher engine, and first flew in 1935 as a spotter for use on the catapults of major warships. Production totalled 746 in two variants: the Walrus Mk I, the standard spotter and air/sea rescue type for the Fleet Air Arm and RAF respectively with the 473kW (635 hp) Bristol Pegasus IIM2 radial engine, and the Walrus Mk II, the RAF's air/sea rescue model with a wooden hull and the 559kW (750 hp) Pegasus VI.

SUPERMARINE WALRUS Mk I

Role: Naval shipborne reconnaissance, later air/sea rescue
Crew/Accommodation: Four
Power Plant: One 750 hp Bristol Pegasus VI air-cooled radial
Dimensions: Span 13.97m (45.83 ft); length 11.45m (37.58 ft); wing area 56.67m^2 (610 sq ft)
Weights: Empty 2,233kg (4,900 lb); MTOW 3,266kg (7,200 lb)
Performance: Maximum speed 217km/h (135 mph) at 1,448m (4,750 ft); operational ceiling 5,639m (18,500 ft); range 966km (600 miles)
Load: Two .303 inch machine guns

SUPERMARINE SPITFIRE and SEAFIRE
(U.K.)

The Spitfire was the British counterpart to the Messerschmitt Bf 109 as its country's main fighter in World War II. Like the German fighter, it remained in production right through the conflict for a total of 20,334 aircraft bolstered by 2,556 new-build Seafire naval fighters. The Spitfire prototype first flew in March 1936 with a 738kW (900 hp) Rolls-Royce Merlin C engine, and despite some fears that the type was not best suited to mass production techniques, was soon ordered into production for its high performance and very great development potential. The Spitfire MkI was powered by the 768kW (1,030 hp) Merlin II and armed with eight 7.7mm (0.303 inch) machine guns or, in the Mk IB variant, four such machine guns and two 20mm cannon: the suffix A indicated eight 7.7mm machine guns, B four such machine guns and two 20mm cannon, C four cannon, and E two cannon and two 12.7mm (0.5 inch) machine guns.

Major fighter variants with the Merlin engine were the initial Mk I, the Mk II with the 876kW (1,175 hp) Merlin XII, the F. and LF.Mks VA, VB and VC in medium- and low-altitude forms with the 1,974kW (1,440 hp) Merlin 45 or 1,096kW (1,470 hp) Merlin 50, the HF.Mk VI high-altitude interceptor with the 1,055kW (1,415 hp) Merlin 47 and a pressurized cockpit, the HF.Mk VII with the two-stage Merlin 61, 64 or 71, the LF, F and HF. Mk VIII with the two-stage Merlin 61, 63, 66 or 70 but an unpressurized cockpit, the LF, F and HF. Mk IX using the Mk V airframe with the two-stage Merlin 61, 63 or 70, the LF and F.Mk XVI using the Mk IX airframe with a cutdown rear fuselage, bubble canopy and Packard-built Merlin 266.

The Spitfire was also developed in its basic fighter form with the larger and more powerful Rolls-Royce Griffon inline engine, and the major variants of this sequence were the LF.Mk XI with the 1,294kW (1,735 hp) Griffon II or IV, and LF the F.Mk XIV with the 1,529kW (2,050 hp) Griffon 65 or 66 and often with a bubble canopy, the F.Mk XVIII with the two-stage Griffon and a bubble canopy, the F.Mk 21 with the Griffon 61 or 64, the F.Mk 22 with the 1,771kW (2,373 hp) Griffon 85 driving a contra-rotating propeller unit, and the improved F.Mk 24. The Spitfire was also used as an unarmed reconnaissance type, the major Merlin-engined types being the Mks IV, X, XI and XIII, and the Griffon-engined type being the Mk XIX.

The Seafire was the naval counterpart to the Spitfire, the main Merlin-engined versions being the Mks IB, IIC and III, and the Griffon-engined versions being the Mks XV, XVII, 45, 46 and 47.

SUPERMARINE SPITFIRE F.Mk XIV E

Role: Fighter
Crew/Accommodation: One
Power Plant: One 2,050 hp Rolls-Royce Griffon 65 water-cooled inline
Dimensions: Span 11.23m (36.83 ft); length 9.96m (32.66 ft); wing area 22.48m^2 (242 sq ft)
Weights: Empty 2,994kg (6,600 lb); MTOW 3,856kg (8,500 lb)
Performance: Maximum speed 721km/h (448 mph) at 7,925m (26,000 ft); operational ceiling 13.106m (43,000 ft); range 740km (460 miles) on internal fuel only
Load: Two 20mm cannon and two .303 machine guns, plus up to 454kg (1,000 lb) of bombs

Supermarine Spitfire Mk.VB

Supermarine Spitfire Mk.LF XVI

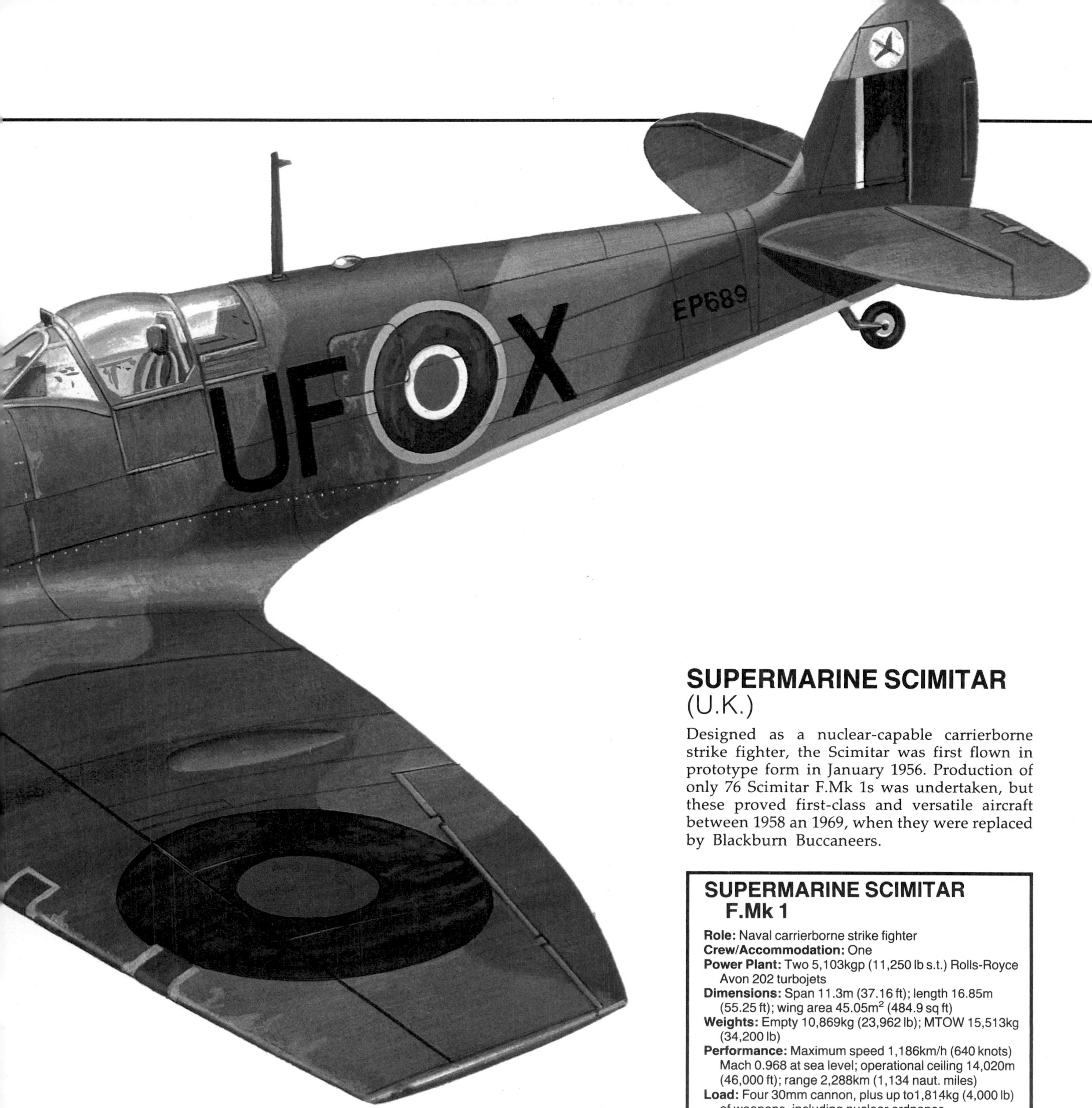

SUPERMARINE SCIMITAR
(U.K.)

Designed as a nuclear-capable carrierborne strike fighter, the Scimitar was first flown in prototype form in January 1956. Production of only 76 Scimitar F.Mk 1s was undertaken, but these proved first-class and versatile aircraft between 1958 an 1969, when they were replaced by Blackburn Buccaneers.

SUPERMARINE SCIMITAR F.Mk 1

Role: Naval carrierborne strike fighter
Crew/Accommodation: One
Power Plant: Two 5,103kgp (11,250 lb s.t.) Rolls-Royce Avon 202 turbojets
Dimensions: Span 11.3m (37.16 ft); length 16.85m (55.25 ft); wing area 45.05m^2 (484.9 sq ft)
Weights: Empty 10,869kg (23,962 lb); MTOW 15,513kg (34,200 lb)
Performance: Maximum speed 1,186km/h (640 knots) Mach 0.968 at sea level; operational ceiling 14,020m (46,000 ft); range 2,288km (1,134 naut. miles)
Load: Four 30mm cannon, plus up to1,814kg (4,000 lb) of weapons, including nuclear ordnance

THOMAS-MORSE MB-3
(U.S.A.)

After developing the S-4 light fighter that was used in World War I as a fighter trainer, Thomas-Morse moved on to the design of a more capable fighter. The company produced two prototypes each of its MB-1, MB-2 and MB-3 designs. The last was powered by the 224kW (300 hp) Wright-Hispano inline engine and seemed worthy of production to the extent of 50 aircraft. Production of 200 improved MB-3A aircraft was entrusted to Boeing, and many of these were later modified as MB-3M trainers.

THOMAS-MORSE MB-3A

Role: Fighter
Crew/Accommodation: One
Power Plant: One 300 hp Wright H-3 water-cooled inline
Dimensions: Span 7.92m (26 ft); length 6.1m (20 ft); wing area 21.27m^2 (229 sq ft)
Weights: Empty 778kg (1,716 lb); MTOW 1,152kg (2,539 lb)
Performance: Maximum speed 227km/h (141 mph) at sea level; operational ceiling 5,944m (19,500 ft); endurance 2.25 hours
Load: Two .303 inch machine guns

TRANSALL C160
(France/West Germany)

So named for the fact that it is a cargo aeroplane with a wing of 160m^2 (645.9 sq ft), the C160 was designed to replace the Noratlas in the air forces of the two sponsoring countries, and first flew in March 1963. Some 169 production aircraft were built for France (50), West Germany (90), Turkey (20) and South Africa (9) before the line was closed in 1972. In 1977 the production programme was revived to produce 29 slightly improved aircraft for France.

TRANSALL C160

Role: Intermediate-range cargo/troop transport
Crew/Accommodation: Four, plus up to 93 troops
Power Plant: Two 6,100 shp Rolls-Royce Tyne Mk.22 turboprops
Dimensions: Span 40m (131.25 ft); length 32.4m (106.29 ft); wing area 160.1m^2 (1,723 sq ft)
Weights: Empty 29,000kg (63,920 lb); MTOW 51,000kg (11,243500 lb)
Performance: Maximum speed 496km/h (308 mph) at 5,500m (18,045 ft); operational ceiling 9,145m (30,000 ft); range 1,850km (1,150 miles) with full payload
Load: Up to 16,000kg (35,270 lb)

A close-up of a TB-3, seen here carrying a parasite I-16 ground attacker.

TUPOLEV ANT-6 (TB-3)
(U.S.S.R.)

In its day this all-metal aeroplane was the world's most advanced four-engined bomber, and a truly remarkable achievement for the Soviet aircraft industry. The prototype first flew in December 1930, and between 1931 and 1937 a total of 818 TB-3 production aircraft was built in variants with the M-17 and several marks of the M-34 inline engines. During World War II most surviving aircraft were converted into G-2 unarmed paratroop or freight transports.

TUPOLEV ANT-6 (B-3)

Role: Heavy bomber
Crew/Accommodation: Six
Power Plant: Four 970 hp M-34RN water-cooled in-lines
Dimensions: Span 40.5m (132.87 ft); length 25.3m (83.01 ft); wing area 230m^2 (2,475.7 sq ft)
Weights: Empty 10,956kg (24,154 lb); MTOW 19,500kg (42,990 lb)
Performance: Maximum speed 288km/h (179 mph) at 3,000m (9,843 ft); operational ceiling 7,740m (25,393 ft); range 3,120km (1,939 miles) with 3,000kg (6,614 lb) warload
Load: Six 7.62mm machine guns, plus up to 5,800kg (12,787 lb) of bombs

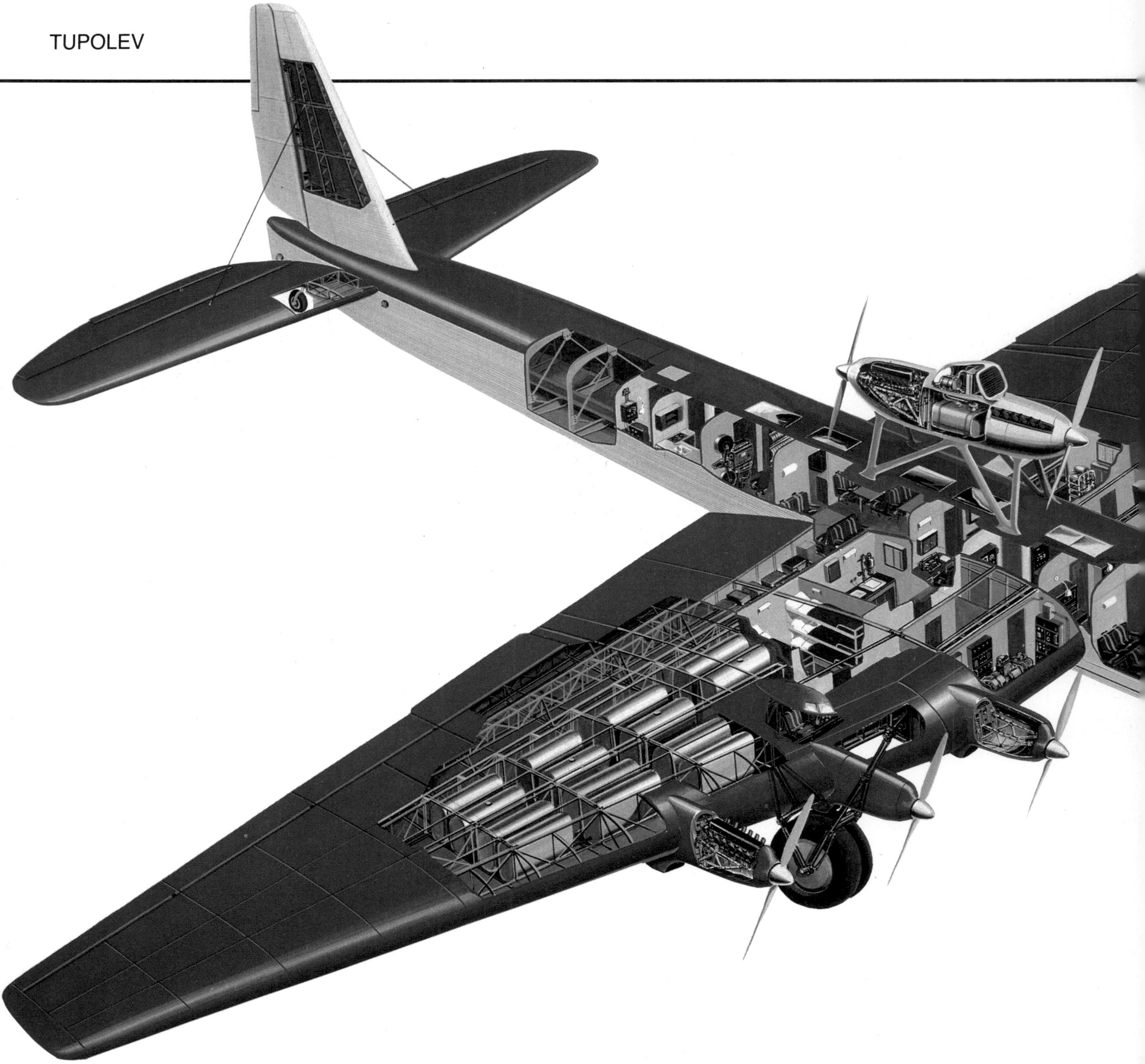

ЦАГИ-АНТ25
CAHI-ANT25
URSS
Сталинский маршрут

TUPOLEV ANT-20
(U.S.S.R.)

In its time this was the world's largest aeroplane, a technology-stretching machine mainly of metal construction and powered by eight inline engines in the passenger-cum-propaganda role. Six of the 671kW (900 hp) AM-34 engines were mounted in the wing leading edges and two as a push/pull pair strut mounted above the centre section. The type first flew in June 1934, and in addition to accommodation for a crew of 20 it had other features such as a printing press, cinema and film laboratory. The aeroplane was lost in a mid-air collision and replaced by an ANT-20bis that flew first in 1939 with six 895kW (1,200 hp) AM-34FRNV engines and accommodation for 64 passengers.

TUPOLEV ANT-20

Role: Heavy transport
Crew/Accommodation: Eight, plus up to 82 passengers
Power Plant: Six 900 hp AM-34 water-cooled in-lines
Dimensions: Span 63m (206.69 ft); length 33m (108.27 ft); wing area 486m^2 (5,231 sq ft)
Weights: Empty 31,950kg (70,438 lb); MTOW 42,000kg (92,594 lb)
Performance: Cruise speed 190km/h (118 mph) at 2,400m (7,874 ft); operational ceiling 4,500m (14,764 ft); range 1,200km (746 miles) with maximum payload
Load: Up to 6,800kg (14,991 lb)

TUPOLEV ANT-25
(U.S.S.R.)

The ANT-25 was designed as a long-distance record breaker, and first flew in June 1933. Two aircraft were built, the second with the 671kW (900 hp) M-34R engine achieving a flight of 9,130km (5,673 miles) in June 1937 and another of 11,500km (7,146 miles) in July 1937. Nothing came of the plan to build 20 long-range bombing research aircraft based on the ANT-25.

TUPOLEV ANT-25

Role: Long-range flight research
Crew/Accommodation: Three
Power Plant: One 950 hp Mikulin M-34R water-cooled inline
Dimensions: Span 34m (111.55 ft); length 13.9m (45.6 ft); wing area 88.20m^2 (949.33 sq ft)
Weights: Empty 4,200kg (9,259 lb); MTOW 11,500kg (25,353 lb)
Performance: Maximum speed 165km/h (103 mph) at 2,000m (6,562 ft); operational ceiling 7,000m (22,966 ft); range 13,000km (8,078 miles)
Load: Up to 7,050kg (15,543 lb) of fuel and lubricant

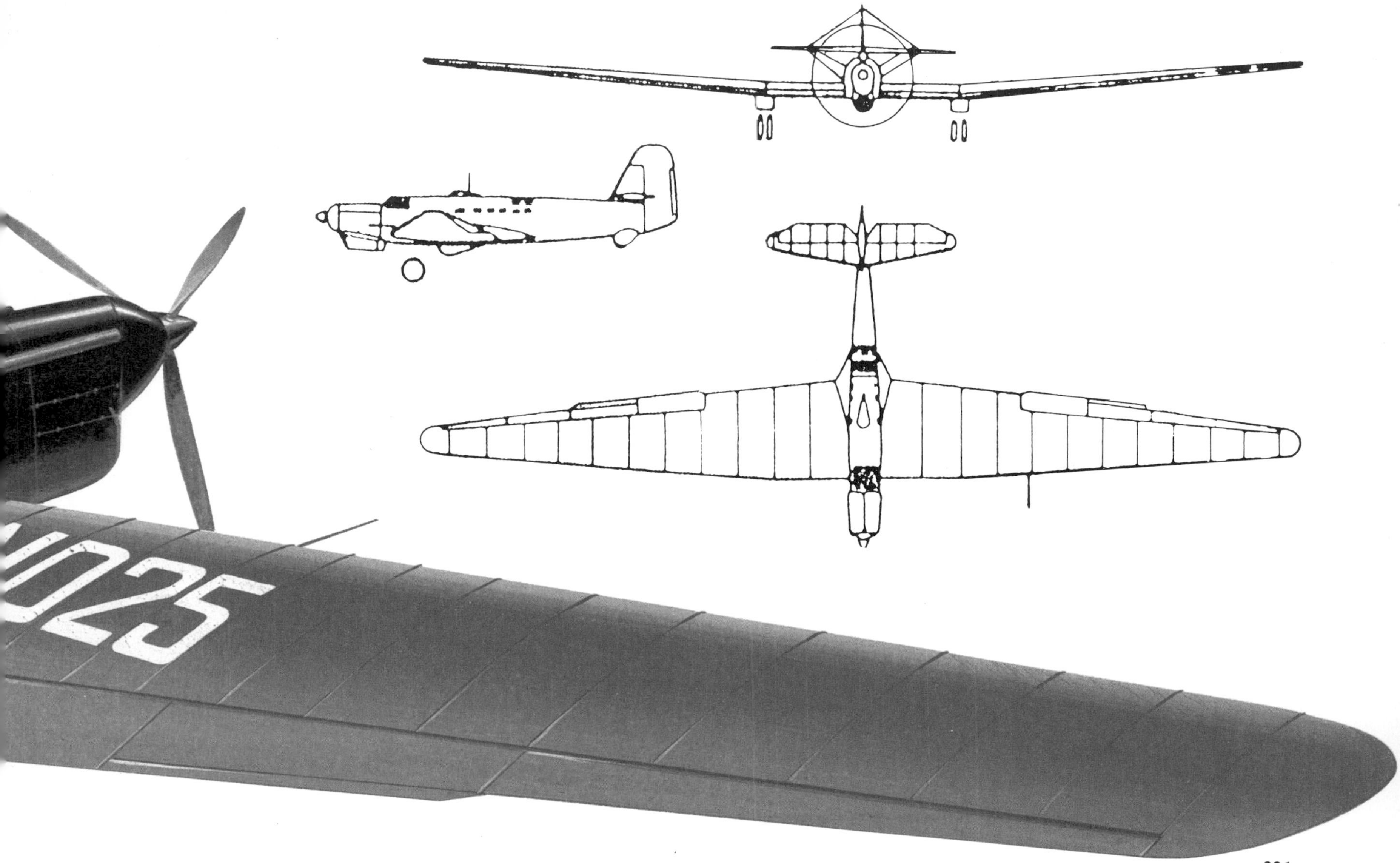

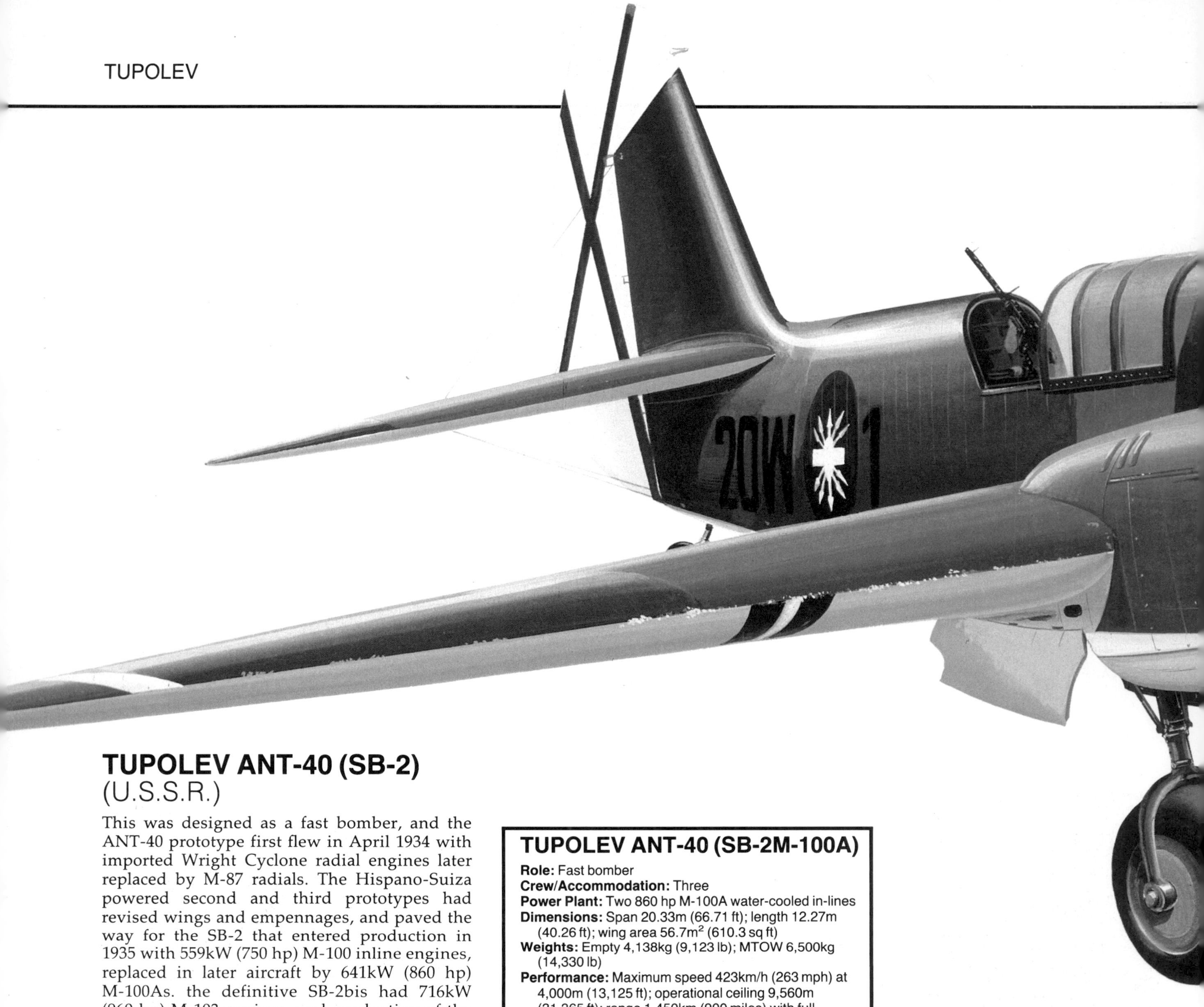

TUPOLEV ANT-40 (SB-2)
(U.S.S.R.)

This was designed as a fast bomber, and the ANT-40 prototype first flew in April 1934 with imported Wright Cyclone radial engines later replaced by M-87 radials. The Hispano-Suiza powered second and third prototypes had revised wings and empennages, and paved the way for the SB-2 that entered production in 1935 with 559kW (750 hp) M-100 inline engines, replaced in later aircraft by 641kW (860 hp) M-100As. the definitive SB-2bis had 716kW (960 hp) M-103 engines, and production of the series totalled 6,656 aircraft by 1940. There were a number of variants of this basic design, the most important of them the PS-40 transport.

TUPOLEV ANT-40 (SB-2M-100A)

Role: Fast bomber
Crew/Accommodation: Three
Power Plant: Two 860 hp M-100A water-cooled in-lines
Dimensions: Span 20.33m (66.71 ft); length 12.27m (40.26 ft); wing area 56.7m^2 (610.3 sq ft)
Weights: Empty 4,138kg (9,123 lb); MTOW 6,500kg (14,330 lb)
Performance: Maximum speed 423km/h (263 mph) at 4,000m (13,125 ft); operational ceiling 9,560m (31,365 ft); range 1,450km (900 miles) with full bombload
Load: Four 7.62mm machine guns, plus up to 600kg (1,323 lb) of bombs

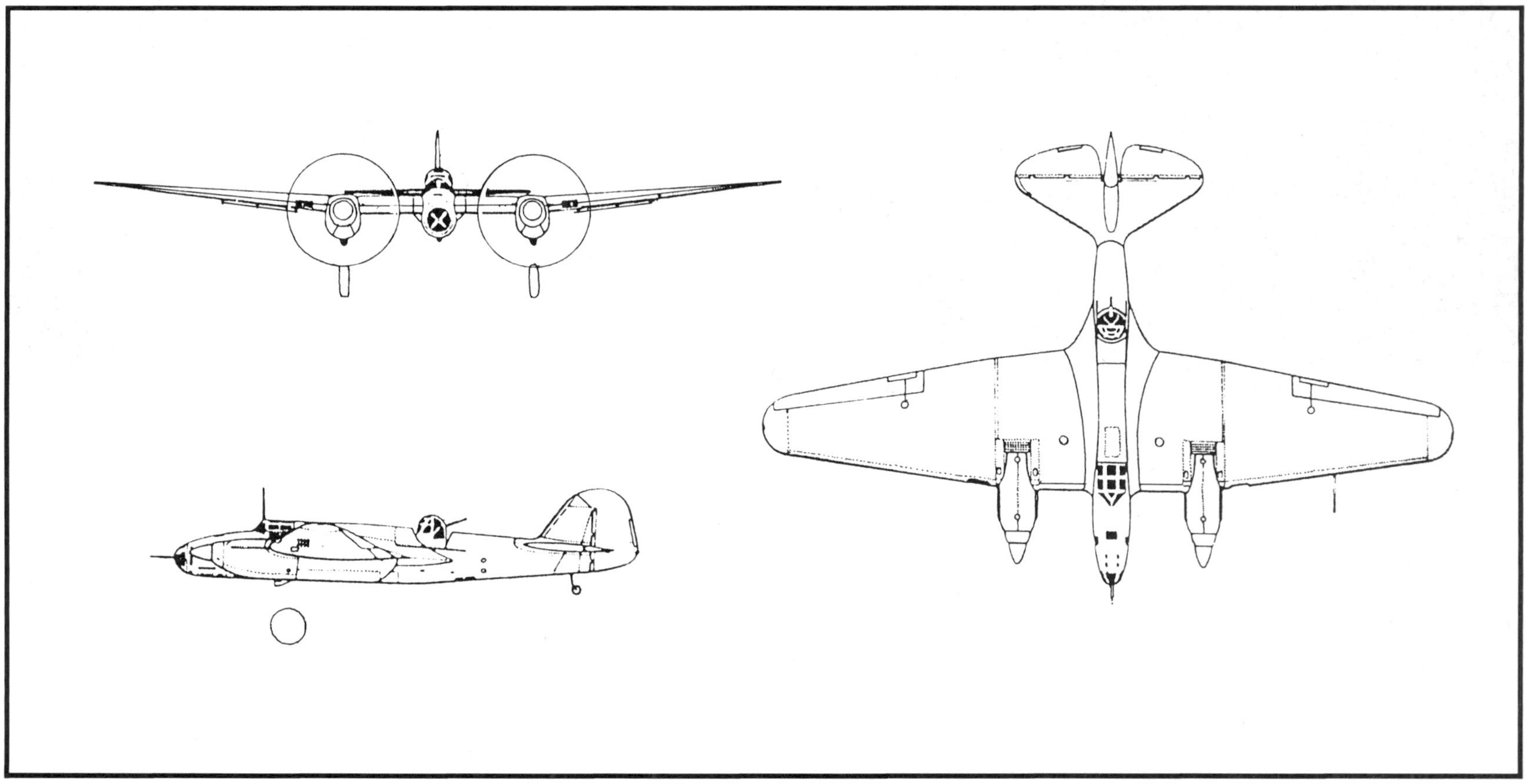

TUPOLEV Tu-2
(U.S.S.R.)

This was the U.S.S.R.'s best medium bomber of World War II, and total production was 2,527. The type began life as the ANT-58 prototype that first flew in January 1941, but was then developed via the ANT-59 that saw the original AM-37 engines replaced by Shvetsov ASh-82 radial engines, the ANT-60 that introduced a simplified structure, and the definitive ANT-61. This entered service in November 1942, but was redesignated Tu-2 in early 1943. The basic production Tu-2 was refined as the Tu-2S, and after World War II the bomber was further developed in specialist types such as a torpedo bomber, escort fighter, long-range bomber and high-speed transport for priority freight.

TUPOLEV Tu-2

Role: Bomber/strike
Crew/Accommodation: Four
Power Plant: Two 1,850 hp Shvetsov ASh-82 FNV air-cooled radials
Dimensions: Span 18.86m (61 ft); length 13.8m (45.25 ft); wing area 48.8m^2 (525.3 sq ft)
Weights: Empty 8,273kg (18,240 lb); MTOW 12,800kg (28,224 lb)
Performance: Maximum speed 560km/h (345 mph) at 5,800m (19,029 ft); operational ceiling 10,980m (36,024 ft); range 1,400km (870 miles) with full bombload
Load: Two 20mm cannon and three 12.7mm machine guns, plus up to 3,000kg (6,614 lb) of bombs

TUPOLEV Tu-16 'BADGER G'

Role: Missile-carrying bomber, reconnaissance, electronic warfare
Crew/Accommodation: Six to nine dependant upon mission
Power Plant: Two 8,750kgp (19,290 lb s.t.) Mikulin AM-3M turbojets
Dimensions: Span 32.93m (108 ft); length 34.8m (114.2 ft); wing area 164.65m^2 (1,772 sq ft)
Weights: Empty 40,000kg (88,185 lb); MTOW 77,000kg (169,756 lb)
Performance: Maximum speed 941km/h (508 knots) at 11,000m (36,090 ft); operational ceiling 12,200m (40,026 ft); radius 2,895km (1,562 naut. miles) unrefuelled with full warload
Load: Three 23mm cannon, plus up to 9,000kg (19,842 lb) of bombs

TUPOLEV Tu-16
(U.S.S.R.)

Another great technical achievement by the Tupolev design bureau, the Tu-16 twin-jet strategic bomber began to enter service in 1953 after the Tu-88 prototype had flown in the winter of 1952. About 2,000 aircraft were produced and though the baseline 'Badger-A' bomber is still in service, other variants are the 'Badger-B', originally with anti-ship missiles but now used as a bomber, the 'Badger-C' anti-ship missile carrier, the 'Badger-D' electronic reconnaissance aeroplane, the 'Badger-E' bomber and photo -reconnaissance aeroplane, the 'Badger-F' variant of the 'Badger-E' but with electronic support measures equipment, the 'Badger-G' improved anti-ship missile carrier, and the improved anti-ship 'Badger-H, J, K and L' electronic countermeasures aircraft. Many of the older aircraft have been converted into either of two types of inflight-refuelling tanker, and the basic type is produced in China as the Xian H-6 bomber.

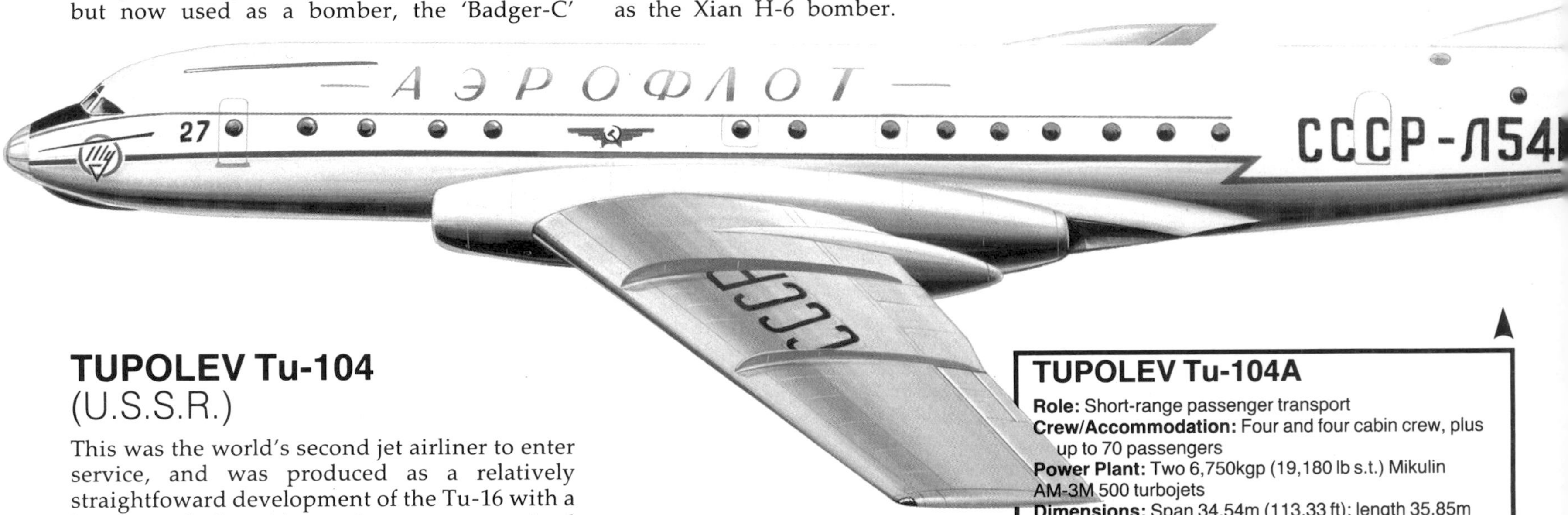

TUPOLEV Tu-104
(U.S.S.R.)

This was the world's second jet airliner to enter service, and was produced as a relatively straightfoward development of the Tu-16 with a low instead of a mid-set wing on a pressurized fuselage. The type first flew in June 1955 and entered service in the summer of 1956 as a 50-passenger type with 6,750kg (14,881 lb) thrust Mikulin AM-3 turbojets. The 70-passenger Tu-104A used the 8,700kg (19,180 lb) thrust AM-3M while the Tu-104B had the same power plant and a lengthened fuselage for 100 passengers. Some Tu-104As were later modified to Tu-104D standard with accommodation for 85 passengers and total production was about 200.

TUPOLEV Tu-104A

Role: Short-range passenger transport
Crew/Accommodation: Four and four cabin crew, plus up to 70 passengers
Power Plant: Two 6,750kgp (19,180 lb s.t.) Mikulin AM-3M 500 turbojets
Dimensions: Span 34.54m (113.33 ft); length 35.85m (117.62 ft); wing area 174.4m^2 (1,887 sq ft)
Weights: Empty 41,600kg (91,710 lb); MTOW 76,000kg (167,550 lb)
Performance: Maximum speed 900km/h (560 mph) at 9,200m (30,184 ft); operational ceiling 11,500m (37,730 ft); range 2,650km (1,645 miles) with full payload
Load: Up to 9,000kg (19,840 lb)

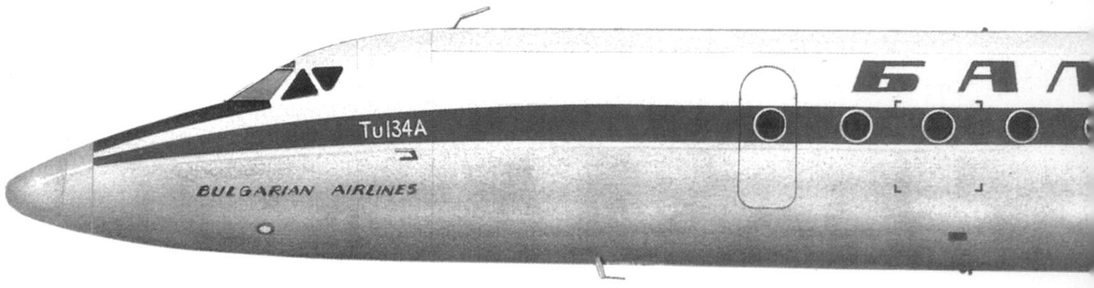

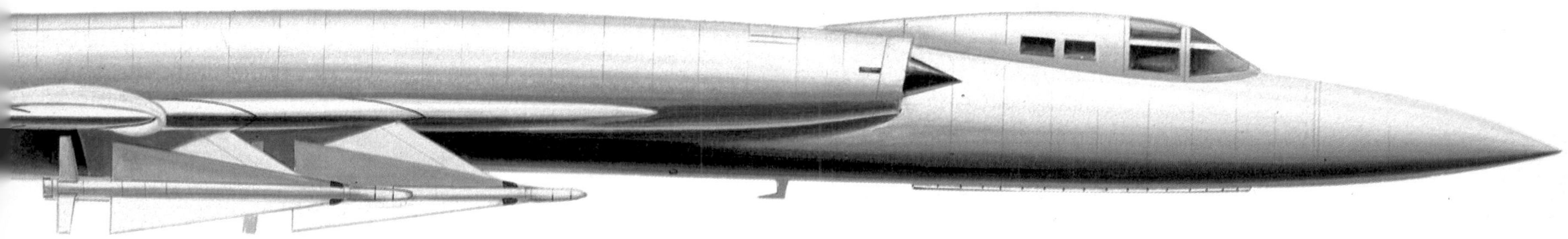

TUPOLEV Tu-26
(U.S.S.R.)

This is the world's largest interceptor, a supersonic type carrying powerful radar and four massive air-to-air missiles in an airframe large enough to accommodate the fuel for long patrols over the U.S.S.R.'s northern and eastern reaches, where airfields are few and far between. The type first flew in Tu-102 prototype form in about 1960, and began to enter service in 1968.

TUPOLEV Tu-28 'FIDDLER B'

Role: Long-range interceptor
Crew/Accommodation: Two
Power Plant: Two 11,795kgp (26,000 lb s.t.) Lyulka AL-7 turbojets
Dimensions: Span 18m (59 ft); length 28m (91.83 ft)
Weights: Empty 35,380kg (78,000 lb); MTOW 43,545kg (96,000 lb)
Performance: Maximum speed 1,745km/h (942 knots) Mach 1.64 at 12,000m (39,370 ft); operational ceiling 20,000m (65,615 ft); radius 1,255km (677 naut. miles) with four missiles
Load: Four long/medium-range air-to-air missiles

TUPOLEV Tu-134
(U.S.S.R.)

In 1962 Tupolev began development of a Tu-124A to replace the Tu-124, the latter having been derived from the TU-16 bomber and therefore lacking many of the desirable features of a purpose-designed airliner. The wings were basically those of the Tu-124 increased in span and area by the use of a new centre section, the fuselage was in essence a lengthened version of the Tu-124's fuselage, and the landing gear was modelled on that of the Tu-124 with strengthened legs. The power plant was completely different, however, in its use of twin turbofans located in rear-fuselage nacelles, and the empennage was new, being of the T-tail variety. The type was redesignated Tu-134 in recognition of these alterations and began to enter service in September 1967 with accommodation for 72 passengers. Further development produced the Tu-134A with longer fuselage for 84 passenges. Conversions of older aircraft produced the Tu-134B with a more advanced cockpit and accommodation for 80 passengers, the Tu-134B-1 with accommodation for 84 or 90 passengers but no galley, and the Tu-134B with more efficient Soloviev D-30-II turbofans.

TUPOLEV Tu-134 'CRUSTY'

Role: Short-range passenger transport
Crew/Accommodation: Three and three cabin crew, plus up to 72 passengers
Power Plant: Two 6,800kgp (14,991 lb s.t.) Soloviev D-30 turbofans
Dimensions: Span 29m (95.14 ft); length 34.95m (114.67 ft); wing area $127.3m^2$ (1,370.3 sq ft)
Weights: Empty 27,500kg (60,627 lb); MTOW 45,000kg (99,200 lb)
Performance: Cruise speed 900km/h (486 knots) at 8,500m (27,887 ft); operational ceiling 12,000m (39,370 ft); range 2,400km (1,295 naut. miles) with 7,000kg (15,432 lb) payload
Load: Up to 7,700kg (16,975 lb)

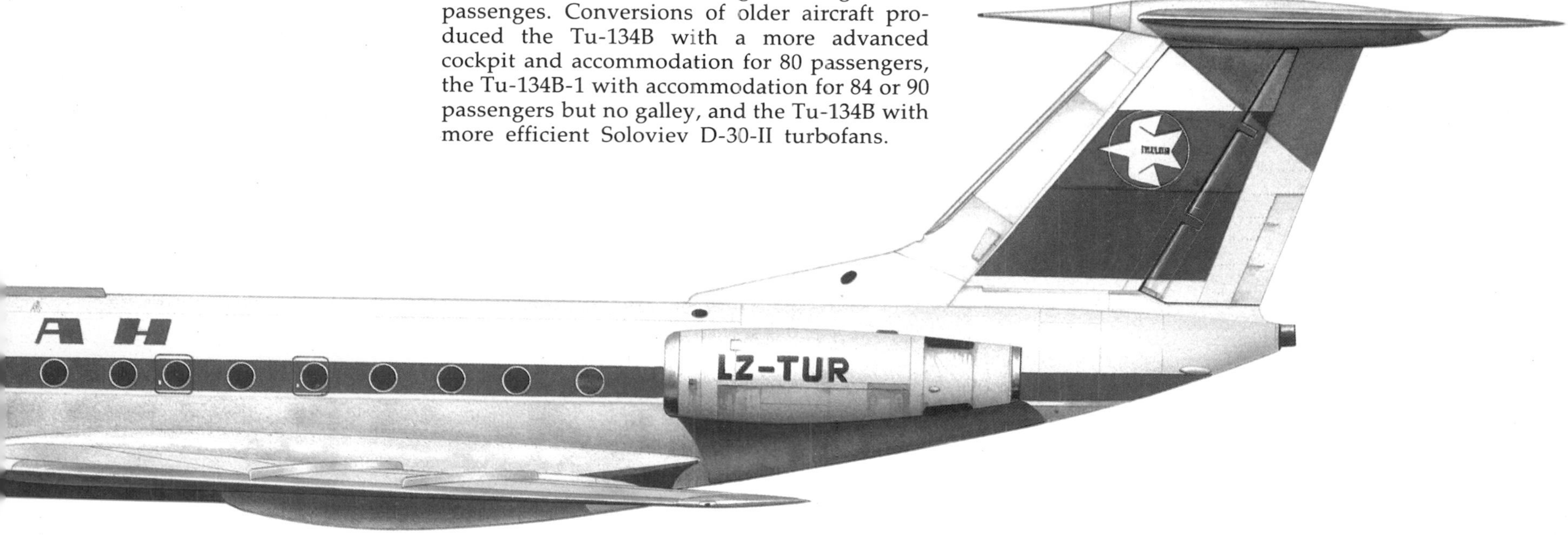

TUPOLEV Tu-26
(U.S.S.R.)

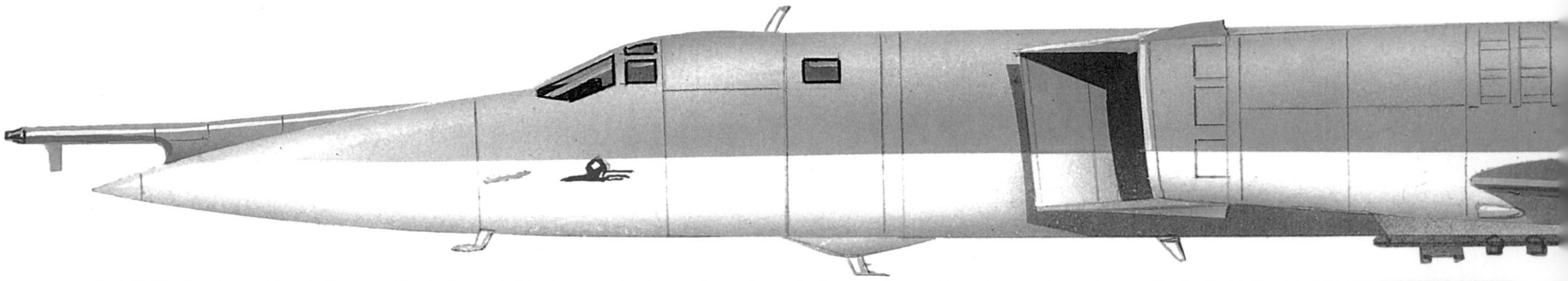

This may be regarded as the successor to the Tu-16 via the interim Tu-22, proffering supersonic performance and, by comparison with the Tu-22, longer range through the adoption of variable-geometry outer wing panels and different engines located in long fuselage trunks rather than in nacelles on the upper rear fuselage. The first Tu-22M 'Backfire-A' aircraft were probably Tu-22 conversions, but the definitive Tu-26 'Backfire-B' with completely revised landing gear and other alterations was a new-build type. The 'Backfire-C' has more advanced inlets suggesting a different engine type and higher performance.

TUPOLEV Tu-26 'BACKFIRE'

Role: Bomber/reconnaissance with variable-geometry wing
Crew/Accommodation Four/five
Power Plant: two 21,000kgp (46,297 lb s.t.) Kuznetsov NK-144 turbojets with reheat
Dimensions: Span spread 34.4 swept 26.2m (spread 112.9 swept 86 ft); length 40.2m (131.9 ft); wing area 170m^2 (1,830 sq ft)
Weights: Empty 47,000kg (103,617 lb); MTOW 122,500kg (270,066 lb)
Performance: Maximum speed 2,126km/h (1,147 knots) Mach 2 at 11,000m (36,090 ft); operational ceiling 17,983+m (59,000+ ft); radius 8,700km (4,695 naut. miles) with one air-to-air refuelling
Load: Two 23mm cannon, plus up to 8,000kg (17,637 lb) of weapons, including one long-range anti-ship missile

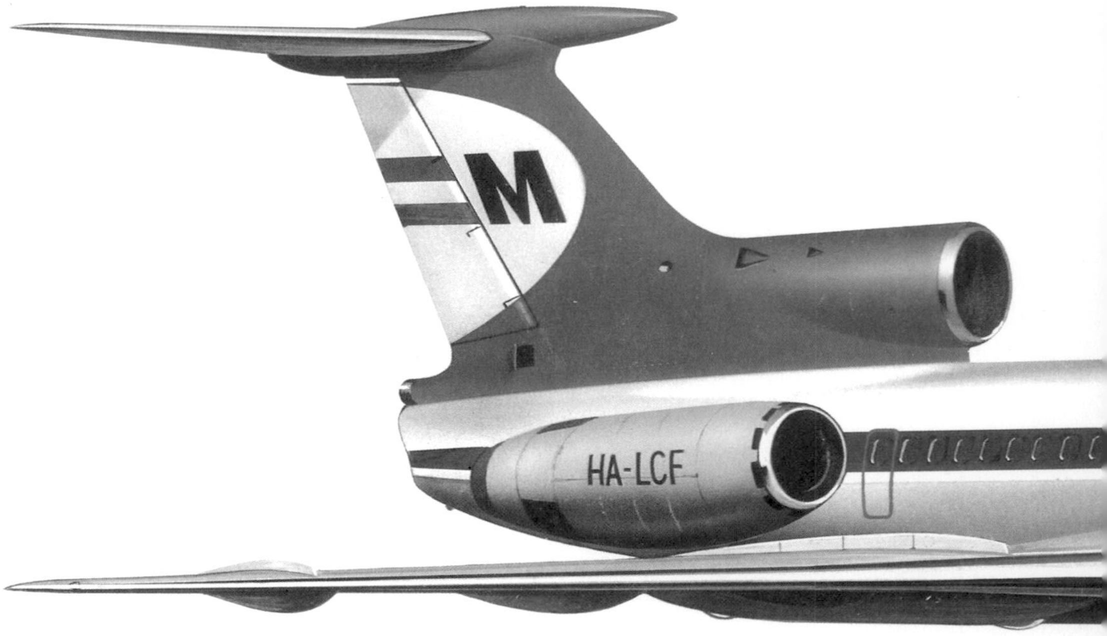

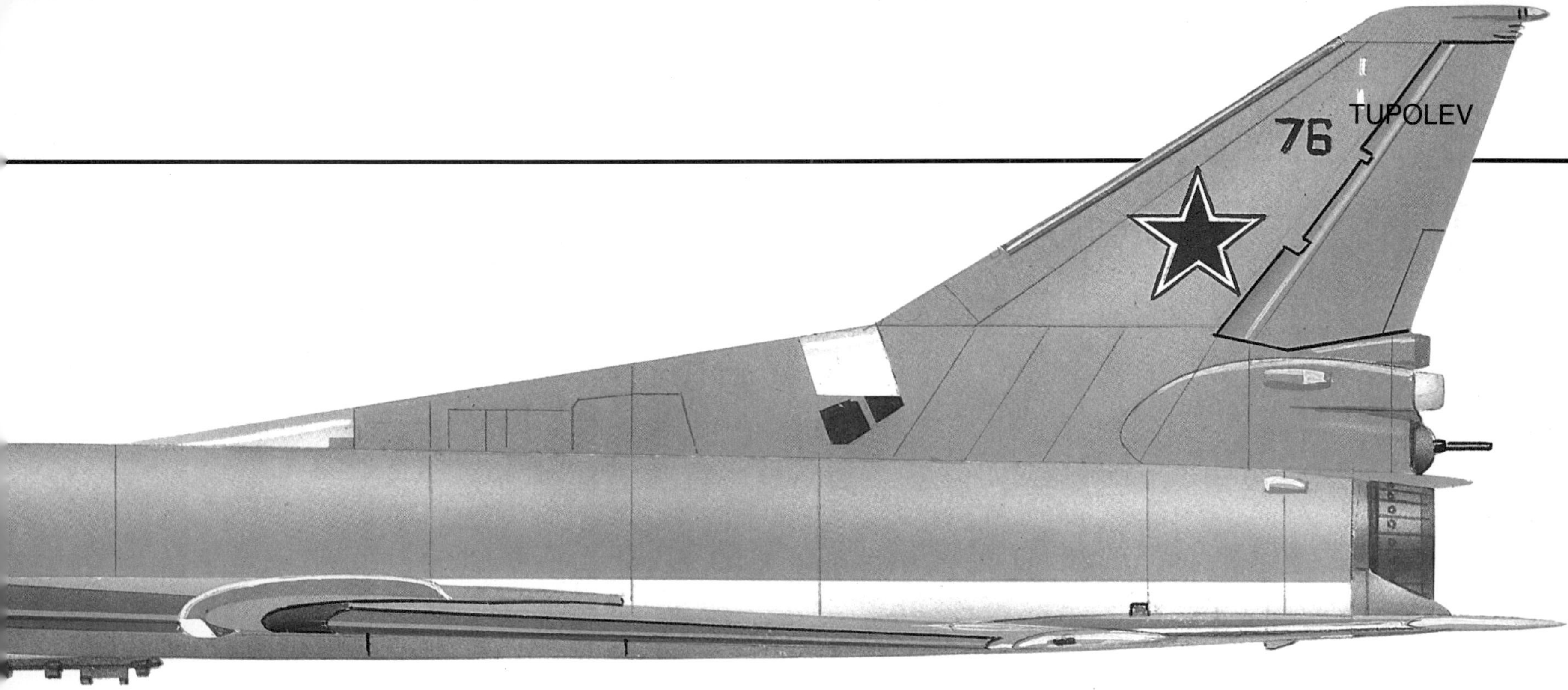

TUPOLEV Tu-154
(U.S.S.R.)

The design of this tri-jet airliner was started in the mid-1960s in response to an Aeroflot requirement that demanded a single type to replace older airliners such as the Antonov An-10, Ilyushin Il-18 and Tupolev Tu-104. The type may be regarded as a scaled-up Boeing Model 727 with considerably more power in the form of three 9,500kg (20,944 lb) static thrust Kuznetsov NK-8-T turbofans. The type first flew in 1971 and began to enter service in 1972 in its original Tu-154 form. Improved versions have been the Tu-154A and Tu-154B with more power, the Tu-154B-2 with more advanced operating equipment, the Tu-154C freighter, and the Tu-154M (later designated Tu-164) with 10,600kg (23,369 lb) static thrust Soloviev D-30KU turbofan engines.

TUPOLEV Tu-154 'CARELESS'

Role: Long/intermediate-range passenger transport
Crew/Accommodation: Three, and three cabin crew, plus up to to 158 passengers
Power Plant: Three 9,500kgp (20,944 lb s.t.) Kuznetsov NK-8-T turbofans
Dimensions: Span 37.55m (123.2 ft); length 47.9m (157.15 ft); wing area 201.45m^2 (2,169 sq ft)
Weights: Empty 43,500kg (95,900 lb); MTOW 90,000kg (198,416 lb)
Performance: Cruise speed 900km/h (486 knots) at 11,000m (36,090 ft); operational ceiling 14,000m (45,932 ft); range 2,500km (1,360 naut. miles) with 158 passengers and 5 tonnes of cargo/mail
Load: Up to 20,000kg (44,090 lb)

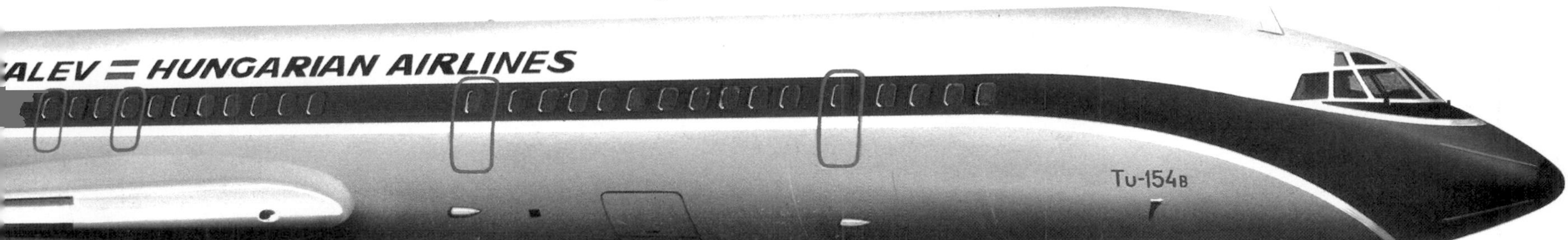

TUPOLEV Tu-160 (U.S.S.R.)

This is the unconfirmed designation of the new strategic bomber known to NATO as the 'Blackjack'. It is a variable-geometry type, larger than the Rockwell B-1B, and probably intended for penetration of enemy airspace at high altitude and supersonic speed for the release of cruise missiles, or at low level and high subsonic speed for the delivery of free-fall weapons.

TUPOLEV Tu-160 'BLACKJACK A'

Role: Long-range variable-geometry bomber reconnaissance
Crew/Accommodation: Three
Power Plant: Four 22,680 kgp (50,000 lb s.t.) Soloviev turbofans with reheat
Dimensions: Span unswept 55.00m, swept 36.75m (unswept 182.00 ft, swept 120.5 ft); length 53.95m (177 ft)
Weights: Empty 115,000kg (253,531 lb); MTOW 267,625kg (590,011 lb)
Performance: Maximum speed 2,443km/h (1,318 knots) Mach 2.3 at 12,000m (40,026 ft); operational ceiling 13,720m (45,000 ft); range 7,300km (4,536 naut. miles)
Load: Up to 16,330kg (36,001 lb) of ordnance or missiles
Note: all data estimated

VICKERS F.B.27 VIMY
(U.K.)

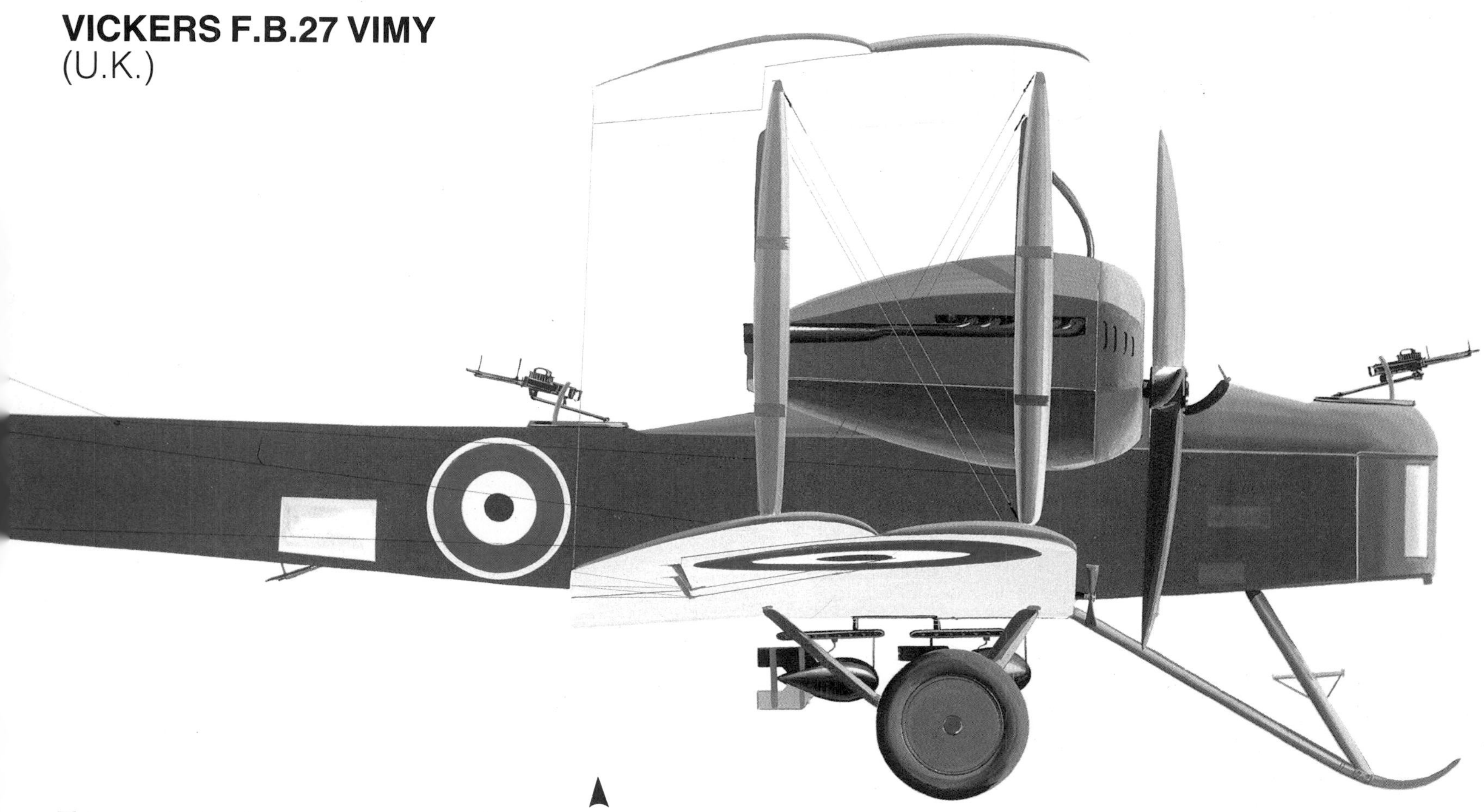

This was the U.K.'s most important bomber of the early 1920s, first flown in November 1917 as one of three British bombers designed to take the bombing war to Germany in the closing stages of World War I. The type secured undying fame as the aeroplane used by Alcock and Brown for the first nonstop crossing of the Atlantic in 1919, and production of the Vimy Mk II service bomber totalled just 230. The basic type was also developed as the Vimy Commercial airliner, of which 40 were ordered by China, and its military transport counterpart the Vernon, of which 55 were used by the RAF.

VICKERS VIMY Mk II

Role: Bomber
Crew/Accommodation: Three
Power Plant: Two 360 hp Rolls-Royce Eagle VIII water-cooled inlines
Dimensions: Span 20.73m (68 ft); length 13.27m (43.54 ft); wing area 123.56m^2 (1,330 sq ft)
Weights: Empty 3,221kg (7,101 lb); MTOW 5,670kg (12,500 lb)
Performance: Maximum speed 166km/h (103 mph) at sea level; operational ceiling 3,658m (12,000 ft); endurance up to 11 hours
Load: Two .303 inch machine guns, plus up to 1,123kg (2,476 lb) of bombs

VICKERS TYPE 132 VILDEBEESTE
(U.K.)

The Vildebeeste (or Vildebeest after 1934) was a biplane torpedo bomber that remained in service up to 1942 for lack of an adequate successor, and accordingly suffered devasting losses in the Japanese onslaught in the Far East from December 1941. The prototype flew in April 1928 with a 343kW (480 hp) Bristol Jupiter VIII radial engine, but the Vildebeest Mk I used the Bristol Pegasus I, the Mk II had the Pegasus IIM3, the Mk III had a revised cockpit allowing a crew of three, and the Mk IV introduced the 615kW (825 hp) Bristol Perseus VIII. Production totalled 189 including 39 aircraft for New Zealand, and Spain built 26 aircraft under licence with Hispano-Suiza inline engines.

VICKERS VILDEBEESTE Mk IV

Role: General-purpose strike (land or water-based)
Crew/Accommodation: Two
Power Plant: One 825 hp Bristol Perseus VIII
Dimensions: Span 14.94m (49 ft); length 11.48m (37.66 ft); wing area 67.63m^2 (728 sq ft)
Weights: Empty 2,143kg (4,724 lb); MTOW 3,856kg (8,500 lb)
Performance: Maximum speed 251km/h (156 mph) at 1,524m (5,000 ft); operational ceiling 5,791m (19,000 ft); range 2,615km (1,625 miles)
Load: Two .303 inch machine guns, plus up to 499 kg (1,100 lb) of bombs

Vickers Armstrong Wellington Mk.II

VICKERS TYPE 271 WELLINGTON SERIES
(U.K.)

This was the U.K.'s most significant medium bomber of World War II, and indeed during the early stages of the war was perhaps the only truly effective night bomber. The type used the form of geodetic construction pioneered in the Wellesley, and was thus immensely strong if somewhat ungainly in appearance. The prototype first flew in June 1936 with 682kW (915 hp) Bristol Pegasus X radial engines, and when production ceased in October 1945, no fewer than 11,461 aircraft had been produced in versions with the Pegasus (Wellington B.Mks I, IA and IC, and GR.Mk VIII), the Bristol Hercules (Wellington B.Mks III, IX and X, and GR.Mks XI, XII, XIII and XIX), the Pratt & Whitney Twin Wasp (Wellington B.Mk IV) and the Rolls-Royce Merlin inline engine (Wellington B.Mks II and VI). Wellingtons were extensively converted later in the type's career so as to operate in alternative roles such as freighting and aircrew training.

VICKERS WELLINGTON B.Mk IC

Role: Heavy bomber
Crew/Accommodation: Five/six
Power Plant: Two 1,050 hp Bristol Pegasus XVIII air-cooled radials
Dimensions: Span 26.27m (86.18 ft); length 19.69m (64.6 ft); wing area 78m^2 (848 sq ft)
Weights: Empty 8,709kg (19,200 lb); MTOW 12,927kg (28,500 lb)
Performance: Maximum speed 378km/h (235 mph) at 1,440m (4,724 ft); operational ceiling 5,486m (18,000 ft); range 2,575km (1,600 miles) with 925kg (2,040 lb) bombload
Load: Six .303 inch machine guns, plus up to 2,041kg (4,500 lb) internally stowed bombload

As an interim transport for use before the first of the Brabazon Committee's post-war airliners entered service, Vickers developed the VC.1 or Type 491, which was later named Viking. This combined the wings and landing gear of the Wellington bomber, with a portly stressed-skin fuselage, and 163 production aircraft were very important in the development of post-war air routes in Europe.

The military equivalent of the Viking was the Valetta, and more than 250 of these were produced in C.Mk 1 general-purpose transport, C.Mk 2 VIP transport, and T.Mk 3 navigator training variants. The Varsity was a specialized crew trainer version of the Valetta, equipped with a nose wheel undercarriage, and 160 such aircraft were built up to 1954.

VICKERS VALETTA C.Mk 1

Role: Military freight/troop transport
Crew/Accommodation: Three, plus up to 34 troops
Power Plant: Two 2,000 hp Bristol Hercules 230 air-cooled radials
Dimensions: Span 27.2m (89.25 ft); length 19.19m (62.95 ft); wing area 81.94m^2 (882 sq ft)
Weights: Empty 11,331kg (24,980 lb); MTOW 16,556kg (36,500 lb)
Performance: Maximum speed 415km/h (258 mph) at 3,048m (10,000 ft); operational ceiling 6,553m (21,500 ft); range 2,350km (1,460 miles) with maximum payload
Load: Up to 3,402 kg (7,500 lb)

VICKERS TYPE 607 VALETTA and TYPE 648 VARSITY (U.K.)

VICKERS TYPE 630 VISCOUNT SERIES
(U.K.)

This was the world's first turboprop-powered airliner to enter service. The aeroplane originated as the VC.2 to meet a requirement issued in World War II by the Brabazon Committee that was charged with assessing the U.K.'s post-war civil air transport needs, and the prototype Type 630 flew in July 1948 with 738kW (990 shp) Rolls-Royce Dart RDa.1 Mk 502 turboprops. The first production version was the Type 700 with accommodation for between 40 and 59 passengers, and a power plant of four 1,044kW (1,400 shp) Dart Mk 506s. Total production was eventually 444 aircraft in a number of increasingly capacious and long-ranged versions that culminated in the Type 810 with provision for 71 passengers.

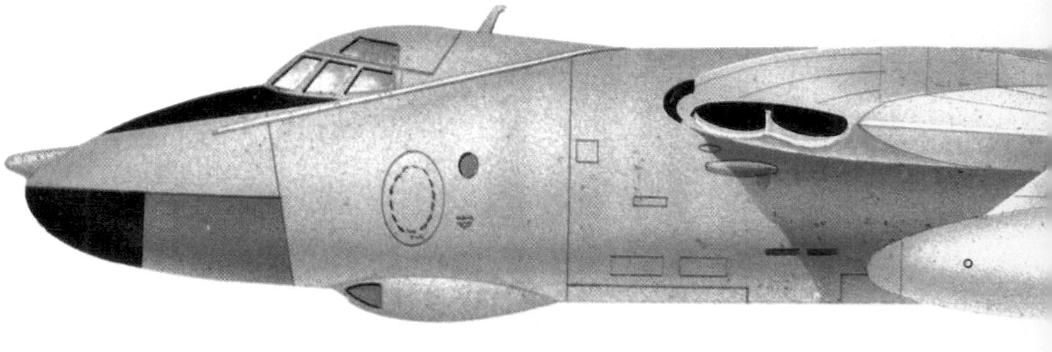

VICKERS VISCOUNT 810

Role: Short-range passenger transport
Crew/Accommodation: Three and two cabin crew, plus up to 71 passengers
Power Plant: Four 2,100 shp Rolls-Royce Dart R.Da. 7/1 Mk. 525 turboprops
Dimensions: Span 28.5m (93.76 ft); length 26.11m (85.66 ft); wing area 89.46m^2 (963 sq ft)
Weights: Empty 18,753kg (41,565 lb); MTOW 32,886kg (72,500 lb)
Performance: Maximum speed 563km/h (350 mph) at 6,100m (20,000 ft); operational ceiling 7,620m (25,000 ft); range 2,775km (1,725 miles) with maximum payload
Load: Up to 6,577 kg (14,500 lb)

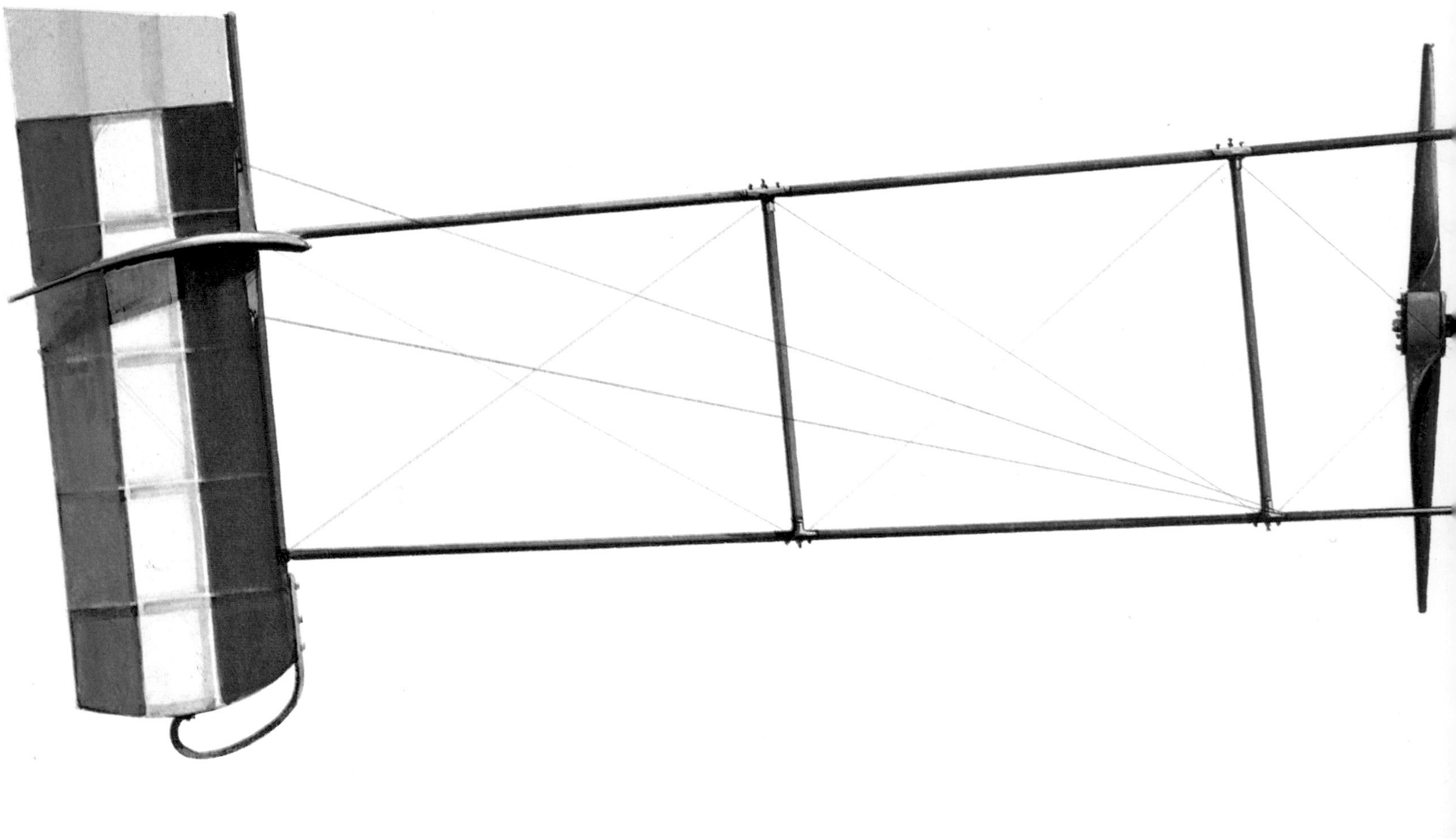

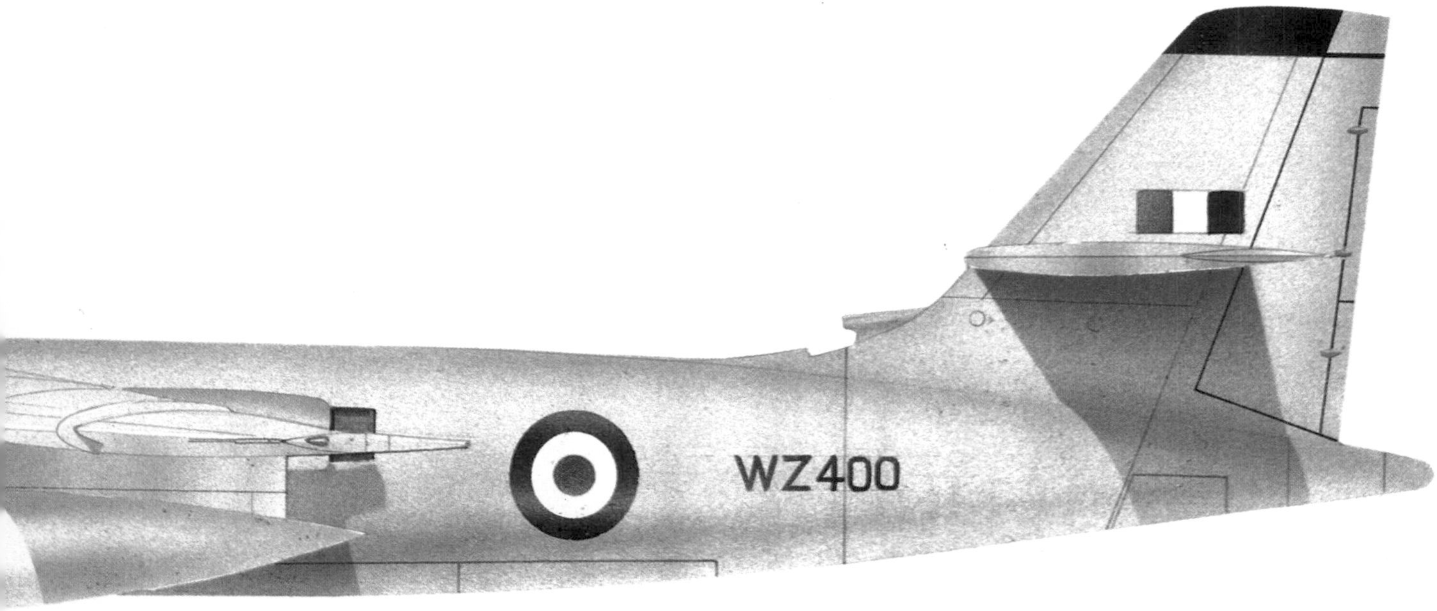

VICKERS TYPE 667 VALIANT
(U.K.)

This was the first of the U.K.'s trio of strategic V-bombers to enter service, and though not as advanced or capable as the later Avro Vulcan and Handley Page Victor, was none the less a worthy aeroplane. The prototype first flew in May 1951 with 2,948kg (6,500 lb) Rolls-Royce Avon RA.3 turbojets, improved to the 3,402kg (7,500 lb) thrust Avon RA.7 in the second prototype. Production totalled 104 aircraft for the RAF in Valiant B.Mk 1 bomber, B(PR).Mk 1 strategic reconnaissance, B(PR)K.Mk 1 bomber, reconnaissance and inflight refuelling tanker, and B(K).Mk 1 tanker variant.

VICKERS VALIANT B.Mk 1

Role: Strategic bomber
Crew/Accommodation: Five
Power Plant: Four 4,536 kgp (10,000 lb s.t.) Rolls-Royce Avon 28 turbojets
Dimensions: Span 34.85m (114.33 ft); length 32.99m (108.25 ft); wing area 219.4m^2 (2,362 sq ft)
Weights: Empty 34,419kg (75,881 lb); MTOW 63,503kg (140,000 lb)
Performance: Maximum speed 912km/h (492 knots) at 9,144m (30,000 ft); operational ceiling 16,459m (54,000 ft); range 7,242km (3,908 naut. miles) with maximum fuel
Load: No defensive armament, but internal stowage for up to 9,525 kg (21,000 lb) of bombs

VOISIN BOMBERS
(France)

Gabriel Voisin was one of France's pioneers of flight, and in World War I produced a number of ungainly pusher biplanes with the payload that allowed them to be used as bombers. Most numerous (built to the extent of more than 1,000 aircraft) was the Type LA (Type III) with the 89kW (120 hp) Salmson (Canton-Unne) M9 engine. The Type LB (Type IV) was generally similar, but carried a 37mm cannon and was powered by a 112kW (150 hp) Salmson engine which was also used in the Types V and VI of 1916. The Type LAP (Type VIII) of autumn, 1916 was a two-seat night bomber with the 164kW (220 hp) Peugeot 8Aa engine, and the Type LAR that appeared in September 1917 was also a night bomber.

VOISIN TYPE III B.2

Role: Bomber/reconnaissance
Crew/Accommodation: Two
Power Plant: One 120 hp Salmson M9 water-cooled inline
Dimensions: Span 14.75m (48.39 ft); length 9.53m (31.27 ft); wing area 49.65m^2 (534.4 sq ft)
Weights: Empty 1,000kg (2,205 lb); MTOW 1,485kg (3,274 lb)
Performance: Maximum speed 122km/h (76 mph) at sea level; operational ceiling 2,980m (9,800 ft); range 465km (289 miles)
Load: One .303 inch machine gun, plus up to 210 kg (463 lb) of bombs

VOUGHT F4U CORSAIR
(U.S.A.)

Vought F4U-1A Corsair

One of several fighters with a claim to having been one of the best fighters of World War II, the Corsair was certainly the best fighter-bomber and a truly distinguished type in this exacting role with cannon, bombs and rockets. The type first flew in May 1940 as the XF4U-1 with the 1,491kW (2,000 hp) Pratt & Whitney XR-2800, and after a troubled development in which the U.S. Navy refused to allow carrier-borne operations until after the British had achieved these on their smaller carriers, the type entered service as the F4U-1.

Total production was 12,571 up to the early 1950s, and the main variants were the baseline F4U-1, the F4U-1C with four 20mm cannon in place of the wing machine guns, the F4U-1D fighter-bomber, the F4U-1P photo-reconnaissance aeroplane, the FG-1 built by Goodyear in three subvariants, the F3A built by Brewster in three subvariants, the F4U-4 with more power, the F2G Goodyear version of the F4U-4, and several F4U-5, F4U-7 and AU-1 post-war models.

VOUGHT F4U-1D CORSAIR

Role: Naval carrierborne fighter bomber
Crew/Accommodation: One
Power Plant: One 2,000 hp Pratt & Whitney R-2800-8 Double Wasp air-cooled radial
Dimensions: Span 12.50m (41 ft); length 10.16m (33.33 ft); wing area 29.17m^2 (314 sq ft)
Weights: Empty 4,074kg (8,982 lb); MTOW 6,350kg (14,000 lb)
Performance: Maximum speed 578km/h (359 mph) at sea level; operational ceiling 11,247m (36,900 ft); range 1,633km (1,015 miles)
Load: Six .5 inch machine guns plus up to 907kg (2,000 lb) of bombs

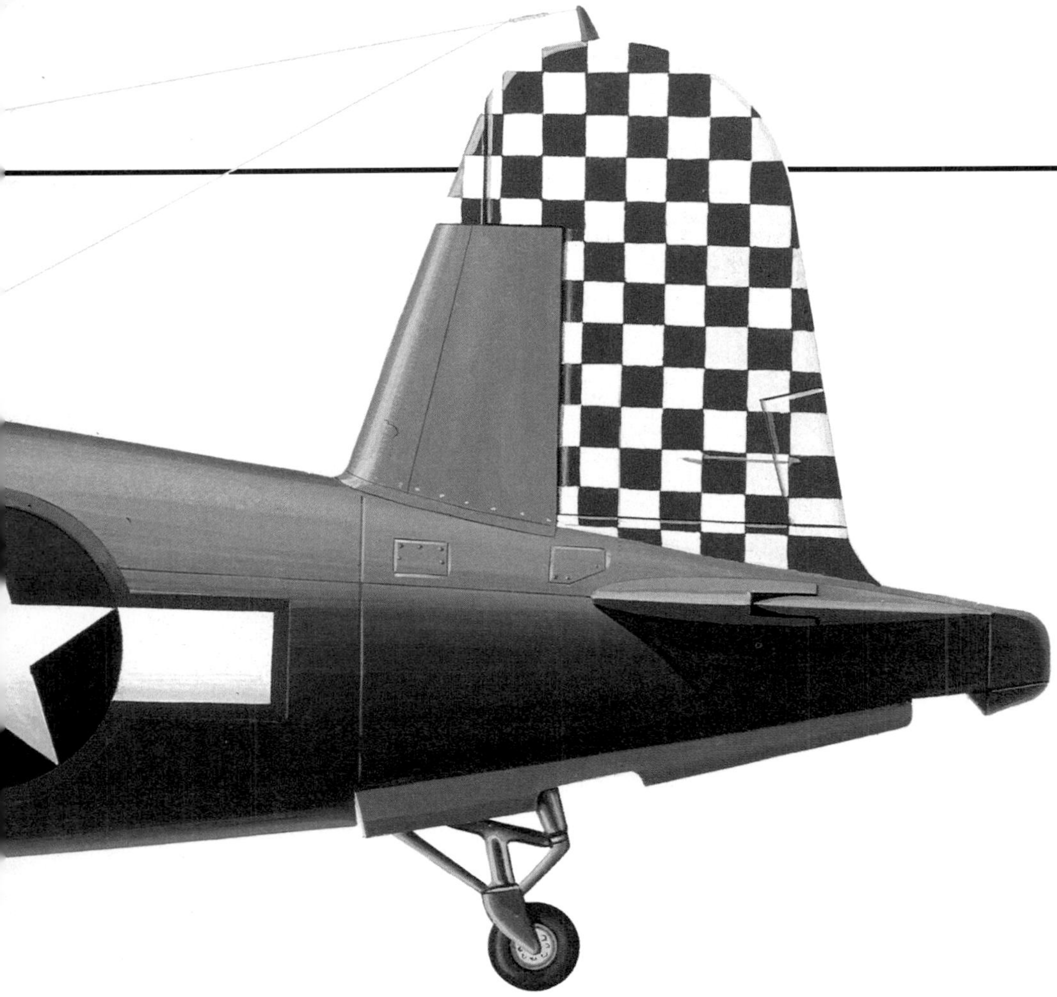

VOUGHT F-8 CRUSADER
(U.S.A.)

A slightly later contemporary of the North American F-100 Super Sabre, the Crusader carrierborne fighter was an altogether more capable machine despite its additional naval equipment. The design's most interesting feature was a variable incidence wing that allowed the fuselage to be kept level during take-off and landing, thereby improving the pilot's field of vision. The XF8U-1 prototype first flew in March 1955 and was followed by a series of progressively more capable production types between the F8U-1 (from 1962 F-8A) and the F8U-2NE (later F-8E. In addition there were F-8H to L rebuilds of older aircraft to an improved standard with a strenghtened airframe and blown flaps. There were also several reconnaissance variants.

VOUGHT F-8E CRUSADER

Role: Naval carrierborne fighter
Crew/Accommodation: One
Power Plant: One 8,165 kgp (18,000 lb s.t.) Pratt & Whitney J57-P-20 turbojet with reheat
Dimensions: Span 10.9m (35.7 ft); length 16.5m (54.2 ft); wing area 34.8m² (375 sq ft)
Weights: Empty 8,960kg (19,750 lb); MTOW 15,420kg (34,000 lb)
Performance: Maximum speed 1,802km/h (973 knots) Mach 1.7 at 12,192m (40,000 ft); operational ceiling 17,374m (57,000 ft); radius 966km (521 naut. miles)
Load: Four 30mm cannon, plus up to 2,268 kg (5,000 lb) of externally carried weapons, which can include four short-range air-to-air missiles

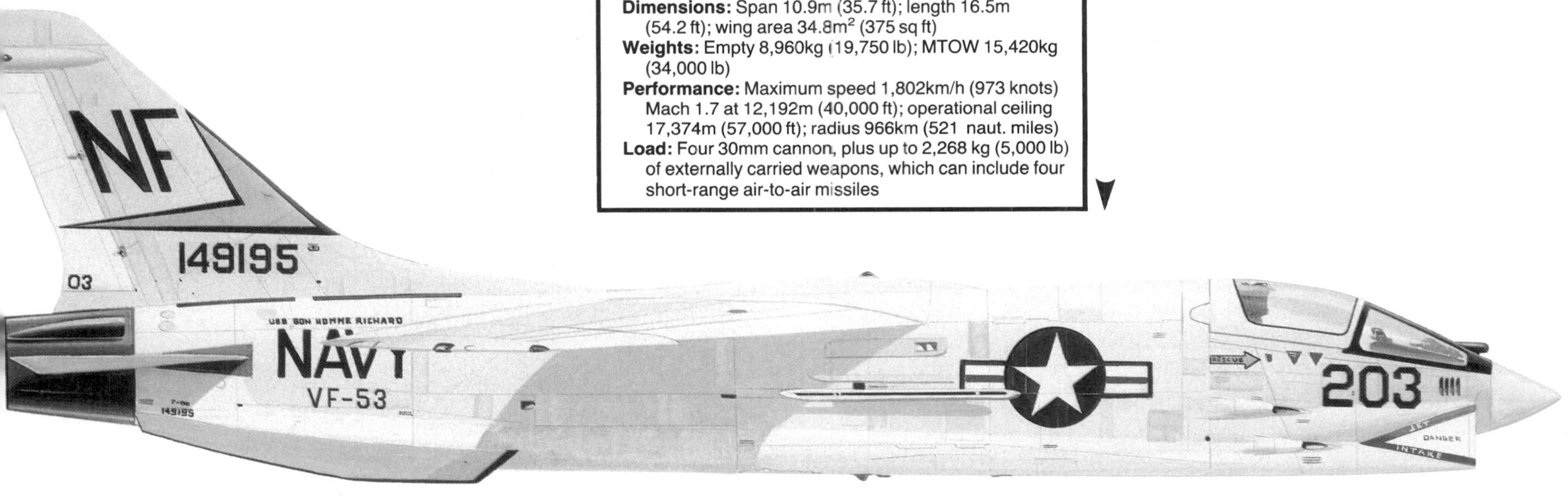

VOUGHT A-7 CORSAIR II
(U.S.A.)

The A-7 was developed with great speed on the aerodynamic basis of the F-8 Crusader to provide the U.S. Navy with a medium-weight replacement for the light-weight McDonnell Douglas A-4 Skyhawk, and first flew in September 1965. The type began to enter service in February 1967 in the form of the A-7A with the 5,148kg (11,350 lb) thrust Pratt & Whitney TF30-P-6 non-afterburning turbofan, replaced by the more powerful TF30-P-8 in the A-7B and by the improved TF30-P-408 in the A-7C. In December 1965 the U.S. Air Force had decided to adopt a version with the Allison TF41-A-1 licence-built Rolls-Royce Spey turbofan, and this A-7D model was mirrored by the Navy's A-7E. There have been some two-seat versions, and Vought is currently preparing an A-7 Plus radical development which will have advanced electronics plus the combination of more power and a revised airframe in order to give supersonic performance.

VOUGHT A-7E CORSAIR II

Role: Naval carrierborne strike
Crew/Accommodation: One
Power Plant: One 6,804 kgp (15,000 lb s.t.) Allison/Rolls-Royce TF41-A-1 turbofan
Dimensions: Span 11.8m (38.75 ft); length 14.06m (46.13 ft); wing area 34.83m^2 (375 sq ft)
Weights: Empty 8,592kg (18,942 lb); MTOW 19,051kg (42,000 lb)
Performance: Maximum speed 1,060km/h (572 knots) at sea level; operational ceiling 13,106m (43,000 ft); range 908km (489 naut. miles) with 2,722 kg (6,000 lb) bombload
Load: One 6-barrel 20mm cannon, plus up to 6,804 kg (15,000 lb) of weapons

WACO CG-4 HADRIAN
(U.S.A.)

This was the U.S.A.'s standard assault glider which saw a lot of action during World War II. Two XCG-4 prototypes were followed by 13,909 CG-4A production gliders, and these were complemented by some 427 CG-15 gliders to a design that was essentially a refined version of that of the CG-4.

WACO CG-4 HADRIAN

Role: Military assault glider
Crew/Accommodation: Two, plus up to 13 troops
Power Plant: None
Dimensions: Span 25.5m (83.66 ft); length 14.73m (48.33 ft); wing area 79.16m^2 (852 sq ft)
Weights: Empty 1,678kg (3,700 lb); MTOW 3,402kg (7,500 lb)
Performance: Maximum gliding speed 193km/h (120 mph); operational ceiling dependent upon tug aircraft
Load: Up to 1,724 kg (3,800 lb)

WESTLAND WAPITI
(U.K.)

The Wapiti was a general-purpose biplane derived ultimately from the de Havilland D.H.9A especially in its flying surfaces. The prototype flew in March 1927, and production of 517 aircraft with successively more powerful variants of the Bristol Jupiter radial engine resulted in the baseline Wapiti Mk I, the Wapiti Mk IA with leading-edge slats, the Wapiti Mk IB with divided main landing gear, the Wapiti Mk II with an all-metal primary structure, and the Wapiti Mk IA main production model with a revised wing structure. Other variants were the Wapiti Mk III South African model with the Armstrong Siddeley Jaguar radial engine, the Wapiti Mk V with a lengthened fuselage, and the Wapiti Mk VI trainer.

WESTLAND WAPITI Mk IA

Role: Military general-purpose/army co-operation
Crew/Accommodation: Two
Power Plant: One 480 hp Bristol Jupiter VIII F air-cooled radial
Dimensions: Span 14.15m (46.42 ft); length 9.65m (31.66 ft); wing area 40.32m^2 (434.0 sq ft)
Weights: Empty 1,320kg (2,910 lb); MTOW 2,223kg (4,900 lb)
Performance: Maximum speed 225km/h (140 mph) at sea level; operational ceiling 6,279m (20,600 ft); range 837m (520 miles)
Load: Two .303 inch machine guns, plus up to 263 kg (580 lb) of bombs

WESTLAND LYNX
(U.K.)

This is one of the most advanced helicopters in the world, and the first of six prototypes flew in March 1971. Extensive production has been undertaken for the British forces and for export, the naval version generally having fixed tricycle landing gear and the army variant possessing twin tubular skids. The type can carry a useful payload, but its high performance and agility suit it better to the armed role with an assortment of disposable stores and advanced sensors. The type is still in production and active development, with more capable land-based variants about to enter service.

WESTLAND LYNX HAS 2

Role: Shipborne anti-submarine/ship helicopter
Crew/Accommodation: Three
Power Plant: Two 1,120 shp Rolls-Royce Gem 4 turboshafts
Dimensions: Rotor diameter 10.62m (34.8 ft); length rotors turning 15.16m (49.75 ft)
Weights: Empty 3,316kg (7,311 lb); MTOW 4,763kg (10,500 lb)
Performance: Maximum speed 271km/h (146 knots) at sea level; operational ceiling 2,575+m (8,450+ ft); endurance 1 hour on station at 113 km (61 naut. miles) from ship with 2 Mk 44 torpedoes
Load: Up to 816 kg (1,800 lb)

WESTLAND SEA KING
(U.K.)

This is the licence-built version of the Sikorsky Sea King, but with a British tactical system that makes the type considerably more capable than its U.S. counterparts. The first British production helicopter, a Sea King HAS.Mk.1, flew in May 1969 and since that time large numbers have been built or converted for the British services (in the anti-submarine, airborne early warning and search and rescue roles) and for export. Power has been increased steadily for use with more advanced weapons. There is also a Commando assault version with fixed landing gear.

WESTLAND-SIKORSKY COMMANDO

Role: All-weather assault/transport helicopter
Crew/Accommodation: Two, plus up to 28 troops
Power Plant: Two 1,660 shp Rolls-Royce Gnome 1400 turboshafts
Dimensions: Overall length rotors turning 22.15m (72.66 ft); rotor diameter 18.90m (62 ft)
Weights: Empty 5,069kg (11,174 lb); MTOW 9,526kg (21,000 lb)
Performance: Cruise speed 207km/h (112 knots) at sea level; operational ceiling 3,048m (10,000 ft); range 445km (240 naut. miles) with full payload
Load: Up to 2,794kg (6,160 lb)

Westland Sea King Mk.5 launching a BAe Sea Eagle anti-ship missile.

Westland Lynx AH Mk.1

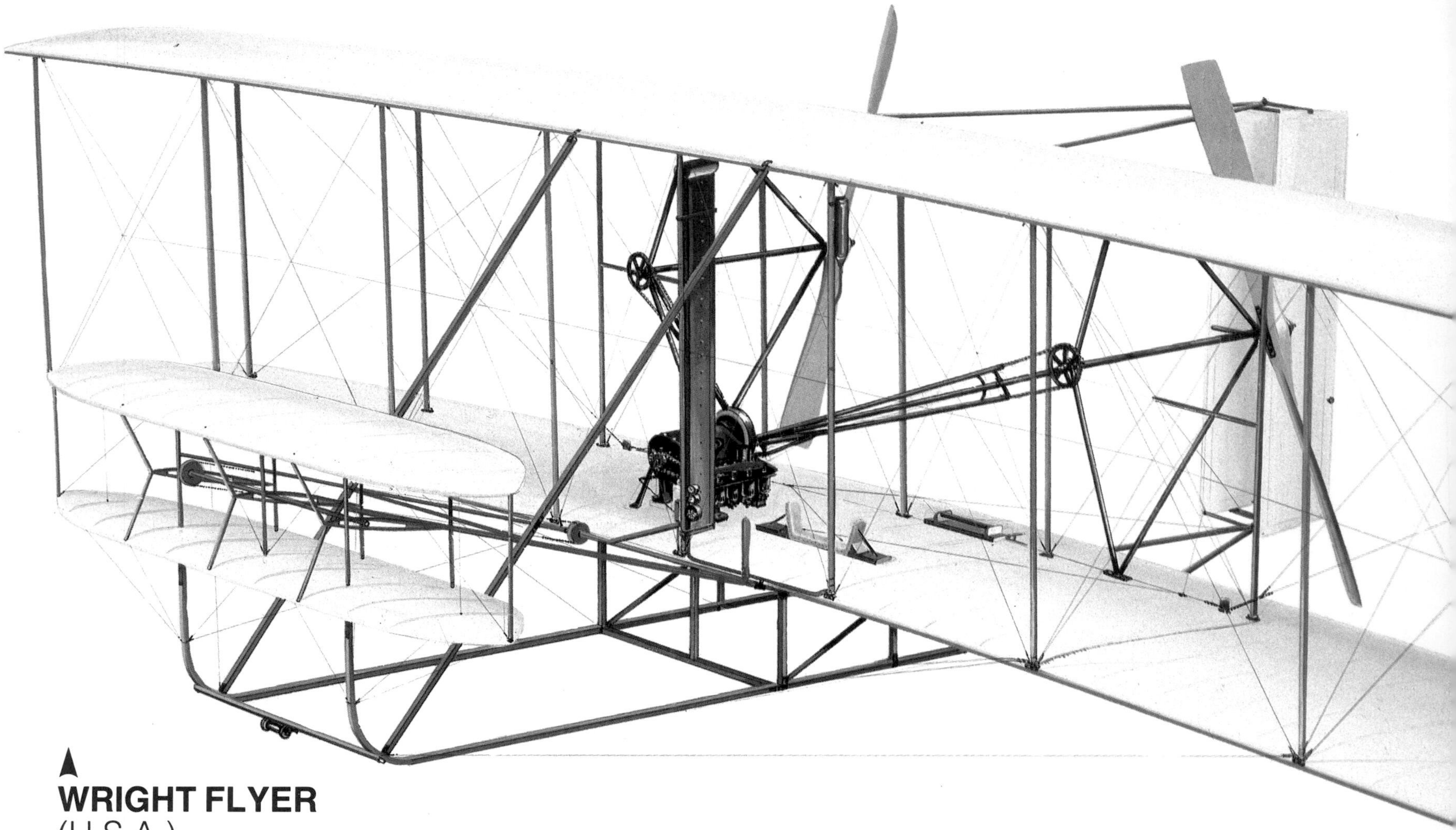

WRIGHT FLYER
(U.S.A.)

With the aeroplane now designated Flyer I, the Wright brothers made the world's first powered, sustained and controlled flights in a heavier-than-air craft on 17 December 1903. This machine was powered by a 9kW (12 hp) Wright engine, and the Flyer II of 1904 had an 11kW (15 hp) engine. In 1905 the brothers produced the world's first really practical aeroplane as the Flyer III with improved controls, but with the engine of the Flyer II.

WRIGHT FLYER 1

Role: Powered flight demonstrator
Crew/Accommodation: One
Power Plant: One 12 hp Wright Brothers' water-cooled inline
Dimensions: Span 12.29m (40.33 ft); length 6.41m (21.03 ft); wing area 47.38m^2 (510 sq ft)
Weights: Empty 256.3kg (565 lb); MTOW 340.2kg (750 lb)
Performance: Cruise speed 48km/h (30 mph) at sea level; operational ceiling 9.14m (30 ft); range 259.7m (852 feet)
Load: None.
Note: the range quoted here was the longest of four flights made by the Wright Brothers on 17 December, 1903

YAKOVLEV Yak-1
(U.S.S.R.)

First flown in January 1940, the I-26 low-wing fighter prototype led to a production series of 8,721 Yak-1 fighters by mid-1943. The main variants of this elegant type were the baseline Yak-1 with the 783kg (1,050 hp) Klimov M-105P inline engine, the Yak-1B with a cutdown rear fuselage to allow incorporation of 360 degree vision canopy, and the YaK-1M with 940kW (1,260 hp) M-105PF engine and structural revisions to reduce weight.

YAKOVLEV Yak-1

Role: Fighter
Crew/Accommodation: One
Power Plant: One 1,050 hp Klimov M-105P water-cooled inline
Dimensions: Span 10m (32.81 ft); length 8.47m (27.79 ft); wing area 17.15m^2 (184.6 sq ft)
Weights: Empty 2,330kg (5,137 lb); MTOW 2,820kg (6,217 lb)
Performance: Maximum speed 580km/h (360 mph) at 5,000m (16,400 ft); operational ceiling 10,000m (32,808 ft); range 850km (528 miles)
Load: One 20mm cannon and two 7.62mm machine guns, plus up to 200 kg (440 lb) of bombs or rockets

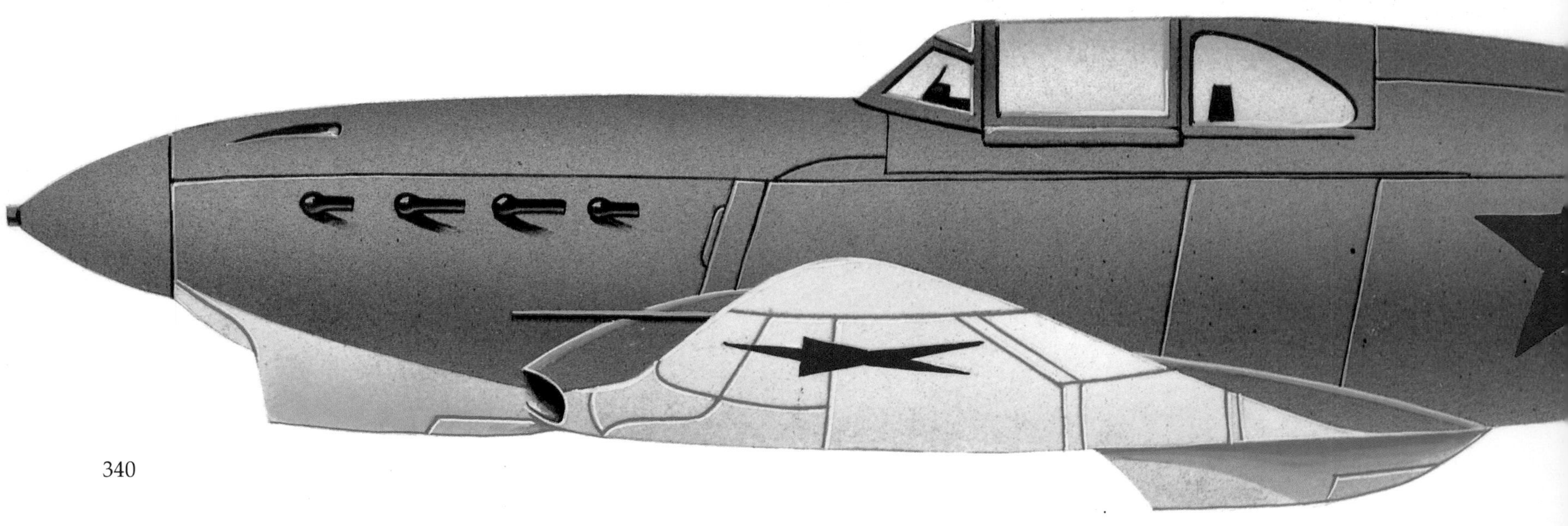

YAKOVLEV Yak-3
(U.S.S.R.)

This was a development of the Yak-1 with initially the Klimov VK-105 and later the VK-107 inline engine in the smallest and lightest airframe able to accommodate it. By comparison with that of the Yak-1 the wing was of shorter span and reduced area, and there were many other detail modifications to the airframe. Problems with the VK-107 were considerable, and though the Yak-3 prototype flew in late 1943 it was July 1944 before the first Yak-3 production aircraft began to reach squadrons, and only then with an interim engine in the form of the 969kW (1,300 hp) Klimov VK-105PF-2. Total production was 4,848 aircraft, and even with the VK-105 the Yak-3 was an exceptional dogfighter. Some aircraft were fitted with the 1,268kW (1,700 hp) VK-107A in 1945, and there were many experimental and development models.

YAKOVLEV Yak-3

Role: Interceptor
Crew/Accommodation: One
Power Plant: One 1,300 hp Klimov VK-105PF-2 water-cooled inline
Dimensions: Span 9.2m (30.18 ft); length 8.49m (27.85 ft); wing area 14.83m^2 (156.6 sq ft)
Weights: Empty 2,105kg (4,641 lb); MTOW 2,660kg (5,864 lb)
Performance: Maximum speed 655km/h (407 mph) at 3,300m (10,827 ft); operational ceiling 10,800m (35,430 ft); range 900km (560 miles)
Load: One 20mm cannon and two 12.7mm machine guns

YAKOVLEV Yak-9
(U.S.S.R.)

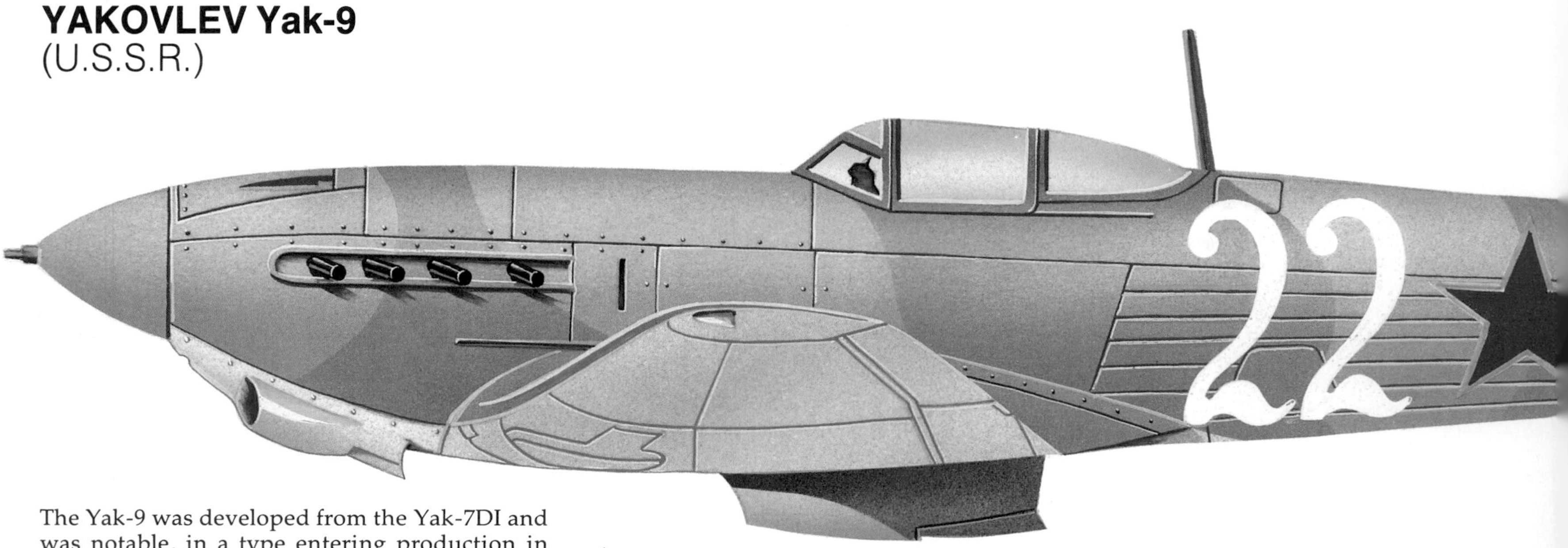

The Yak-9 was developed from the Yak-7DI and was notable, in a type entering production in mid-World War II, for its mixed wood and metal primary structure. Production lasted to 1946 and totalled 16,769 aircraft in several important and some lesser variants. These included the original Yak-9, the Yak-9M with revised armament, the Yak-9D long-range escort fighter, the Yak-9T anti-tank variant with 20, 23 or even 37mm cannon, the Yak-9K heavy anti-tank fighter with 45mm cannon, the Yak-9B high-speed light bomber with internal provision for four 100kg (220 lb) bombs, the Yak-9MPVO night-fighter, the Yak-9DD very long-range escort fighter, the Yak-9U conversion trainer in three subvariants, the Yak-9P with two fuselage-mounted 20mm cannon, and the Yak-9R reconnaissance aeroplane.

YAKOVLEV Yak-9D

Role: Fighter
Crew/Accommodation: One
Power Plant: One 1,360 hp Klimov VK-105PF-3
Dimensions: Span 9.74m (32.03 ft); length 8.55m (28.05 ft); wing area 17.1m^2 (184.05 sq ft)
Weights: Empty 2,770kg (6,107 lb); MTOW 3,080kg (6,790 lb)
Performance: Maximum speed 602km/h (374 mph) at 2,000m (6,560 ft); operational ceiling 10,600m (34,775 ft); range 1,410km (876 miles)
Load: One 20mm cannon + one 12.7mm machine gun

YAKOVLEV Yak-25
(U.S.S.R.)

The Yak-25 was developed as a swept-wing all-weather interceptor, and first flew in about 1953 with 2,200kg (4,850 lb) Mikulin AM-5 turbojets. Production of about 1,000 aircraft began in 1954, the initial 'Flashlight-A' having a crew of two and being followed by the 'Flashlight-B' reconnaissance prototypes with the second crew member in a revised nose, and the 'Flashflight-C' tactical reconnaissance aeroplane with longer-span wings and powered, like late-production 'Flashlight-As', by 2,800kg (6,173 lb) thrust Tumanskii RD-9 turbojets. A 'Mandrake' high-altitude reconnaissance version had straight long-span wings.

YAKOVLEV Yak-25 'FLASHLIGHT A'

Role: All-weather/night interceptor
Crew/Accommodation: Two
Power Plant: Two 2,200kgp (4,850 lb s.t.) Mikulin AM-5 turbojets
Dimensions: Span 12.34m (40.49 ft); length 16.65m (54.68 ft); wing area 37.12m^2 (399.6 sq ft)
Weights: Empty 9,850kg (21,716 lb); MTOW 16,000kg (35,274 lb)
Performance: Speed 1,030km/h (556 knots) Mach 0.97 at 11,000m (36,090 ft); operational ceiling 15,500m (50,853 ft); radius 1,250km (674 naut. miles)
Load: Two 37mm cannon

YAKOVLEV Yak-28
(U.S.S.R.)

This resembles the Yak-25 in general configuration, but is more highly swept for true supersonic performance. The type first flew in about 1960 as the 'Brewer -A' light bomber and began to enter production in 1962. There have been several variants including the 'Brewer-B' strike bomber, 'Brewer-C' strike bomber with a lengthened fuselage, 'Brewer-D' multi-sensor reconnaissance aeroplane, 'Brewer-E' electronic escort aeroplane, 'Firebar' all-weather fighter and 'Maestro' conversion trainer.

YAKOVLEV Yak-28P 'FIREBAR'

Role: All-weather/night interceptor
Crew/Accommodation: Two
Power Plant: Two 4,600 kgp (13,670 lb) Tumansky R-11 turbojets with reheat
Dimensions: Span 12.5m (41.01 ft); length 21.95m (72.01 ft); wing area 35m^2 (376.7 sq ft)
Weights: Empty 17,000kg (37,480 lb); MTOW 18,500kg (40,785 lb)
Performance: Maximum speed 1,998km/h (1,078 knots) Mach 1.88 at 11,000m (36,090 ft); operational ceiling 16,765m (55,000 ft); radius 900km (485 naut. miles) with missiles
Load: Two medium-range air-to-air missiles

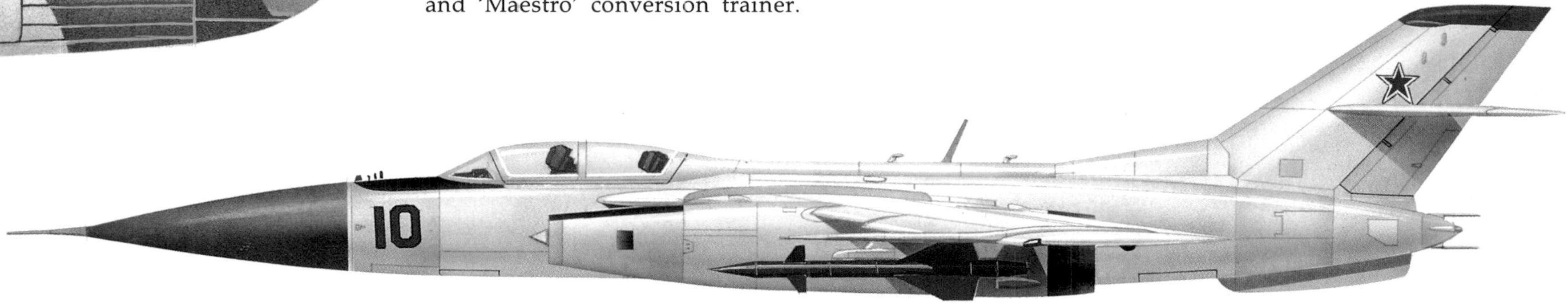

YAKOVLEV Yak-38 (U.S.S.R.)

This naval strike fighter is believed to have been developed as an interim warplane pending the availability of more advanced aircraft, and is now known to be a STOVL type rather than the purely VTOL aeroplane it was originally considered to be. This had led to an upward revision of the Yak-38's capabilities. The type entered service in 1976 and it is notable for its power plant, which comprises two lift turbojets in the forward fuselage and a vectored-thrust turbojet farther aft. The 'Forger-A' is the single-seat model, and the 'Forger-B' the combat-capable two-seater.

YAKOVLEV Yak-38 'FORGER-A'

Role: Vertical take-off and landing naval strike fighter
Crew/Accommodation: One
Power Plant: One 8,200 kgp (18,078 lb s.t.) Lyulka vectored thrust turbofan and two 4,100 kgp (9,039 lb s.t.) Koliesov lift turbojets
Dimensions: Span 7.5m (24.6 ft); length 16m (52.5 ft); wing area 15.5m^2 (167 sq ft)
Weights: Empty 5,500kg (12,125 lb); MTOW 9,980kg (22,002 lb)
Performance: Maximum speed 1,164km/h (628 knots) Mach 0.95 at sea level; operational ceiling 11,887m (39,000 ft); radius 371km (200 naut.miles)
Load: Up to 1,000 kg (2,205 lb) of externally carried weapons, including two medium-range air-to-air missiles

YAKOVLEV Yak-42
(U.S.S.R.)

This was designed to replace the Ilyushin Il-18 and Tupolev Tu-134, and is therefore a medium-range type with accommodation for a maximum of 120 passengers. The type first flew in March 1975, and two prototypes were used for the evaluation of 11-degree and 23-degree swept wings, the latter being selected for the production model.

YAKOVLEV Yak-42 'CLOBBER'

Role: Intermediate/short-range passenger transport
Crew/Accommodation: Two/three and four cabin crew, plus up to 120 passengers
Power Plant: Three 6,500 kgp (14,330 lb s.t.) Lotarev D-36 turbofans
Dimensions: Span 34.88m (114.44 ft); length 36.38m (119.36 ft); wing area 150m^2 (1,615 sq ft)
Weights: Empty 32,500kg (71,650 lb); MTOW 54,000kg (119,049 lb)
Performance: Cruise speed 810km/h (503 mph) at 7,620m (25,000 ft); operational ceiling 13,000+m (42,650 ft); range 1,740km (1,081 miles) with full payload
Load: Up to 10,800 kg (23,810 lb)

YOKOSUKA D4Y SUISEI (COMET)
(Japan)

This carrier-based bomber was unusual among Japanese warplanes of World War II in being planned round an inline engine, the Aichi Astuta version of the Daimler-Benz DB 601A. The prototype flew with an imported DB 600G and revealed good performance leading to large-scale production. The first of an eventual 2,038 aircraft entered service in the autumn of 1942, and the main variants were the D4Y1 with the 895kW (1,200 hp) AE1A Astuta 32 in reconnaissance, dive-bomber and catapult-launched models, the D4Y2 with the 1,044 (1,400hp) AE1P Astuta 32 in reconnaissance, catapult-launched and night-fighter models, and the D4Y3 with the 1,163kW (1,560 hp) Mitsubishi MK8P Kinsei radial engine, of which many were converted to D4Y4 kamikaze aircraft.

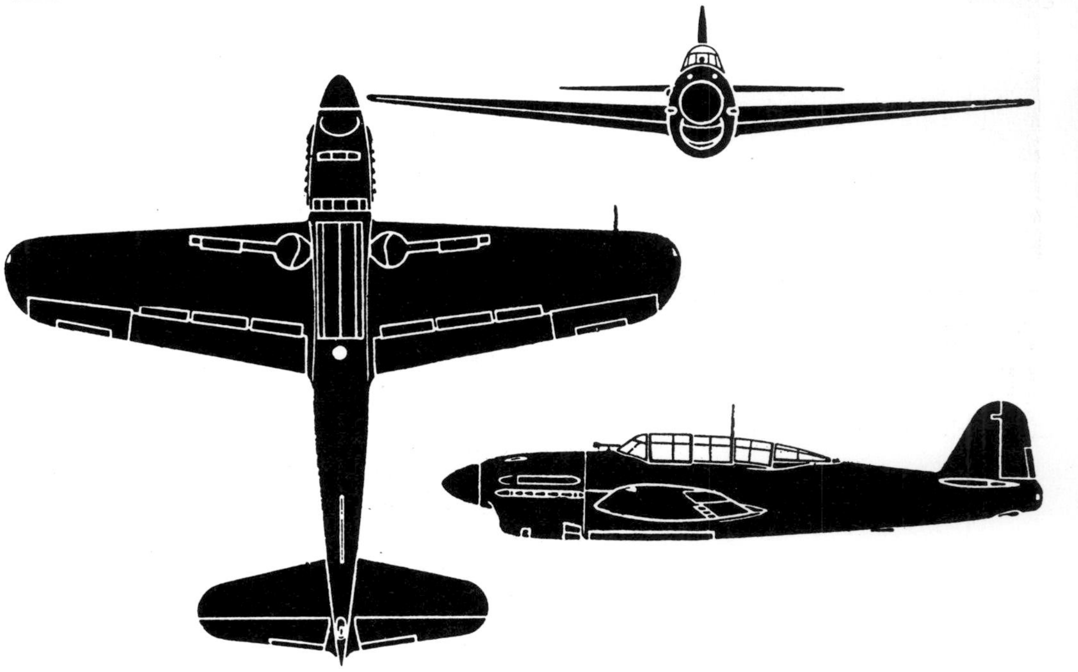

YOKOSUKA D4Y3 SUISEI 'JUDY'

Role: Naval carrierborne strike
Crew/Accommodation: Two
Power Plant: One 1,560 hp Mitsubishi MK8P Kinsei 62 air-cooled radial
Dimensions: Span 11.5m (37.73 ft); length 10.22m (33.53 ft); wing area 23.6m^2 (254 sq ft)
Weights: Empty 2,501kg (5,514 lb); MTOW 4,657kg (10,267 lb)
Performance: Maximum speed 574km/h (357 mph) at 6,050m (19,849 ft); operational ceiling 10,500m (34,449 ft); range 1,520km (944 miles) with full warload
Load: Three 7.7mm machine guns, plus 750 kg (1,653 lb) of bombs

YOKOSUKA P1Y GINGA (MILKY WAY)
(Japan)

This low-altitude aeroplane was conceived for the level bombing, dive-bombing and torpedo bombing roles, and first flew in August 1943. Production totalled 1,098 aircraft, and though performance was generally good these aircraft were hampered by their chronic maintenance problems. The P1Y1 initial production model was powered by 1,361kW (1,825 hp) Nakajima NK9C Homare 12 radial engines in bomber, attack and night-fighter subvariants, and the P1Y2 had 1,380kW (1,850 hp) Mitsubishi MK4T-A Kasei 25A radial engines in night-fighter and bomber subvariants.

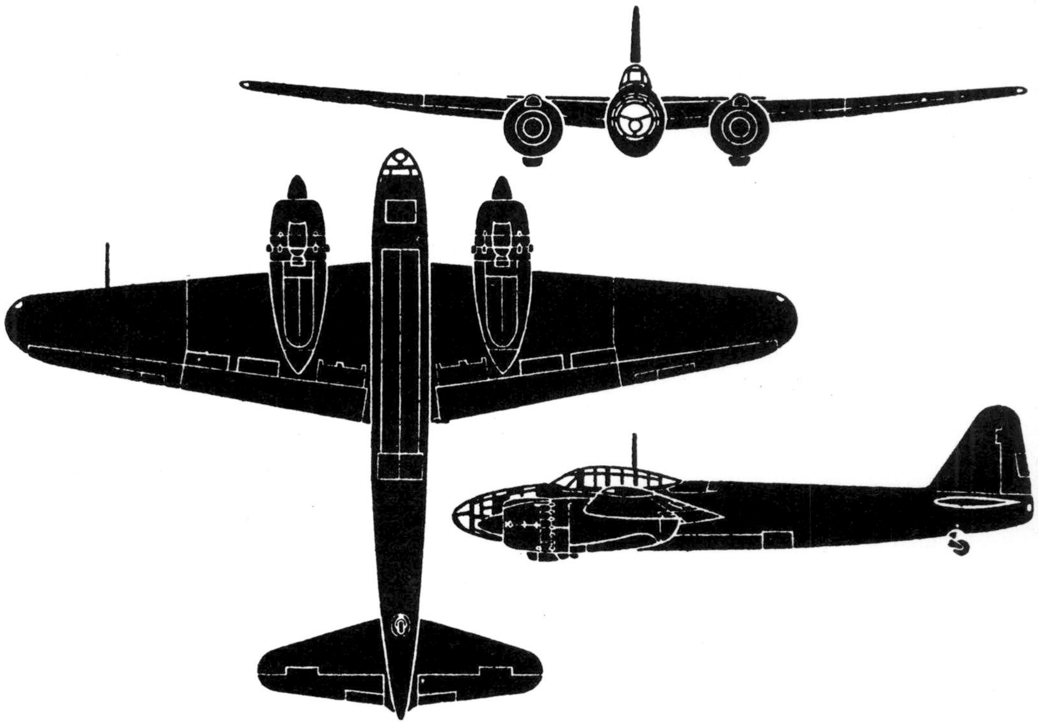

YOKOSUKA P1Y1 GINGA 'FRANCES'

Role: Bomber
Crew/Accommodation: Three
Power Plant: Two 1,825 hp Nakajima NK9C Homare 12 air-cooled radials
Dimensions: Span 20m (65.62 ft); length 15m (49.21 ft); wing area 55m^2 (592.0 sq ft)
Weights: Empty 6,690kg (14,749 lb); MTOW 13,500kg (29,976 lb)
Performance: Maximum speed 555km/h (345 mph) at 5,900m (19,355 ft); operational ceiling 10,220m (33,530 ft); range 1,245km (774 miles) with torpedo
Load: One 20mm cannon and one 12.7mm machine gun, plus up to 1,000kg (2,205 lb) of bombs or a 800kg (1,764 lb) torpedo

ZLIN TRENER (TRAINER) SERIES
(Czechoslovakia)

The Trener series began in response to a Czech Air Force requirement of 1947 for a primary trainer, and in 1948 the Zlin 26 was selected in preference to the Praga E.112. The type entered production with the 78kW (105 hp) Walter Minor 4-III inline engine, and this was the precursor of a long series of trainer and aerobatic aircraft. Next came the Zlin 126 Trener II with the same engine, then the Zlin 226T Trener-6 with the 119kW (160 hp) Walter Minor 6-III, and finally among the trainer series, the Zlin 326 Trener-Master version of the Trener-6 with retractable landing gear.

ZLIN Z226

Role: Basic trainer
Crew/Accommodation: Two
Power Plant: One 160 hp Walter Minor 6-III air-cooled in-line
Dimensions: Span 10.6m (34.75 ft); length 7.8m (25.59 ft); wing area 15.45m^2 (166.3 sq ft)
Weights: Empty 680kg (1,499 lb); MTOW 975kg (2,150 lb)
Performance: Maximum speed 243km/h (151 mph) at sea level; operational ceiling 5,000m (16,350 ft); range 980km (610 miles) with tip tank fuel
Load: None

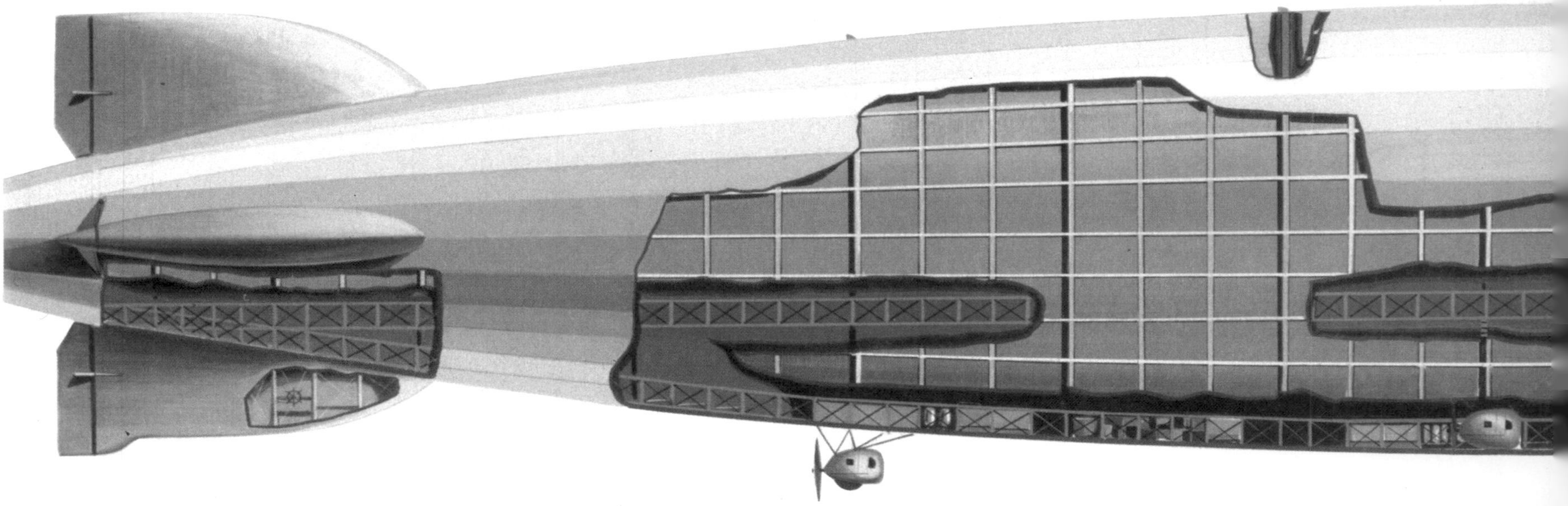

ZEPPELIN LZ 127 'Graf Zeppelin' (Germany)

First flown in September 1928, the LZ 127 was the most successful passenger-carrying airship ever built. In technical terms the type was a logical successor to the company's rigid airships used in World War I for maritime reconnaissance and bombing, though its payload of 15,000kg (33,070 lb) was concentrated on the accommodation and facilities for a maximum of 24 passengers. Between October 29, and November 1, 1928 the ship set a dirigible straight-line distance record of 6,384.5 km (3,967.3 miles) that still stands. Between August 8 and 29 of the following year, still captained by Dr Hugo Eckener, LZ 127 achieved the first circumnavigation of the Earth by an airship. The ship was later used for the first airline services across the North and South Atlantics, and before being scrapped at the beginning of World War II had carried 13,100 passengers and logged more than 1.6 million km (1 million miles).

ZEPPELIN LZ 127 'Graf Zeppelin'

Role: Passenger transport airship
Crew/Accommodation: Twenty-six, including cabin crew, plus 24 passengers
Power Plant: Five 580 hp Maybach VL2 water-cooled in-lines
Dimensions: Maximum diameter 30.50m (100.10 ft); length 236.60m (776.25 ft)
Gas volume: 105,000m^3 (3,708,040 cu ft)
Gross lift: 111,000kg (244,713 lb)
Performance: Cruise speed 110km/h (68 mph) at sea level; range 12,000km (7,456 miles) maximum in still air
Payload: Up to 15,000kg (33,070 lb), including passengers

Middle: The LZ 127 being 'walked out' of its airship shed.
Lower: LZ 127's command and crew section.

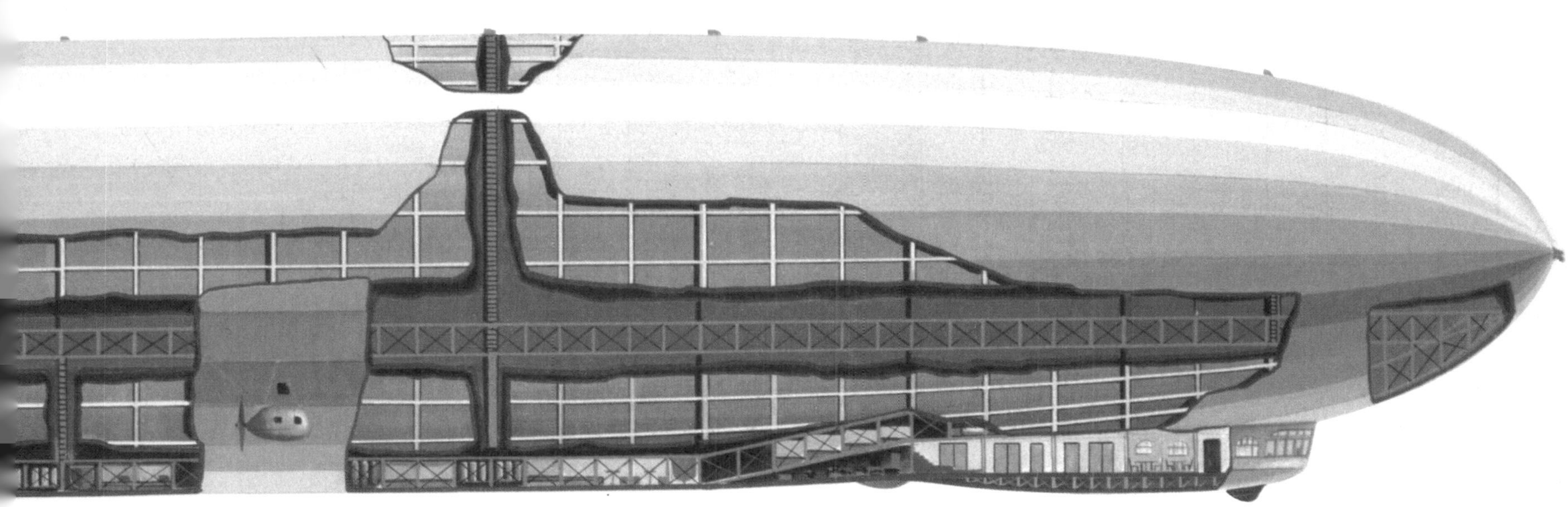

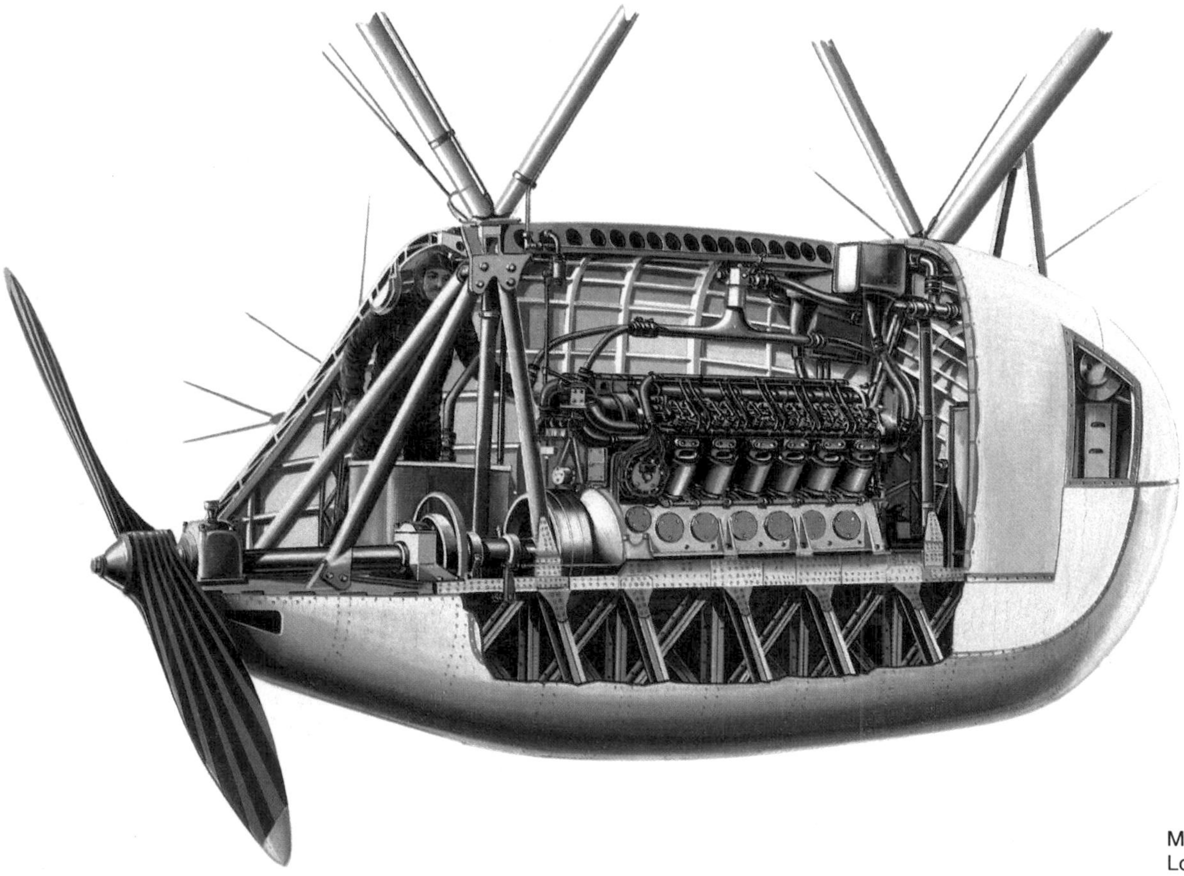

Middle: The rearmost of the LZ 127's five engine cabs.
Lower: The luxurious passenger salons and berths.

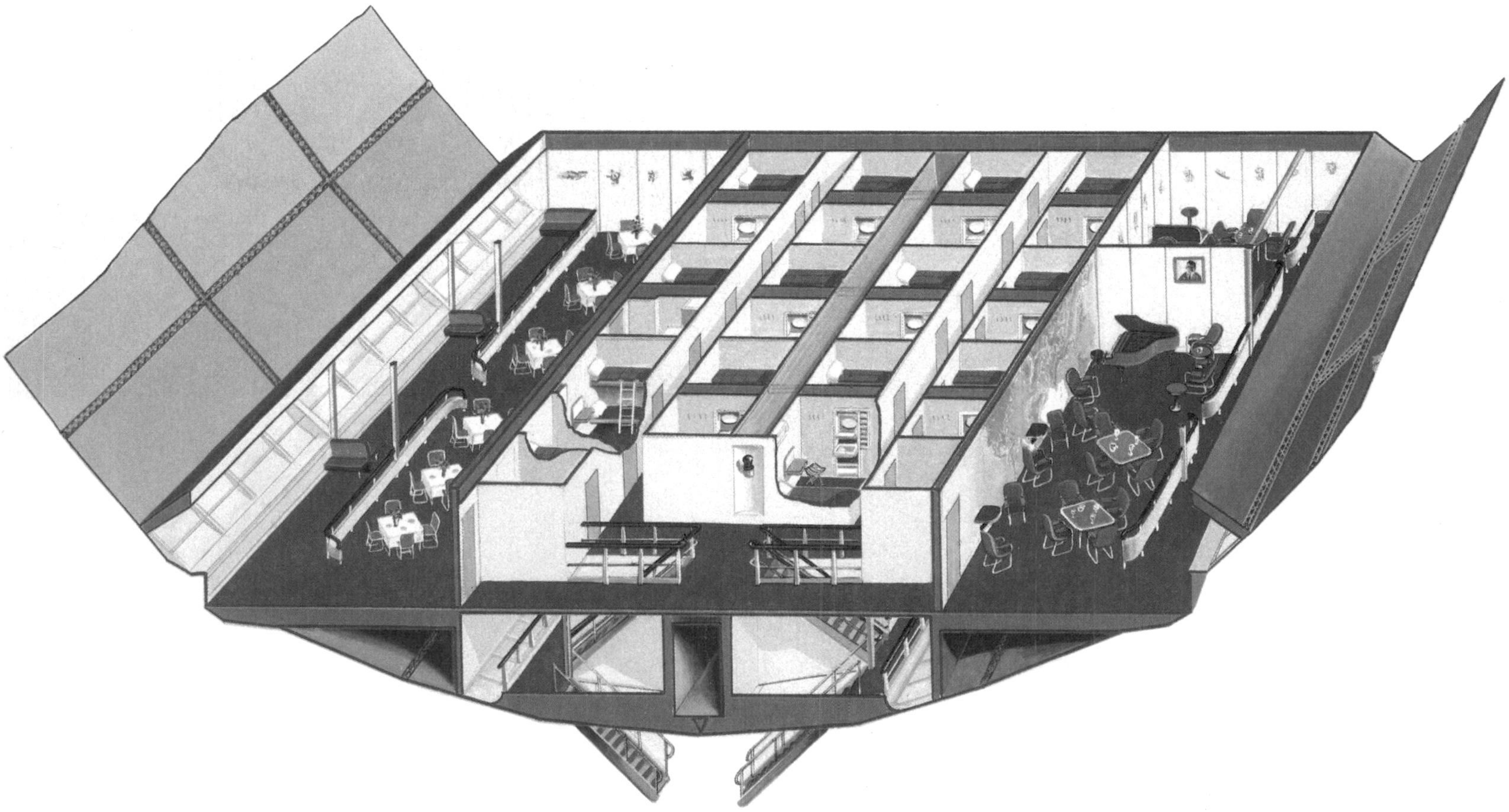

INDEX